Springer Desktop Editio

L. Brandsma, S. F. Vasilevsky, H. D. Verkruijsse
Application of Transition Metal Catalysts in Organic Synthesis
ISBN 3-540-65550-6

H. Driguez, J. Thiem (Eds.)
Glycoscience, Synthesis of Oligosaccharides and Glycoconjugates
ISBN 3-540-65557-3

H. Driguez, J. Thiem (Eds.)
Glycoscience, Synthesis of Substrate Analogs and Mimetics
ISBN 3-540-65546-8

H. A. O. Hill, P. J. Sadler, A. J. Thomson (Eds.)
Metal Sites in Proteins and Models, Iron Centres
ISBN 3-540-65552-2

H. A. O. Hill, P. J. Sadler, A. J. Thomson (Eds.)
Metal Sites in Proteins and Models, Phosphatases, Lewis Acids and Vanadium
ISBN 3-540-65553-0

H. A. O. Hill, P. J. Sadler, A. J. Thomson (Eds.)
Metal Sites in Proteins and Models, Redox Centres
ISBN 3-540-65556-5

A. Manz, H. Becker (Eds.)
Microsystem Technology in Chemistry and Life Sciences
ISBN 3-540-65555-7

P. Metz (Ed.)
Stereoselective Heterocyclic Synthesis
ISBN 3-540-65554-9

H . Pasch, B. Trathnigg
HPLC of Polymers
ISBN 3-540-65551-4

T. Scheper (Ed.)
New Enzymes for Organic Synthesis, Screening, Supply and Engineering
ISBN 3-540-65549-2

Springer
Berlin
Heidelberg
New York
Barcelona
Hong Kong
London
Milan
Paris
Singapore
Tokyo

H. Driguez, J. Thiem (Eds.)

Glycoscience

Synthesis of Substrate Analogs and Mimetics

Dr. Hugues Driguez
Centre de Recherches sur les
Macromolécules Végétales (CERMAV)
601, Rue de la Chimie, B. P. 53
F-38041 Grenoble Cedex 9, France
E-mail: HDriguez@cermav.grenet.fr

Prof. Dr. Joachim Thiem
Institut für Organische Chemie
Universität Hamburg
Martin-Luther-King-Platz 6
D-20146 Hamburg, Germany
E-mail: thiem@chemie.uni-hamburg.de

Description of the Series

The Springer Desktop Editions in Chemistry is a paperback series that offers selected thematic volumes from Springer chemistry series to graduate students and individual scientists in industry and academia at very affordable prices. Each volume presents an area of high current interest to a broad non-specialist audience, starting at the graduate student level.

Formerly published as hardcover edition in the review series
Topics in Current Chemistry (Vol. 187) ISBN 3-540-62032-X

Cataloging-in-Publication Data applied for

ISBN 3-540-65546-8
Springer-Verlag Berlin Heidelberg New York

Printed in Germany

Cover: design & production, Heidelberg
Typesetting: Fotosatz-Service Köhler OHG, Würzburg
SPIN: 10711912 02/3020 - 5 4 3 2 1 0 - Printed on acid-free paper

Preface

Within recent years and after a first coverage of carbohydrate chemistry in the series Topics in Current Chemistry (Vol. 154 in 1990) this field has undergone something like a "quantum jump" with regard to interest for a wider community of scientists. Apparently, for most areas of the natural sciences in general and in particular for those bordering natural products chemistry, progress in the saccharide field has attracted considerable attention and, in fact, many bridging collaborations have resulted. Glycoscience is becoming a term to cover all sorts of activities within or at the edge of carbohydrate research.

It was within the course of a sabbatical collaboration that the editors – both "born" carbohydrate chemists – felt it highly appropriate and desirable to compile a number of contemporary reviews focussing on developments in this field. A number of colleagues actively pursueing outstanding research in the saccharide field agreed to discuss topical issues to which they with their groups have contributed significantly. The result is a fresh and contemporary coverage of selected topics including future outlooks in glycoscience.

In this current volume, particular emphasis has been placed on the demanding synthetic approaches to and on the biological implications of carbohydrate-derived modulators or inhibitors. The sophisticated and now already highly developed synthetic approaches to complex C-glycosides is highlighted by *J.-M. Beau* and *T. Gallagher* and the synthesis of the more biological revelant C-glycosides by *F. Nicotra*. *H. Driguez* focusses on the synthesis and application of thiooligosaccharides for inhibitory studies. The previously nicely developed sugar lactone chemistry allows its synthetic wealth to flow into very attractive syntheses of functionalized heterocycles as described by *I. Lundt*. Further glycosidase inhibitors have been elaborated as outlined by *A. de Raadt*, *C.W. Ekhart*, *M. Ebner* and *A.E. Stütz*. Approaches to inhibitors for other glycanohydrolyses are reported by *R.V. Stick*, and the synthetic development of heparinoid mimetics is elaborated by *H.P. Wessel*. The increasing interest and recent development of multivalent glycoconjugates is covered by *R. Roy*, and in his contribution on amphiphilic sugar derivatives *P. Boullanger* allows us to imagine the supramolecular features embedded in carbohydrate structures.

The authors, the editors and the publisher hope that, by reading this volume, many scientists in the life sciences area will have been able to acquire a taste for the subject and that they will join the growing glycoscience community and actively contribute to this most fascinating research.

Grenoble and Hamburg
December 1996

Hugues Driguez
Joachim Thiem

Contents

Nucleophilic C-Glycosyl Donors for C-Glycoside Synthesis

Jean-Marie Beau[1] and Timothy Gallagher[2]

[1] Institut de Chimie Moléculaire, Université de Paris-Sud, F-91405 Orsay, France
[2] School of Chemistry, University of Bristol, Bristol, BS8 1TS, United Kingdom

This chapter reviews the methods available for the synthesis of C-glycosides that involve the use of carbohydrate units expressing nucleophilic reactivity at C(1) (the anomeric site). A wide variety of "anionic" (organometallic derivatives) and neutral (free radical) C(1) nucleophiles are known and examples of all structural types are presented. Coverage includes methods for generating effective C(1)-nucleophiles, their reactivity and utility in C-glycoside synthesis and, where appropriate, any associated limitations.

Table of Contents

Topics in Current Chemistry, Vol. 187

Symbols and abbreviations

Ac	acetyl
Ac_2O	acetic anhydride
AIBN	1,1-azobisisobutyronitrile
Bn	benzyl
Bz	benzoyl
CSA	camphorsulfonic acid
dba	dibenzylidene acetone
DBU	1,8-diazabicyclo[5.4.0]undec-7-ene
DEAD	diethyl azodicarboxylate
Dibal-H	diisobutyl aluminium hydride
DIPS	diisopropylsulfide
DMSO	dimethyl sulfoxide
dppf	1,1′-bis(diphenylphosphino)ferrocene
HMPA	hexamethylphosphoramide
LCIPA	lithium cyclohexylisopropylamide
LDA	lithium diisopropylamide
LDBB	lithium di-*tert*-butylbiphenylide
LN	lithium naphthalenide
LTMP	lithium 2,2,6,6-tetramethylpiperidide
LUMO	lowest unoccupied molecular orbital
mCPBA	meta chloroperoxybenzoic acid
MOM	methoxymethyl
NIS	*N*-iodosuccinimide
PCC	pyridinium chlorochromate
PDC	pyridinium dichromate
py	pyridine
TBS	*tert*-butyldimethylsilyl
Tf	trifluoromethane sulfonyl
TFA	trifluoroacetic acid
TMS	trimethylsilyl
TPS	triisopropylsilyl

1 Introduction

The synthesis of C-glycosides has experienced a rapid and widespread evolution over the last two decades [1–4]. While a variety of methods exist for establishing C-glycosyl linkages, conventional carbohydrate reactivity, that is the addition of an external nucleophile to an electrophilic C(1) (anomeric) center (Scheme 1), is most widely employed. Access to this reactivity is governed by the presence of the ring-oxygen atom, but this residue will also support the establishment of nucleophilic reactivity at C(1). Such nucleophilicity, though not readily exploited in *O*-glycosylation, does provide a versatile and complementary approach to the synthesis of C-glycosides (Scheme 2).

Defining the boundaries associated with nucleophilic reactivity is, however, less clear cut. With this in mind (and to allow the reader to make an informed assessment of the chemistry that is now available), we have avoided making our classifications too rigid.

In the following we have broadly divided the available C(1) nucleophiles according to structure and reactivity. C(1)-Nucleophiles based on metallation at an sp^3-center are categorized in terms of the presence and, as appropriate, the nature of a hetero atom (O or N) substituent at C(2). C(1)-Lithiated glycals correspond to nucleophilic sp^2-centers, and variations on this theme (concerning the nature of the metal component) are covered in depth.

Enolates represent a special class of sp^2-hybridized C(1) nucleophiles. This is an important and developing aspect of carbohydrate reactivity and the carbonyl function can either be located at C(2), or be external to the sugar ring (as in sialic acid derivatives). Although the latter are not strictly C(1) nucleophiles, both their value in C-glycoside synthesis and relationship to other topics covered in this chapter merit inclusion of these systems.

Carbon-based radicals stabilized by oxygen, though electronically neutral, can exhibit nucleophilic reactivity. C-Glycoside synthesis based on anomeric

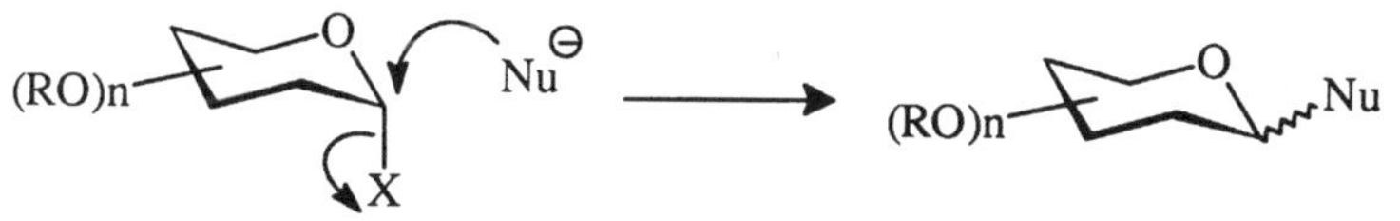

Electrophilic *C*-Glycosyl Donor

Scheme 1

Nucleophilic *C*-Glycosyl Donor

Scheme 2

radicals is very important, but the space available here precludes a comprehensive coverage of the topic. Key features, which accentuate the nucleophilic aspects of anomeric radicals will, however, be addressed.

2 Nucleophilic sp^3-Anomeric Centers

2.1 C(1)-Metallated 2-Deoxypyranosyl Compounds

2.1.1 *Reductive Metallation Processes*

The first example of the non-stabilized C(1)-organolithium nucleophile was provided by reductive lithiation of 2-deoxy-D-glucopyranosyl chloride **2** or phenylsulfides **3** [5] with lithium naphthalenide (LN), a process discovered by Cohen and Matz [6] on cyclic α-alkoxysulfides, to give the C(1)-lithiated 2-deoxypyranoside **4**. Lithio reagent **4** reacts with electrophiles in moderate yields to provide selectively the corresponding α-D-C-glycosides **5** (Scheme 3). The stereoselectivity results from a two-step single-electron transfer mechanism. In the first step, transfer of one electron occurs from LN to the LUMO of X (X = Cl, SPh) to provide anion radical **6** which cleaves to generate axial radical **7** (stabilized by the anomeric effect) (Scheme 4). A second electron transfer to **7** then leads, under kinetic control, to the organolithium **4**, which is configurationally stable at −78 °C and reacts with electrophiles with retention of configuration. This method has been used as a key step in a synthesis of **11**, a monocyclic analogue of compactin, by reductive lithiation of thioglycoside **8** with lithium di-*tert*-butylbiphenylide (LDBB) at low tem-

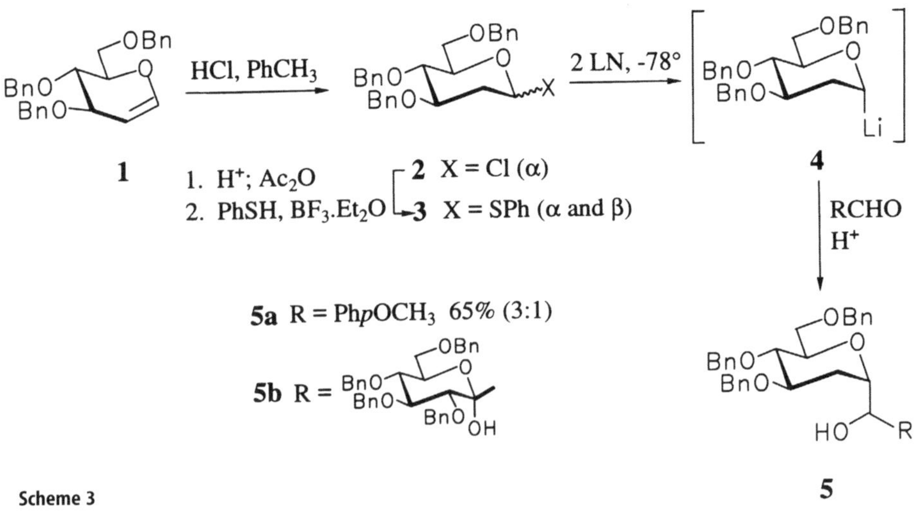

HCl, $PhCH_3$

2 LN, -78°

1

1. H^+; Ac_2O

2. PhSH, $BF_3.Et_2O$

2 X = Cl (α)

3 X = SPh (α and β)

4

RCHO

H^+

5a R = Ph*p*OCH_3 65% (3:1)

5b R =

5

Scheme 3

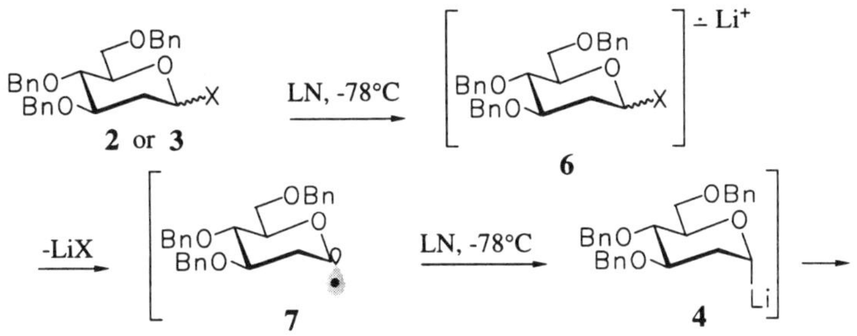

Scheme 4

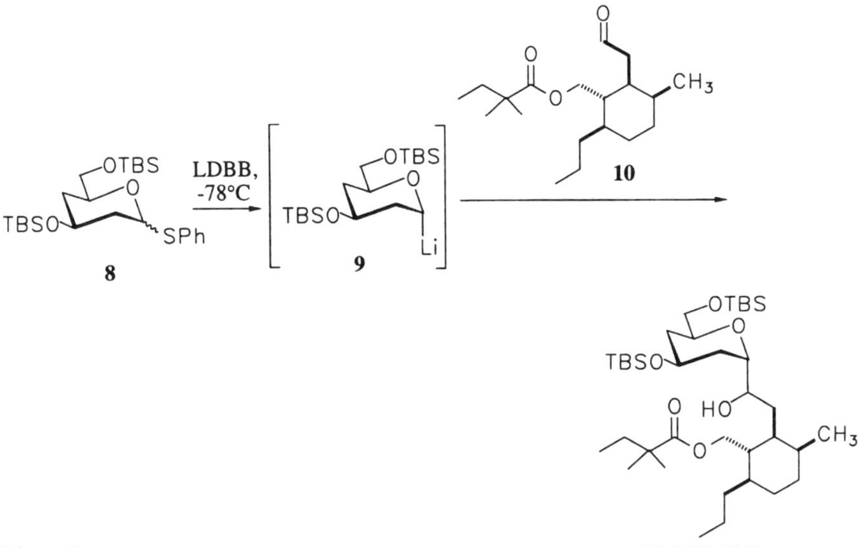

11 84% (7:1)

Scheme 5

perature followed by addition of the α-lithio reagent **9** to aldehyde **10** [7] (Scheme 5).

All attempts to transform the α-lithio reagent **4** to the more stable β-anomer by warming to −20 °C, a process that is successful with less functionalized substrates [8], failed, though C(1) isomerization was achieved when the hydroxyl groups of the sugar are protected as methyl ethers [9]. Thus, reductive lithiation (LDBB) of sulfide **12** gives the α-lithio reagent **13** at −78 °C which reacts with acetone to furnish the α-C-glycoside **15** (Scheme 6). When the lithiated reagent **13** is warmed to −20 °C for 45 min, essentially complete isomerization takes places leading to the β-lithio reagent **14** which was trapped by acetone to give β-C-glycoside **16**. The lower yield observed with **16** (vs. **15**) is attributed to competing protonation during isomerisation but this study also provided a detailed analysis of the stereoselectivity of this type of process.

When electrophiles other than carbonyl compounds or alkyl halides are employed, the basic α-lithio reagents need to be converted to other organo-

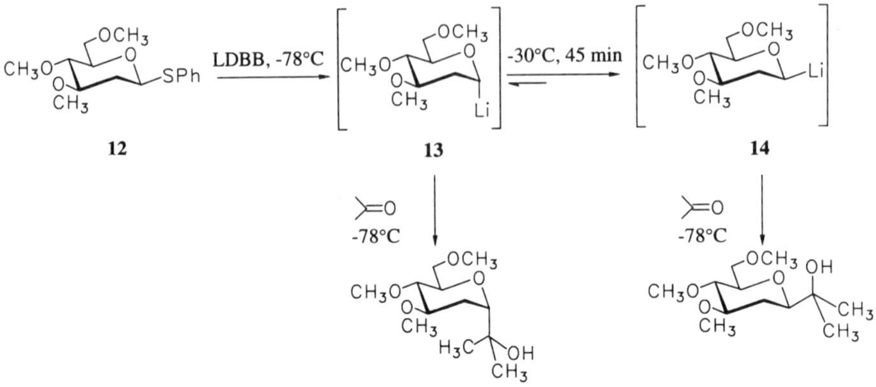

Scheme 6

12 → LDBB, -78°C → [13] ⇌ -30°C, 45 min → [14]

13 → acetone, -78°C → **15** 81% (α:β, 98:2)

14 → acetone, -78°C → **16** 59% (α:β, 1:99)

metallic nucleophiles. Access to cuprate reagents is possible, which occurs with retention of configuration, and an example is shown in Scheme 7. Reductive lithiation of the α-chloride (obtained by hydrochlorination of glycal **17**) provides the α-lithio reagent **18** which was converted into the mixed cuprate reagent **19** [10]. Reaction of **19** with racemic epoxide **20** requires the presence of an excess of BF_3. Et_2O, and leads to the α-C-glycosides **21** (as a mixture of diastereomers) (Scheme 7). This transformation has been used as a key step in the stereochemical assignment of the C(1)-C(10) fragment of nystatin A1 [10].

The reductive lithiation of α-alkoxy phenylsulfides is a slow process (typically 0.5–1 h at –78 °C) and lowering the LUMO of the electron acceptor by using, for example, an anomeric sulfone, leads to a much faster electron transfer [11]. Reductive lithiation of sulfone **22** is fast (less than 1 min) and leads to similar α-lithio reagents to those described above and Scheme 8 shows examples of simple α-C-2-deoxyglycosides **23** and **24** prepared by this protocol. The most interesting feature of anomeric sulfones is that alkylation *prior to* the reductive desulfonylation event is achievable. In this way, a one-pot four-step sequence

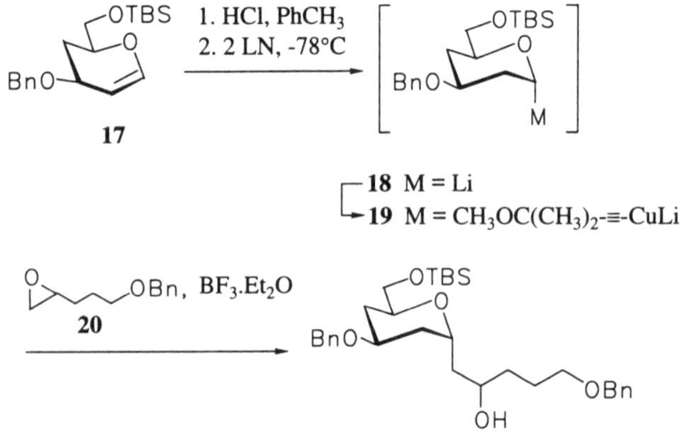

17 → 1. HCl, $PhCH_3$; 2. 2 LN, -78°C →

18 M = Li

19 M = $CH_3OC(CH_3)_2$-≡-CuLi

→ **20**, $BF_3.Et_2O$ →

Scheme 7

21 (36% from **17**)

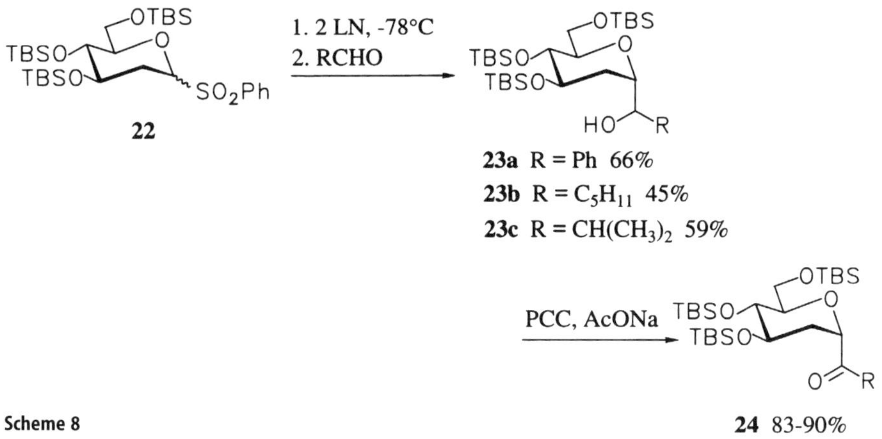

TBSO, TBSO, OTBS, O, SO_2Ph **22** — 1. 2 LN, -78°C 2. RCHO → HO, R

23a R = Ph 66%
23b R = C_5H_{11} 45%
23c R = $CH(CH_3)_2$ 59%

PCC, AcONa → O, R

24 83-90%

Scheme 8

involving sequential sulfone deprotonation-electrophile (El^+) addition-reductive lithiation-protonation provides a stereoselective entry to β-C-2-deoxyglycosides **28** from sulfone **22** [12] (Scheme 9). Again, homolytic C–S bond cleavage of the initial alkylated intermediate **25** provides an axial radical **26** that, after a second electron transfer, generates the configurationally stable lithium reagent **27** which is protonated, using methanol, to give **28** with retention of configuration.

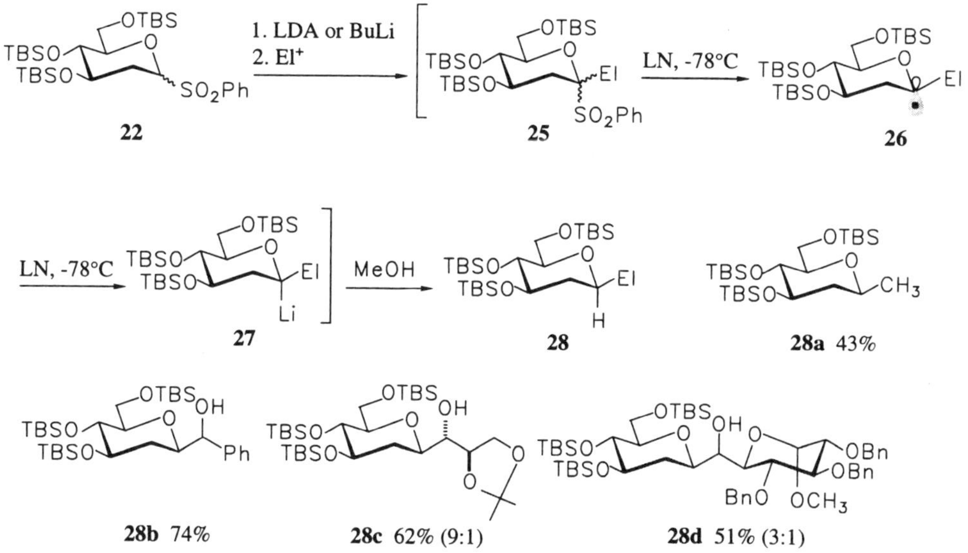

22 — 1. LDA or BuLi 2. El^+ → [**25** (El, SO_2Ph) — LN, -78°C → **26** (El) — LN, -78°C → **27** (El, Li)] — MeOH → **28** (El, H)

28a (CH_3) 43%

28b (OH, Ph) 74%

28c 62% (9:1)

28d 51% (3:1) (OBn, OBn, BnO, OCH_3)

Scheme 9

If dimethyl carbonate or phenyl esters (e.g. phenyl 1,2-*O*-isopropylidene-D-glycerate **29**) are used as electrophiles towards **22**, then α-C-2-deoxyglycosides are obtained selectively [13] (Scheme 10). In this case, the intermediate acylated

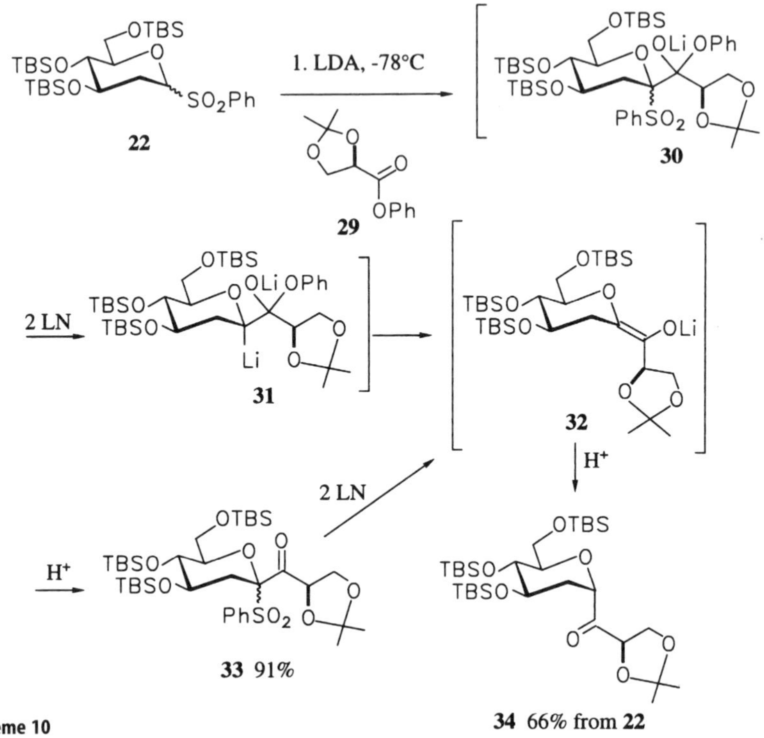

Scheme 10

sulfones (e.g. **33**) are stable (often crystalline) and a one-pot (as above) or a two-step procedure (sulfone deprotonation-acylation then reductive lithiation-protonation) can be conducted at will. The selective formation of α-C-glycosides is explained in Scheme 10. Reductive lithiation of either adduct **30** (one-pot procedure via **31**) or ketosulfone **33** leads to enediolate **32** that undergoes kinetic protonation from the *exo* face to give the α-product **34** (α/β ratio, 22:1). Finally, alkylation rather than protonation can be exploited as illustrated in Scheme 11.

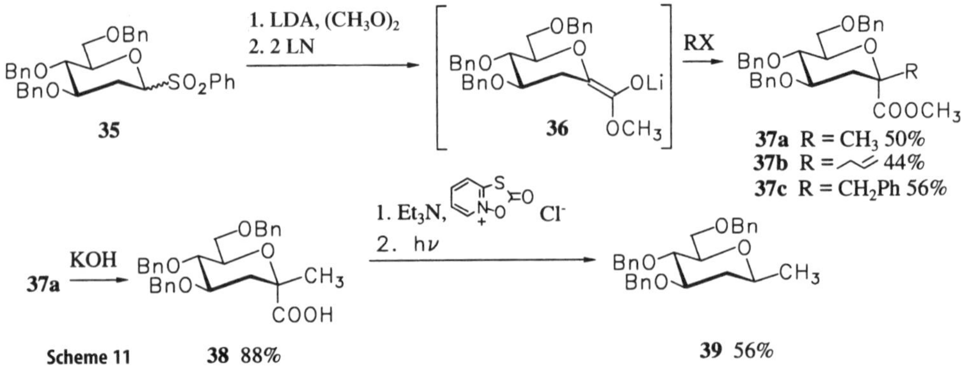

Scheme 11

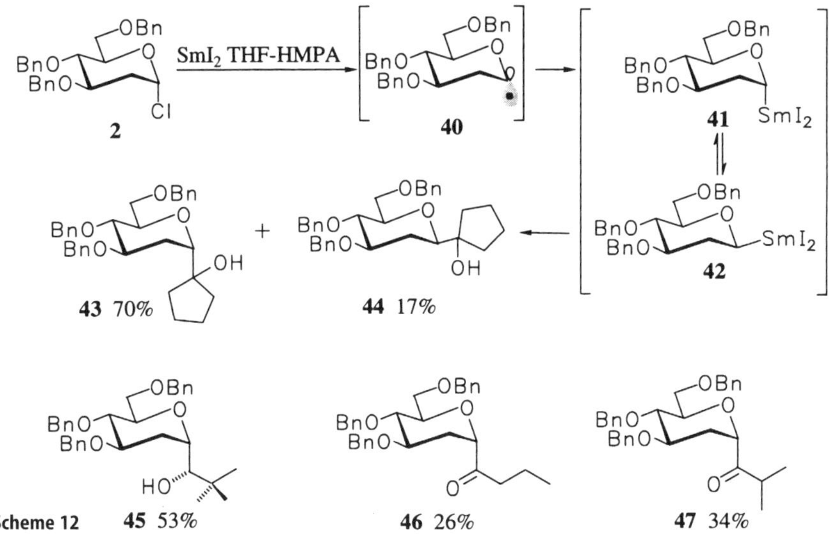

Scheme 12

Using sulfone **35**, the standard sequence provides enediolate **36** which is alkylated from the β-face to give bis-C-glycosides **37** [14]. A Hunsdiecker-type decarboxylation of C-glycoside **37a** (via carboxylic acid **38**) gives β-C-glycoside **39** [14]. Access to such simple C-glycosides is, however, more straightforward using the one-pot sequence shown in Scheme 9.

Reductive samariation of 2-deoxypyranosyl chlorides or arylsulfones by samarium diiodide, a mild but powerful single-electron transfer reagent [15], is a useful alternative for preparing C(1)-metallated anomeric species. Reaction of chloride **2** at room temperature with the SmI_2-THF-HMPA system in the presence of cyclopentanone, gives a mixture of α-C-glycoside **43** and β-C-glycoside **44** [16] (Scheme 12). This process displays two major differences to reductive lithiation which explain the modest α/β selectivity that is observed. Reductive samariation is necessarily conducted at room temperature to achieve a reasonable rate and, secondly, Barbier conditions are needed (the ketone must be present during the reductive metallation step) to avoid excessive protonation of the intermediate glycosyl samarium(III) species **41** or **42** by THF. Under these conditions, the kinetic nucleophile **41**, produced by reduction of the anomeric radical **40**, undergoes partial isomerization to the β-species **42** before being trapped by the electrophile (cyclopentanone). Aldehydes, as more reactive electrophiles, give only α-C-glycosides **45**, **46**, and **47** (after a PDC oxidation) but are inefficient substrates because of competing carbonyl reduction by SmI_2 [16]. This is a consequence of the presence of HMPA which is required to enhance the reducing power of SmI_2 for efficient single-electron transfer [17].

Anomeric 2-pyridylsulfones, such as **48**, were introduced by Skrydstrup and Beau [18] and have a LUMO energy level low enough that reductive samariation is possible in THF alone (no HMPA needed). Under Barbier conditions, good yields of C-glycosides **50–53** are available with aldehydes and ketones (Scheme 13) [19]. The low selectivity associated with **50–53** is again a consequence

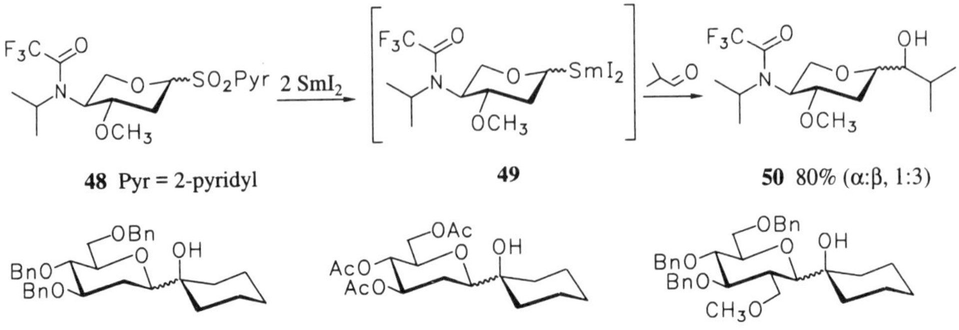

48 Pyr = 2-pyridyl **49** **50** 80% (α:β, 1:3)

51 82% (α:β, 1:1) **52** 88% (α:β, 1:1) **53** 73% (α:β, 1:4)

Scheme 13

of isomerization giving both anomers of the organosamarium species **49** at room temperature. The notable differences in selectivities between the two procedures (compare **43/44**, Scheme 12 and **51**, Scheme 13) involving formally the same anomeric organosamarium species may well reflect the role of HMPA.

Nonstereoselective reductive samariations on 2-deoxy glycosides, though of limited synthetic value, are nevertheless of mechanistic interest. However, this type of process does have utility in the preparation of C(2)-oxygenated or C(2)-aminated C-glycosides which is described in Sect. 2.2.3.

2.1.2
Deprotonation Processes

The anomeric proton of *O*-glycosides and *S*-glycosides is not sufficiently acidic to be deprotonated by strong bases without affecting the other functional groups present in sugars. To facilitate deprotonation, an anion stabilizing group must first be introduced at the anomeric center. This group then needs to be removed after carbon-carbon bond formation has been completed and the stereoselectivity of the final C-glycoside is most often defined during this removal step. Anomeric nitro, sulfone and ester groups have been used as anion stabilizers, although it should be appreciated that the resulting "anions", because of the nature of the stabilizing moiety, are essentially sp^2-centers.

Anomeric triphenylphosphonium salts have been used as well as phenylsulfides, but in the latter case extra stabilization is necessary (see below). Anomeric nitrosugars, which have been extensively studied in C-glycosylation reactions by Vasella, will be covered in Sect. 2.2.1 and ester enolates derived from 3-deoxy-2-ketoulosonic acids (sialic acid and KDO derivatives), which bear a structural similarity to 2-deoxy pyranosides, will be covered in Sect. 4.4. Deprotonation of anomeric phenylsulfones has been discussed in Sect. 2.1.1 and additional transformations on closely related compounds are presented in Scheme 14 [20]. Alkylation of phenylsulfone **54** with epoxide **55** provides adduct **56** which eliminates benzenesulfinic acid at room temperature to give the C(1)-alkylated glycal **57**; a similar elimination is also observed with adducts derived from

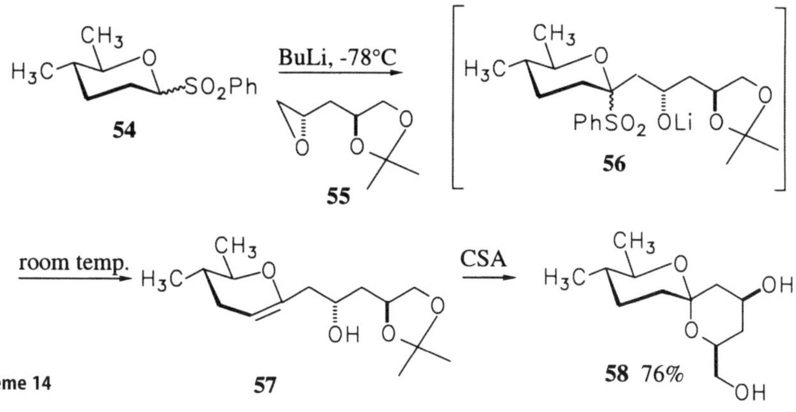

Scheme 14

aldehydes [21]. Enol ether **57** cyclizes to spiro ketal **58** on reaction with camphorsulfonic acid (CSA).

Phosphorus-based reagents have also been prepared. Glucal **1** reacts cleanly with triphenylphosphine hydrobromide to provide the anomeric triphenylphosphonium salt **59** [22] (Scheme 15). Phosphonium ylide formation and condensation with octanal gives the exocyclic enol ether **60**. The β-C-glycoside **61** is obtained exclusively when hydrogenation over Pd/C is carried out in the presence of 1% triethylamine and further reduction leads to the β-C-glycoside triol **62**. Similar Wittig condensations, though not involving sugars,

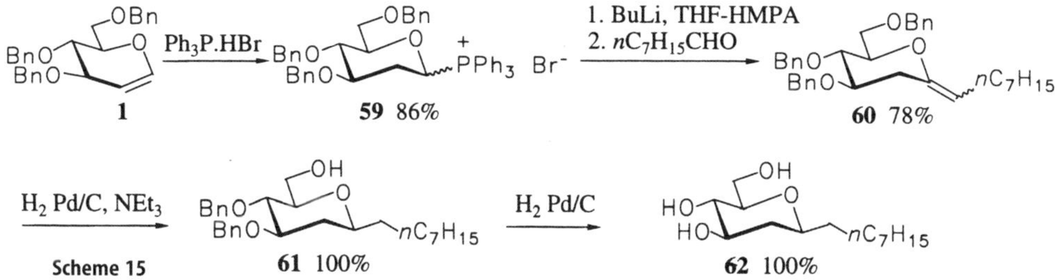

Scheme 15

have been applied to the construction of spiroketals [22]. In an independent report [23], 2-(diphenylphosphinoxy)tetrahydropyran **64** was prepared from phosphonium salt **63** (Scheme 16). Deprotonation of phosphine oxide **64** with LDA and coupling with aldehyde **65** provides a mixture of *endo* and *exo* enol ethers **66** that cyclize to spiroketal **67** upon acid treatment.

The anomeric proton in phenylthio 2,3-dideoxy-hex-2-enopyranosides is sufficiently acidic to enable deprotonation and, for example, reaction of *S*-glycoside **68** with BuLi leads to dianion **69** which alkylates or acylates at C(1) from the α-face to give C-glycosides **70** (Scheme 17) [24]. Similarly, deprotonation of hex-2-enopyranoside **71** and coupling with cyclic sulfate **72** provides alkylated product **73** which, upon hydrolysis at pH > 3.5, undergoes a loss of sulfate and a

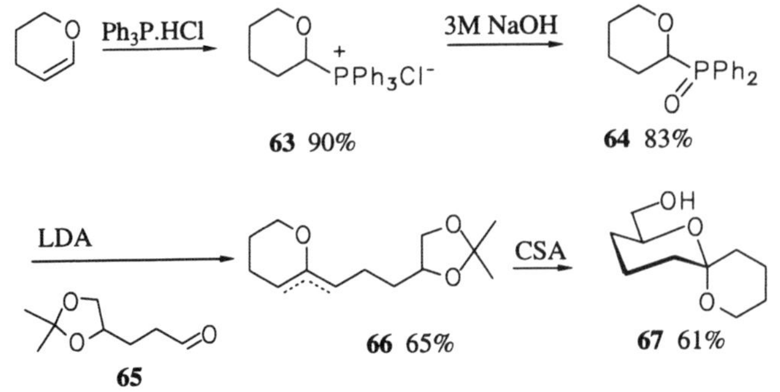
Ph3P.HCl
3M NaOH
PPh3Cl−
PPh2
63 90%
64 83%
LDA
CSA
OH
65
66 65%
67 61%

Scheme 16

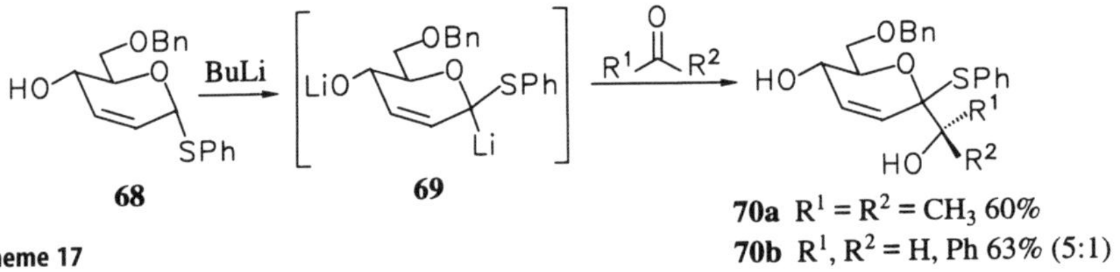
OBn
HO
SPh
BuLi
LiO
Li
R1
R2
HO
68
69
70a R1 = R2 = CH3 60%
70b R1, R2 = H, Ph 63% (5:1)

Scheme 17

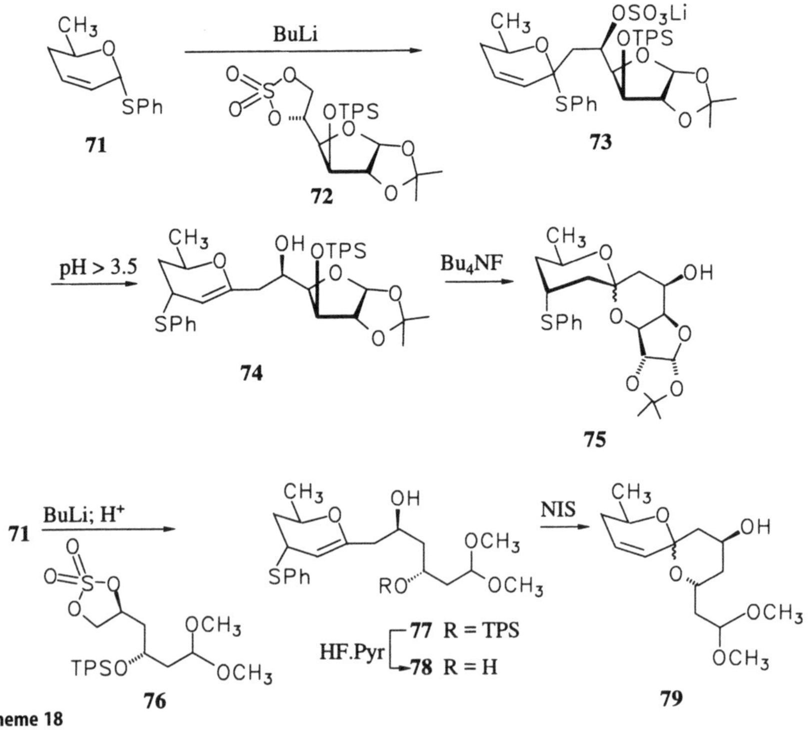
CH3
SPh
BuLi
OTPS
OSO3Li
71
72
73
pH > 3.5
OH
Bu4NF
74
75
71
BuLi; H+
NIS
OCH3
TPSO
RO
76
HF.Pyr
77 R = TPS
78 R = H
79

Scheme 18

Ferrier-type rearrangement to give glycal **74** [25] (Scheme 18). Spiroketals **75** are then obtained after desilylation. A similar sequence of reactions based on cyclic sulfate **76** furnishes adduct **77** which, after desilylation to give **78**, provides spiroketals **79** upon *N*-iodosuccinimide (NIS) treatment.

2.1.3
Trialkylstannyl-Based Transmetallations

In reductive lithiation processes, the stereoselectivity of the C–C bond forming reaction results from reduction of a transient anomeric radical under kinetic control to an anomeric anionic species that is stereochemically stable at –78 °C. Stereochemical information can also be stored prior to the lithiation event, as with stereodefined anomeric trialkylstannanes. α-Alkoxy organostannanes are known to be useful precursors of α-alkoxy organolithium compounds through a fast, low-temperature exchange reaction with alkyllithium reagents [26]. Tin-lithium exchange occurs with retention of configuration and the sp^3-organolithium reagent that is formed also reacts with electrophiles with retention of configuration [27]. Isomeric tributyl-(2-deoxy-D-*arabino*-hexopyranosyl)stannanes **80** and **83** are both produced from pyranosyl chloride **2**, prepared by reaction of tri-*O*-benzyl-D-glucal **1** with HCl in toluene [28] (Scheme 19). Treatment of chloride **2** with tributylstannyl lithium [26] in THF at 0 °C provides the tributyl-β-D-glycosylstannane **80** as a result of an S_N2 reaction (stereoselectivity of ca. 60:1). The isomeric tributyl-α-D-glucosylstannane **83** is obtained from chloride **2** by reductive lithiation with 2 equiv. of LN in THF at –78 °C followed by addition of tributyltin chloride. Transmetallation of the

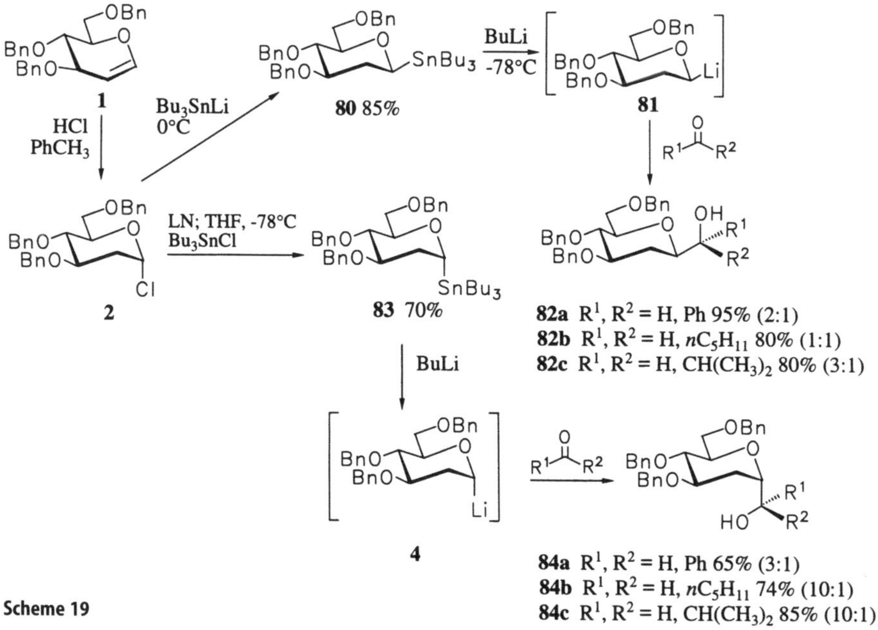
OBn
O
BnO
BnO
1
HCl
PhCH3
Bu3SnLi
0°C
SnBu3
80 85%
BuLi
-78°C
Li
81
R1
R2
2
Cl
LN; THF, -78°C
Bu3SnCl
83 70%
OH
BuLi
4
HO
82a R1, R2 = H, Ph 95% (2:1)
82b R1, R2 = H, nC5H11 80% (1:1)
82c R1, R2 = H, CH(CH3)2 80% (3:1)
84a R1, R2 = H, Ph 65% (3:1)
84b R1, R2 = H, nC5H11 74% (10:1)
84c R1, R2 = H, CH(CH3)2 85% (10:1)

Scheme 19

isomeric stannanes **80** and **83** leads to the configurationally stable organolithium reagents **81** and **4** (see Sect. 2.1.1) respectively which are trapped both stereospecifically and efficiently by aldehydes to give C-glycosides **82** and **84**; competing protonation is the only significant side reaction observed [28]. These anomeric organolithiums add to enones (e.g. 2-cyclohexenone) in a 1,2-fashion but reactions involving alkyl iodides lead to poor yields of alkylated products.

As a consequence of the asymmetric nature of these C(1) nucleophiles, additions of e.g. organolithium **4** to prochiral carbonyl compounds proceeds with facial selectivity, with the *syn*-isomer **86** being favored over the *anti*-isomer **88** (Scheme 20). Although the β-lithio species **81** shows modest *syn*-selectivity towards aldehydes, higher diastereoselectivity is obtained with the α-lithio variant **4** (*syn*: *anti* ca. 10:1) [29]. This facial discrimination can be explained by considering the two possible diastereomeric transition state structures (shown in Scheme 20 for the α-series) in which the major path is the one where the R group of the aldehyde and C(2) of the sugar are in an *anti* arrangement (**85** preferred to **87**). Taking C(2) as the largest carbanion substituent, the empirical model proposed by Bassindale et al. [30] also predicts the same sense of selectivity when chelation is not a major contributing factor. Of course, the influence of the other asymmetric centers in the sugar substrate cannot be ignored which may account for the differences observed between the α- and the β-series.

These C(1) organolithium reagents can also be converted to the organocopper reagents, such as **89** and **92**, by treatment with the copper(I) bromide-dimethyl sulfide complex in diisopropyl sulfide-THF (1:1, v/v) at −78 °C [31] (Scheme 21). Boron trifluoride etherate-promoted conjugate additions of these modified reagents to representative enones, using a version of Yamamoto's procedure [32], produces the corresponding C-glycosides **90**–**94** with retention of configuration at the anomeric center. Hindered enones, such as mesityl oxide, are less useful as electrophiles because higher temperatures (≥ 0 °C) are required and the cuprate reagents show limited stability under these conditions. Epoxides can be used as electrophilic partners when cyanocuprate reagents [33] (available from the organolithium derivative and lithium 2-thienylcyanocuprate

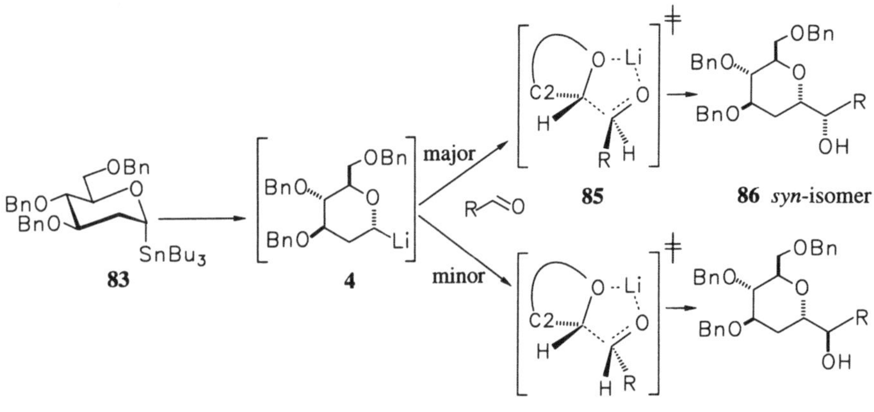

Scheme 20

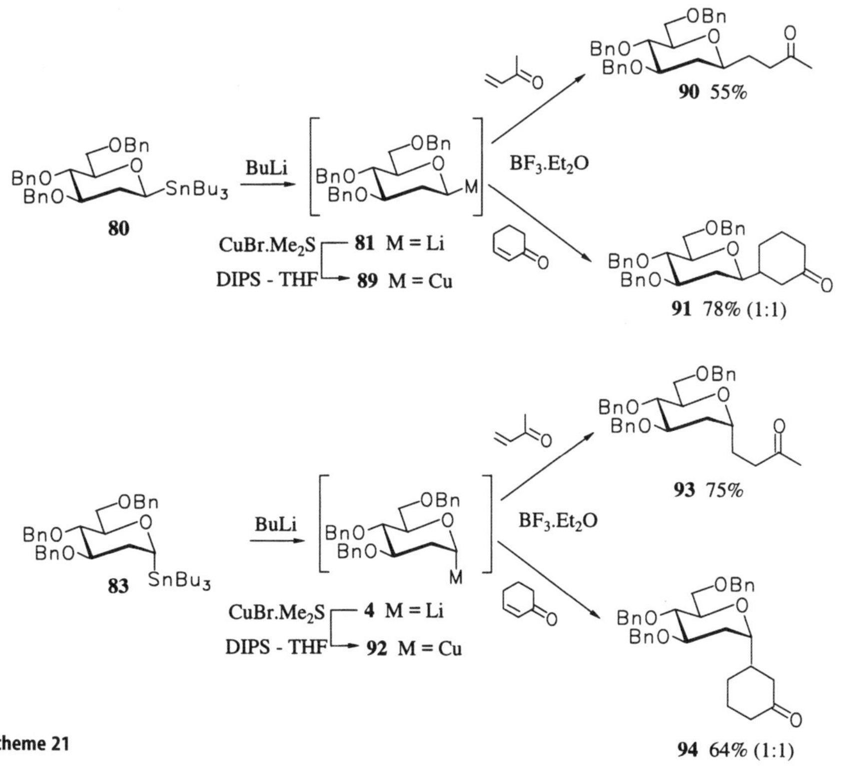

OBn
BnO O
BnO SnBu3
80
BuLi
OBn
BnO O
BnO M
$CuBr.Me_2S$ **81** M = Li
DIPS - THF **89** M = Cu
$BF_3.Et_2O$
OBn
BnO O
BnO O
90 55%
OBn
BnO O
BnO O
91 78% (1:1)

OBn
BnO O
BnO
83 SnBu3
BuLi
OBn
BnO O
BnO
M
$CuBr.Me_2S$ **4** M = Li
DIPS - THF **92** M = Cu
$BF_3.Et_2O$
OBn
BnO O
BnO O
93 75%
OBn
BnO O
BnO O
94 64% (1:1)

Scheme 21

OSi*t*BuMe2
Ph2*t*BuSiO O
SnBu3
95
1) BuLi, THF, -78°
2) Cu(2-thienyl)CNLi
OSi*t*BuMe2
Ph2*t*BuSiO O
CuCNLi2
S
96
(±) O OBn
$BF_3.Et_2O$
71%
OSi*t*BuMe2
O OH
OBn
O
Ph2*t*BuSi
+
OSi*t*BuMe2
O OH
OBn
O
Ph2*t*BuSi
97a 2 : 1 **97b**

OBn
BnO O
BnO
OBn
OH
98 71%
OBn
BnO O
BnO
OBn
OH
99 67%
OBn
BnO O
BnO
OBn
OH
100 63%

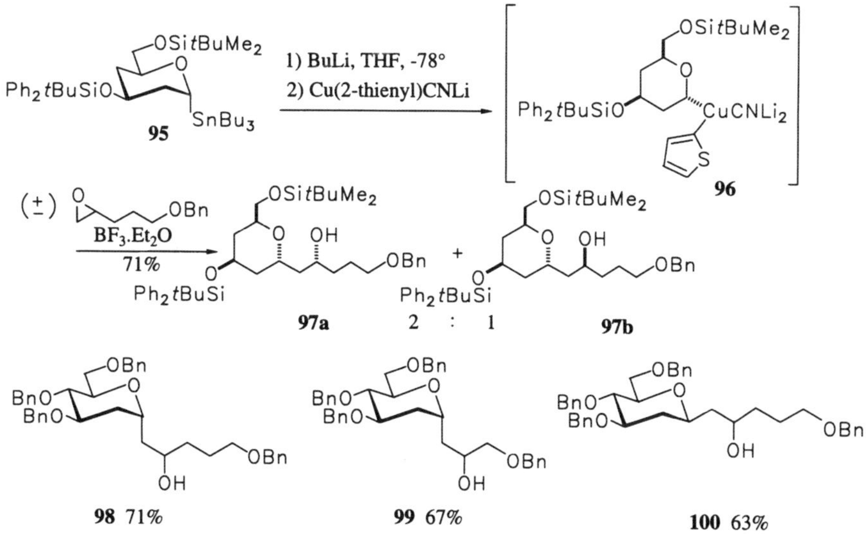

Scheme 22

[34]) are employed (Scheme 22). Alkylation of cyanocuprate **96** using an epoxide requires an excess of $BF_3 . Et_2O$ but good yields of C-glycosides **97–100** have been reported.

2.2
C(1)-Metallated 2-Hydroxy(alkoxy) Sugars

Much more challenging is the production of C(1)-metallated sugars that incorporate a heteroatom (*O*, *N*) at C(2). These targets have greater biological relevance than do the 2-deoxy series – the relationship to e.g. D-glucose, D-glucosamine is obvious – but one must contend with (and prevent) facile *β*-elimination of the C(2) substituent. The first solution to this problem was reported by Vasella who used the weakly basic anions derived from C(1)-nitro sugars which tolerate the presence of a heteroatom at C(2). Methods have now been developed that extend the applicability of this concept to reductive metallation and tin-based transmetallation processes, leading to nonstabilized 2-oxy and 2-amino substituted C(1)-nucleophiles.

2.2.1
Anomeric Nitro Sugars

Anomeric nitro sugars [35] are most efficiently prepared by ozonolysis of the *N*-glycosylnitrones obtained by reaction of aromatic aldehydes with the corresponding sugar oximes via the tautomeric *N*-glycosylhydroxylamines [36]

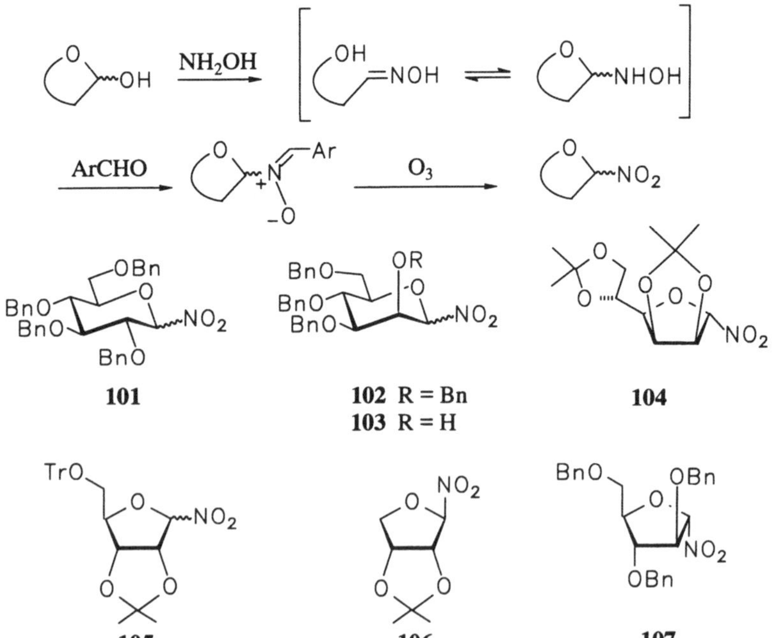

NH_2OH ArCHO O_3

101 102 R = Bn 103 R = H 104 105 106 107

Scheme 23

(Scheme 23). A variety of anomeric nitro sugars has been generated in this manner in both the pyranoid (**101**–**103**) and furanoid (**104**–**107**) series [36, 37]. These molecules have been applied to chain elongation processes via Henry and Michael reactions with a subsequent solvolysis of the anomeric nitro group being used to introduce a hydroxy group [38]. Within the context of C-glycoside synthesis, the nitro group has to be replaced by a hydrogen atom once C–C bond formation is complete. Baumberger and Vasella [39] have shown that denitration under radical conditions (with Bu_3SnH) in the pyranose series proceeds under stereoelectronic control, with axial hydrogen transfer, to provide exclusively the β-D-C-glycopyranosides (Scheme 24). Reaction of the potassium nitronate derived from nitro sugars **101** (*gluco* series) and **102** (*manno* series) with an excess of paraformaldehyde followed by acetylation provide the addition products **108/109** which, on reduction, gave the β-D-C-glycosides **110/111** in good overall yield. In the *gluco* series, the reduction step proceeds via the anomeric radical **112** which undergoes hydrogen transfer from the α-face. Further examples of this reduction methodology (**113** → **114**; **115** → **116**; **117** → **118**) are shown in Scheme 24. The lack of selectivity associated with reduction of nitro

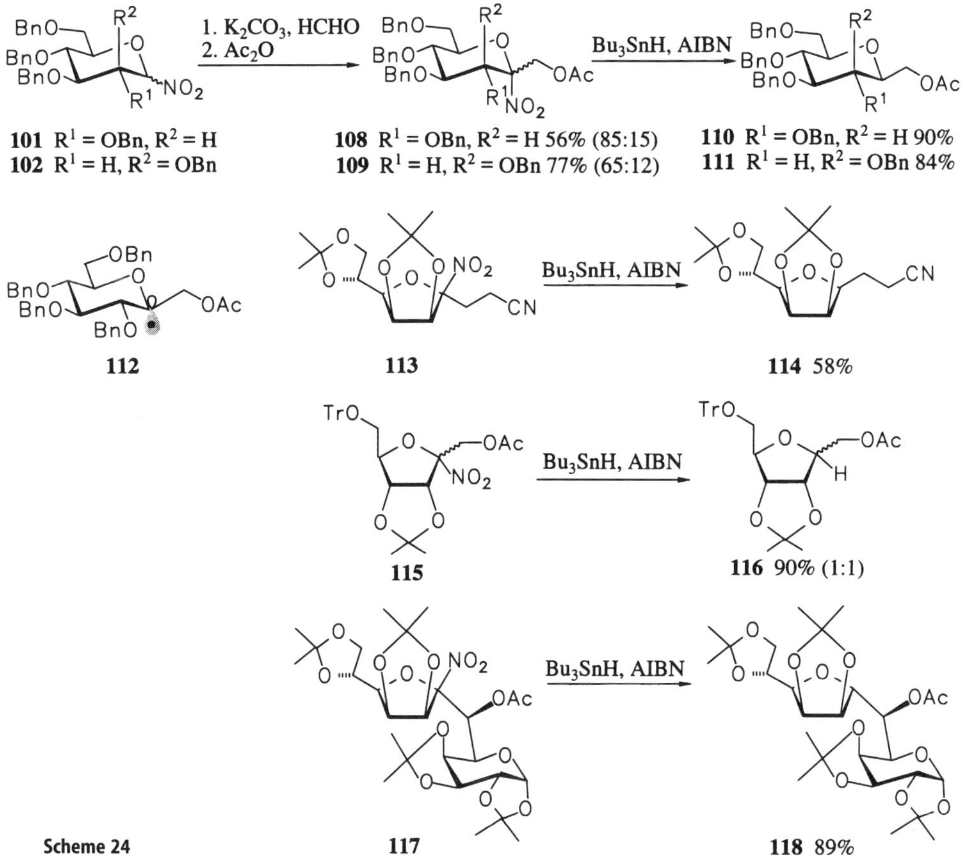

1. K_2CO_3, HCHO
2. Ac_2O

Bu_3SnH, AIBN

101 R^1 = OBn, R^2 = H
102 R^1 = H, R^2 = OBn

108 R^1 = OBn, R^2 = H 56% (85:15)
109 R^1 = H, R^2 = OBn 77% (65:12)

110 R^1 = OBn, R^2 = H 90%
111 R^1 = H, R^2 = OBn 84%

112

113 → Bu_3SnH, AIBN → **114** 58%

115 → Bu_3SnH, AIBN → **116** 90% (1:1)

117 → Bu_3SnH, AIBN → **118** 89%

Scheme 24

sugar **115** can be explained by antagonistic electronic (*H*-transfer favored from the α-face) and steric (*H*-transfer favored from the β-face) effects.

Rather than a hydrogen atom transfer, the intermediate nucleophilic anomeric radical (see Sect. 5) can be trapped by an electron-deficient alkene [40]. Treatment of alkylated nitro sugar **113** with an excess of acrylonitrile in the presence of Bu_3SnH, provides the bis-C-glycoside **122** (Scheme 25). ESR spectroscopy demonstrated that a single-electron transfer from $Bu_3Sn^{\cdot}$ to the nitro group provides radical anion **119** which cleaves to the anomeric radical **120**. Acrylonitrile reacts with **120** from the *exo*-face to produce, via radical **121**, bis-C-glycoside **122** [40]. Utilizing the same procedure with glucose-derived **108** gave bis-C-glycoside **123**.

Another interesting aspect of C(1)-nitro sugars is that the nitro group, having served as an anion-stabilizing group (to allow chain elongation without competitive β-elimination), can behave as a leaving group with appropriate nucleophiles. For example, formylation and protection of the D-arabinofuranoside **107** gives C-glycoside **124** which, on treatment with the sodium salt of nitromethane, provides bis-C-glycofuranosides **125** (Scheme 26) [37]. These derivatives have, in turn, been converted to the mono-phosphonate analogues of β- and α-D-fructose 2,6-bisphosphate, **126** and **127** respectively. Finally, anomeric nitro sugars can be partners in condensation reactions leading to bis-(1,1)-linked carbohydrates. Reaction of the readily available nitro mannosyl bromide **128** [35] with the lithium nitronate derived from **104** furnishes the bis-α,α-C-linked dimer **129**. Subsequent reduction (using Na_2S) provides *Z*-enediol **130** which upon hydrogenation gives the α,β-C-linked dimer **131** [41] (Scheme 27).

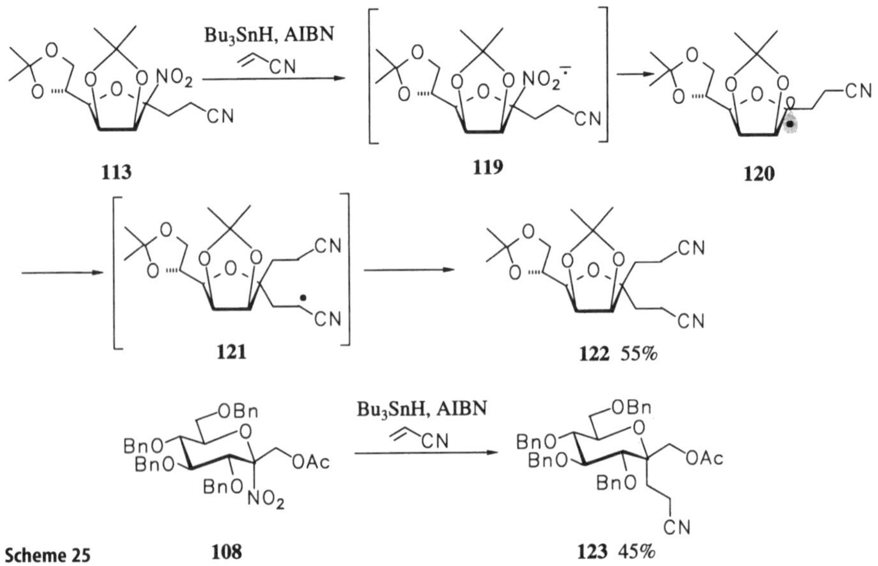

Bu₃SnH, AIBN

113 **119** **120**

121 **122** 55%

108 **123** 45%

Scheme 25

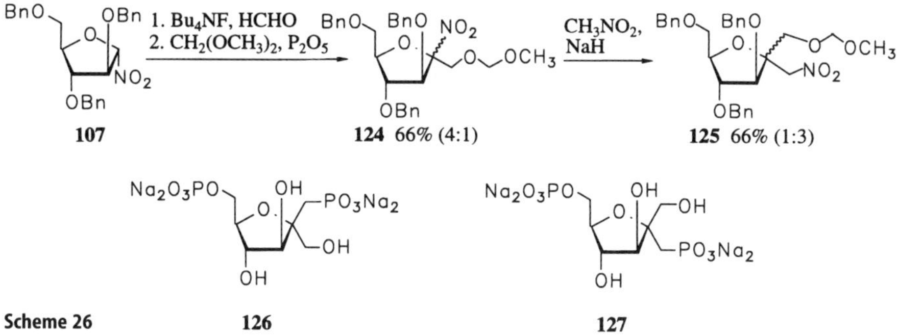

1. Bu_4NF, HCHO
2. $CH_2(OCH_3)_2$, P_2O_5

CH_3NO_2, NaH

107 **124** 66% (4:1) **125** 66% (1:3)

126 **127**

Scheme 26

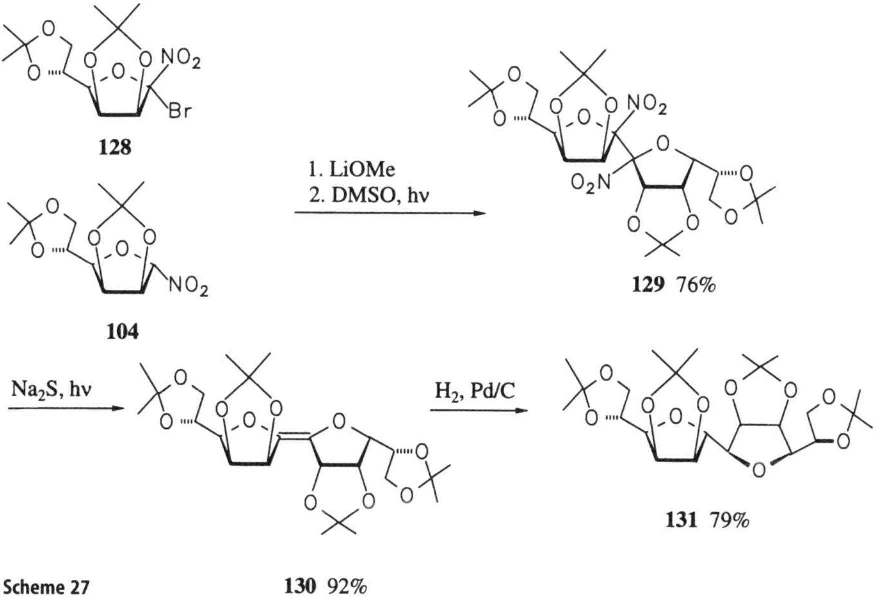

1. LiOMe
2. DMSO, hν

128 **104** **129** 76%

Na_2S, hν

H_2, Pd/C

131 79%

Scheme 27 **130** 92%

2.2.2
Reductive Lithiation and Tin-Based Transmetallations

These routes to C-glycosides were originally developed on 2-deoxy sugars because of the issues associated with facile β-elimination of a C(2)-subsitutent. This problematic elimination can be prevented by a protective metallation of the C(2) residue prior to metallation at C(1) i.e a dianion is involved. Wittmann and Kessler have shown [42] that treatment of 3,4,6-tri-*O*-benzyl-α-D-glucopyranosyl chloride **132** with butyl lithium at −78 °C (to produce the lithium alkoxide) followed by LN at −100 °C generates the dilithio species **133** that reacts with simple electrophiles (aldehydes and MeI in the presence of CuI) to give C-glycosides **134** with retention of configuration (Scheme 28). Interestingly, use of the corresponding α-stannane **135** is compromized by the poor yields

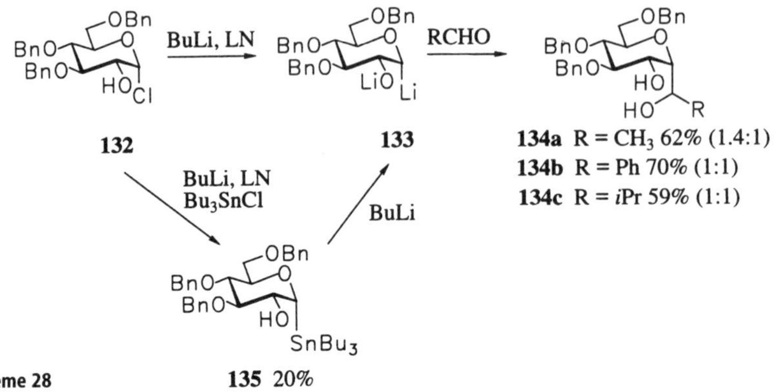

BuLi, LN
RCHO
132
133
134a R = CH_3 62% (1.4:1)
134b R = Ph 70% (1:1)
134c R = *i*Pr 59% (1:1)
BuLi, LN
Bu_3SnCl
BuLi
135 20%

Scheme 28

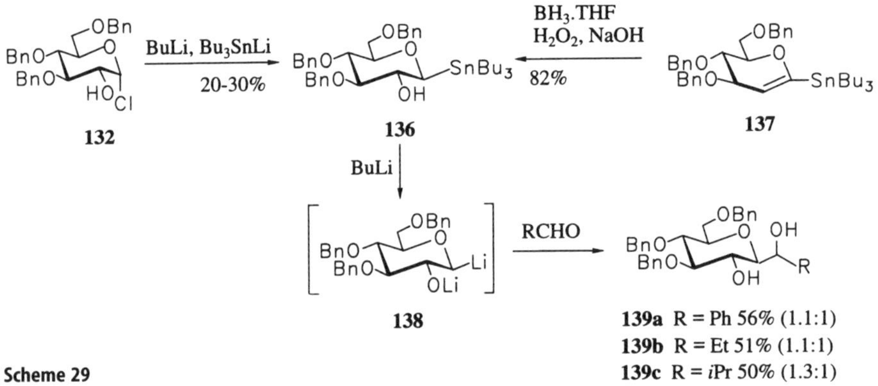

BuLi, Bu_3SnLi
20-30%
BH_3.THF
H_2O_2, NaOH
82%
132
136
137
BuLi
RCHO
138
139a R = Ph 56% (1.1:1)
139b R = Et 51% (1.1:1)
139c R = *i*Pr 50% (1.3:1)

Scheme 29

associated with the preparation of this intermediate from chloride **132** [29]. A similar problem is encountered in the preparation of the isomeric *β*-stannane **136** [29] (Scheme 29). Here, a more reliable, although lengthier route to stannane **136** relies on a hydroboration-oxidation sequence [29] based on vinyl stannane **137** (see Sect. 3.2). Hydroxyl deprotonation of **136** followed by transmetallation generates the *β*-dilithio reagent **138** which reacts with aldehydes to give C-glycosides **139**. Opening of glucal epoxide **140** [43] with triphenylstannyl lithium provides an efficient route to triphenylstannane **141** [44] (Scheme 30). However, conversion of **141** to the *β*-dilithio reagent **138** requires a large (10 equiv.) excess of butyl lithium as a consequence of phenyl/butyl exchange around the tin center prior to the transmetallation event [44]. Acylation of **138** is only possible with the less reactive benzonitrile providing, after hydrolytic workup, C-benzoylated product **142**. Reactivity profiles are not, however, straightforward and simple C-alkylation (using dimethyl sulfate) to give **143** requires conversion of dianion **138** to a cyanocuprate reagent.

Opening of epoxide **140** with tributylstannylmethyl magnesium is probably the best available route to stannane **136** [45] (Scheme 31). After protection of the C(2)–OH as a MOM ether, stannane **144** undergoes a facile cross-coupling reac-

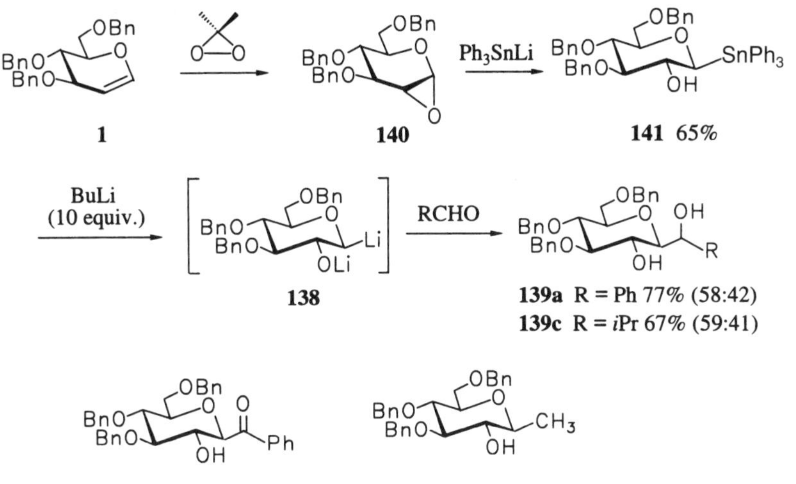

1 → 140 → (Ph3SnLi) 141 65%

BuLi (10 equiv.) → [138] → RCHO →

139a R = Ph 77% (58:42)
139c R = *i*Pr 67% (59:41)

Scheme 30 142 84% 143 77%

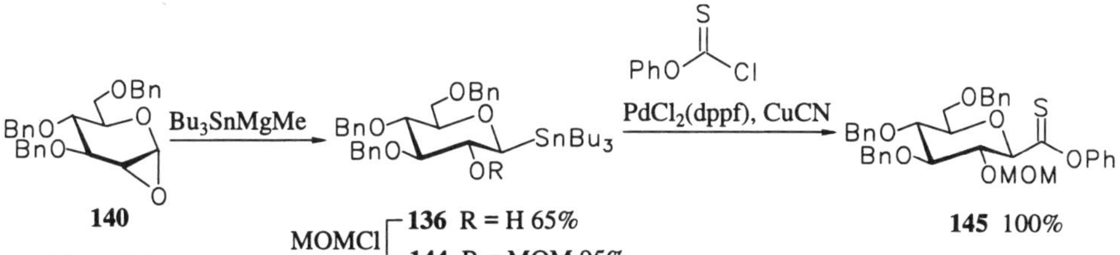

140 → ($Bu_3SnMgMe$) → 136 R = H 65% → (MOMCl) 144 R = MOM 95% → (PhOC(S)Cl, $PdCl_2$(dppf), CuCN) → 145 100%

Scheme 31

tion with thioformates (e.g. phenyl chlorothionoformate) under Pd(0)–Cu(I) catalysis providing the corresponding *O*-phenyl thioester **145** in quantitative yield [45] (Scheme 31). Additional aspects of transition metal-mediated synthesis of C-glycosides are dealt with in Sect. 2.4.

2.2.3
Reductive Samariation

Derivatives of 2-alkoxy (or acyloxy) glycopyranosides are unsuitable as substrates for reductive lithiation because of competitive (and comprehensive) *β*-elimination. Even a trimethylsilyloxy group at C(2) as in **147** will either rearrange to the C(1)-silane **148** in the *gluco* or *galacto* series or, as in the *manno* series **149**, undergo elimination (to give **151**) [46] (Scheme 32). In 1988, Prandi and Beau [47] showed that reductive samariation of acetobromo-D-galactose **152** in the presence of heptanal (Barbier conditions) did produce *β*-D-C-glycoside **153** in which acetate migration from C(2) to the exocyclic hydroxyl was observed (Scheme 33). Yields are poor but two features are noteworthy. First, in an intermolecular samarium-mediated Barbier reaction, the coupling event must proceed, at least in part, through an anionic mechanism via a transient

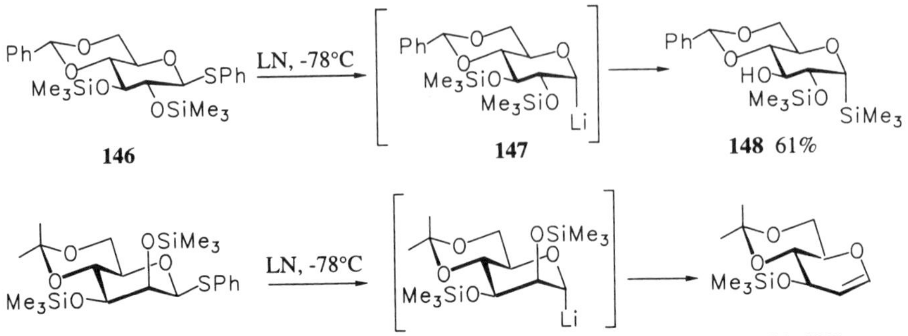

LN, -78°C

146 **147** **148** 61%

LN, -78°C

Scheme 32 **149** **150** **151** 68%

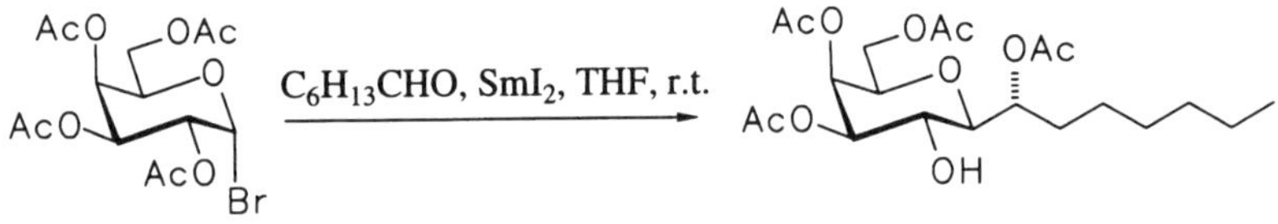

$C_6H_{13}CHO$, SmI_2, THF, r.t.

Scheme 33 **152** **153** 15%

SmI_2, THF-HMPA

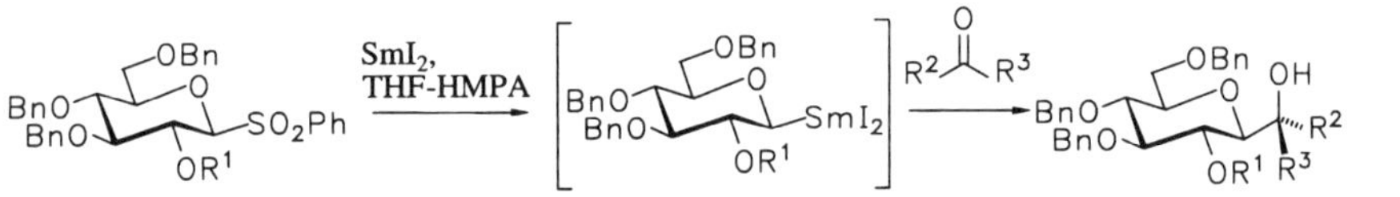

154 R^1 = Bn
155 R^1 = Na

156

157a R^1 = Bn, R^2, R^3 = cyclopentyl 18%
157b R^1 = Bn, R^2 = *t*Bu, R^3 = H 24%
157c R^1 = H, R^2, R^3 = H, *t*Bu 16% (1:1)

Scheme 34

β-anomeric organosamarium reagent [48]; an alternative radical process would be expected to lead to an α-C-glycoside. Secondly, the intermediate anionic species does not simply undergo elimination but is stable enough, at room temperature, to enable a productive carbon-carbon bond formation. Sinaÿ and coworkers demonstrated that reductive samariation of phenylsulfone **154**, with the SmI_2-THF-HMPA system and in the presence of aldehydes or ketones, gives the β-D-C-glycosides **157** in low yields, along with varying amounts of glucal **1** [16] (Scheme 34). In an effort to improve the yield of C-glycoside **157c**, reductive samariation of the C(2) sodium alkoxide **155** was examined but proved fruitless. Contrary to expectation, Skrydstrup and Beau observed that reductive samariation (SmI_2, THF, no HMPA) of α-mannopyranosyl 2-pyridyl sulfone **158** (under Barbier conditions) provides the α-C-mannopyranosides **159** very efficiently and with excellent stereoselectivity and, under these conditions, formation of glucal **1** is reduced [19] (Scheme 35). Similar results are obtained with other C(2)–OH protecting groups such as Me, Bn or MEM and with the C(2)–$OSiMe_2$ *t*-Bu (TBS), no elimination product was detected [49]. In contrast, β-C-glucopyranosides **161** are formed exclusively from the corresponding β-gluco-

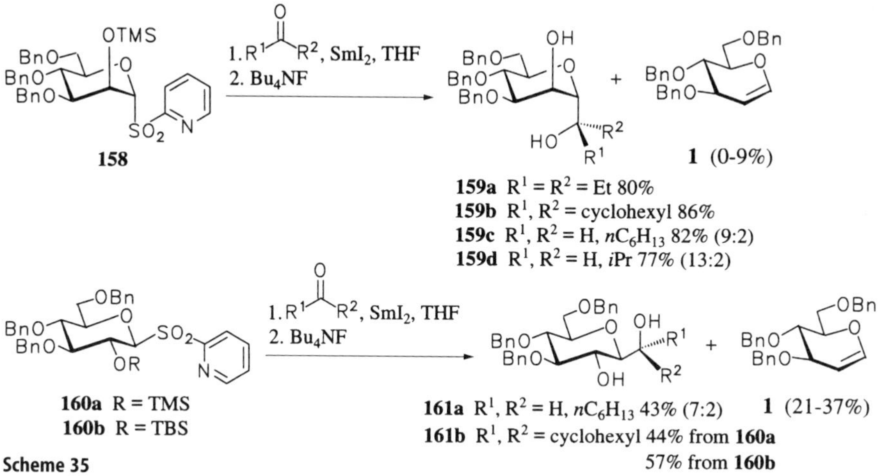

OTMS
BnO
BnO
BnO
-O
SO_2 N
158

1. $R^1C(O)R^2$, SmI_2, THF
2. Bu_4NF

OH
BnO
BnO
BnO
-O
HO
R^2
R^1
+
OBn
BnO
BnO
O
1 (0-9%)

159a $R^1 = R^2$ = Et 80%
159b R^1, R^2 = cyclohexyl 86%
159c R^1, R^2 = H, nC_6H_{13} 82% (9:2)
159d R^1, R^2 = H, *i*Pr 77% (13:2)

OBn
BnO
BnO
O
SO_2
OR
N

1. $R^1C(O)R^2$, SmI_2, THF
2. Bu_4NF

OBn
BnO
BnO
O
OH
R^1
OH
R^2
+
OBn
BnO
BnO
O

160a R = TMS
160b R = TBS

161a R^1, R^2 = H, nC_6H_{13} 43% (7:2) **1** (21-37%)
161b R^1, R^2 = cyclohexyl 44% from **160a**
57% from **160b**

Scheme 35

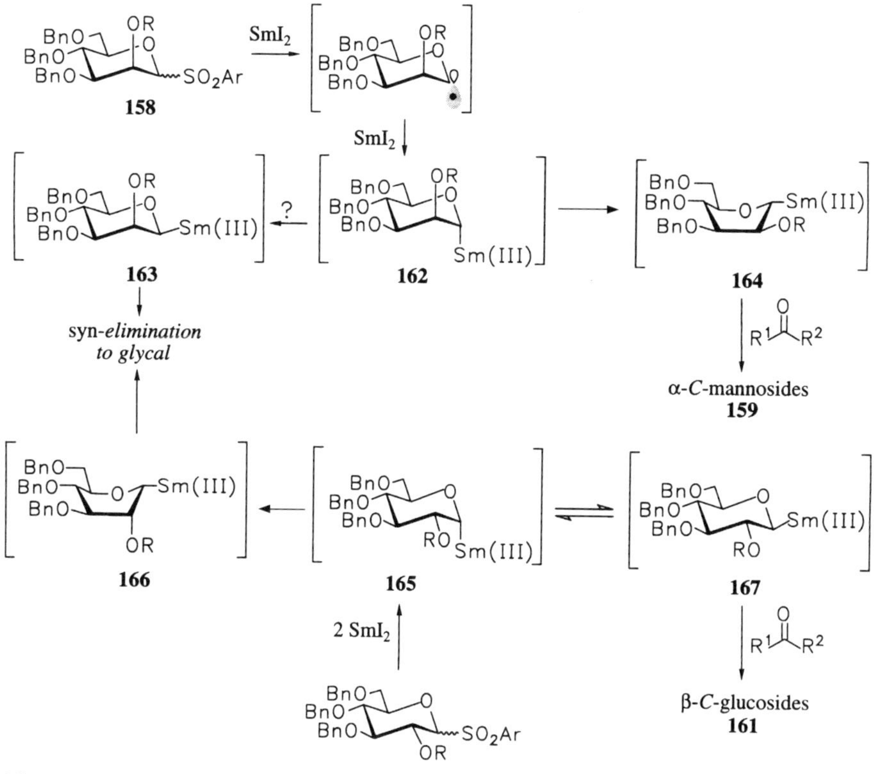

BnO
BnO
BnO
OR
-O
SO_2Ar
158

SmI_2

BnO
BnO
BnO
OR
-O
O

SmI_2

BnO
BnO
BnO
OR
-O
Sm(III)
163

?

BnO
BnO
BnO
OR
-O
Sm(III)
162

BnO
BnO
BnO
O
Sm(III)
OR
164

syn-*elimination*
to glycal

$R^1C(O)R^2$
α-*C*-mannosides
159

BnO
BnO
BnO
O
Sm(III)
OR
166

BnO
BnO
BnO
O
RO
Sm(III)
165

BnO
BnO
BnO
O
Sm(III)
RO
167

$R^1C(O)R^2$
β-*C*-glucosides
161

2 SmI_2

BnO
BnO
BnO
O
SO_2Ar
OR
160

Scheme 36

pyranosyl 2-pyridyl sulfones **160** Yields are moderate because of competitive elimination. Better yields of β-C-glucosides are obtained when the protecting group at *O(2)* is bulkier (TMS *vs.*TBS) as in **160a/b.**

It is informative to compare these results with those available in the 2-deoxy series (see Sect. 2.1.1) and the mechanistic pathways are summarized in Scheme 36. Sulfone reduction of **158** (*manno* series) or **160** (*gluco* series) leads to the kinetically favored α-organosamarium (III) species **162** or **165** respectively via reduction of the corresponding anomeric radicals. In the *manno* series, epimerization to give the β-species **163** is slow compared to a facile conformational change to generate **164** which places the C(1) and C(2) substituents in more stable equatorial positions, ultimately leading to α-C-mannosides **159.** In the *gluco* series, this conformational change (**165** to give **166**) is energetically more demanding and epimerization to the β-species **167** dominates which then leads to the corresponding β-C-glucosides **161**. The exclusive formation of the 1,2-*trans* C-glycosides is explained by a facile *syn*-elimination of the samarium reagents **163** and **166** at room temperature. In short, only the 1,2-*trans* organosamarium species **164** and **167** lead to a productive C–C bond formation.

This samarium-based method is also a useful route for a rapid assembly of C-disaccharides. Reductive samariation of sulfone **158** in the presence of

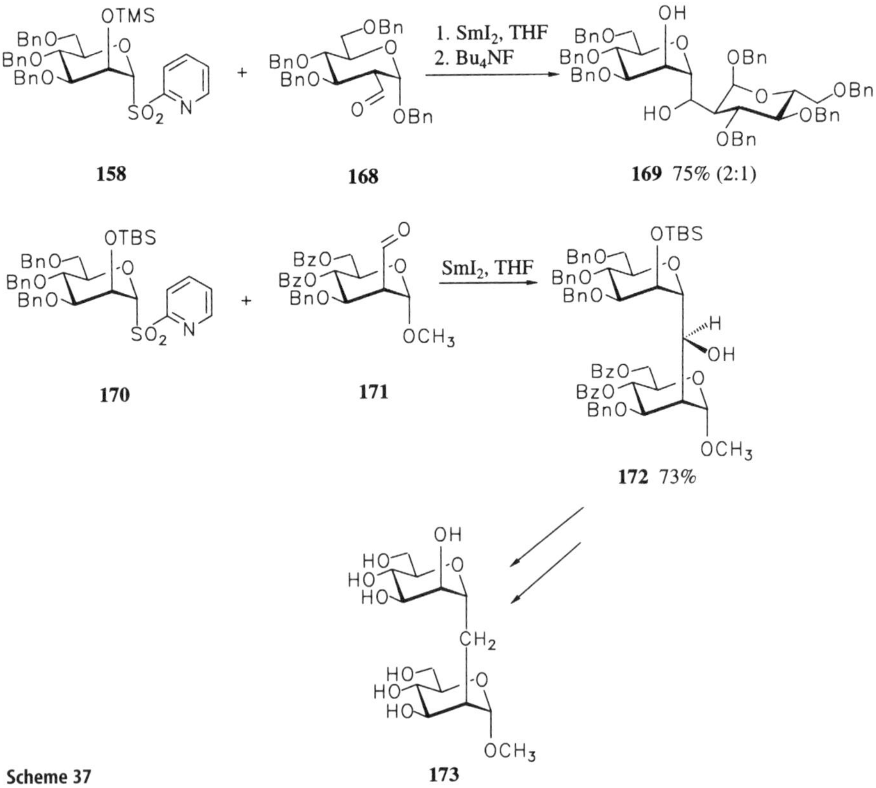

158 + **168** → 1. SmI_2, THF; 2. Bu_4NF → **169** 75% (2:1)

170 + **171** → SmI_2, THF → **172** 73% → **173**

Scheme 37

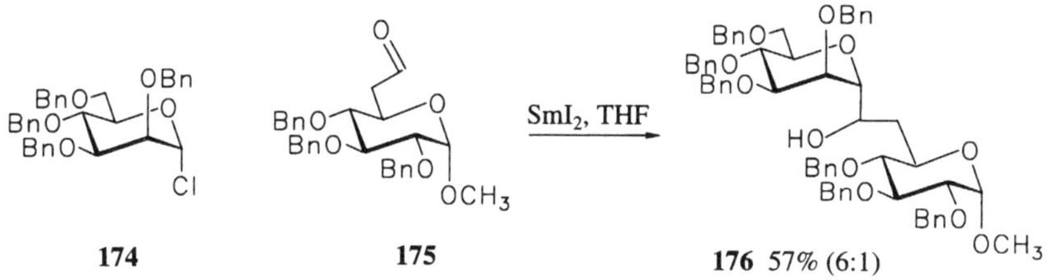

174 175 176 57% (6:1)

SmI_2, THF

Scheme 38

aldehyde **168** gives the α-C-disaccharide **169**, a derivative of an α-C-mannopyranosyl(1 → 2)-D-glucopyranose [19] (Scheme 37). Similarly, coupling of sulfone **170** and aldehyde **171** furnishes α-C-disaccharide **172** which was transformed to methyl α-C-mannobioside **173**, a C-glycoside analogue of the *Mycobacterium tuberculosis* capping disaccharide [50]. This unique behavior of the α-D-mannopyranosyl samarium(III) species was recently exploited in the synthesis of a C-linked 1,6-disaccharide **176** by reductive samariation of mannopyranosyl chloride **174** in the presence of aldehyde **175** [51] (Scheme 38). In general terms, one should expect that any substituent at the anomeric center possessing an appropriate low energy LUMO will lead efficiently to an anomeric samarium (III) nucleophile suitable for C-glycoside synthesis.

2.3
C(1)-Metallated 2-Amino-2-deoxypyranosides

Vasella has applied the concept of anomeric anion stabilization by a nitro group to the β-D-*N*-acetyl-D-glucosamine derivative **177**, available in four steps from *N*-acetyl-D-glucosamine [52] (Scheme 39). Reaction of the tetraethylammonium nitronate derived from **177** with aldehyde **178** provides *anti*-**179** which then undergoes stereoselectively reduction (see Sect. 2.2.1) to provide β-C-glycoside **180**, intermediate in a synthesis of *N*-acetyl-neuraminic acid.

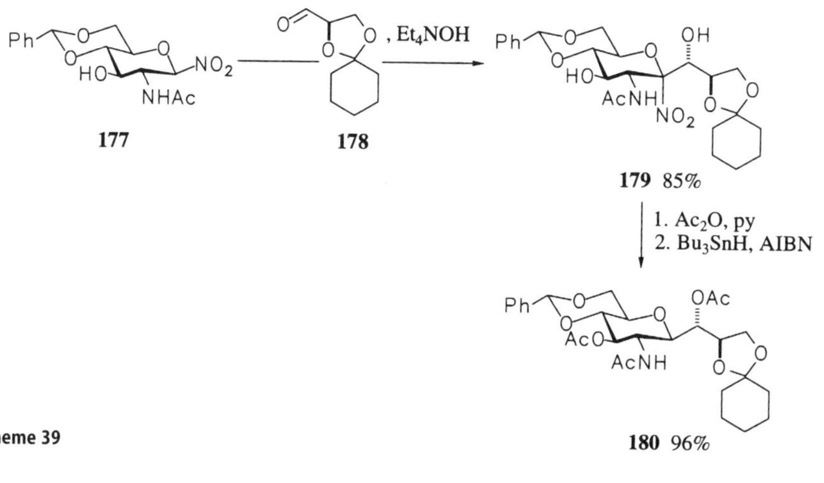

177 178 179 85%

, Et_4NOH

1. Ac_2O, py
2. Bu_3SnH, AIBN

180 96%

Scheme 39

In contrast to the extensive studies that have focused on C(1) nucleophiles carrying a 2-alkoxy/2-hydroxy moiety (see Sect. 2.2), the chemistry of the corresponding 2-amino-2-deoxy variants is less well explored. Kessler has described the generation of the dilithio species **181** (by halogen-lithium exchange) and **183** (by tin-lithium exchange) as well as the reactivity of these C(1) nucleophiles towards simple electrophiles (aldehydes and Me_2SO_4) as a route to 2-acetamido-2-deoxy-C-glycosides (Scheme 40) [53]. The two key transformations (generation of the dilithiated species and electrophilic trapping) are stereospecific, thereby providing access to the α- and β-C-glycosides **182** and **184** respectively.

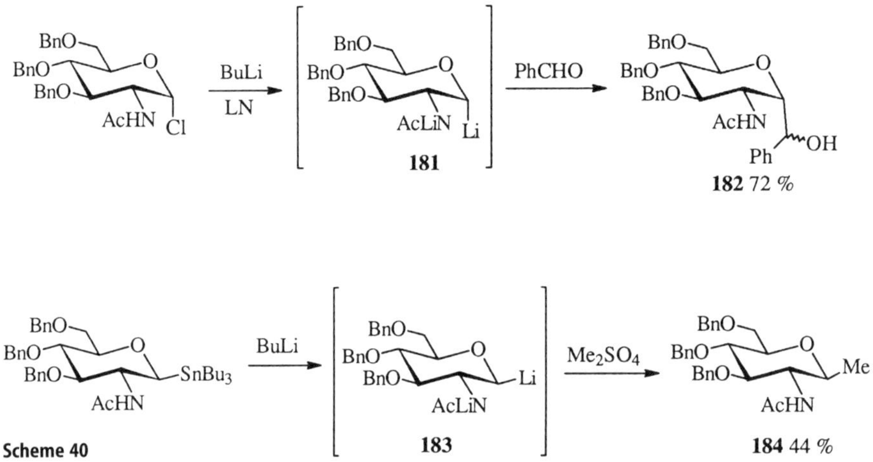

Scheme 40

Reductive samariation has also found application in this aspect of C-glycoside synthesis. An exception to the 1,2-*trans* "rule" associated with 2-alkoxy (or silyloxy) glycopyranosyl 2-pyridylsulfones (see Sect. 2.2.3) was observed by Skrydstrup and Beau in the reductive samariation of anomeric pyridyl sulfones derived from *N*-acetyl-D-galactosamine [54]. Reductive samariation of 2-pyridyl sulfone **185** in the presence of carbonyl compounds does not provide the anticipated β-products but rather α-C-glycosides **186** (Scheme 41). This transformation is only stereoselective (α/β ratios, 9–20:1) because competitive elimination does not occur (as is observed in the 2-oxo series). This unusual result is ascribed to a strong complexation of the anomeric samarium moiety by the acetamido group which appears to stabilize the α-anomeric configuration, either as conformer **187** or with samarium occupying an "equatorial" position as in **188** [54]. This stereocontrolled synthesis of α-D-C-galactosamine derivatives is representative of a chelation-controlled C-glycosylation.

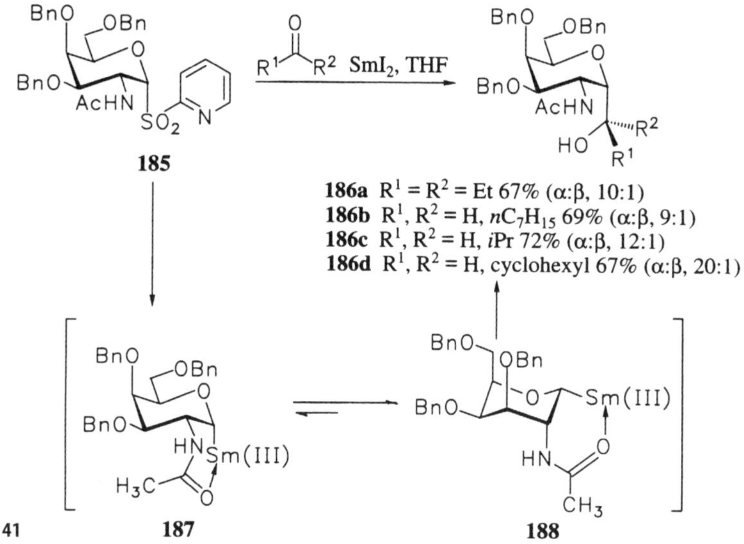

$R^1C(O)R^2$, SmI_2, THF

185

186a $R^1 = R^2 =$ Et 67% (α:β, 10:1)
186b R^1, R^2 = H, $n$$C_7H_{15}$ 69% (α:β, 9:1)
186c R^1, R^2 = H, *i*Pr 72% (α:β, 12:1)
186d R^1, R^2 = H, cyclohexyl 67% (α:β, 20:1)

Scheme 41 **187** **188**

2.4
Transition Metal-Mediated C(1) Nucleophilic Reactivity

While transition metals have been used extensively in C-glycoside construction, assigning a precise reactivity profile to C(1) can be difficult and this section will cover cases where nucleophilic character is apparent, albeit loosely defined. The use of palladium-mediated synthesis of C-glycosides has been reported and this topic, which involves an anomeric tin intermediate, has been covered in Sect. 2.2.2. The insertion of carbon monoxide into a carbon-metal (C(1)-metal) bond is, however, an important aspect of transition metal-mediated C-glycoside formation, and this process has been observed and exploited in a number of situations.

2.4.1
Manganese Reagents

Stable C(1) organomanganese complexes, based on both furanosides and pyranosides, are available by displacement of the corresponding anomeric bromides with sodium pentacarbonylmanganate (I); for example, reaction of **189** leads to complex **190** [55, 56]. In the presence of carbon monoxide, facile insertion into the C–Mn bond of **190** gives an acyl manganese intermediate **191**, which may be trapped by an alkoxide (or thiolate) nucleophile to give the corresponding carboxylic acid derivative **192** (Scheme 42). Alternatively, a three-component coupling sequence involving complex **193** and a reactive alkene or alkyne may also be effected leading to more highly functionalized C-glycosides **194** and **195** (Scheme 43) [57]. This process requires use of high pressure conditions as well as a subsequent protolytic or photolytic step to achieve demetallation of the initially-formed addition products.

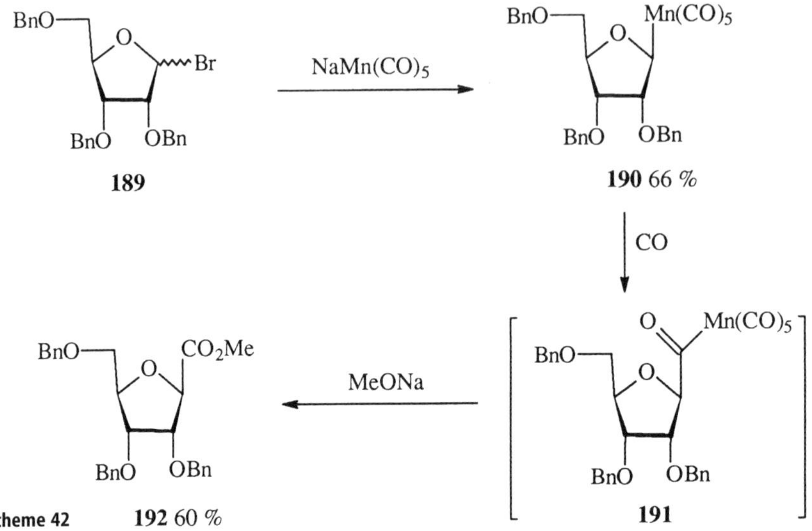

189

NaMn(CO)$_5$

190 66 %

CO

191

MeONa

192 60 %

Scheme 42

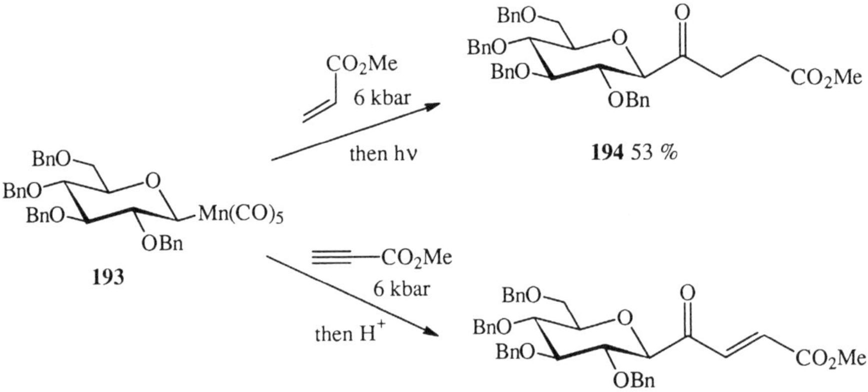

193

CO$_2$Me 6 kbar then hν

194 53 %

≡—CO$_2$Me 6 kbar then H$^+$

195 52 %

Scheme 43

2.4.2
Cobalt and Iron Reagents

Cobalt-mediated reductive siloxymethylation of anomeric acetates provides a related approach to C-glycosides (Scheme 44) [58]. Again, initial reaction involves a metal-centered "nucleophile" with the C-glycosides **196** and **197** being set up in a subsequent CO-insertion/reduction sequence. Good levels of stereocontrol have been reported at C(1) and a role for the C(2)-OAc (as a participating group) has been suggested. Interestingly, the closely related C(1) iron complex **198** [59] is stable under the conditions used for CO-insertion, and iron-based methods have not found utility in C-glycoside synthesis.

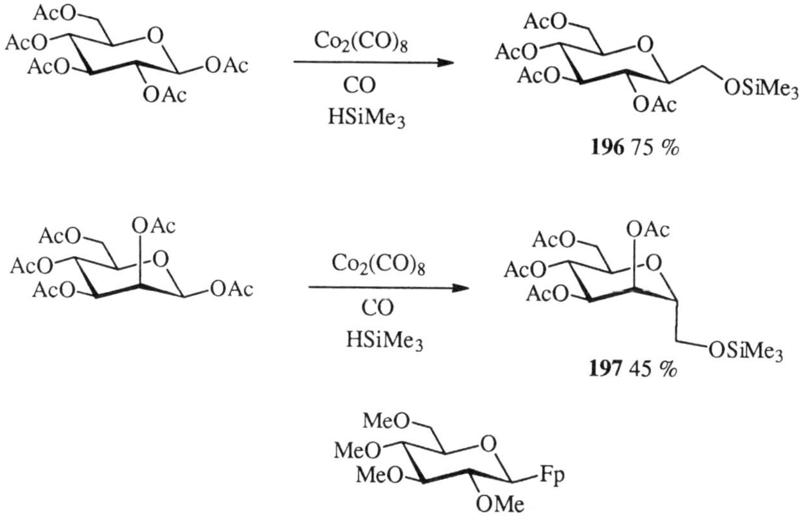

$Co_2(CO)_8$, CO, $HSiMe_3$

196 75 %

$Co_2(CO)_8$, CO, $HSiMe_3$

197 45 %

198 [Fp = (η_5–C_5H_5)Fe$(CO)_2$]

Scheme 44

3
Nucleophilic *sp*²-Anomeric Centers

Another way to facilitate metallation at C(1) is to work with sugars possessing a vinylic anomeric hydrogen, that is to use glycals as substrates. Direct deprotonation with strong bases is possible – with certain limitations – and the C(1) stannylated glycals provide the basis of useful routes to C(1)-substituted glycals and C-glycosides, particularly when harnessed to palladium-catalyzed coupling reaction with organic halides.

3.1
Deprotonation Processes

Three independent papers in 1986 reported that glycals e.g. **199** are deprotonated by *tert*-butyl lithium [60, 61] or butyl lithium-potassium *tert*-butoxide [62] (Schlosser's base) provided that silyl ethers are used as protecting groups (Scheme 45). The lithiated glucal **200** reacts with various electrophiles to give the corresponding C(1)-substituted glycals **201** [60], with transmetallation by catalytic amounts of copper iodide aiding the C-allylation reaction to give **201b** [61]. Trapping of **200** with a tin-based electrophile is an important process that will be discussed in Sect. 3.2. C(1)-Lithiated glycals e.g. **202** (from L-rhamnal) react with quinone derivatives, such as **203**, in a 1,2-fashion to provide quinone ketal **204** [63] (Scheme 46). Adduct **204** provides a route to the C-aryl glycoside **206** by reductive aromatization with diisobutylaluminium hydride and stereoselective hydration of the intermediate glycal **205**. An alternative method for reductive aromatization of a quinone adduct, e.g. **207** derived from lithio glycal **202**, uses sodium dithionite [64] (Scheme 47). Similar chemistry has also been applied to furanoid glycals [65]. Lithiation of glycal **208** followed by reaction with benzo-

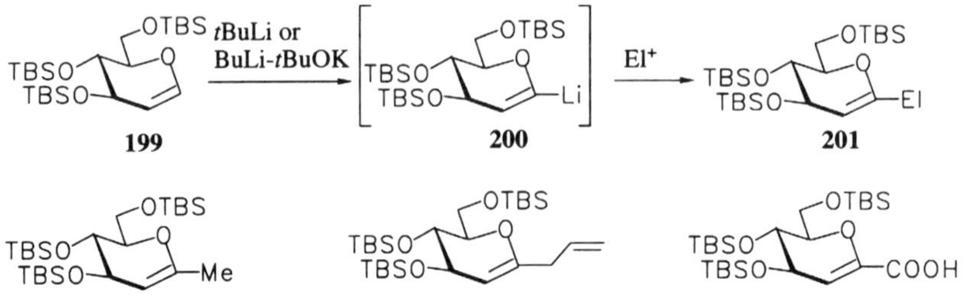

199 200 201

201a 86% 201b 70% 201c 60%

Scheme 45

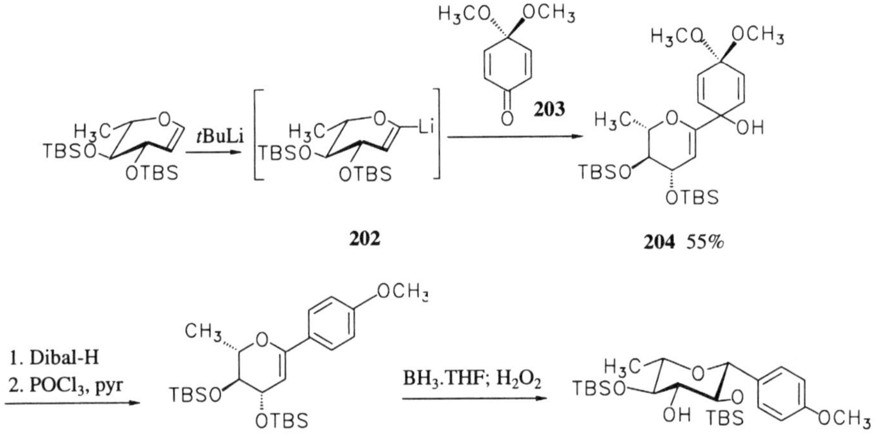

202 203 204 55%

205 94% 206 51%

Scheme 46

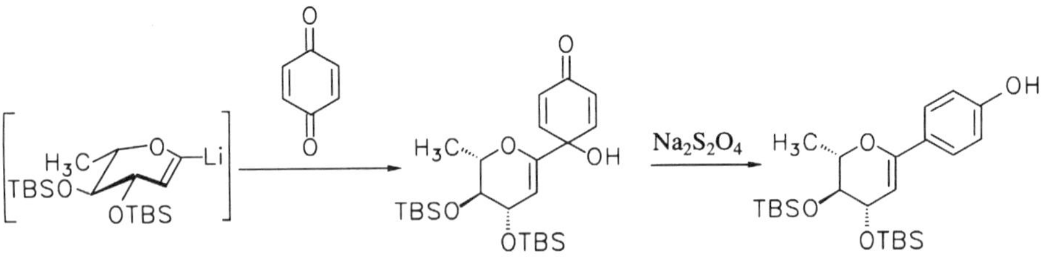

202 207 91% 84%

Scheme 47

1,4-quinone leads to adduct **209** which has been converted directly to the C-aryl glycofuranoside **210** by a hydroboration-oxidation sequence, a procedure not successful within the pyranoid series. An application of this strategy to the synthesis of bis-C-aryl glycosides **216** based on the kidamycin antibiotics is shown in Scheme 49. 1,2-Addition of lithiated rhamnal **202** to 1,4-naphthoquinone provides the C-naphthoquinol glycal **211** [66] (Scheme 49). Silylation to give **212**

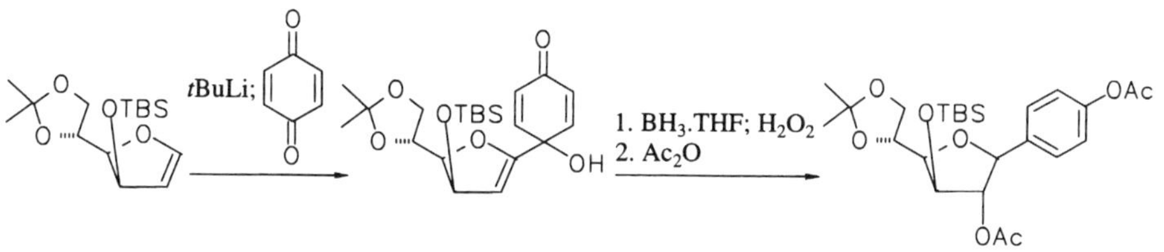

208 **209** 76% **210** 54%

Scheme 48

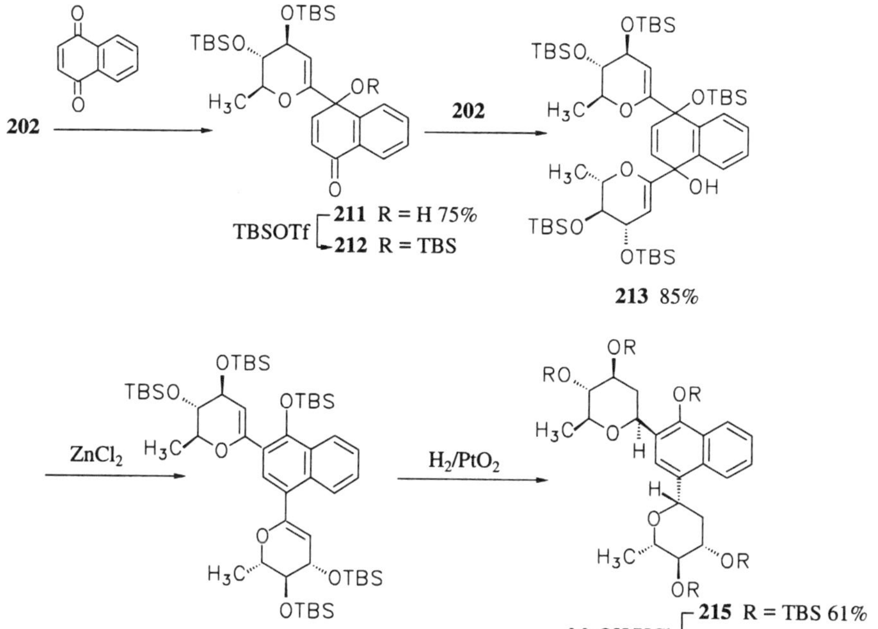

211 R = H 75%
212 R = TBS
213 85%
214 92%
215 R = TBS 61%
216 R = H 69%

Scheme 49

followed by reaction with a second equivalent of **202** results in the the 1,4-bis adduct **213** which undergoes regioselective Lewis acid-mediated rearrangement to the 1,3-bisaryl glycal **214**. Finally, hydrogenation over platinum oxide affords 2,4-bis-C-glycosylated naphthol **215** which was desilylated to give **216**.

Conversion of the lithiated glycals to the more stable organozinc reagents is an attractive option. In a synthesis of the methyl ester of vineomycinone B2 **221**, Tius and coworkers [67] found that coupling of the chlorozinc derivative **217** with the iodoanthracenyl derivative **218**, in the presence of a palladium(0) catalyst [$Pd(PPh_3)_2Cl_2$, i-Bu_2AlH], provided efficient access to the C-aryl glycal **219** (Scheme 50). Subsequent cyanoborohydride reduction at a controlled pH provides the β-D-C-glycoside **220** as a single stereoisomer.

Appropriate substituents at C(2) can facilitate direct C(1)-lithiation of glycals. Schmidt and coworkers [68] demonstrated that Schlosser's base achieves de-

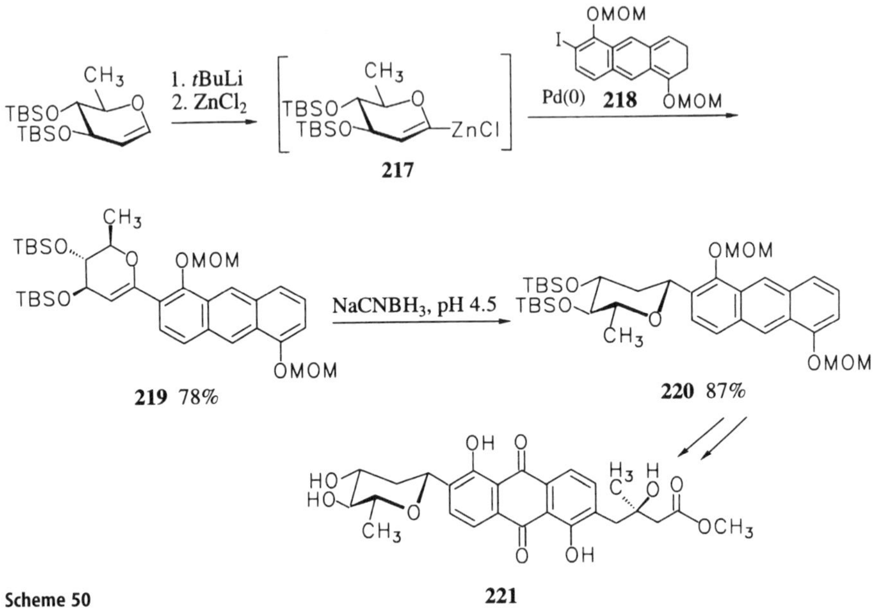

Scheme 50

protonation of the *O*-benzyl-protected 2-alkoxy-D-glucal **222**, leading to a C(1)-lithio species **223** that can be trapped by a limited range of simple electrophiles to give C-glycals **224** (Scheme 51). With a 2-(phenylthio) [58] or even better, a 2-(phenylsulfinyl) substituent at C(2) [69], LDA is a sufficiently strong base to permit a clean C(1)-lithiation. Scheme 52 illustrates the use of 2-(phenylsulfinyl)glucal **225** which is readily available from glucal **1** (PhSCl then DBU elimination and oxidation): Deprotonation of **225** with LDA cleanly gives lithiated glucal **226** which reacts with aldehydes as shown. Removal of the phenylsulfinyl group from adduct **227 a** was carried out using Raney nickel and stereospecific hydration (as earlier) leads to β-D-C-glucoside **228**. This chemistry has been successfully applied to the synthesis of a novel β-glucosidase inhibitor **229** [70] and the C-linked cellobiose derivative **230** [71].

Schmidt has used the same strategy to construct the bicyclic structure **235** of ezomycin A [72]. C(1) Lithiation of 2-(phenylsulfinyl)-D-galactal **231** followed by reaction with glyoxal monoacetal gives adduct **232** (Scheme 53). The (*R*)-isomer of **233** was then carried through to the β-D-C-galactose **234** which was ultimately converted, under Vorbrüggen conditions, to the target nucleoside **235**.

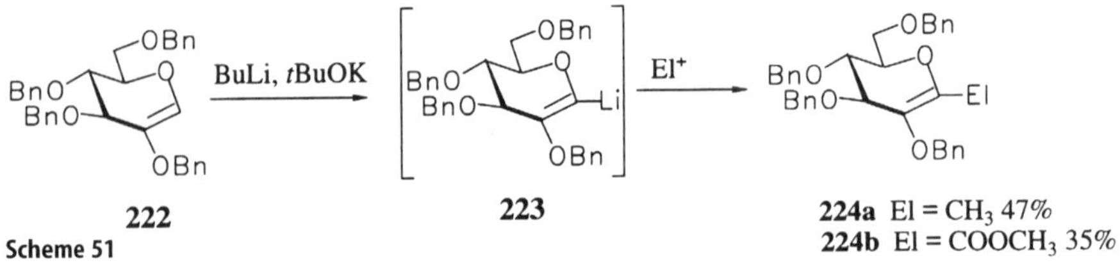

Scheme 51

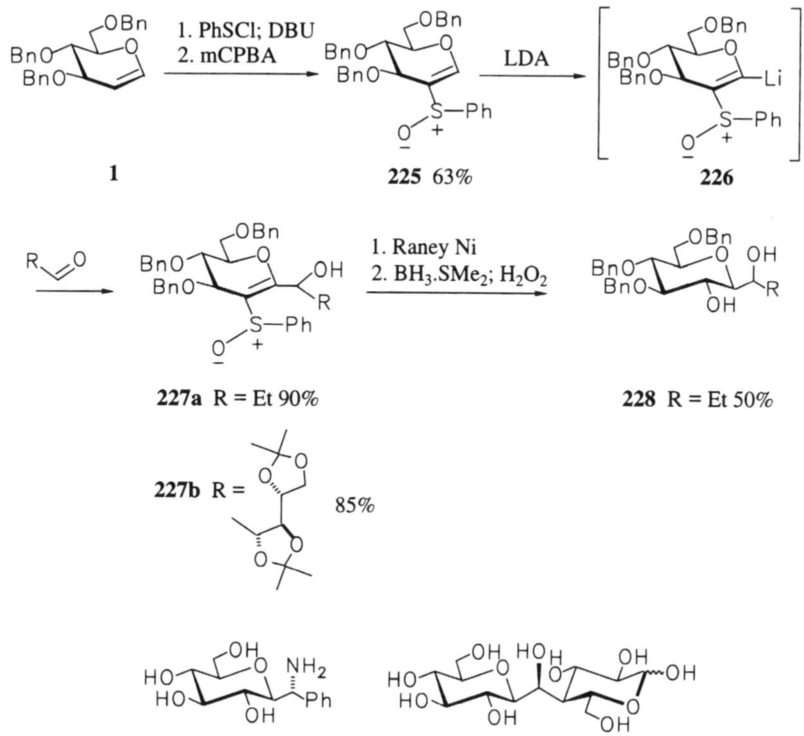

1. PhSCl; DBU
2. mCPBA

LDA

1

225 63%

226

1. Raney Ni
2. $BH_3.SMe_2$; H_2O_2

227a R = Et 90%

228 R = Et 50%

227b R = 85%

229

230

Scheme 52

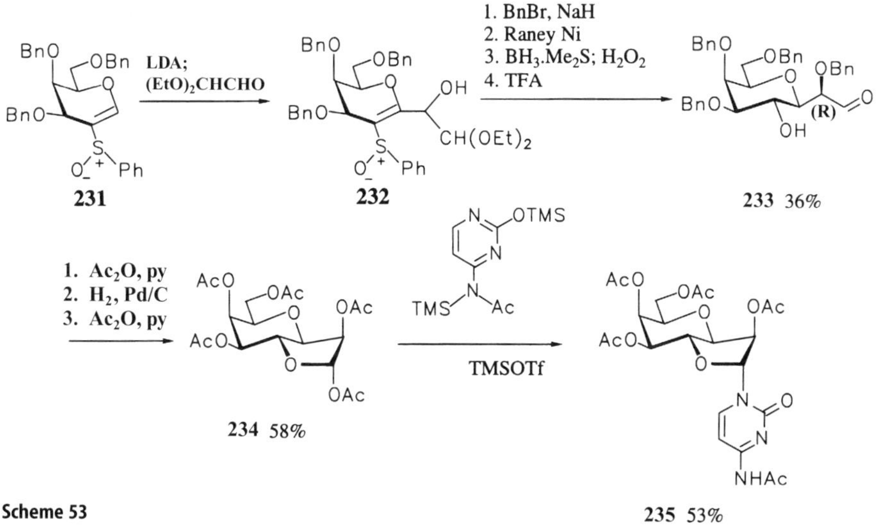

LDA;
$(EtO)_2CHCHO$

1. BnBr, NaH
2. Raney Ni
3. $BH_3.Me_2S$; H_2O_2
4. TFA

231

232

233 36%

1. Ac_2O, py
2. H_2, Pd/C
3. Ac_2O, py

TMSOTf

234 58%

235 53%

Scheme 53

3.2
C(1)-Stannylated Glycals

In the preceding section, the preparation and reactivity of 1-tributylstannyl glycals (e.g. **236**) was omitted, although these reagents do provide access to the same series of C(1)-lithiated glycals via facile tin-lithium exchange. In addition, vinyl stannanes of this type undergo efficient Pd(0)-catalyzed cross-coupling reactions (Stille reaction [73]) which extend the usefulness of this class of glycals in C-glycoside synthesis.

The first preparation of C(1)-(tributylstannyl)-D-glucal **236** by stannylation of the C(1)-lithiated species was reported by Hanessian [62] (Scheme 54). After exchange of the protecting groups, tin-lithium exchange of **237** cleanly provides the lithio reagent **238** which reacts with methyl iodide and aldehydes, including a 6-aldehydo-D-glucoside, in good yields. The advantage of this process over the routes described in Sect. 3.1 is that the lithiated glycal is produced under mild conditions. However, the synthesis of C(1)-stannylated glycals does necessitate use of a strong base which means that certain protecting groups, typically benzyl ethers or benzylidene residues, are compromized. Silyl protection is usually acceptable, but even here competitive metallation α to silicon may be encountered [74]. In 1986, Beau and Sinaÿ [60] proposed a preparation of the C(1)-tributylstannyl glycals based on the corresponding *S*-glycosides (Scheme 55). Oxidation of **239** provides anomeric sulfones which, upon treatment with base, eliminate to the vinyl sulfones **240**. A radical-mediated addition-elimination sequence then converts vinyl sulfone **240** to the corresponding vinyl stannanes **241** [60, 75–79]. An added advantage is that, under these conditions, benzyl, benzylidene, amidyl or free hydroxyl groups are tolerated and the range of vinyl stannanes available by this method are shown in Scheme 55. Incomplete stannylation of the vinyl sulfones can be a problem that may be overcome by use of a low concentration of the starting sulfone and larger amounts of Bu_3SnH/AIBN, conditions which are diagnostic of a premature quenching of the radical chain (Scheme 56). C(1)-Lithiated glycals prepared via tin-lithium exchange, react with epoxides that have been activated by boron trifluoride etherate and, contrary to the constraints associated with the related sp^3-C(1)-nucleophiles (see Sect. 2.1.3), prior conversion to the cyanocuprate reagents is not required [33].

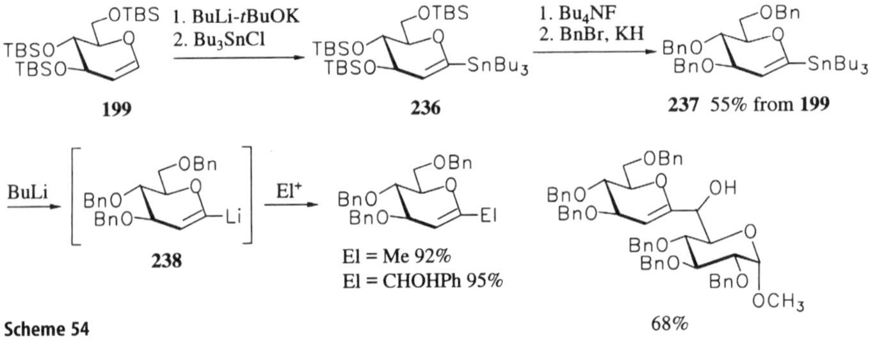

Scheme 54

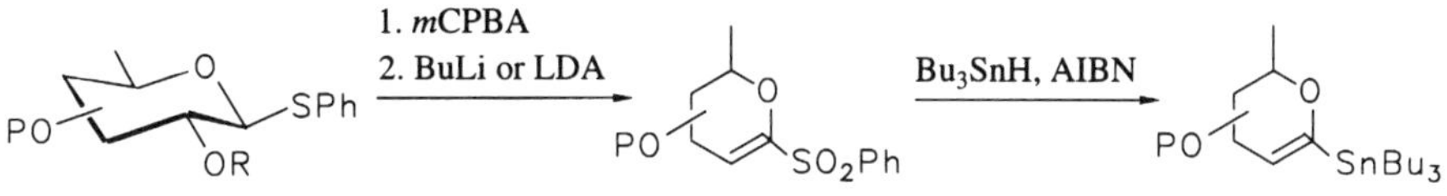

1. *m*CPBA
2. BuLi or LDA

Bu_3SnH, AIBN

239 **240** **241**

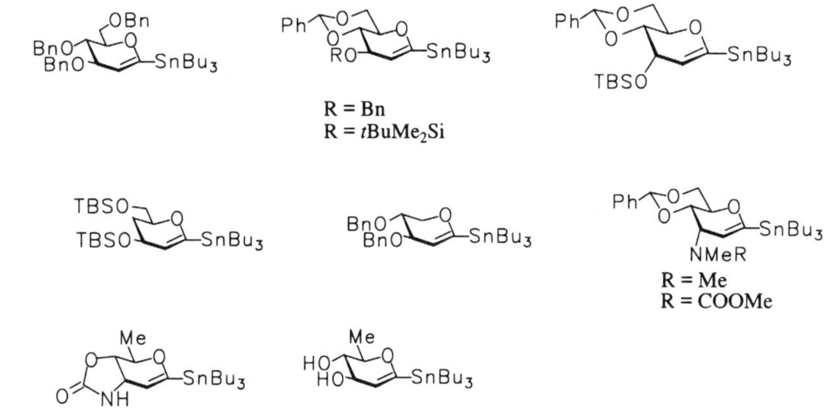

R = Bn
R = $tBuMe_2Si$

R = Me
R = COOMe

Scheme 55

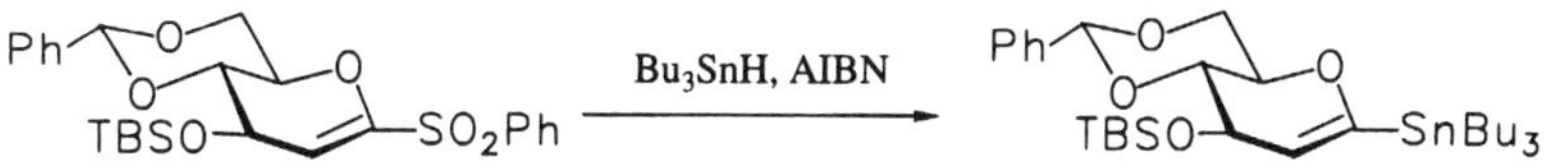

Bu_3SnH, AIBN

Starting material concentration	Bu_3SnH	AIBN	Yield
0.1 M	2.5 equiv.	0.1 equiv	71%
0.02 M	3 equiv.	0.6 equiv.	93%

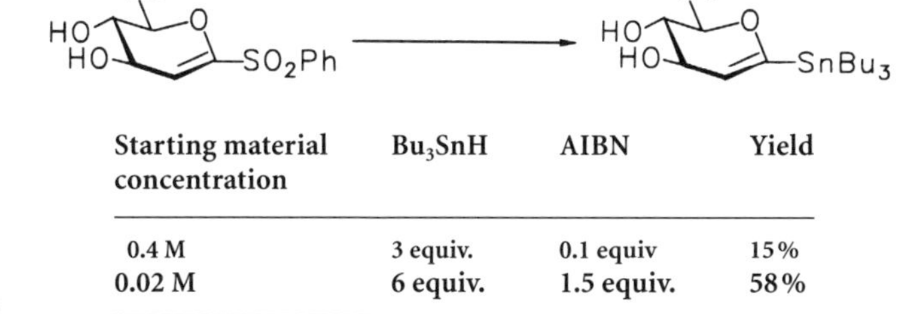

Starting material concentration	Bu_3SnH	AIBN	Yield
0.4 M	3 equiv.	0.1 equiv	15%
0.02 M	6 equiv.	1.5 equiv.	58%

Scheme 56

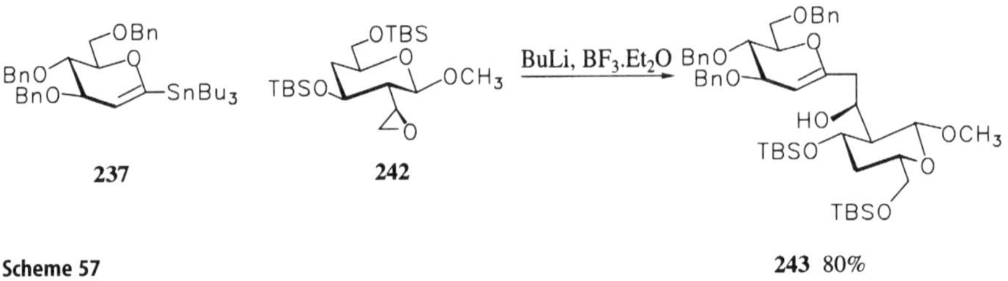
OBn
BnO
BnO
O
SnBu3
237
OTBS
O
TBSO
OCH3
O
242
BuLi, BF3.Et2O
OBn
BnO
BnO
O
HO
TBSO
OCH3
O
TBSO
243 80%

Scheme 57

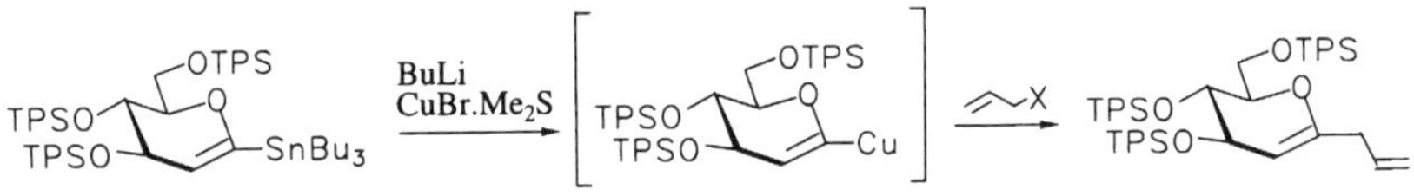
OTPS
TPSO
TPSO
O
SnBu3
BuLi
CuBr.Me2S
OTPS
TPSO
TPSO
O
Cu
X
OTPS
TPSO
TPSO
O

Scheme 58 244

Scheme 57 illustrates an example of this process: coupling of the lithio reagent derived from vinyl stannane **237** with epoxide **242** leads to the C-disaccharide derivative **243** in good yield. A double transmetallation to give an organocopper reagent **244** can facilitate the C-glycosylation reaction with allyl halides [80] (Scheme 58).

The added advantage of the C(1)-stannylated glycals is their ability to participate in palladium-catalyzed coupling reactions with organic halides, a process independently reported by Beau [75] and Friesen [81]. Vinyl stannane **237** can be benzylated, allylated or acylated provided that appropriate catalysts are used [75, 77] and representative examples are given in Scheme 59. The C-arylation of

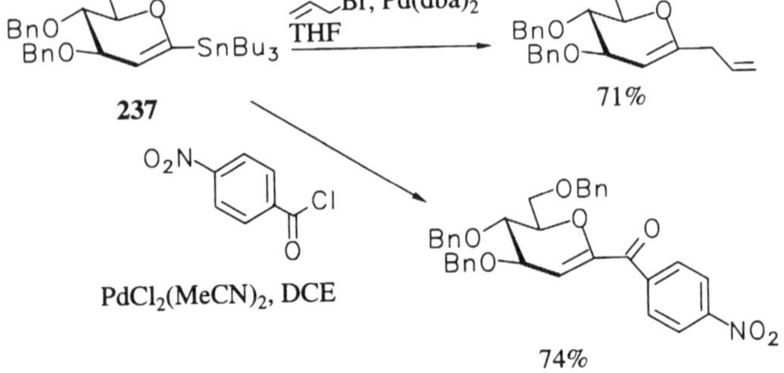
OBn
BnO
BnO
O
BnBr, Pd(Ph3P)4
PhCH3
74%
OBn
BnO
BnO
O
SnBu3
Br, Pd(dba)2
THF
OBn
BnO
BnO
O
71%
237
O2N
Cl
O
PdCl2(MeCN)2, DCE
OBn
BnO
BnO
O
O
NO2
74%

Scheme 59

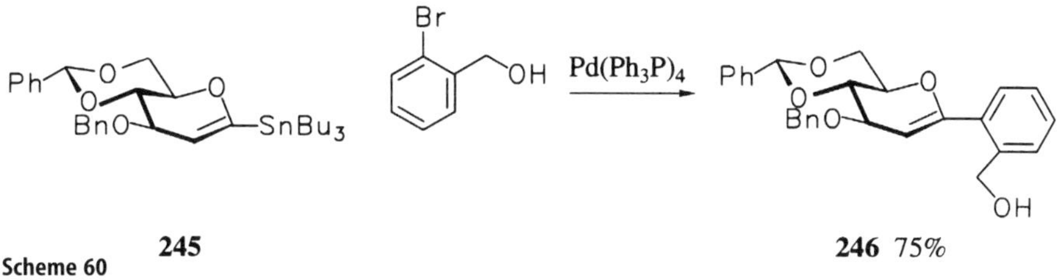

Br

OH $Pd(Ph_3P)_4$

SnBu$_3$

245 **246** 75%

Scheme 60

C(1)-stannylated glycals is certainly an important transformation and use of freshly prepared tetrakis(triphenylphosphine)palladium(0) [$Pd(Ph_3P)_4$] in toluene provides efficient conditions for couplings involving aryl bromides [75, 77]. The C(1)-stannyl-D-glucal **245** couples to 2-bromobenzyl alcohol to give the C-arylglycoside **246** (Scheme 60) although Friesen and Sturino [81] preferred to use bis(triphenylphosphine)palladium dichloride [$Pd(Ph_3P)_2Cl_2$] with a range of *o*- and *p*-substituted bromobenzenes (Scheme 61).

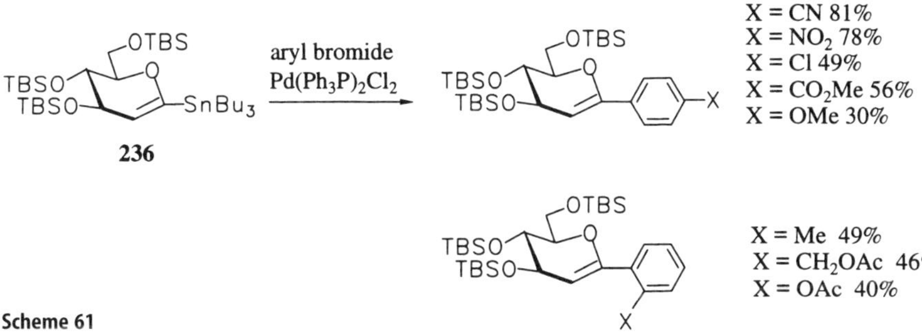

aryl bromide $Pd(Ph_3P)_2Cl_2$

236

X = CN 81%
X = NO_2 78%
X = Cl 49%
X = CO_2Me 56%
X = OMe 30%

X = Me 49%
X = CH_2OAc 46%
X = OAc 40%

Scheme 61

Dubois and Beau have extended the scope of this reaction to the di- and tri-C-glycosylation of aromatic compounds [75, 77], which provide models for the synthesis of di-C-glycosylated antitumor compounds. Coupling of 1,3-dibromobenzene with 2 equivalents of the tin reagent **247** gives the symmetrical 1,3-di-C-glycoside **248** (Scheme 62). Sequential C-glycosylation is also possible by reaction of **247** with an excess of 1,3-dibromobenzene which provides the mono-coupled product **249**. Coupling of **249** with a different tin reagent **237** yields the disymmetrical 1,3-diglycosylbenzene **250**. Likewise the di- **251** or tri-C-glycosyl benzene **252** is obtained by treatment of stannane **237** with 1,4-di- or 1,3,5-tribromo benzene (Scheme 63).

To fully exploit this chemistry requires, after the key C–C bond formation, a stereoselective functionalization of the glycal double bond. Scheme 64 summarizes transformations on C(1)-phenyl-D-glucal **253** that are all stereoselective. Hydroboration-oxidation of C(1)-phenyl glycal **254** to give C-glycoside **255** emphasizes the fact that the initial attack of the reagent on the C=C bond of the glycal always occurs *anti* to the substituent at C(3).

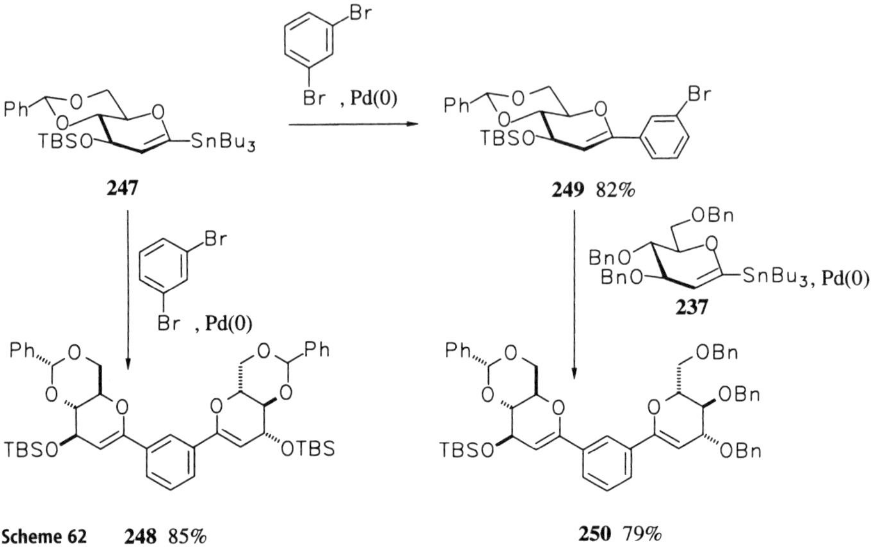

Scheme 62 **248** 85% **250** 79%

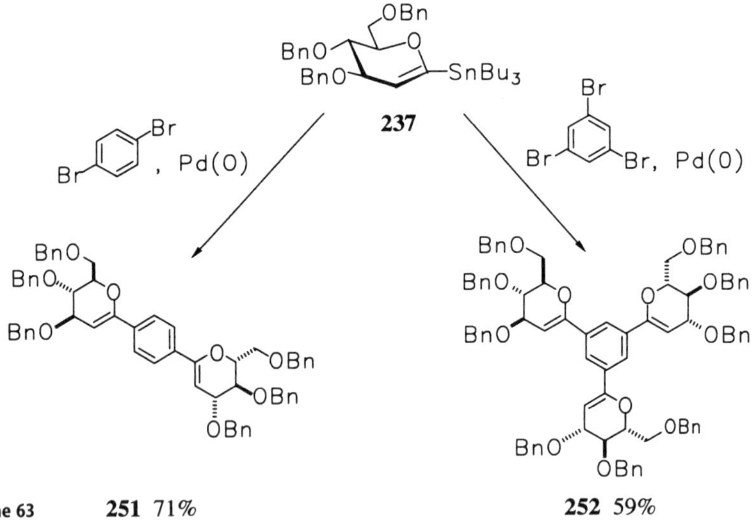

Scheme 63 **251** 71% **252** 59%

Both Beau [76, 78] and Friesen [82, 83] have applied this palladium-catalysed approach to C-glycosides to the synthesis of the structurally related antifungal antibiotics, papulacandin and chaetiacandin. The C(1)-aryl glycals **256** [76, 78] and **257** [82, 83] are key components which are both obtained by C(1)-arylation of the corresponding stannylated glycals **247** and **236** respectively showing that the Pd(0)-coupling process is efficient even with sterically congested aryl bromides.

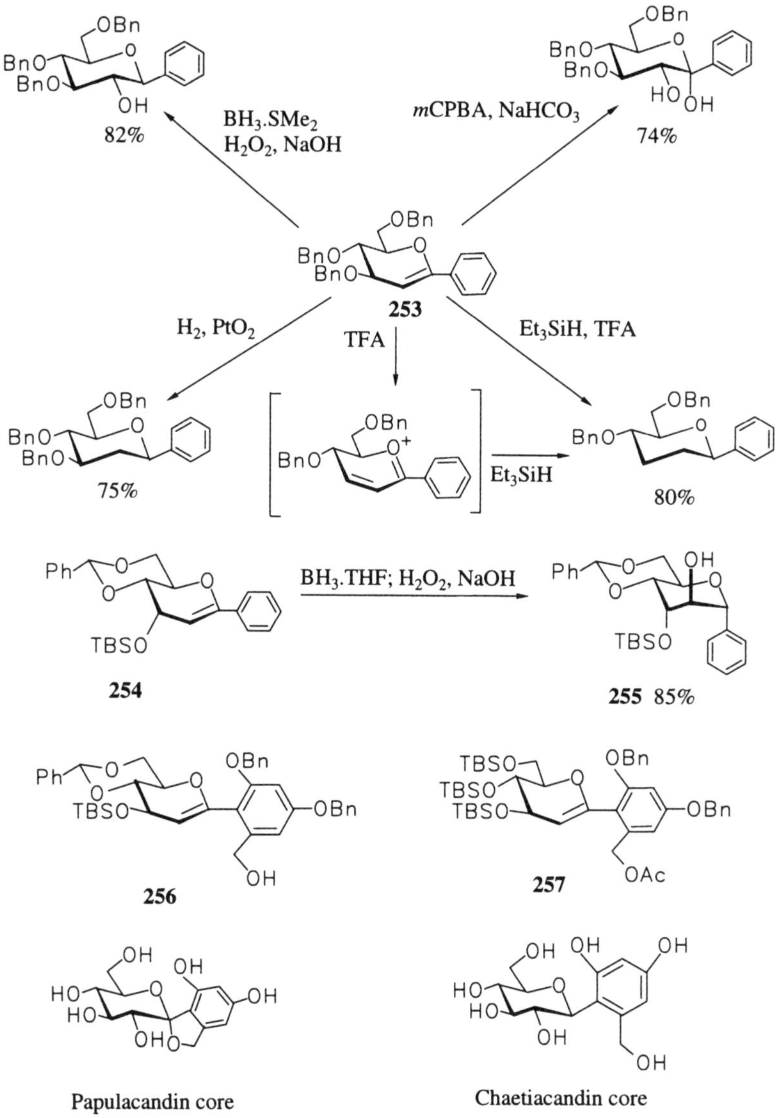

Scheme 64

4
Enolates and Related Systems as C(1) Nucleophiles

4.1
The Synthetic Potential of Enolates as C(1) Nucleophiles

Within the broader context of organic synthesis, one of the most powerful methods for establishing C–C bonds is based on exploiting the versatile nucleo-

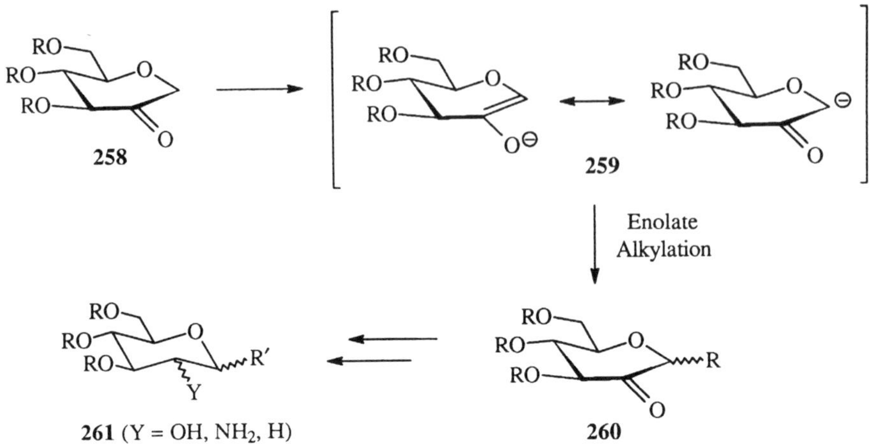

258 259 260 261 (Y = OH, NH_2, H)

Enolate Alkylation

Scheme 65

philic reactivity that is associated with enolates, silyl enol ethers and enamines. Such reactivity is also very attractive in the field of carbohydrate chemistry and C-glycoside synthesis, and C(1) nucleophiles based on 2-ketosugars offer, in principle, one of the most adaptable methods for gaining access to a wide range of C-glycosyl variants.

However, our ability to express a useful nucleophilic character at C(1) (via organometallic derivatives) is frequently constrained by the presence of an oxygen or nitrogen substituent at C(2) (Sect. 2.1). Such C(2) substitution patterns are, nonetheless, important but this proximity of a potential leaving group to the nucleophilic (anionic) C(1)-site frequently results in competitive, if not exclusive, β-elimination leading to glycals. While methods are now available to circumvent this conundrum (see Sects. 2.2 and 2.3), a 2-ketosugar **258** also offers an attractive solution to this problem (Scheme 65). Indeed, the very presence of a C(2) carbonyl moiety within a carbohydrate skeleton actually promotes the formation of, and concomitantly stabilizes, C(1) nucleophilicity i.e. via enolate **259**. Once the C-glycosyl linkage has been introduced by alkylation of the enolate intermediate, the residual C(2) ketone can play a further role. This functional group is now available for stereoselective manipulation of the initially formed adduct **260** (e.g. by ketone reduction, reductive amination, or used as a branching point for homologation) to provide access to a wide range of interesting and, more importantly, C(2)-substituted C-glycoside variants **261**.

4.2 Regiochemical Issues and Problems Associated with Enolate Generation

Given the attractive scenario outlined in Scheme 65, one must ask why 2-ketosugars have not yet emerged as important precursors to C(1) nucleophiles and fulfilled their latent potential. Several factors seem to contribute to this, with the most important being our ability to control the regioselectivity

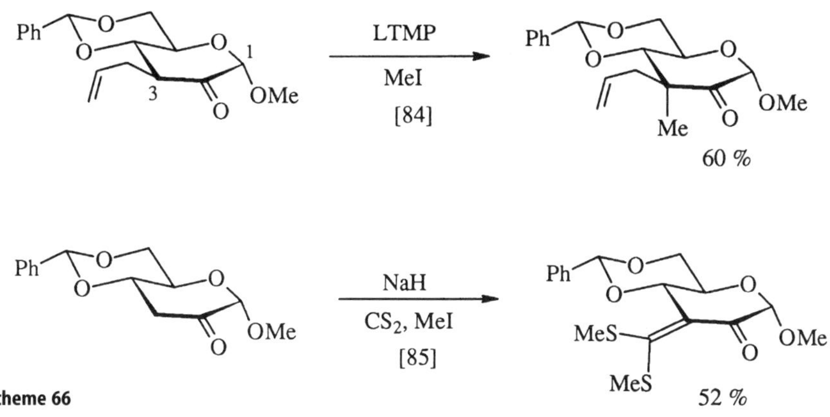

Scheme 66

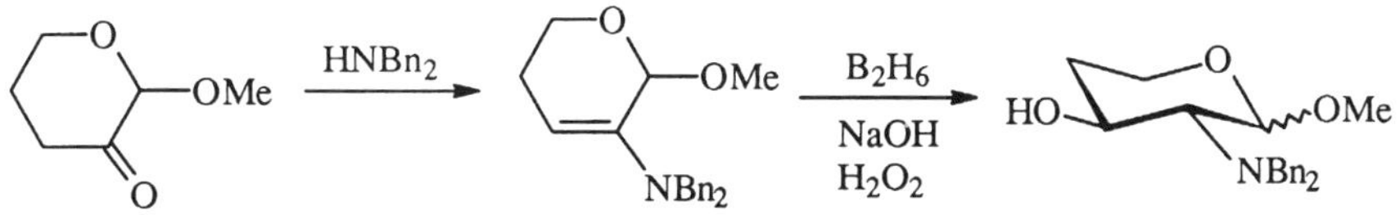

262 263 60 % 264 87 %

Scheme 67

associated with the key enolate formation step. Scheme 66 illustrates two examples [84, 85] where the preferred mode of enolization of the 2-keto-pyranoside is towards C(3) rather than C(1). Other studies involving enolization of keto sugars have been reported [86–91] and, though perhaps less directly relevant to this discussion, serve to illustrate some of the constraints that are encountered. Enamine formation, albeit using a simple substrate **262**, follows a similar preference [92] leading to the C(3)-isomer **263** which was used to prepare the dideoxypentose **264** (Scheme 67) [93, 94]. Stereoelectronic factors [95] contribute to the kinetic and thermodynamic preferences associated with enolization of heterocyclic ketones of this general type and, while a more extensive discussion is not appropriate here, a destabilizing interaction between the developing carbanion character at C(1) and the lone pairs of the adjacent (and conformationally constrained) ring-oxygen must play a significant role [96, 97].

4.3 Carbohydrate Enolates Derived from C(2) Ketones

4.3.1 *Deprotonation as a Source of Enolate Reactivity*

The generation of an enolate (as a C(1) nucleophile) by a deprotonation step was achieved when the alternative mode of enolization towards C(3) was

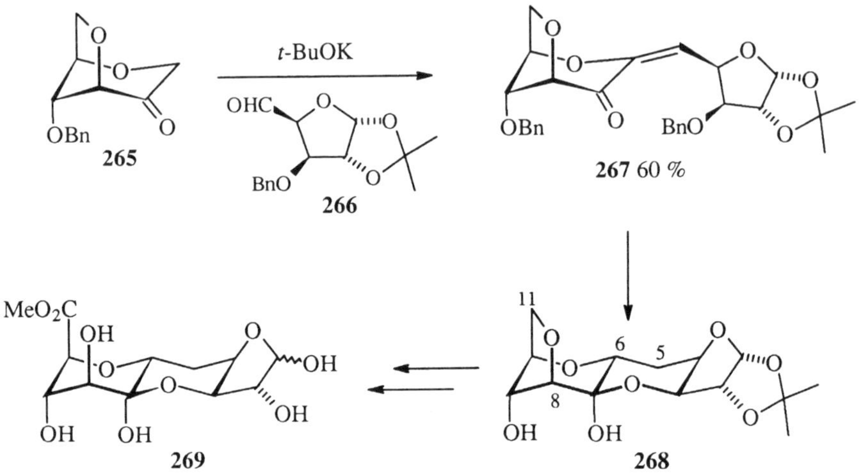

Scheme 68

blocked. This has been achieved using the bicyclic ketone **265**, where C(3) is now a bridgehead position [98]. C-Glycoside synthesis has been carried out via aldol condensations using aldehydes as the electrophilic partner and this chemistry has been exploited to construct C-glycoside **269**, the undecose glycosyl component of the herbicidin class of nucleosides (Scheme 68) [99]. Enolization of ketone **265** was carried out using potassium *tert*-butoxide as the base and in the presence of aldehyde **266**. The resulting enone **267** was obtained as a single double bond isomer and it is important to appreciate that a wide variety of alternative bases was evaluated for this aldol process, but without success. The stereochemistry at the key C-glycosyl linkage [C(5)–C(6) in **269**] was set by hydrogenation of enone **267**, which also served to establish the basic tricyclic herbicidin skeleton, and the 8,11-anhydro bridge associated with **268** was cleaved by a process involving regioselective radical bromination [100].

Two other aspects of the chemistry of ketone **265** merit discussion. Conventional enolate alkylation reactions at C(1) proved to be unworkable, even with reactive alkyl halides (Scheme 69) [101]. Silyl enol ether formation to give **270**, though available, proved to be of limited value. Attempts to utilize **270** in Mukaiyama-type processes (using either aldehydes or α-chlorosulfides in the presence of Lewis acid activator) failed and instead, silyl enol ether **270** underwent rapid Lewis acid-mediated rearrangement to give levoglucosenone **272** [102]. The ease of this process (which occurs using $LiClO_4$ at $-78\,°C$) is attributed to a favorable stereoelectronic relationship (as in **271**) between one of the lone pairs of *O(5)*, the π-system of the silyl enol ether and the axial oxygen bond at C(3). Alternative C(3)-blocked ketones **273** and **274** have been prepared [98] but offer little advantage. The preparation of the acetal-bridged derivative **273** involves a multistep and inefficient route and lactone **274**, though readily accessible, is exceptionally labile with respect to lactone cleavage which severely limits any further application.

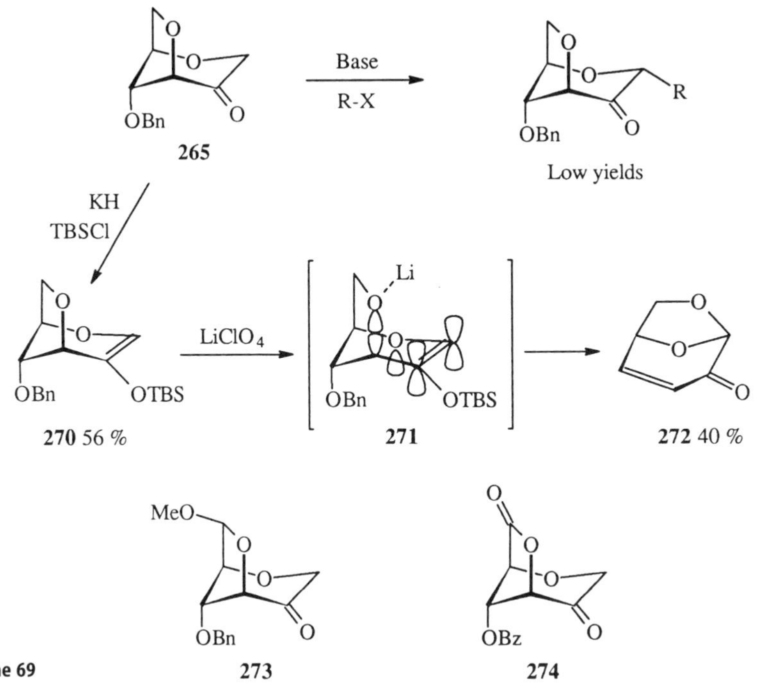

Scheme 69

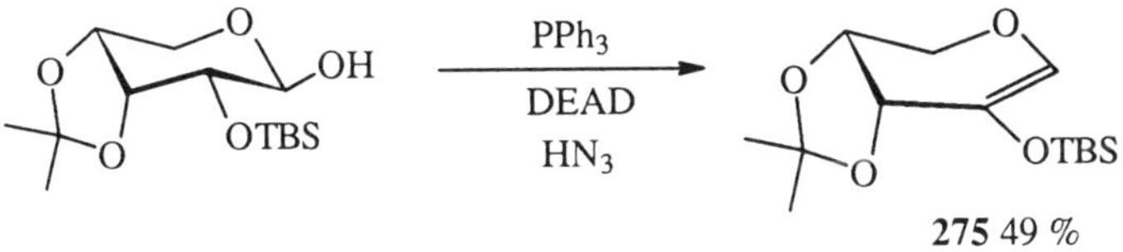

Scheme 70

Interestingly, a simpler silyl enol ether variant **275** has been prepared [103] (Scheme 70) but, to date, no reports have appeared relating to the ability of **275** to function as a C(1) nucleophile.

4.3.2
Reductive Cleavage as a Source of Enolate Reactivity

An alternative approach to a C(1) nucleophile based on a 2-ketosugar is available. Lichtenthaler has shown [104] that 3,4,6-tri-*O*-benzoyl-α-D-*arabino*-hexosulosyl bromide **276** (which is readily prepared from the corresponding 2-*O*-benzoyl-D-glycal) undergoes reductive cleavage using zinc to give a Reformatsky-type nucleophile **277** which was trapped by simple aldehydes, e.g. methanal, to give 2-keto-C-glycosides **278** (Scheme 71). This report represents the first example of a viable C(1) nucleophile based on an enolate, and is attractive because the methodology exploits "conventional" carbohydrate

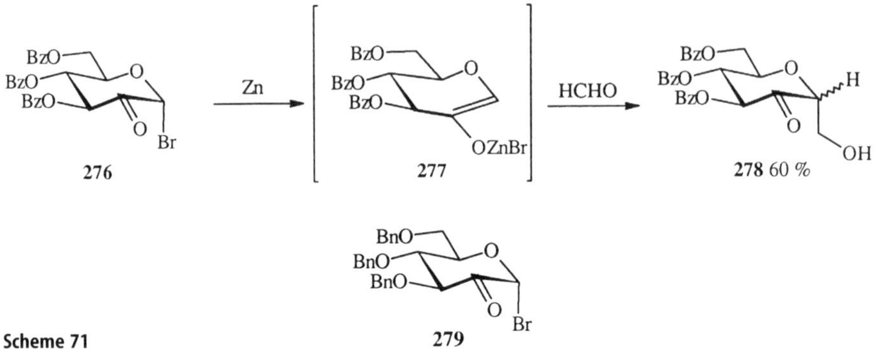

Scheme 71

reactivity (via a glycosyl bromide) [105, 106]. The tri-*O*-benzyl variant **279** is also available and although this carbohydrate unit has not been used in C-glycoside synthesis, important applications in the *O*-glycoside area have been developed [107].

This approach to C(1) nucleophiles has been recently extended. The Lichtenthaler zinc enolate reacts efficiently with more demanding aldehyde electrophiles to provide C-disaccharides [108, 109], and activation of the C(1) bromide **276** can also be carried out using $CeCl_3$/NaI (Scheme 72). The latter method is based on the earlier work of Fukuzawa [110] and others [111], although the mechanism of this cerium-mediated reaction has yet to be fully understood.

Bromide **280** (derived by bromination of silyl enol ether **270**) undergoes both zinc- and cerium-mediated cleavage under mild and essentially neutral conditions, and was used to prepare the nucleoside-containing C-glycoside **282** (Scheme 73) [112, 113]. The aldehyde **281** used in this transformation was exceptionally sensitive to basic conditions which completely precluded use of a conventional enolate obtained by deprotonation of ketone **265** (Sect. 4.3.1).

In summary, enolates derived from 2-ketosugars may serve as progenitors to C-glycosides, but this topic is still at an early phase of its development and warrants more extensive investigation and evaluation.

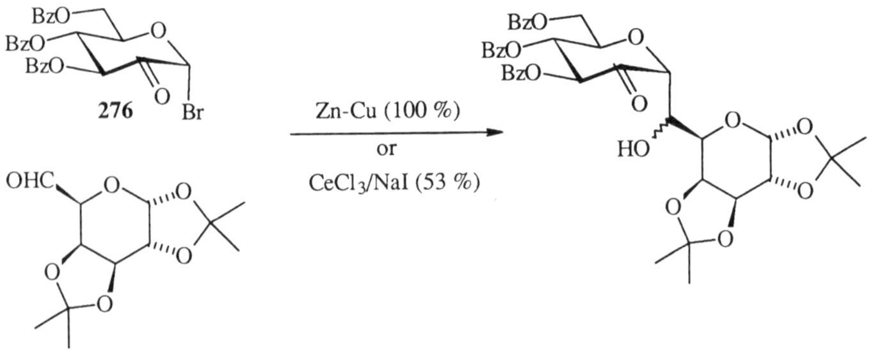

Scheme 72

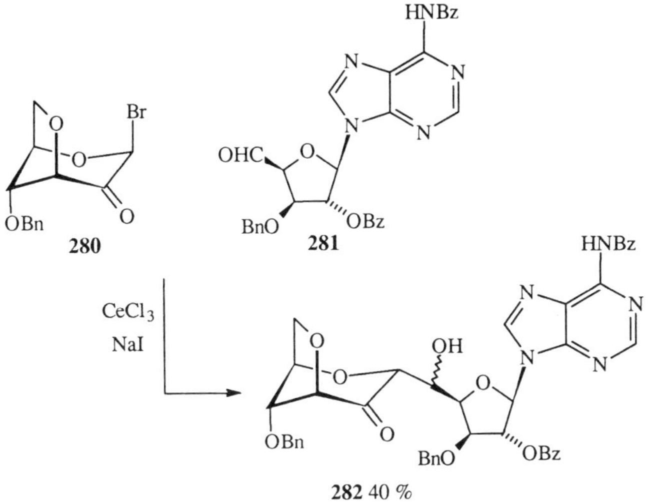

280

281

$CeCl_3$
NaI

282 40 %

Scheme 73

4.4
Enolate Reactivity at the Anomeric Site Based on Ester Stabilization

It is also appropriate to recognize the role of enolates stabilized by an exocyclic carbonyl function in C-glycoside synthesis. The use of LN to achieve reductive cleavage of anomeric sulfones to provide access to an ester-stabilized enolate and, ultimately, 2-deoxy-β-C-glycosides has already been illustrated in Scheme 11 (Sect. 2.1.1).

The enolate reactivity associated with this approach to C-glycoside synthesis was, however, first developed with octulosonic acid derivatives, such as **283** [114], **284** [115], and **285** [116], and a series of examples involving aldehyde and halide-based electrophiles are shown in Scheme 74. Related studies involving stereoselective protonation of this class of exocyclic enolate have also been described [13] and Scheme 75 illustrates this with an example of reductive samariation (using ethane-1,2-diol, but no HMPA) of an anomeric acetate **286** [117].

5
Anomeric Radicals as C(1) Nucleophiles

The recognition that oxygen-stabilized carbon centered radicals, though electronically neutral, display nucleophilic reactivity opens a number of opportunities for the synthesis of C-glycosides [118–120]. C–C Bond formation via radical reactions is also especially appropriate to the anomeric site of carbohydrates since many of the methods available for radical generation require homolysis of a C–X bond (X= oxygen, halogen, sulfur etc.) and introduction of this type of heteroatom substituent at C(1) is straightforward. Additionally, radical-mediated C-glycoside synthesis offers advantages associated with mild reaction conditions, no special requirements for protecting groups and, furthermore, good levels of anomeric stereocontrol are also observed.

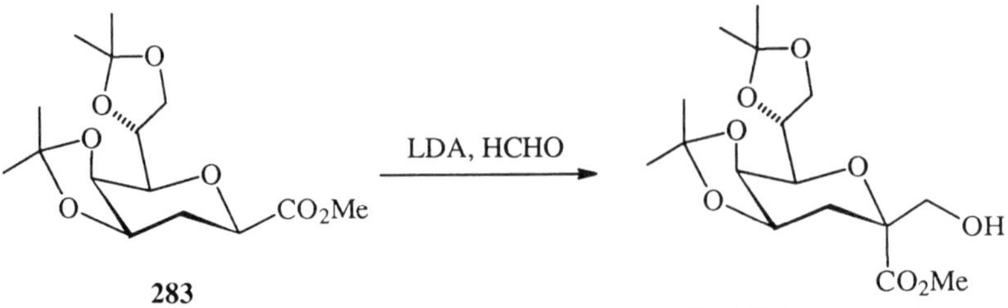

283

LDA, HCHO

CO_2Me

91 % (β-isomer predominates)

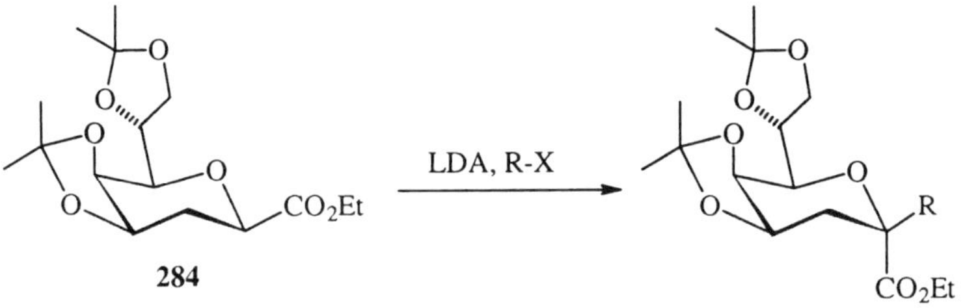

284

LDA, R-X

CO_2Et

R=Me 50 %
R=CN 55 %
R=$CH_2C{\equiv}CH$ 50 %

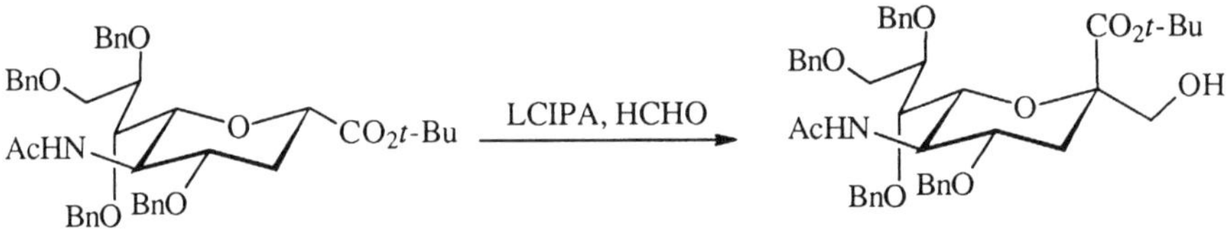

285

LCIPA, HCHO

CO_2t-Bu

46 % (1:3 mixture of α/β isomers

Scheme 74

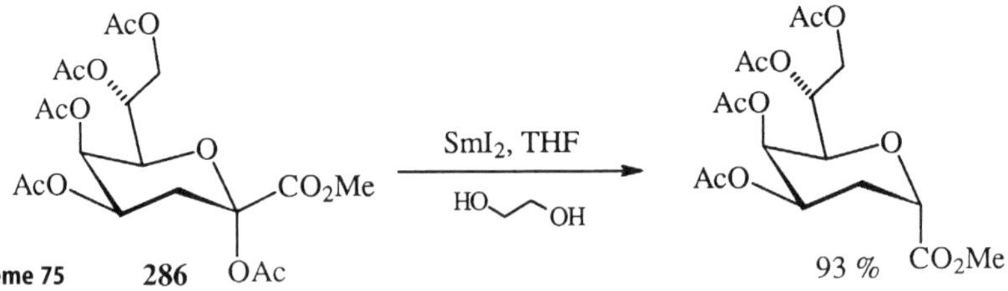

286

SmI_2, THF
HO‿‿OH

93 %

Scheme 75

C-Glycoside synthesis may be achieved in two ways. Intermolecular radical addition reactions are observed with *(i)* polarized, electron-deficient alkenes, *(ii)* alkenes that provide a high level of stabilization to the initial radical adduct and *(iii)* substrates that undergo a facile fragmentation (e.g. allyl stannanes). Additions to “less reactive” substrates, though not favored for intermolecular processes, are observed if the two components are tethered in an intramolecular array.

5.1
Intermolecular Additions

While a full discussion of radical-mediated C-glycoside synthesis is beyond both the space available and the general remit of this chapter, illustrative examples of the nature of the transformations available will be described. Comprehensive reviews are, however, available [3, 4] and should be consulted for more detailed accounts.

Anomeric radicals may be generated via homolytic cleavage of a variety of carbon-heteroatom residues. Most often anomeric bromides [120] are used for this purpose but other halides (fluorides [121], chlorides [122] and iodides [123]), xanthates [124], sulfides [125] (also selenides [126] and tellurides [127]), the nitro grouping [40] (Sect. 2.2.1) and Hunsdieker decarboxylation (under Barton conditions) of a C(1) carboxylate [14] also provide access to anomeric radicals. Electrochemical cleavage (the Kolbe reaction) has been utilized as a method for chain extension of a pre-formed C-glycoside [128]. An alternative approach to generating an anomeric radical has been devised by Fraser-Reid, based on an intramolecular addition to a glycal. Tin-mediated cleavage to the tethered bromoacetal **287** provided a primary radical that cyclized onto the glycal to give the anomeric radical **288.** This nucleophilic intermediate was efficiently trapped *inter alia* by acrylonitrile to give the C(2)-branched C-glycoside **289** (Scheme 76) [129, 130].

C-Glycoside synthesis via the intermolecular addition of anomeric radicals to activated alkenes has been pioneered by Giese and co-workers [120]. A wide variety of acceptors have been used in this area and both simple [131, 132] and more complex (a C-disaccharide **290**) [133] examples are shown (Scheme 77).

Invariably, the predominant (or exclusive) product has α-stereochemistry at the newly formed C-glycosyl linkage (see Scheme 77), and disubstitution at C(1) has also been achieved (Scheme 25, Sect. 2.2.1). Allylstannanes serve as radical acceptors in intermolecular processes (Scheme 78) [122, 134, 135] and, under appropriate circumstances, dimerization of the anomeric radicals is observed, as in C-disaccharide **291** (Scheme 79) [136].

While numerous applications of anomeric radicals exist, three examples serve to illustrate possible future applications and directions for this chemistry. Barton has developed a short and convergent synthesis of the C-nucleoside showdomycin **292** (Scheme 80) [127]. Kessler has prepared an interesting C-linked glycopeptide unit **294** via radical addition to a dehydroaminoacid acceptor **293** (Scheme 81) [137]. In the latter example, good anomeric stereo-

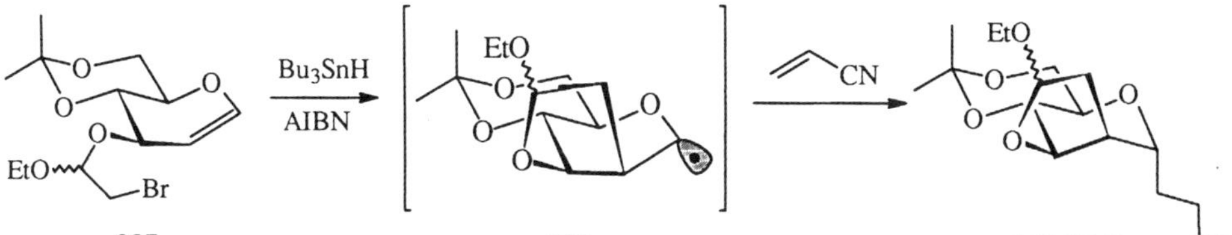

Bu_3SnH, AIBN; CN (acrylonitrile)

287 **288** **289** 87 %

Scheme 76

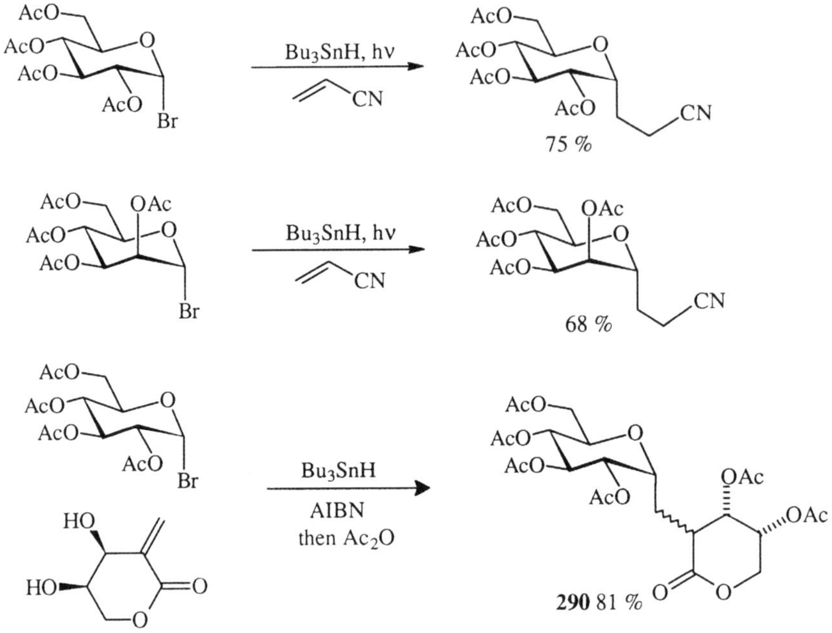

AcO AcO AcO AcO O Br Bu_3SnH, hν CN AcO AcO AcO AcO O CN 75 %

AcO OAc AcO AcO O Br Bu_3SnH, hν CN AcO OAc AcO AcO O CN 68 %

AcO AcO AcO AcO O Br HO HO O O Bu_3SnH AIBN then Ac_2O AcO AcO AcO AcO O OAc OAc O O **290** 81 %

Scheme 77

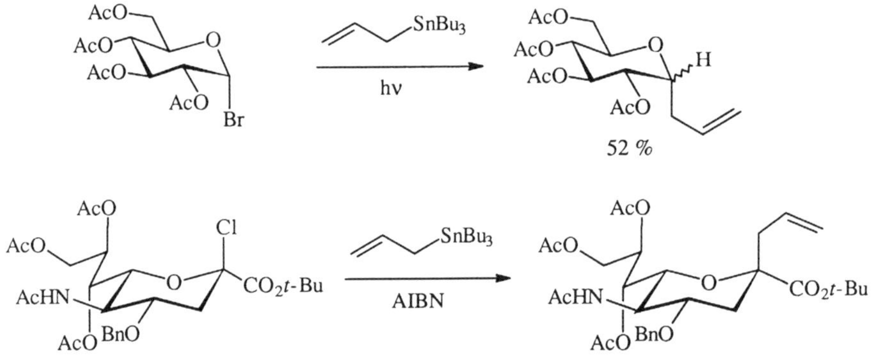

AcO AcO AcO AcO O Br SnBu3 hν AcO AcO AcO AcO O H 52 %

AcO AcO Cl AcHN O CO_2t-Bu AcO BnO SnBu3 AIBN AcO AcO AcHN O CO_2t-Bu AcO BnO 92% 1:1.8 mixture of isomers

Scheme 78

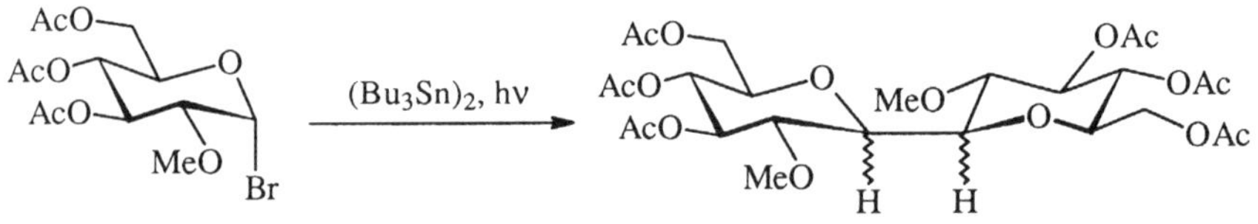

AcO AcO AcO MeO O Br $(Bu_3Sn)_2$, hν AcO AcO AcO MeO O H H MeO O OAc OAc OAc

Scheme 79

291 32 % (3 diastereomers)

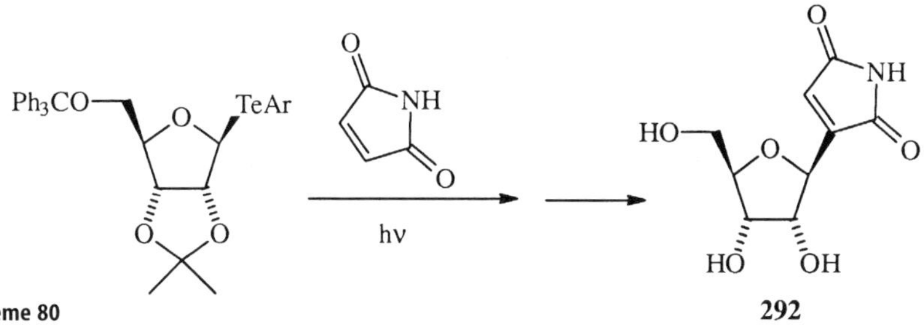
Ph3CO
O
TeAr
O
O
O
NH
O
hν
HO
O
O
NH
O
HO
OH

Scheme 80 **292**

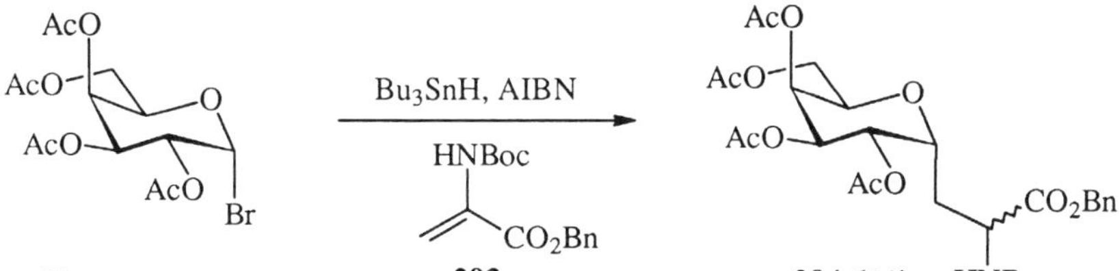
AcO
AcO
O
AcO
AcO
Br
Bu3SnH, AIBN
HNBoc
CO2Bn
AcO
AcO
O
AcO
AcO
CO2Bn
HNBoc

Scheme 81 **293** **294** 61%

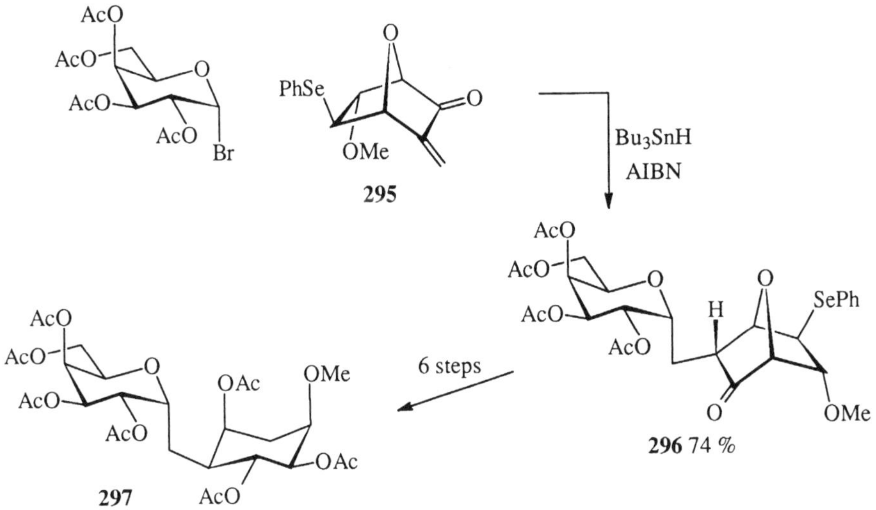
AcO
AcO
O
AcO
AcO
Br
PhSe
O
O
OMe
Bu3SnH
AIBN
AcO
AcO
O
AcO
AcO
H
O
SePh
O
OMe
6 steps
AcO
AcO
O
AcO
AcO
OAc
OMe
OAc
AcO

295

296 74 %

297

Scheme 82

control was observed but the α-amino acid center was produced as a mixture of epimers. Carbasugars (similar to a monosaccharide unit but containing no ring oxygen) have recently attracted significant interest and Vogel [138] has adapted his "naked sugar" methodology to provide a radical acceptor **295** which undergoes addition of a galactosyl radical to give adduct **296.** This intermediate has then been transformed to the C-disaccharide mimic **297** (Scheme 82).

5.2
Intramolecular Additions

Intramolecular radical coupling reactions have also been developed and this strategy is particularly useful when only weakly or non-electrophilic acceptor substrates are involved. This is a developing topic and one example will serve to illustrate the current status of this aspect of C-glycoside synthesis. Sinaÿ has exploited the well-established concept of a removable tether as the basis of his approach to the synthesis of C-disaccharide **300** (Scheme 83) [139]. The mixed silyl ether **298** serves to hold donor and acceptor components in close proximity and the requisite radical was generated by samarium-mediated cleavage of an anomeric sulfone (see Sect. 2.2.3). The coupling reaction, which is a 9-*endo-trig* process, was done at high dilution (5×10^{-3} M) using a syringe pump and adduct **299** was converted to the C-disaccharide **300** in 50% yield from **298**. In addition to silyl ethers, ketals also serve as viable tethers in these types of coupling reactions [140].

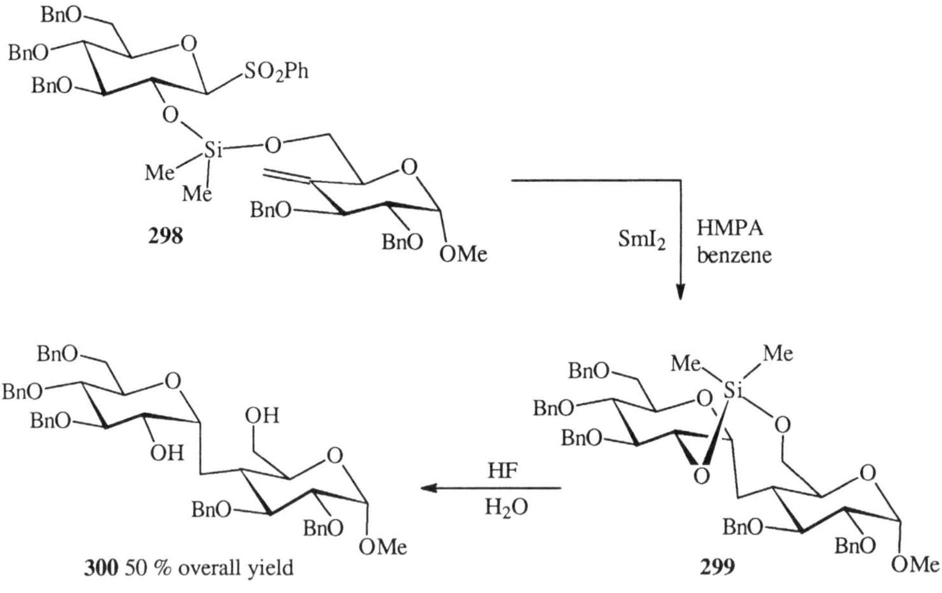

Scheme 83

6
Conclusions

In this review we have presented a complete and, as far as possible, up to date account of how C-glycosides may be synthesized using C(1) nucleophiles. Free radical based methods, which have provided numerous and useful procedures for C-glycoside synthesis, were beyond the scope of this review and only a few representative examples are mentioned.

One of the interesting features of C(1) nucleophiles is that, for the most part, these reagents deviate from the "traditional" carbohydrate reactivity profile (see Schemes 1 and 2). Consequently, this is an area of chemistry that should not be regarded as "specialized" and many of the reactions that we have described have much more general applications in synthetic organic chemistry. Indeed, the advances that have been made in C-glycoside chemistry owe much to the work of noncarbohydrate chemists, but this is also an area that has served as a test bed in its own right and many of the transformations that we have described have also gone on to find utility elsewhere. We have pointed out problems and areas that are deficient. Clearly many opportunities exist and we are confident that these will be brought to fruition as interest in carbohydrate mimics continues to mount.

References

1. Postema MHD (1992) Tetrahedron 48:8545
2. Herscovici J, Antonakis K (1992) In: Atta-ur- Rahman (ed) Studies in Natural Products Chemistry, vol 10, Elsevier, Amsterdam
3. Postema MHD (1995) C-Glycoside synthesis. CRC Press, London
4. Levy DE, Tang C (1995) The chemistry of C-glycosides, Pergamon, Oxford
5. Lancelin JM, Morin-Allory L, Sinaÿ P (1984) J Chem Soc Chem Commun 355
6. Cohen T, Matz JR (1980) J Am Chem Soc 102:6900
7. Ermolenko MS, Olesker A, Lukacs G (1994) Tetrahedron Lett 35:715
8. Cohen T, Lin M-T (1984) J Am Chem Soc 106:1130
9. Rychnovsky SD, Mickus DE (1989) Tetrahedron Lett 30:3011
10. Prandi J, Beau J-M (1989) Tetrahedron Lett 30:4517
11. Beau J-M, Sinaÿ P (1985) Tetrahedron Lett 26:6185
12. Beau J-M, Sinaÿ P (1985) Tetrahedron Lett 26:6189
13. Beau J-M, Sinaÿ P (1985) Tetrahedron Lett 26:6193
14. Crich D, Lim LBL (1990) Tetrahedron Lett 31:1897
15. Girard P, Namy J-L, Kagan HB (1980) J Am Chem Soc 102:2693
16. de Pouilly P, Chénedé A, Mallet J-M, Sinaÿ P (1993) Bull Soc Chim Fr 130:256
17. Inanaga J, Ishikawa M, Yamaguchi M (1987) Chem Lett 1485
18. Mazéas D, Skrydstrup T, Doumeix O, Beau J-M (1994) Angew Chem Int Ed Engl 33:1383
19. Mazéas D, Skrydstrup T, Beau J-M (1995) Angew Chem Int Ed Engl 34:909
20. Greck C, Grice P, Ley SV, Wonnacott A (1986) Tetrahedron Lett 27:5277
21. Ley SV, Lygo B, Wonnacott A (1985) Tetrahedron Lett 26:535
22. Ousset JB, Mioskowski C, Yang Y-L, Falk JR (1984) Tetrahedron Lett 25:5903
23. Ley SV, Lygo B (1984) Tetrahedron Lett 25:113
24. Valverde S, Garcia-Ochoa S, Martin-Lomas M (1987) J Chem Soc Chem Commun 383
25. Gomez AM, Valverde S, Fraser-Reid B (1991) J Chem Soc Chem Commun 1207
26. (a) Still WC (1978) J Am Chem Soc 100:1481 (b) McGarvey GJ, Kimura M (1982) J Am Chem Soc 104:5422
27. (a) Still WC, Sreekumar C (1980) J Am Chem Soc 102:1201 (b) Sawyer JS, MacDonald TL, McGarvey GJ (1984) J Am Chem Soc 106:3376
28. (a) Lesimple P, Beau J-M, Sinaÿ P (1985) J Chem Soc Chem Commun 894 (b) Lesimple P, Beau J-M, Sinaÿ P (1987) Carbohydr Res 171:289
29. Lesimple P, Beau J-M (1994) Bioorg Med Chem 2:1319
30. Bassindale AR, Ellis RJ, Lau JC-Y, Taylor PG (1986) J Chem Soc Chem Commun 98
31. Hutchinson DK, Fuchs PL (1987) J Am Chem Soc 109:4930
32. (a) Yamamoto Y, Maruyama K, (1978) J Am Chem Soc 100:3240 (b) Yamamoto Y, Yamamoto S, Yatagi H, Ishihara Y, Maruyama K (1982) J Org Chem 47:119

33. Prandi J, Audin C, Beau J-M (1991) Tetrahedron Lett 32:769
34. Lipshutz BH, Koerner M, Parker DA (1987) Tetrahedron Lett 28:945
35. Aebischer B, Vasella A, Weber H-P (1982) Helv Chim Acta 65:621
36. Aebischer B, Vasella A (1983) Helv Chim Acta 66:789
37. Meuwly R, Vasella A (1986) Helv Chim Acta 69:751
38. Aebischer B, Bieri JH, Prewo R, Vasella A (1982) Helv Chim Acta 65:2251
39. Baumberger F, Vasella A (1983) Helv Chim Acta 66:2210
40. Dupuis J, Giese B, Hartung J, Leising M, Korth H-G, Sustmann R (1985) J Am Chem Soc 107:4332
41. Aebischer B, Meuwly R, Vasella A (1984) Helv Chim Acta 67:2236
42. Wittmann V, Kessler H (1993) Angew Chem Int Ed Engl 32: 1091
43. Halcomb RL, Danishevsky SJ (1989) J Am Chem Soc 111: 6661
44. Frey O, Hoffmann M, Wittmann V, Kessler H, Uhlmann P, Vasella A (1994) Helv Chim Acta 77:2060
45. Belosludtsev YY, Bhatt RK, Falk JR (1995) Tetrahedron Lett 36:5881
46. Pedretti V, Veyrières A, Sinaÿ P (1990) Tetrahedron 46:77
47. Prandi J, Beau J-M unpublished results
48. (a) Namy J-L, Collin J, Bied C, Kagan HB (1992) Synlett 733 (b) Curran DP, Fevig TL, Jasperse CP, Totleben ML (1992) Synlett 943 (c) Molander GA, McKie JA (1991) J Org Chem 56:4112
49. Jarreton O, Skrydstrup T, Beau J-M, unpublished results
50. Jarreton O, Skrydstrup T, Beau J-M (1996) J Chem Soc Chem Commun 1661
51. Hung S-C, Wong C-H (1996) Tetrahedron Lett 37:4903
52. Julina R, Müller I, Vasella A, Wyler R (1987) Carbohydr Res 164:415
53. Hoffmann M, Kessler H (1994) Tetrahedron Lett 35:6067
54. Urban D, Skrydstrup T, Riche C, Chiaroni A, Beau J-M (1996) J Chem Soc Chem Commun 1883
55. Deshong P, Slough GA, Elango V, Trainor GL (1985) J Am Chem Soc 107:7788
56. Deshong P, Slough GA, Elango V (1987) Carbohydr Res 171:342
57. (a) Deshong P, Slough GA, Rheingold AL (1987) Tetrahedron Lett 28:2229 (b) Deshong P, Sidler DR, Slough GA (1987) Tetrahedron Lett 28:2233
58. Chatani N, Ikeda T, Sano T, Sonoda N, Kurosawa H, Kawasaki Y, Murai S (1988) J Org Chem 53:3387
59. Trainor GL, Smart BE (1983) J Org Chem 48:2447
60. Lesimple P, Beau J-M, Jaurand G, Sinaÿ P (1986) Tetrahedron Lett 27:6201
61. Nicolaou KC, Hwang C-K, Duggan ME (1986) J Chem Soc Chem Commun 925
62. Hanessian S, Martin M, Desai RC (1986) J Chem Soc Chem Commun 926
63. Parker KA, Coburn CA (1991) J Am Chem Soc 113:8516
64. Parker KA, Coburn CA (1992) J Org Chem 57:5547
65. Parker KA, Su D-S (1996) J Org Chem 61:2191
66. Parker KA, Koh Y-H (1994) J Am Chem Soc 116:11149
67. (a) Tius MA, Gu Y, Gomez-Galeno J (1990) J Am Chem Soc 112:8188 (b) Tius MA, Gomez-Galeno J, Gu X, Zaidi JH (1991) J Am Chem Soc 113:5775
68. Schmidt RR, Preuss R, Betz R (1987) Tetrahedron Lett 28:6591
69. Preuss R, Schmidt RR (1989) Liebigs Ann Chem 429
70. Schmidt RR, Dietrich H (1991) Angew Chem Int Ed Engl 30:1328
71. Schmidt RR, Preuss R (1989) Tetrahedron Lett 30:3409
72. Maier S, Preuss R, Schmidt RR (1990) Liebigs Ann Chem 483
73. Stille JK Angew Chem Int Ed Engl (1986) 25:508
74. (a) Friesen RW, Sturino CF, Daljeet AK, Kolaczewska AE (1991) J Org Chem 56:1944 (b) Friesen WF, Trimble LA (1996) J Org Chem 61:1165
75. Dubois E, Beau J-M (1990) J Chem Soc Chem Commun 1191
76. Dubois E, Beau J-M (1990) Tetrahedron Lett 31:5165
77. Dubois E, Beau J-M (1992) Carbohydr Res 228:103
78. Dubois E, Beau, J-M (1992) Carbohydr Res 223:157

79. Barbaud C, Skrydstrup T, Beau J-M, unpublished results
80. Grondin R, Leblanc Y, Hoogsteen K (1991) Tetrahedron Lett 32:5021
81. Friesen RW, Sturino CF (1990) J Org Chem 55:2572
82. Friesen RW, Sturino CF (1990) J Org Chem 55:5808
83. Friesen RW, Daljeet AK (1990) Tetrahedron Lett 31:6133
84. Wood AJ, Jenkins PR, Fawcett J, Russell DR (1995) J Chem Soc Chem Commun 1567
85. Kobayashi H, Shibata N, Watanabe M, Komido M, Hashimoto N, Hisamichi K, Suzuki S (1992) Carbohydr Res 231:317
86. Klemer A, Thiemeyer H (1984) Liebigs Ann Chem 1094
87. Klemer A, Wilbers H (1987) Liebigs Ann Chem 815
88. Klemer A, Beerman H (1983) J Carbohydr Chem 2:457
89. Klemer A, Wilbers, H (1985) Liebigs Ann Chem 2328
90. Klemer A, Stegt H (1985) J Carbohydr Chem 4:205
91. Rocherolle V, Christobal Lopez JC, Olesker A, Lukacs G (1988) J Chem Soc Chem Commun 513
92. Descours D, Picq D, Anker D, Pacheco H (1982) Carbohydr Res 105:9
93. Eiden F, Wanner KT (1984) Liebigs Ann Chem 1759
94. Eiden F, Wanner KT (1985) Arch Pharm 318:207
95. (a) Hine J, Malone LG, Liotta CL (1967) J Am Chem Soc 89:5911 (b) Hine J, Dalsin PD (1972) J Am Chem Soc 94:6998 (c) Lehn J-M, Wipff G (1976) J Am Chem Soc 98:7498
96. (a) Hirsch JA, Wang XL (1982) Synth Commun 12: 333 (b) Goldsmith DJ, Dickinson CM, Lewis AJ (1987) Heterocycles 25:291
97. Cox PJ, Griffin AM, Newcombe NJ, Lister S, Ramsay MVJ, Alker D, Gallagher T (1994) J Chem Soc Perkin Trans 1 1994:1443
98. Griffin AM, Newcombe NJ, Alker D, Ramsay MVJ, Gallagher T (1993) Heterocycles 35:1247
99. Newcombe NJ, Mahon MF, Molloy KC, Alker D, Gallagher T (1995) J Am Chem Soc 115:6430
100. (a) Ferrier RJ, Furneaux RH (1980) Aust J Chem 33:1025 (b) Somsak L, Ferrier RJ (1991) Adv Carb Chem Biochem 49:37
101. Griffin AM (1995) PhD Dissertation University of Bristol
102. Griffin A, Newcombe NJ, Gallagher T (1994) In: Witczak ZJ (ed) Levoglucosenone and levoglucosans, chap 3, ATL Press, Mount Prospect
103. Schörkhuber W, Zbiral E (1980) Liebigs Ann Chem 1455
104. Lichtenthaler FW, Schwidetzky S, Nakamura K (1990) Tetrahedron Lett 31:71
105. Kaji E, Lichtenthaler FW (1993) Trends Glycosci Glycotechnol 5:121
106. Lichtenthaler FW, Kläres U, Lergenmüller M, Schwidetzky S (1992) Synthesis 179
107. (a) Lichtenthaler FW, Schneider-Adams T (1994) J Org Chem 59:6728 (b) Lichtenthaler FW, Schneider-Adams T, Immel S (1994) J Org Chem 59:6735
108. Schwidetzky S (1988) PhD Dissertation Technische Hochschule Darmstadt
109. Binch HM, Griffin AM, Schwidetsky S, Ramsay MVJ, Gallagher T, Lichtenthaler FW (1995) J Chem Soc Chem Commun 967
110. (a) Fukuzawa S, Fujinami T, Sakai S (1985) J Chem Soc Chem Commun 777 (b) Fukuzawa S, Tsuruta T, Fujinami T, Sakai S (1987) J Chem Soc Perkin Trans 1 1987: 1473
111. (a) Imamoto T, Kusumoto T, Tawarayama Y, Sugiura Y, Mita T, Hatanaka Y, Yokoyama M (1984) J Org Chem 49:3904 (b) Imamoto T, Kusumoto T, Yokoyama M (1983) Tetrahedron Lett 24:5233
112. Binch HM, Griffin AM, Gallagher T (1996) Pure Appl Chem 68:589
113. Binch HM, Gallagher T (1996) J Chem Soc Perkin Trans 1 1996:401
114. Norbeck DW, Kramer JB, Lartey PA (1987) J Org Chem 52:2174
115. (a) Luthman K, Orbe M, Wåglund T, Claesson A (1987) J Org Chem 52:3777 (b) Wåglund T, Luthman K, Orbe M, Claesson A (1990) Carbohydr Res 206:269 (c) Orbe M, Luthman K, Wåglund T, Claesson A, Csöregh I (1991) Carbohydr Res 211:1
116. Wallimann K, Vasella A (1991) Helv Chim Acta 74:1520

117. Hanessian S, Girard C (1994) Synlett 863
118. Descotes G (1988) J Carbohydr Chem 7:1
119. Dupuis J, Giese B, Rüegge D, Fischer H, Korth HG, Sustmann R (1984) Angew Chem Int Ed Engl 23:896
120. Giese B (1985) Angew Chem Int Ed Engl 24:553
121. Nicolaou KC, Dolle RE, Chucholowski A, Randall JL (1984) J Chem Soc Chem Commun 1153
122. Keck GE, Enholm EJ, Yates JB, Wiley MR (1985) Tetrahedron 41:4079
123. Giese B, Gilges S, Gröninger KS, Lamberth C, Witzel T (1988) Liebigs Ann Chem 615
124. Araki Y, Endo T, Tanji M, Nagasawa J, Ishido Y (1988) Tetrahedron Lett 29:351
125. Keck GE, Enholm EJ, Kachensky DF (1984) Tetrahedron Lett 25:1867
126. Adlington RM, Baldwin JE, Basek A, Kozyrod RP (1983) J Chem Soc Chem Commun 944
127. Barton DHR, Ramesh M (1990) J Am Chem Soc 112:891
128. Harenbrock M, Matzeit A, Schäfer HJ (1996) Liebigs Ann Chem 55
129. Lopez JC, Fraser-Reid B (1989) J Am Chem Soc 111:3450
130. Lopez JC, Gomez AM, Fraser-Reid B (1995) J Org Chem 60:3871
131. Dupuis J, Giese B (1983) Angew Chem Intl Ed Engl 22:622
132. Giese B, Dupuis M, Leising M, Nix M, Lindner HJ (1987) Carbohydr Res 171:329
133. Giese B, Witzel T (1986) Angew Chem Intl Ed Engl 25:450
134. Keck GE, Yates JB (1982) J Am Chem Soc 104:5829
135. Paulsen H, Matschulat P (1991) Liebigs Ann Chem 487
136. Giese B, Rückert B, Gröninger KS, Muhn R, Lindner HJ (1988) Liebigs Ann Chem 997
137. Kessler H, Wittmann V, Köck M, Kottenhahn M (1992) Angew Chem Int Ed Engl 31:902
138. (a) Ferritto R, Vogel P (1995) Tetrahedron Lett 36: 3517 (b) Bimwala RM, Vogel P (1992) J Org Chem 57:2076
139. Chénedé A, Perrin E, Rekai ED, Sinaÿ P (1994) Synlett 420
140. Vauzeilles B, Cravo D, Mallet JM, Sinaÿ P (1993) Synlett 522

Synthesis of C-Glycosides of Biological Interest

Francesco Nicotra

Dipartimento di Chimica Organica e Industriale, Università degli Studi di Milano,
Via Venezian, 21 I-20133 Milano, Italy

This review article deals with C-glycosides, their biological interest and the synthetic methods for their preparation. C-Glycosides are carbohydrate analogs in which a carbon atom substitutes the oxygen involved in the glycosidic linkage. This modification compromises the anomeric reactivity of the sugar, giving rise to antimetabolites that can inhibit carbohydrate processing enzymes. The synthesis of C-glycosides can be performed from natural sugars, or from non-carbohydrate chiral starting materials. In the first case, the most common, the electrophilic character of the anomeric carbon is in general exploited, in reactions with proper carbon-nucleophiles. In this review, the different C-glycosylation procedures will be classified, and the stereochemical outcome of the reactions highlighted and explained. Particular attention will be paid to the uneasy syntheses of C-glycosides of aminosugars and of non-reducing disaccharides, and to the syntheses of C-glycosides of specific biological interest, such as the phosphono analogs of glycosyl phosphates.

Table of Contents

Topics in Current Chemistry, Vol. 187

1 Introduction

Carbohydrate analogs in which a carbon atom substitutes the glycosidic oxygen (or nitrogen in the case of nucleosides) are defined as C-glycosides (Fig. 1).

The interest in C-glycosides was originally only academic. In 1945, Hurd et al. obtained the first C-glucoside by reaction of 2,3,4,6-tetra-*O*-acetyl-α-D-glucopyranosyl chloride **1** with a Grignard reagent [1] (Scheme 1). Studies on the synthesis of these compounds, starting from different glycosyl halides and using different organometallic reagents, followed this observation [2]. In 1961 Helferich and Bettin obtained the unexpected 2,3,4,6-tetra-*O*-acetyl-β-D-glucopyranosyl cyanide **4** submitting 2,3,4,6-tetra-*O*-acetyl-α-D-glucopyranosyl bromide **3** to the glycosylation procedure which employs mercuric cyanide. The cyanide anion acted as nucleophile instead of the alcohol, affording **4** [3] (Scheme 1).

The interest in C-glycosides increased greatly after 1970, when the pharmacological properties of some C-nucleosides (Fig. 2) of natural origin were observed [4].

Later on, in the 1980s, a renewed impulse in the synthesis of C-glycosides came from the observation that many macrolidic structures of biological inte-

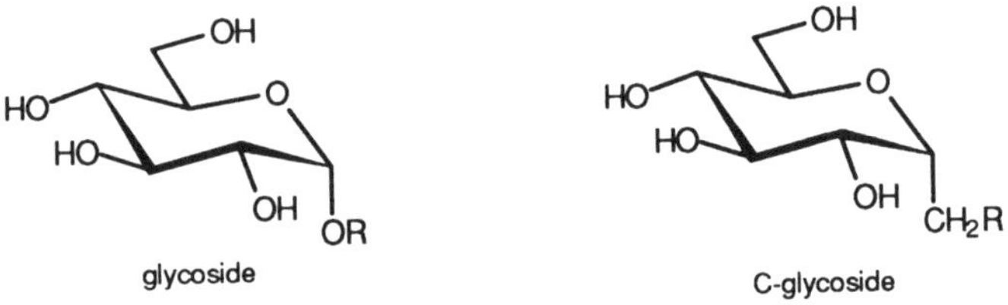

glycoside

C-glycoside

Fig. 1

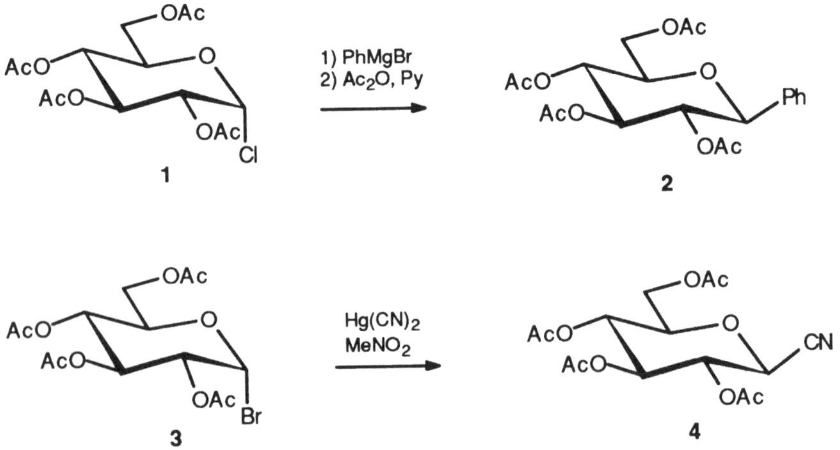

Scheme 1. Early syntheses of C-glycosides

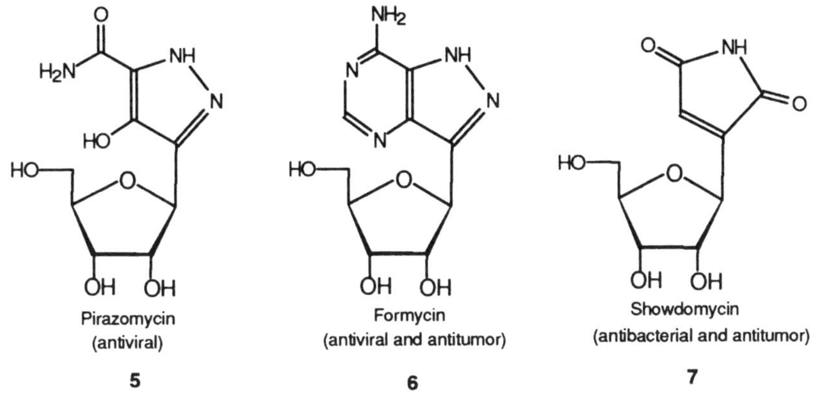

Pirazomycin
(antiviral)

5

Formycin
(antiviral and antitumor)

6

Showdomycin
(antibacterial and antitumor)

7

Fig. 2. Some C-nucleosides

rest, such as palytoxin (Fig. 3) are structurally related to C-glycosides. In this period some synthetic chemists, involved in projects devoted to the synthesis of these macrocycles, focused their attention on carbohydrate chemistry and in particular on the synthesis of C-glycosides. The aim was the use of C-glycosides as chiral building blocks in the synthesis of macrocyclic structures. This interest enriched the chemistry of carbohydrates with new methodologies and reagents

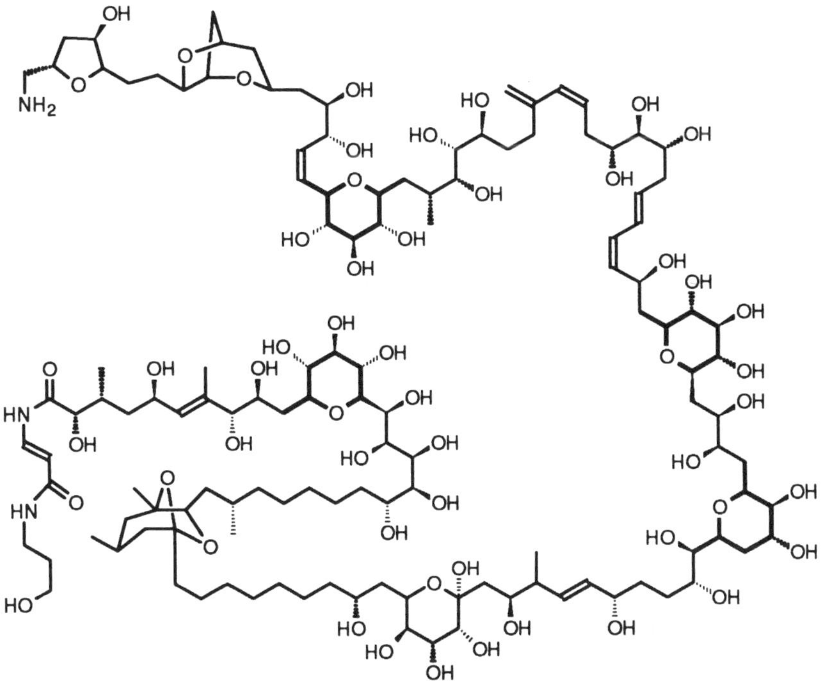

Fig. 3. Palytoxin; the C-glycosidic building blocks are emphasized

used in non-carbohydrate synthetic chemistry. New and more efficient C-glycosylation procedures were proposed, and the stereochemical outcome of the reactions were studied in depth.

More recently, in the 1990s, the explosive growth of knowledge in glycobiology, the science that studies the multivarious biological roles of carbohydrates, renewed the interest in carbohydrate mimics, and among them C-glycosides. It became clear that most of the biological recognition phenomena involve the glycidic part of cell-wall glycoconjugates. In particular, the glycoforms of glycoproteins and glycolipids were shows to be involved in phenomena of great pharmacological interest, such as the adhesion of the cells with viruses, bacteria, toxins and tumor cells. C-Glycosides then became attractive as potential inhibitors of carbohydrate processing enzymes and as stable analogs of the glycoforms involved in the above-mentioned recognition phenomena.

2 Biological Interest

2.1 Interfering in Carbohydrate Metabolism

The glycoforms of glycoproteins and glycolipids are involved in many cell-cell and cell-pathogen adhesion phenomena. As an example, the HIV infection cycle starts with the recognition between the glycidic part of the viral *N*-linked glycoprotein gp120, and the specific CD4 receptor of T-lymphocytes. Inhibitors of the biosynthesis of *N*-linked glycoproteins showed anti-HIV activity [5].

Another pharmacologically important process, in which the glycoforms and their processing enzymes are involved, is tumor cell growth and invasion. Tumor cell invasion of base membrane requires a preliminary interaction between an oligosaccharide chain located on the surface of metastatic cells, and E-selectins (endothelial leukocyte adhesion molecules, ELAM-1) which are expressed on the surface of the endothelial cells during inflammation. The oligosaccharide chain responsible for this interaction is NeuAcα2-3Galβ1-4GlcNAc(α1-3-Fuc)-, defined sialyl Le^x because it ends with a sialic acid unit (*N*-acetylneuraminic acid, Neu5Ac). It was observed that tumor cell glycoproteins are heavily glycosylated, and the glycidic components are further branched and sialylated, generally they are rich in the sialyl Le^x oligosaccharide [6]. Furthermore, tumor cells show an abnormal abundance of glycosidases, the enzymes that hydrolyse the glycosidic bond [7]. These observations suggest that inhibitors of glycoprotein processing enzymes, and in particular inhibitors of glycosidases, should be anticancer agents. Recent findings indicate that this is really the case [8].

2.2 Inhibition of *N*-Linked Glycoprotein Biosynthesis

Both in viruses and tumors, specific *N*-linked glycoproteins are involved in the recognition phenomena responsible for the pathology. The biosynthesis of these glycoproteins follows a common route (Scheme 2) in which *N*-acetyl-α-D-glucosamine 1-phosphate (GlcNAc-1P) is converted into the UDP derivative (UDP–GlcNAc), that in turn reacts with a dolichyl phosphate affording a GlcNAc–P–P–dolichyl. Then a second unit of GlcNAc is linked to the first by another GlcNAc-transferase, and finally different mannosyltransferases and glucosyltransferases add 8 mannose and 3 glucose units, affording a common dolichyl-P-P-oligosaccharide. This oligosaccharide is now ready to be transferred to an asparagine residue of the protein. At this point, a process defined as "trimming" cuts off sequentially the glucose units, employing α-glucosidases (glucosidase I and II), to afford a "high-mannose" oligosaccharide. To obtain glycoproteins of complex type, such as those of tumors and viruses, further trimming occurs, which involves mannosidases IA, IB and II. Now, on this common oligosaccharidic precursor, different glycosyltransferases build up the specific oligosaccharide.

The possibility of interfering in these processes, employing stable analogs of the metabolic intermediates reported in Scheme 2, or mimics of the receptor recognized glycoforms, have stimulated interest in the synthesis of related C-glycosides.

2.3 Stable Analogs of Glycosyl Phosphates, the Main Glycosylating Agents and Metabolic Regulators

Particular efforts have been devoted to the synthesis of stable C-glycosidic analogs of glycosyl phosphates. Glycosyl phosphates are in fact the main metabolic precursors and glycosylating agents in the biosynthesis of glycoconjugates. Furthermore some glycosyl phosphates are important metabolic regulators; this is the case for example of fructose 1,6-bisphosphate which is a potent activator of the glycolysis and inhibitor of the gluconeogenesis [9]. Mimics of *N*-acetyl-α-D-glucosamine 1-phosphate are probably the most attractive synthetic targets. This sugar phosphate in fact is not only the metabolic precursor of *N*-linked-glycoproteins (Scheme 2), but it is also the precursor of the bacterial cell-wall components teichoic acid and mureine (Fig. 4). Furthermore *N*-acetyl-α-D-glucosamine 1-phosphate is involved in glycosylation-deglycosylation of some proteins, an abundant and dynamic process the role of which is not yet clear [10], and still under investigation.

3 Synthesis

C-Glycosides are usually synthesized starting from the parent sugar, properly protected and functionalized, in a process generally defined C-glycosylation. Examples of C-glycoside syntheses starting from non-carbohydrate substrates, mainly Diels-Alder adducts, are also known, but less common.

The C-glycosylation procedures generally exploit the electrophilic character of the anomeric carbon of the sugar. The new C–C bond can be formed by reaction of the aldehydic function of the sugar with an organometallic or a Wittig

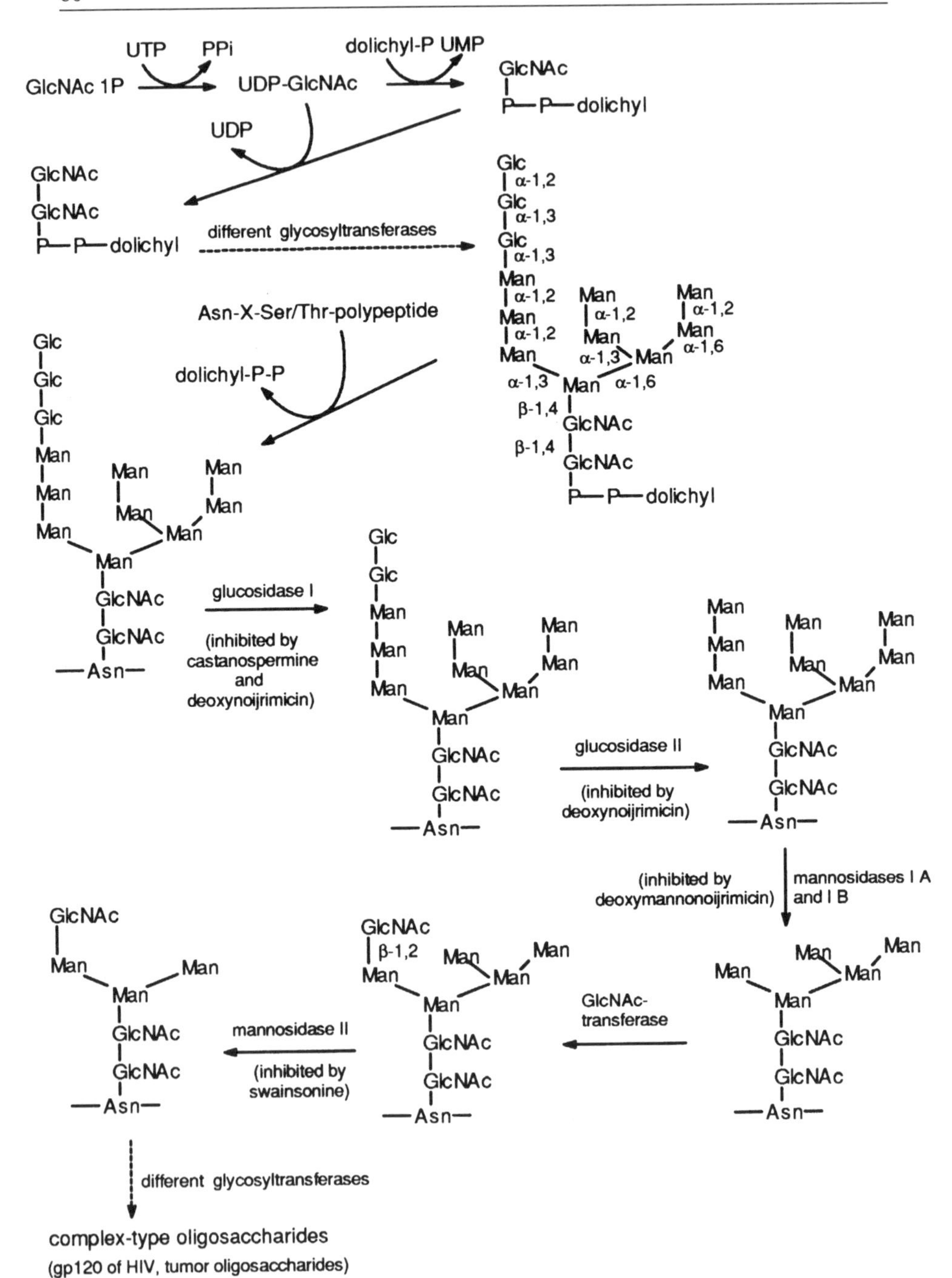

Scheme 2. The biosynthesis of *N*-linked glycoproteins

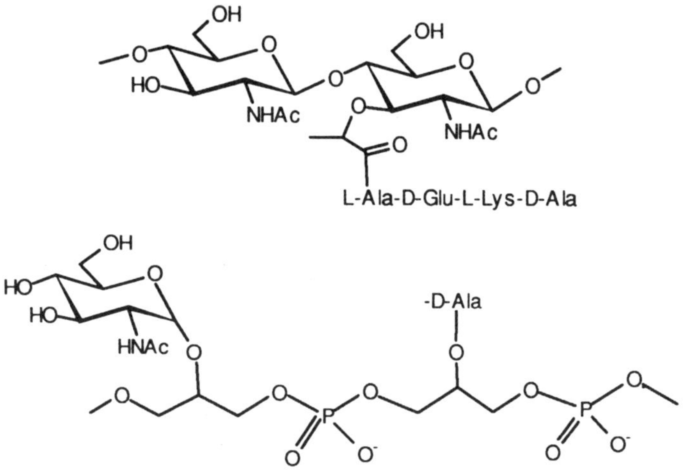

Fig. 4. Repeating unit of mureine (above) and theichoic acid (below)

reagent. Alternatively the anomeric carbon can be converted into a halide, which reacts with organometallic reagents or carbanions. The electrophilic character of the anomeric carbon can also be enhanced by Lewis acids, catalysing its conversion into a carbocation which reacts with electron rich carbon atoms, in a Friedel-Crafts type reaction. The anomeric carbon of the sugar can be converted into a radical, which reacts with unsaturated carbon atoms, and its polarity can be reversed by conversion into a carbanion, which reacts with electrophiles such as aldehydes. Scheme 3 describes in a very concise form all these possibilities. Further anomeric functionalizations, such as the formation of lactones, epoxides, and carbon-carbon double bonds, have also been exploited in the synthesis of C-glycosides.

In C-glycosylation reactions a new stereocenter is formed, and in general only one of the two stereoisomers is desired. Therefore, it is quite important to develop stereoselective C-glycosylation procedures, in which the stereochemical outcome can be predicted.

This article will describe our efforts in the stereoselective synthesis of C-glycosides of biological interest, giving, in the meantime, an overview of the main C-glycosylation procedures and explaining their stereochemistry.

3.1 Reaction of Glycosyl Halides with Organometallics and Carbanions

A protected glycosyl halide reacts with a Grignard reagent affording a C-glycoside [1] (Scheme 1). In the presence of an acyl participating group at C-2, which directs the attack of the nucleophile from the opposite side (as in O-glycosylations), the stereochemistry of the reaction is predictable. In the absence of a participating group, the C-glycosylation occurs mainly with inversion of the anomeric configuration, the reaction intermediate being probably an ion-pair. So, the reaction of the 2,3,4,6-tetra-*O*-benzyl-α-D-glucopyranosyl chloride with

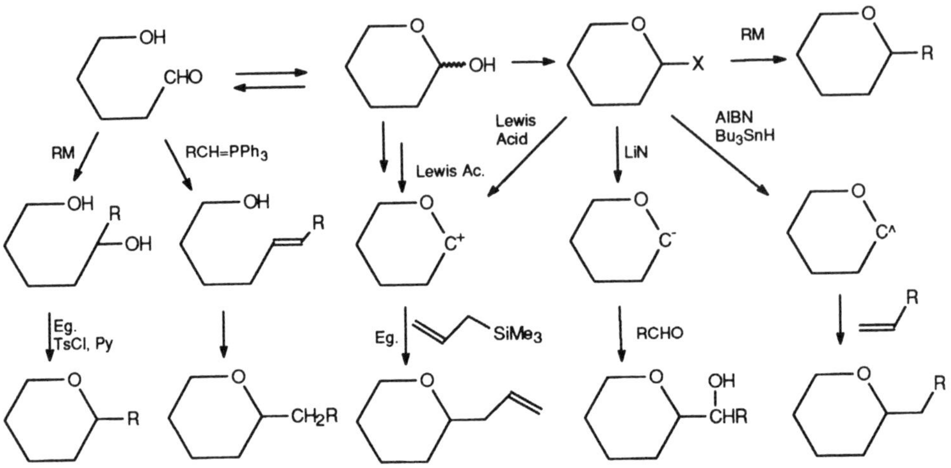

Scheme 3. General scheme of C-glycosylation procedures (LiN = lithium naphthalide)

2,4-dimethoxyphenyl magnesium bromide affords the corresponding C-glucosyl derivative in a β: α ratio of 4:1 [11].

The C-glycosylation reaction has been tested with other organometallic reagents such as alkyl- and aryllithium, alkyl- and arylcadmium, alkyl- and arylzinc, lithiumarylcuprates, alkylaluminum and silver acetylides, and some results are reported in Table 1. It seems in general that if the nucleophile is too hard, such as in the case of alkyllithium reagents, the elimination competes with the substitution reaction. This is the case for the reaction of $LiCH_2PO(OMe)_2$ with 2,3,4,6-tetra-*O*-benzyl-α-D-glucopyranosyl bromide **8**, that we have effected in the attempt to synthesize the phosphono analog of glucose 1-phosphate. As shown in Scheme 4, the reaction afforded **9**, the product of elimination of HBr.

Stabilized carbanions, such as malonyl carbanions, which have a soft character, react with glycosyl halides affording the corresponding C-glycosyl malonates. Once more, the presence of a participating group at C-2 directs the attack of the nucleophile from the opposite side. An example of the application

Table 1. Reaction of glycosyl halides with different organometallic reagents

RM	Leaving group	Yields %
RMgBr, RMgCl	Cl, Br	50–85
PhLi	Cl	27
Ar_2Cd	Cl	20–83
Ph_2Zn	Cl	72
R_2CuLi	Br	40–60
$AgC{\equiv}CCO_2Et$	Cl	42
furanylHgCl	Br	58
Me_3Al	F	95

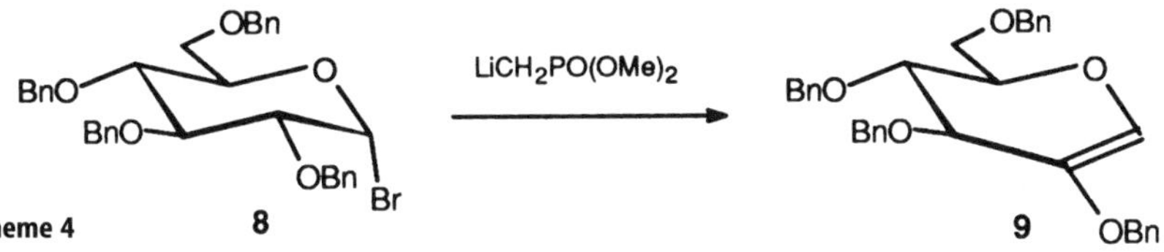

Scheme 4

of this C-glycosylation procedure is reported in the stereoselective synthesis of the phosphono analog **15** of *β*-D-glucose 1-phosphate (Scheme 5) [12]. In this synthesis 2,3,4,6-tetra-*O*-acetyl-α-D-glucopyranosyl bromide **10** was reacted with the anion of dibenzyl malonate to afford stereoselectively the *β*-C-glycosyl malonate **11**, taking advantage of the presence of the acetyl participating group at C-2. The debenzylation of the malonate, followed by acid catalyzed decarboxylation, afforded 2-(2,3,4,6-tetra-*O*-acetyl-*β*-D-glucopyranosyl)acetic acid

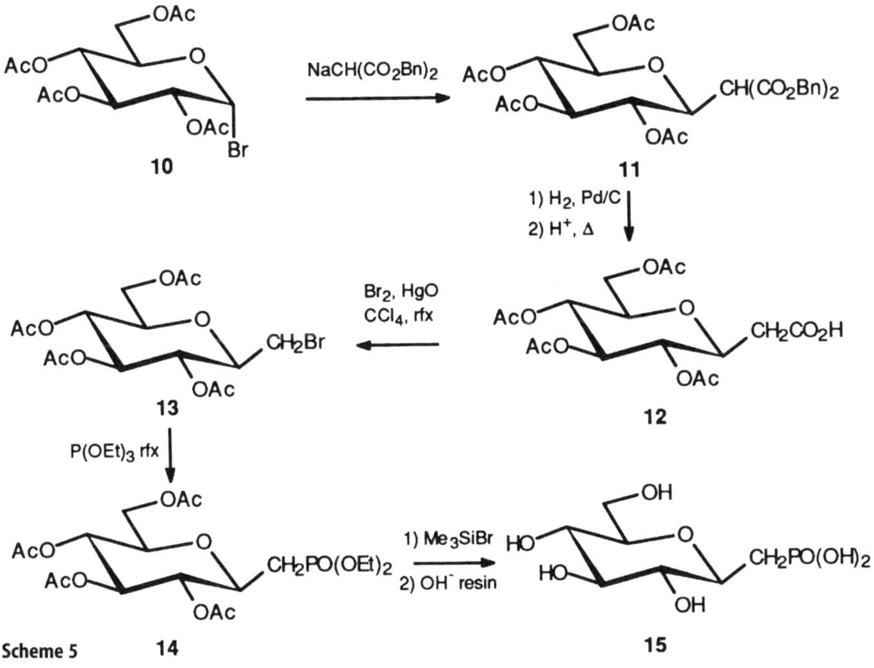

Scheme 5

12, which was then submitted to bromodecarboxylation, Arbuzov reaction and deprotection, to give the phosphono analog **15** of *β*-D-glucose 1-phosphate. Any attempt to apply this method to the synthesis of C-glycosides of D-glucosamine was unfruitful. Glycosyl halides of differently protected D-glucosamine do not react with malonyl carbanions or Grignard reagents, or react in very poor yields [13] (Scheme 6).

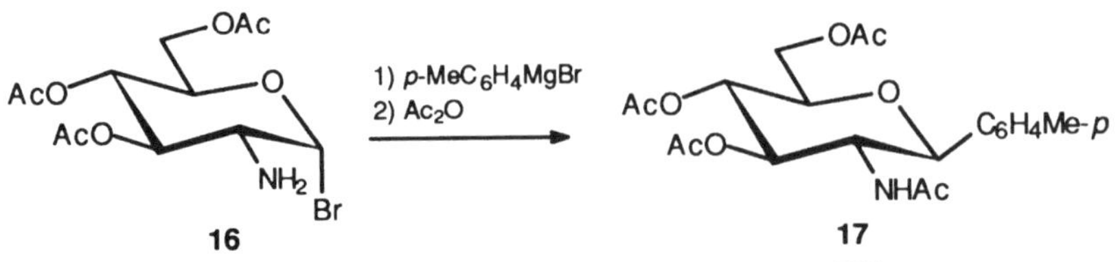

Scheme 6

3.2 Lewis Acid Catalyzed C-Glycosylations

The electrophilic character of the anomeric carbon of a sugar can be enhanced by treatment with a Lewis acid. This treatment affords a stabilized anomeric carbocation intermediate which easily reacts with a nucleophile. The reaction can be effected starting from glycosyl halides, glycosyl acetates, glycosides and also from aldoses. The better the leaving group, the easier the reaction. The stabilized anomeric carbocation intermediate reacts in a Friedel-Crafts type electrophilic substitution with aromatic compounds, alkenes, enolethers, enamines, allylsilanes, propargylsilanes, trimethylsilyl cyanide and alkinylstannanes (Scheme 7). The stereochemistry of the reaction depends on the nature of the substrate, and to some extent also on the experimental conditions. In the case of pyranosides the anomeric effect exerts a clear influence, the nucleophile attaches the anomeric center from the side opposite to the axial ion-pair of the pyranosidic oxygen. So, D-glucopyranosides and D-mannopyranosides afford α-C-glycosides with excellent stereoselection [14]. In the case of furanosides the

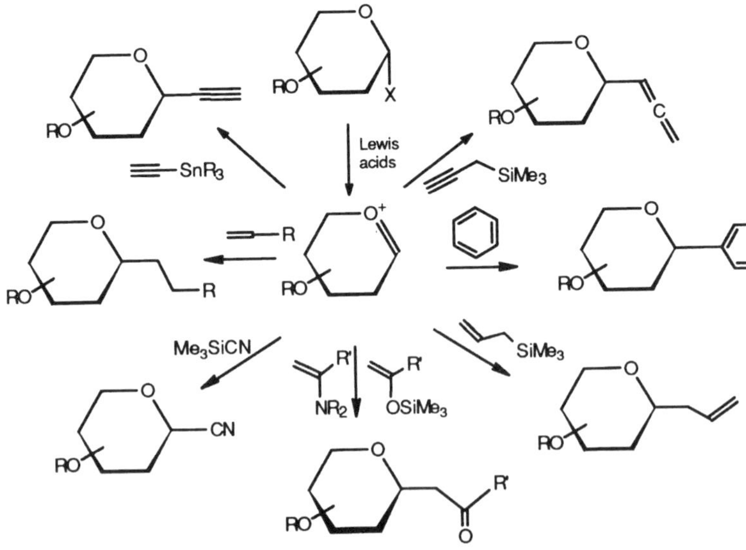

X=Cl, F, OAc, OMe, thiopyridyl, trifluoroimidate, *p*-nitrobenzoate, OH

Scheme 7. Lewis acid catalyzed C-glycosylations

stereochemical results are less clear, the preferred conformation of the intermediate carbocation being less defined. In these cases the experimental conditions, such as solvents and temperatures, play a relevant role.

The Lewis acid catalyzed C-glycosylation method is probably one of the most efficient to prepare C-glycosides of ketoses, the anomeric carbocation intermediate being in this case more stabilized by a further substituent. We applied this procedure to synthesize C-fructosides [15] (Scheme 8). The reaction of methyl

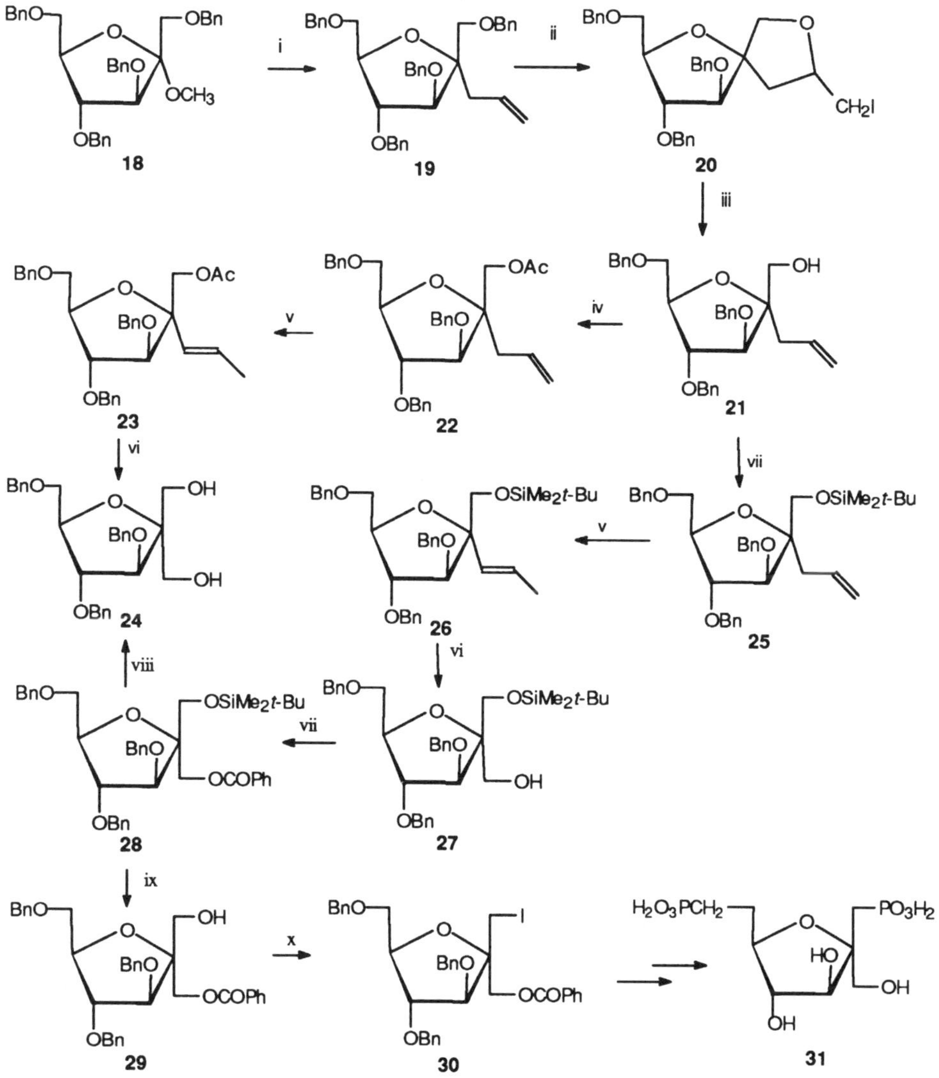

Scheme 8. i, $Me_3SiCH_2CH{=}CH_2$, Me_3SiOTf MeCN, 0 °C; ii, I_2, THF; iii, Zn, AcOH; iv, Ac_2O-Py; v, $PdCl_2(MeCN)_2$, PhH, rfx; vi, O_3, CH_2Cl_2, –78 °C, then $Zn(BH_4)_2$; vii, *t*-$BuMe_2SiCl$-TEA-DMAP, CH_2Cl_2; viii, Bu_4NF; ix, AcOH, H_2O, THF; x, I_2, Im, Ph_3P

1,3,4,6-tetra-*O*-benzyl-α-D-fructofuranoside **18** with allyltrimethylsilane and the Lewis acid trimethylsilyltriflate stereoselectively afforded the α-allyl C-glycoside **19** in about 60% d.e., independent of the nature of the solvent and the temperature. The reaction was unexpectedly stereoselective, taking into account that it occurs on a furanosidic ring. The manipulation of the α-allylic appendage so introduced allowed us to obtain different α-C-furanosides. On the other hand, we were also able to apply this reaction to the synthesis of β-C-fructosides. To do that we exploited the double bond of the allylic appendage of **19**, which on treatment with iodine effects the selective debenzylation of the C-1 hydroxyl group, with formation of the cyclic iodoether **20**. Treatment of **20** with zinc and acetic acid resulted in a reductive elimination, with formation of **21**. The molecule **21** corresponds to **19**, deprotected at C-1. Once deprotected, this hydroxyl group of **21** can be functionalized in different ways, so acting as a β-oriented C-glycosidic substituent. The α-oriented allylic anomeric appendage of **21** can in turn be converted into a hydroxymethyl group, taking the place of the original C-1 of the fructose skeleton. The conversion of the allylic substituent into a hydroxymethyl group was effected by isomerization of the double bond (**23** or **26**), ozonolysis and reduction of the ozonide to a hydroxyl group (**24** or **27**). The procedure required a delicate choice of protecting groups and experimental conditions, as the proximity of the two appendages can cause reciprocal interference. For example, protecting the β-oriented hydroxyl group as an acetate (**22**), and effecting the ozonolysis of **23** with reduction of the ozonide with $NaBH_4$ (iv in Scheme 8), causes a deacetylation, giving rise to formation of the diol **24**. In the hypothesis that this hydrolysis is due to the basic conditions generated by the reductant, the neutral $Zn(BH_4)_2$ was used, but once more the diol **24** was the main product. The protecting group was then changed, turning to the *tert*-butyldimethylsilyl ether which is stable under basic conditions. However, also with this protecting group, the reduction of the ozonide with $LiAlH_4$ was unexpectedly accompanied by extensive hydrolysis of the *tert*-butyldimethylsilyl ether despite the known stability under basic conditions. To obtain the desired product **27**, it was necessary to reduce the ozonide of **26** with triphenylphosphine, then using the neutral reductant $Zn(BH_4)_2$ to reduce the aldehyde obtained. In addition, the protection of the newly formed α-oriented hydroxyl group of **27**, and the subsequent hydrolysis of the *tert*-butyldimethylsilyl ether to restore the β-oriented hydroxymethyl group required many different attempts to find the right experimental conditions. For example, treatment of **27** with 2-methoxyethoxymethyl chloride resulted in extensive desilylation. The benzoate was the protecting group of choice, provided that acetic acid, and not fluoride anions, is used for the subsequent deprotection of the silyl ether of **28**. In fact the use of tetrabutylammonium fluoride caused debenzoylation with formation of the diol **24**.

The above-mentioned procedure for the synthesis of β-C-fructosides has been used to synthesize the bisphosphono analog of β-D-fructose 2,6-bisphosphate [16], which is, as reported in Sect. 2.3, an important activator of glycolysis and inhibitor of gluconeogenesis. To prepare the target molecule we first attempted the conversion of the free hydroxyl group of **21** into an iodide which in turn can be easily converted into a phosphonate. However, this conversion occurred in very low yields, so a more complex procedure was required. Instead of converting directly the hydroxyl group of **21** into a phosphonate, and then its allylic substituent into an hydroxymethyl group, the following sequence of reaction was required. Protection of the hydroxyl group on the β-appendage of **21** to afford **25**; three-step conversion of the α-appendage of **25** into a protected hydroxymethyl group of **28**; deprotection of the hydroxyl group of the β-appendage of **28** to afford **29**; transformation of **29** into the iodide **30** by treatment with triphenylphosphine, iodine and imidazole; and finally conversion of the iodide **30** into the corresponding phosphonate by reaction with triethylphosphite. To complete the synthesis a second methylphosphonate was subsequently introduced in place of the hydroxyl group at C-6 of the fructose skeleton. This was effected by selectively deprotecting the benzyl group at C-6 by controlled catalytic hydrogenation, oxidizing the obtained hydroxyl group to an aldehyde, and reacting the aldehyde with $CH_2[PO(OEt)_2]_2$ in presence of a base. This last reaction was attempted in different experimental conditions, the best results being obtained with DBU, LiCl in acetonitrile.

The C-fructoside **29** was used in a project devoted to the synthesis of C-sucrose. The idea was to convert the free hydroxyl group of **29** into a functional group which could be attacked by a glucosyl radical, so allowing the linkage of a glucose moiety to a carbon linked at the anomeric center of the fructose moiety, with the correct β-orientation. To do that the free hydroxyl group of **29** was converted into the aldehyde **32**, which on reaction with carboethoxymethylenetriphenylphosphorane in refluxing acetonitrile afforded the bicyclic C-glycoside **33** (Scheme 9) as the result of an intramolecular transesterification. The α,β-unsaturated ester **33** was tested as scavenger in a reaction with a glucosyl radical, but unfortunately the reaction was unsuccessful (Scheme 9).

BnO OH O BnO OCOPh OBn **29** PCC BnO O O BnO OCOPh OBn **32** $Ph_3P{=}CHCO_2Et$ BnO O BnO O O OBn **33** OBn O BnO BnO OBn

Scheme 9. (PCC = pyridinium chlorochromate)

The synthesis of C-glycosyl derivatives of non-reducing disaccharides, such as C-sucrose, is quite difficult. It requires the linking between the anomeric centers of two sugars through a carbon atom bridge. Kishi and Dyer [17] and Nicotra et al. [18] developed different synthetic procedures allowing the preparation of C-disaccharides related to sucrose building up ex novo the fructosidic moiety of the molecule. In our case, we exploited the Lewis acid catalyzed C-glycosylation procedure for the stereoselective introduction of an α-oriented allylic appendage at the anomeric center of 2,3,4,6-tetra-*O*-benzyl-D-glucopyranose

34 (Scheme 10). The allylic appendage of the obtained 1-(2,3,4,6-tetra-*O*-benzyl-α-D-glucopyranosyl)-2-propene **35** was converted into the stabilized Wittig reagent **38** which was then reacted with differently protected enantiomerically pure D-glyceraldehyde derivatives to afford the (E)-olefin **39**. The osmylation of the double bond of **39**, stereo-directed by the allylic oxygen, afforded the diol **40**, the newly formed stereocenters of which have the configurations required for the target molecule. The deprotection of **40** allows the formation of an hemiketal structure which mimics the furanosidic moiety of sucrose. D-Glyceraldehydes protected as 2,3-*O*-isopropylidene and as 2-benzyl-3-benzyloxysuccinoyl derivatives were employed in this synthesis. In the first case, an equilibrium between the furanosidic and pyranosidic hemiketal ring is observed, whereas in the second case the deprotection by catalytic hydrogenation gives rise to a molecule in which the hydroxyl group at the furanosidic end is succinylated.

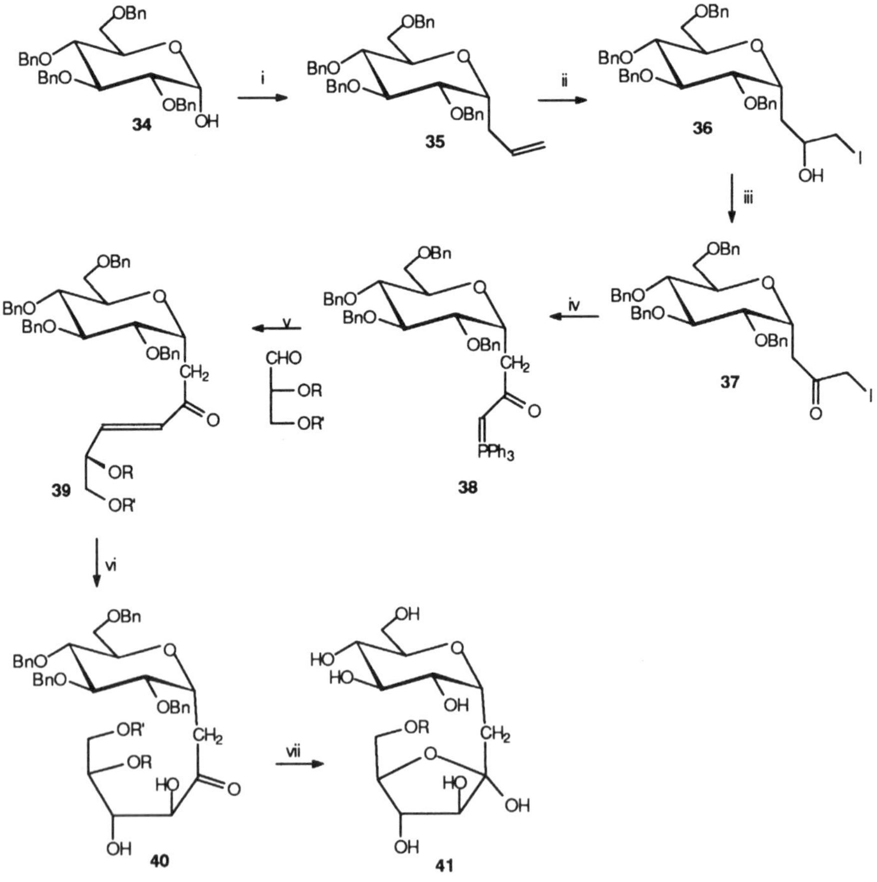

OBn BnO BnO OBn OH 34 i OBn BnO BnO OBn 35 ii OBn BnO BnO OBn 36 OH I
iii
OBn BnO BnO OBn CH2 O 39 OR OR' v CHO OR OR' OBn BnO BnO OBn CH2 O PPh3 38 iv OBn BnO BnO OBn 37 O I
vi
OBn BnO BnO OBn OR' CH2 OR HO O OH 40 vii OH HO HO OH OR CH2 O HO OH OH 41

Scheme 10. i, allyltrimethylsilane-trimethylsilyltriflate; ii, $Hg(OAc)_2$, acetone-water; then I_2, THF; iii, PCC, CH_2Cl_2; iv, Ph_3P, then $NaHCO_3$; v, MeCN; vi, OsO_4 cat, *N*-methylmorpholine-*N*-oxide, acetone-water; vii, deprotection

This succinyl group avoids the formation of the pyranosidic ring and acts as a spacer to conjugate the mole-cule to resins or biopolymers. The succinylated analog of sucrose **41** (R= $HO_2CCH_2CH_2CO$-) is not sweet, and it is unable to inhibit invertase. It is however an inhibitor of α-glucosidase from baker's yeast.

An interesting result, in which the Lewis acid catalyzed C-glycosylation procedure is involved, has been obtained treating 1-deoxy-3,4,6-tetra-*O*-benzyl-D-arabinohexulose **42** with Lewis acids. A dimerization occurs, with formation of the C-disaccharidic structure **43**, in which both the anomeric centers are involved in the C-glycosidic linkage [19] (Scheme 11). The reaction probably involves an enolether intermediate which acts as a nucleophile on the oxonium ion intermediate from which it is probably generated. The attack occurs stereoselectively from the β-face affording the C-disaccharide **43** (mixture of anomers). To confirm the hypothesis that the enolether **44** is the intermediate in this reaction, it was synthesized from the corresponding lactone (with titanocene dichloride-trimethylaluminum complex, the Tebbe reagent), and treated with

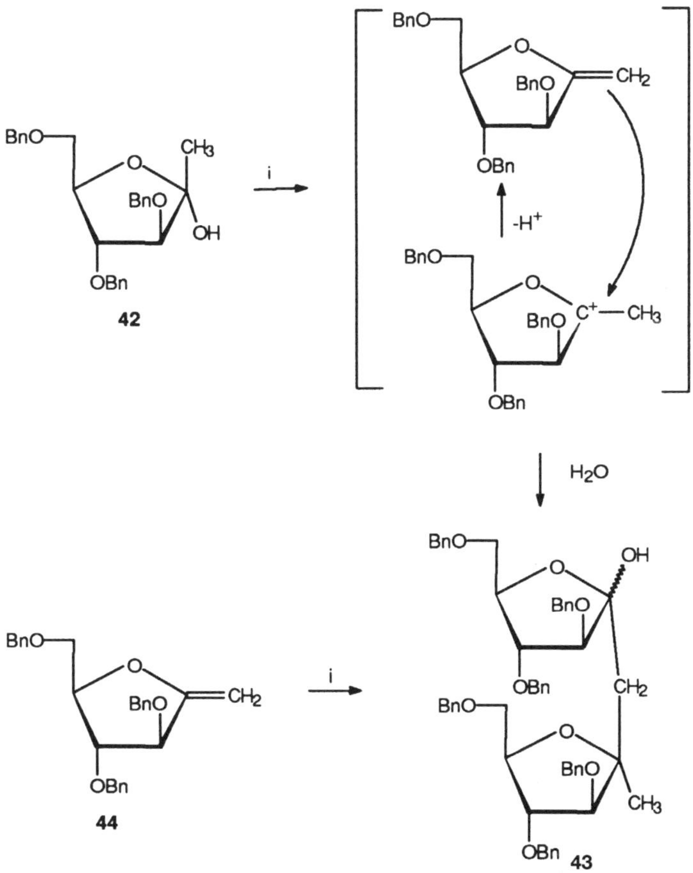

Scheme 11. i, BF_3-OEt_2, MeCN, 0 °C

Lewis acids. As expected, the same C-disaccharide **43** was obtained. The procedure was extended to glucose. In this case however the 1-deoxy-glucoheptulose **45** does not dimerizes by treatment with Lewis acids, whereas the enolether **46** does, affording the C-disaccharide **47** [20] (Scheme 12). Once more the attack of the nucleophile on the oxonium ion is stereoselective, but in this case it occurs from the α-face, as expected for a glucopyranosidic structure. Attempts have also been made to react the enolethers **44** and **46** with different glycosyl cation intermediates, in various experimental conditions. In all the cases in which a Lewis acid was used to generate the glycosyl cation, the product of dimerization of the enolether was obtained. Employing non-acidic anomeric activation methods, such as the reaction of a thiophenyl or a thiopyridyl glycoside with methyl iodide, no reaction occurred or, eventually, under drastic conditions, the glycosyl carbocation underwent elimination.

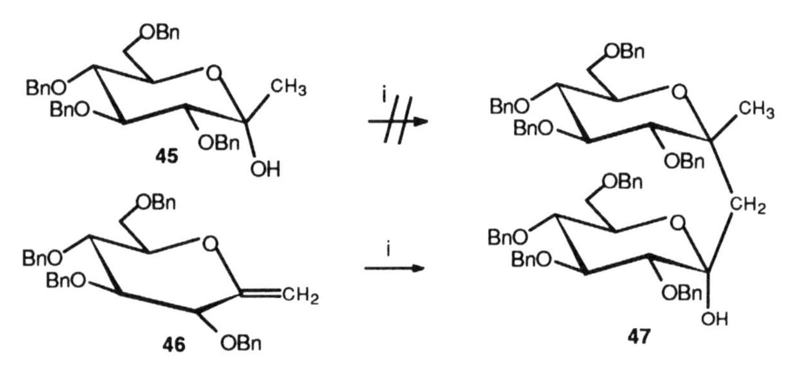

Scheme 12. i, BF_3OEt_2, MeCN, 0 °C

The Lewis acid catalyzed C-glycosylation procedure has never been applied to 2-aminosugars, the nucleophilic nitrogen interfering in the reaction. C-Glycosides of D-glucosamine and D-galactosamine were however obtained by applying the procedure to 2-azido-2-deoxy-glycopyranosides **49**, which are obtained in turn from the corresponding glycals **48** (Scheme 13). Alternatively C-glycosides of 2-aminosugars could be easily obtained provided that the hydroxyl group at C-2 of a C-glycoside could be selectively deprotected in a simple manner. Once deprotected, this hydroxyl group can be converted into an amino group with

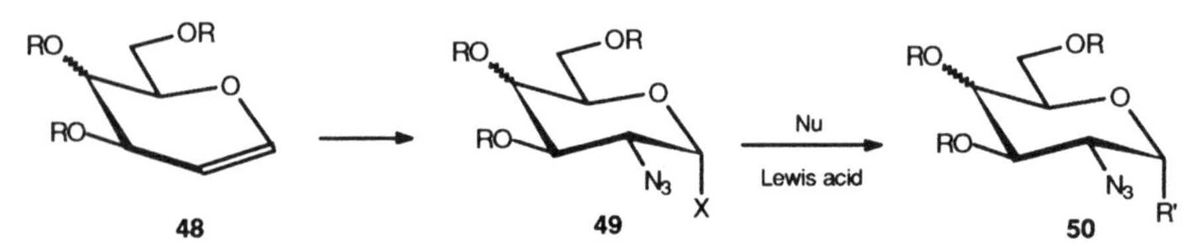

Scheme 13. X=$-ONO_2$, $-Br$ or $-OC(NH)CCl_3$; Nu=$Me_3SiCH_2CH_2$, $Bu_3SnC{\equiv}CR$ or Me_3SiCN; R'=$-CH_2CH{=}CH_2$, $C{\equiv}CR$ or $-CN$

inversion or alternatively with retention of configuration. To obtain α-C-glycosides of D-glucosamine and D-mannosamine following this approach, we started from the easily available 1-(2,3,4,6-tetra-*O*-benzyl-α-D-glucopyranosyl)-2-propene **51**. In fact, the selective deprotection of the benzyl protecting group at C-2 of **51** can be effected taking advantage of the fact that this oxygen is the only one in the molecule that can intramolecularly attack the activated (as iodonium ion) double bond of the glycosidic appendage. The attack results in a cyclization and debenzylation (Scheme 14), and the obtained cyclic α-iodoether **52** treated with zinc and acetic acid affords **53**, in which the original double bond is regenerated and the C-2 hydroxyl group is deprotected. The whole deprotection process occurs with 68% overall yield [21]. Once deprotected, the hydroxyl group of **53** can be oxidized and then converted into an oxime, the reduction of which, either by catalytic hydrogenation or by reduction with $LiAlH_4$, affords the C-glucosamine **55**. Alternatively, the sequence triflate, azide and reduction affords the C-mannosamine **54**.

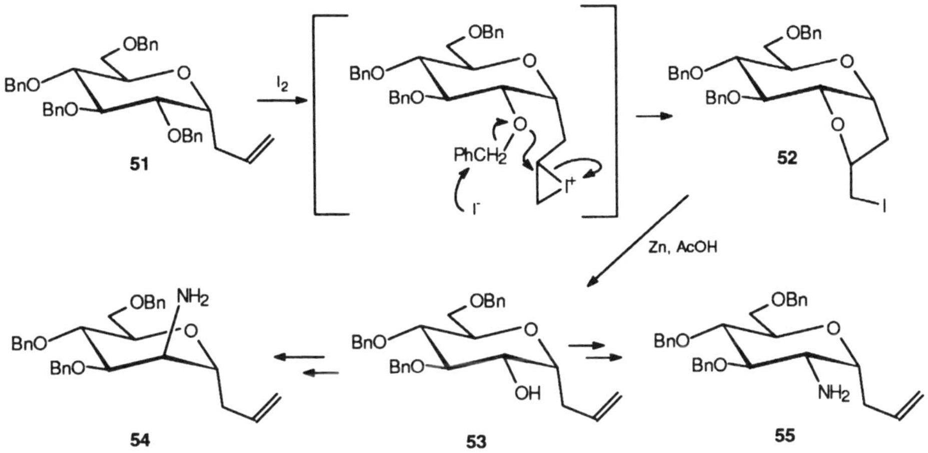

Scheme 14

3.3 C-Glycosides from Glyconolactones

The lactonic carbon of a sugar is harder than the corresponding glycosyl halide. This allows an efficient addition of organolithium reagents to glyconolactones with the formation of the corresponding lactol (Scheme 15). The reaction stops in general at this level, and the lactol can be converted into a C-glycoside by a Lewis acid catalyzed reduction with triethylsilane [14]. The reaction mechanism of this reduction is similar to that of Lewis acid catalyzed C-glycosylations, the nucleophile being changed from a carbon atom to a hydride. Also the stereochemistry of the process is the same of that reported in Lewis acid catalyzed C-glycosylations, and is influenced by the anomeric effect. In the case of D-glucopyranoses and D-mannopyranoses, the hydride attacks the oxonium ion from the α-face giving

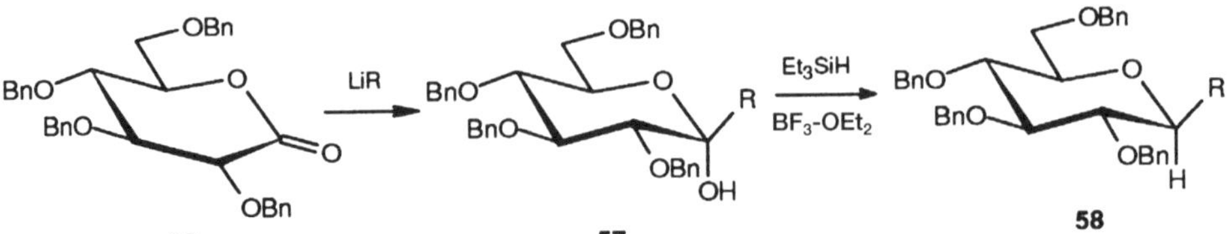

Scheme 15

rise to β-C-glycosides. It is worth noting that the procedure is complementary, from the stereochemical point of view, to that described in Sect. 3.2.

This procedure could in principle open the way, inter alia, to the phosphono analog of β-D-glucose 1-phosphate simply by reacting the gluconolactone **56** with $LiCH_2PO(OMe)_2$ and then reducing the obtained lactol **59**. Unfortunately the lactol phosphonate **59** was inert to this reduction (Scheme 16).

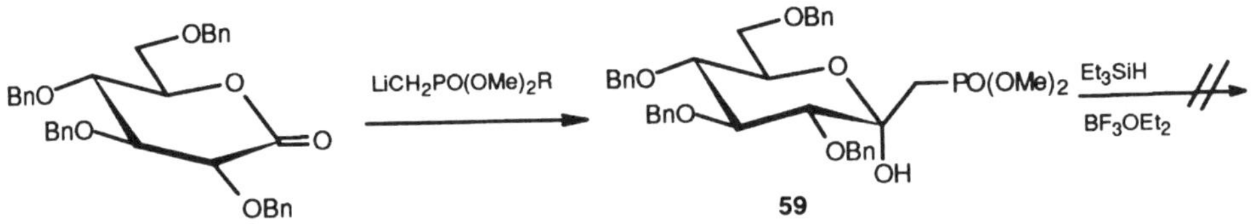

Scheme 16

Glyconolactones undergoes methylenation [22], difluoromethylenation [23], and dichloromethylenation [24] to afford glycoexoenitols (Scheme 17) which can be manipulated to afford different C-glycosides. We studied the possibility of obtaining C-glycosides using the electron-rich double bond of glycoexoenitols (**60** and its galacto isomer) as electron-poor radical scavengers. The malonyl radical obtained from bromomalonate and Bu_3SnH reacted with different hexoenitols to afford the corresponding C-glycosyl methylmalonates (**61** and its galacto isomer) (Scheme 18) [25]. In the case of the D-gluco and D-galacto derivatives, the β-C-glycosides were the only detectable products, the reaction being highly stereoselective. Once more the anomeric effect is responsible for the

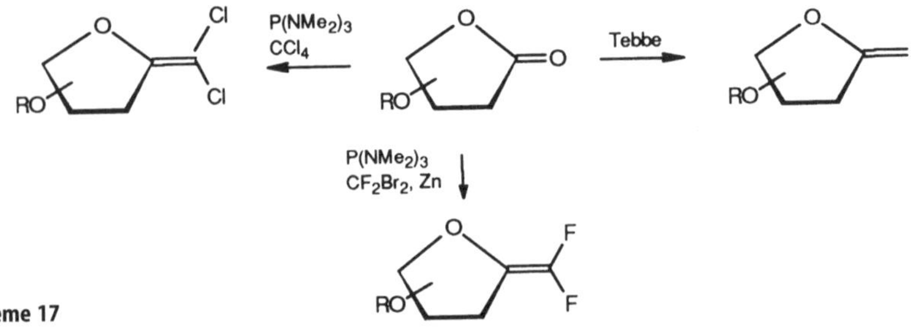

Scheme 17

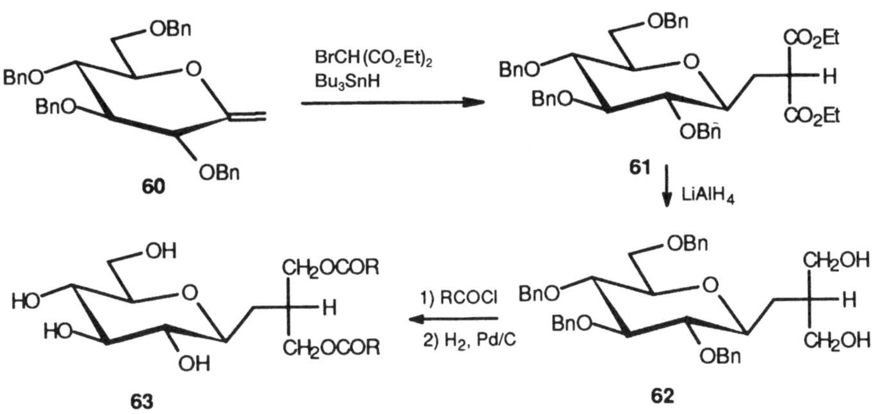

Scheme 18

stereochemical outcome of the reaction, since the α-oriented glycopyranosyl radical intermediate is more stable and then reduced by Bu_3SnH. The malonyl group of **61** (and its galacto isomer) can be reduced to afford stable analogs of glycosyl glycerols, in which the oxygen that links the sugar to the glycerol is substituted by a carbon atom. Furthermore, the esterification of the hydroxyl group of the glycerol moiety gives rise to stable analogs of glycolipids (**63**). It is noteworthy that natural occurring glycosyl glycerols and their diacyl derivatives show antitumor activities [26]; they are however labile to glycosidases, whereas this is not the case for stable C-glycosidic analogs.

3.4 Wittig Reaction – Cyclization Procedure

Aldoses react with Wittig reagents affording an open chain product defined as glycoenitol. If the free hydroxyl group of the glycoenitol effects a nucleophilic addition to the double bond (which must be activated), an a C-glycoside is formed. The double bond can be activated by conjugation with an electron withdrawing group (Scheme 19, path a), or by reaction with an electrophile, E^+ (Scheme 19, path b).

When stabilized ylids such as $Ph_3P{=}CHCO_2Et$ are used, the cyclization of the open chain product to afford a C-glycoside can be effected by a base catalyzed

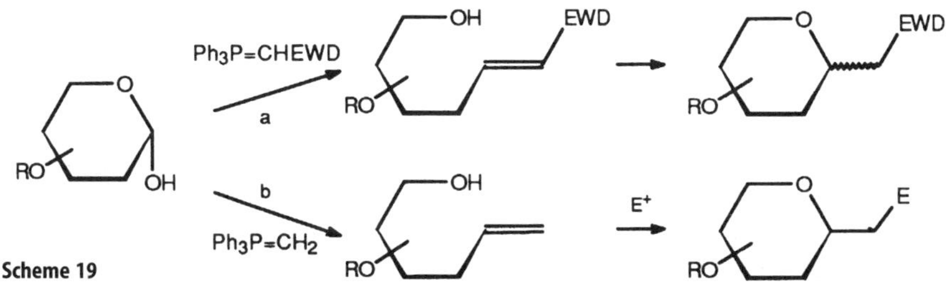

Scheme 19

intramolecular Michael reaction. It has been shown that in some cases the Michael cyclization occurs spontaneously during the Wittig reaction. This occurs when furanoses with an isopropylidene protecting group (such as a 2,3-isopropylidene-5-triphenylmethyl-D-ribofuranose **67** in Scheme 20) are used [27], or in general at high reaction temperatures, such as in refluxing *N,N*-dimethylformamide [28]. Furthermore, it has been shown that the presence of a partici-

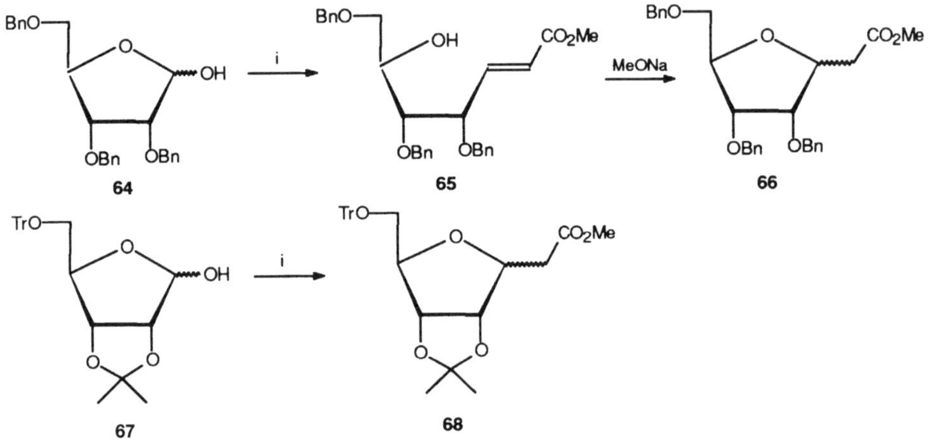

Scheme 20. i, $Ph_3P{=}CHCO_2Me$, MeCN, rfx

pating group at C-2 of the starting sugar favors the spontaneous cyclization during the Wittig reaction [29]. This is the case with the reaction of 2-acetamido-2-deoxy-4,6-*O*-benzylidene-D-glucopyranose **69**, which on treatment with $Ph_3P{=}CHCO_2Et$ affords directly and stereoselectively, the α-C-glycopyranoside **70** (Scheme 21) [30], one of the few examples of C-glycosides of glucosamine.

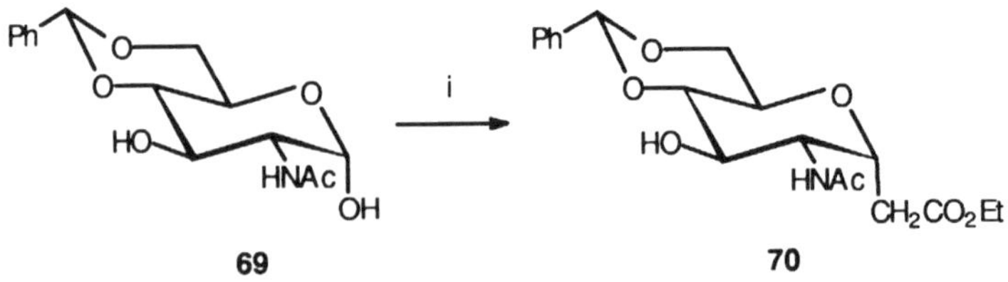

Scheme 21. i, $Ph_3P{=}CHCO_2Et$, MeCN, rfx

The stereochemistry of the C-glycosidic products, obtained with the abovementioned procedure, depends on the structure of the starting sugar and on the experimental conditions. In general, base catalyzed cyclizations afford a mixture of anomers, which on prolonged reaction time is converted into the thermodynamic product. For example, in the case of 4,6-*O*-isopropylidene-D-glucose, the mixture of anomers originally formed by reaction of the open chain intermediate with KOH, is gradually converted into the β-C-glucopyranose, the thermodynamic product [31]. Interestingly, in the case of 4,6-*O*-benzylidene-*N*-acetyl-D-glucosamine **69**, which cyclizes spontaneously during the reaction with $Ph_3P{=}CHCO_2Et$ (Scheme 21), the α-C-glucopyranoside **70**, which is the kinetic product, is formed. This result can be explained in the light of the spontaneous cyclization of the open chain intermediate, which occurs in neutral non-equilibrating conditions favored by the presence of the *N*-acetyl group.

A more efficient stereocontrol in the cyclization of glycoenitols can be obtained by activation of the double bond with an electrophile. In this case the presence of an electron withdrawing group conjugated with the double bond is not necessary. Sinaÿ et al. proposed the use of the simple ylid $Ph_3P{=}CH_2$, which reacts with 2,3,4,6-tetra-*O*-benzyl-D-glucopyranose **71** affording the glucoheptenitol **72**. Treatment of **72** with mercuric acetate gives rise to a stereoselective cyclization with formation of the α-C-glucoside **73** (Scheme 22, first line) [32]. The stereochemical outcome of the reaction has been studied [33], and the results clearly indicate that it depends on the orientation of the allylic alkoxy substituent, the 1,2-*cis* product being favored. Scheme 22 shows selected examples of this study. Starting from 2,3,4,6-tetra-*O*-benzyl-D-glucopyranose **71**, the α-C-glucoside **73** is formed as the only product.

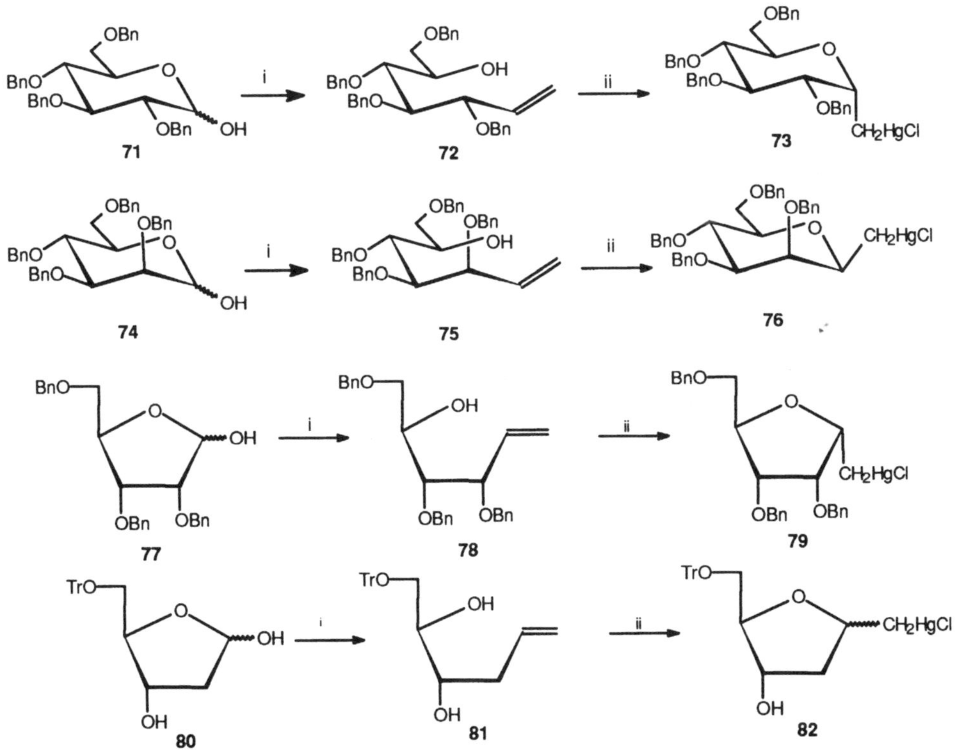

Scheme 22. i, $Ph_3P{=}CH_2$, THF; ii, $Hg(OAc)_2$, THF, then KCl

Changing the structure of the sugar, but not the stereochemistry at C-2, such as in the case of 2,3,5-tri-*O*-benzyl-D-ribofuranose **77**, the stereochemistry of the reaction does not changes, the α-C-riboside **79** being formed in 70% d.e. Changing the stereochemistry at C-2 of the starting sugar, as in the case of 2,3,4,6-tetra-*O*-benzyl-D-mannopyranose **74**, the stereochemistry of the reaction changes, the β-C-mannoside **76** being the predominant product (40% d.e.). In the absence of substituents at C-2, as in the case of 5-triphenylmethyl-D-deoxyribofuranose **80**, no stereoselection is observed, the α- and β-C-2-deoxyribosides being obtained in equal amounts (**82**). The lower stereoselection observed in the formation of the β-C-mannoside can be explained by taking into account the fact that the β-mannosidic structures are strongly disfavored for stereoelectronic reasons, and are very difficult to obtain. These stereochemical results can be explained in the light of the observation that the more stable conformation of allylic alcohols (and ethers) is that in which the heteroatom lies in the same plane as the double bond. The attack of the electrophile then occurs from the less hindered side as shown in Fig. 5.

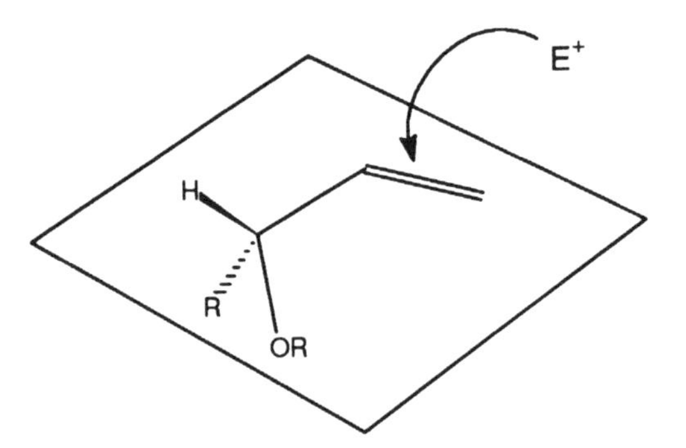

Fig. 5

Different electrophiles, such as I_2, Br_2, NBS, NIS, PhSeCl or metachloperoxybenzoic acid have been used in the cyclization of glycoenitols [34], but the best stereochemical results have been obtained with mercuric salts and iodine. Iodine however can give rise to a debenzylation with formation of the more favored 5-member cyclization product. This is the case for **72** which, when treated with iodine, affords the C-fructosidic structure **83** [35] (Scheme 23).

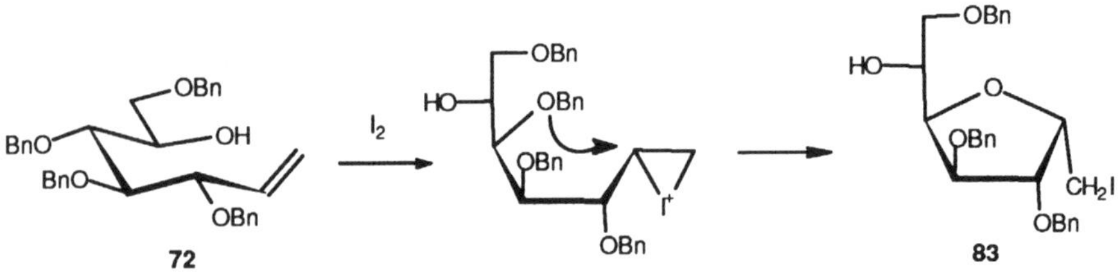

Scheme 23

Following the Wittig cyclization C-glycosylation procedure, the phosphono analog of α-D-glucose 1-phosphate, one of the main carbohydrate metabolites, has been stereoselectively synthesized according to Scheme 24 [12]. The mercurioderivative **73** was converted into the corresponding bromide **84** (or iodide) which in turn was submitted to the Arbuzov reaction in refluxing triethylphosphite to afford the corresponding phosphonate **85**. Treatment of the phosphonate **85** with iodotrimethylsilane effected the concomitant deprotection of the benzyl groups and the hydrolysis of the phosphonic ester, giving rise to **86**, the phosphono analog of α-D-glucose 1-phosphate. In a similar manner the phosphono analogs of α- and β-D-mannose 1-phosphate and of α-D-ribose 1-phosphate have been prepared [36, 37]. On the other hand, all attempts to apply this procedure to synthesize the phosphono analog of α-D-glucosamine 1-phosphate, which is probably the most important phosphono analog from a pharmacological point of view, failed.

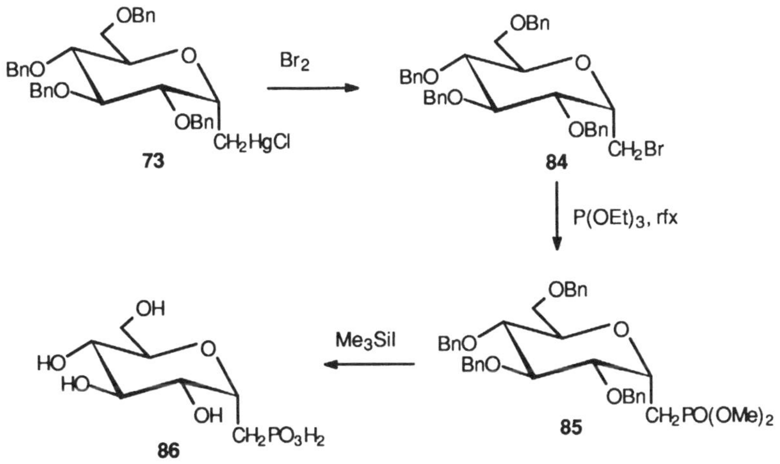

Scheme 24. i, $Ph_3P{=}CH_2$, THF; ii, $Hg(OAc)_2$, THF, then KCl; iii, Br_2, THF; iv, $P(OEt)_3$, rfx; v, Me_3SiI

The Wittig-cyclization C-glycosylation procedure is applied with difficulty to sugars different from glucose or arabinose. In fact, in the case of other sugars, the yields of the Wittig reaction are often very low. In particular, all attempts to apply this reaction to differently protected glucosamines failed. Furthermore, the isolation of the Wittig products from triphenylphosphine oxide, the side-product of the reaction, is often tedious and time consuming. To overcome these difficulties, an alternative way to obtain the glycoenitols, the cyclization of which affords C-glycosides, has been developed. It consists in the vinylation of a sugar with one carbon atom less [38] (Scheme 25), a reaction that generates a new stereocenter, the stereochemistry of which must be controlled. The application of this procedure to different protected pentoses indicates that the addition of vinylmagnesium bromide to the carbonyl group in equilibrium with the hemiacetal form is not stereoselective, even at very low temperature. On the other hand, divinylzinc, which is a more chelating reagent, stereoselectively affords the *threo*-products, in accordance with the Cramchelation model. The glycoenitols obtained by stereoselective vinylation of aldoses were mercuriocyclized to afford stereoselectively the C-glycosyl mercurio derivatives with the 1,2-*cis* relationship, according to the previously described stereochemical outcome of the mercuriocyclization reaction (Scheme 25). C-Glycosides obtained by this method have an interesting property – the hydroxyl group at C-2, which originates from the carbonyl group of the starting sugar, is the only one deprotected. This fact allows the selective manipulation of the deprotected hydroxyl group to obtain, for example, 2-amino-, 2-deoxy- and 2-fluoro-C-glycosides or simply to invert the stereochemistry of the C-2 hydroxyl group.

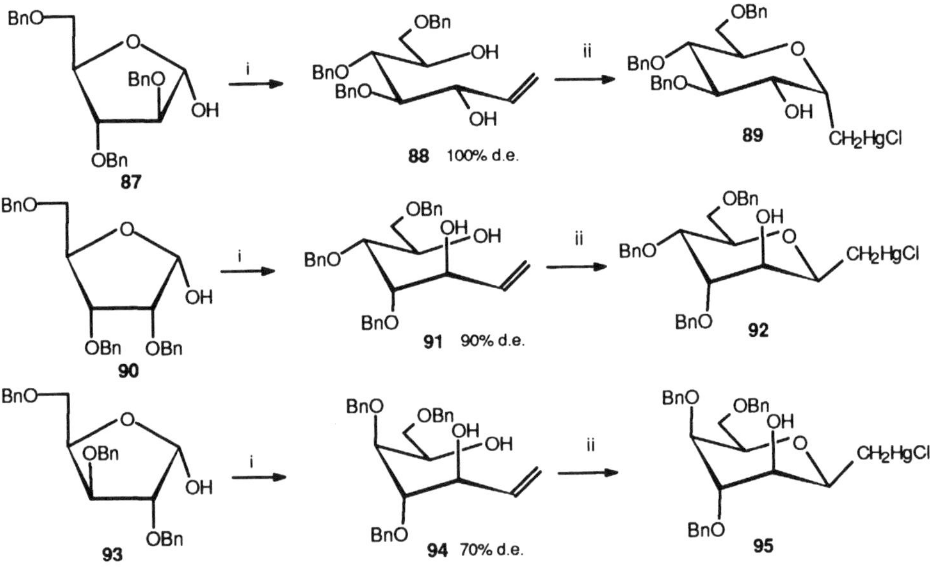

Scheme 25. i, $Zn(CH{=}CH_2)_2$, ii, $Hg(OAc)_2$, THF, then KCl

The procedure now described could afford C-glycosides of D-glucosamine, provided that the whole process could be applied to *N*-benzyl-2,3,5-tri-*O*-benzyl-D-arabinofuranosylamine **96** (Scheme 26), in equilibrium with the corresponding imine **97**. Divinylzinc, which stereoselectively reacts with aldoses, was unable to react with **96**; on the contrary, the more reactive vinylmagnesium bromide, which was not stereoselective with aldoses, reacted smoothly with the glycosylamine **96** affording the desired aminoglucoenitol **98** in excellent stereoselection (88% d.e.) (Scheme 26) [39]. Once more the Cramchelation model can be invoked to explain the stereoselection. The mercuriocyclization of **98** stereoselectively afforded the α-C-glycoside **99** of D-glucosamine in about 60% diastereomeric excess.

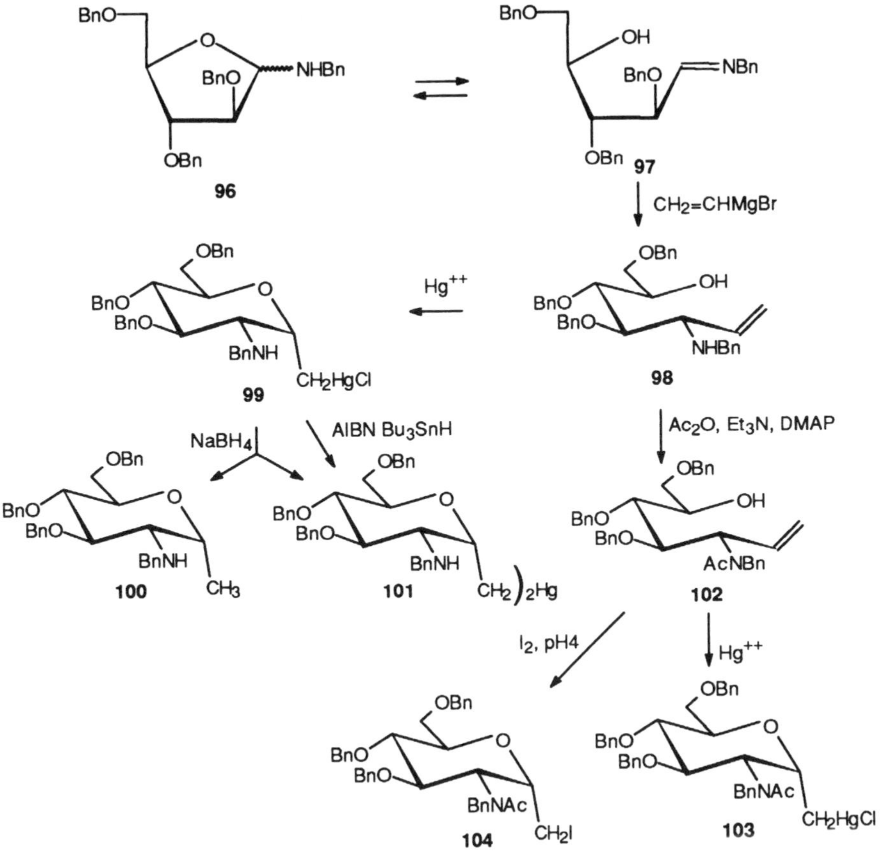

Scheme 26. (DMAP = dimethylaminopyridine)

The elaboration of the C-glycosidic appendage of **99** was troublesome [40]. The reduction with $NaBH_4$ afforded a mixture of the 1-(glycopyranosyl)methane **100** and the dimeric structure **101**, whereas the reduction with $LiAlH_4$ gave the starting open chain product **98**. The formation of the dimeric product **101** can be optimized treating **99** with AIBN and tributyltin hydride. All attempts to convert the C-Hg linkage of **99** into an electrophilic function such as a halide failed. For example, treatment of **99** with I_2 or with Br_2 gave a complex mixture of products. Under the hypothesis that the nucleophilic character of the aminic nitrogen of **99** could interfere with the formation of the adjacent electrophile, we decided to lower the nucleophilicity of the nitrogen by acetylation. The acetylation of **99** was however ineffective, even when drastic conditions such as the use of dimethylaminopyridine were employed. The formation of the *N*-acetyl derivative **103** required the acetylation of the less hindered open chain intermediate **98**, followed by mercuriocyclization of the acetylated product **102**. Once more, all attempts to convert **103** into a halide by different halodemercuriation methods were unsuccessful. It was possible however to obtain the iodide **104** by direct iodocyclization of **102** with iodine at pH 4 under carefully controlled experimental conditions. The iodide **104** was, however, very labile, and was inadequate inter alia to synthesize the phosphono analog of *N*-acetyl-α-D-glucosamine 1-phosphate.

The synthesis of the phosphono analog of *N*-acetyl-α-D-glucosamine 1-phosphate, the importance of which has been underlined in Sect. 2.3, requires in conclusion a synthetic procedure in which the introduction of the amino function is done as the last step. This was effected exploiting the vinylation – electrophilic cyclization procedure described above, that allows to prepare C-glycosides with the hydroxyl group at C-2 deprotected (Scheme 27). The α-C-glucopyranosyl mercurioderivative **89** obtained from 2,3,5-tri-*O*-benzyl-D-arabinose **87** according to Scheme 25, was converted into the iodide **105**, the free hydroxyl group of which required protection before effecting the Arbuzov reaction. Otherwise the cyclic phosphonate **106** is formed. Protection of the hydroxyl group of **105** as a *tert*-butyldimethylsilyl ether, Arbuzov reaction and subsequent deprotection, allowed production of the phosphonate **109** with the hydroxyl group at C-2

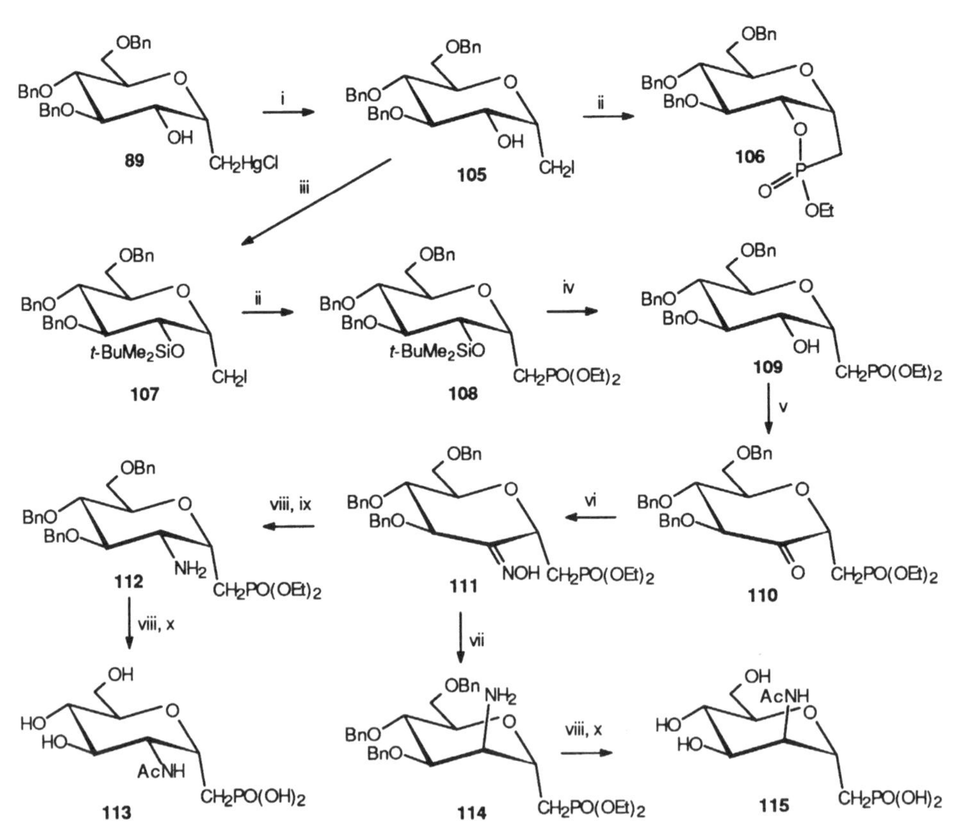

Scheme 27. i, I_2, THF; ii, $P(OEt)_3$ rfx; iii, *t*-$BuMe_2SiCl$, imidazole, DMF; iv, CF_3COOH–H_2O; v, Ac_2O-DMSO; vi, NH_2OH, MeOH-THG; vii, H_2, Ni-Raney; viii, Ac_2O-dimethylaminopyridine-CH_2Cl_2; ix, B_2H6-THF; x, Me_3SiI

selectively deprotected. The conversion of this molecule into the phosphono analog of D-mannosamine was effected by oxidation of the free hydroxyl group, conversion into the oxime and reduction. Interestingly, reduction by Ni-Raney catalyzed hydrogenation afforded the manno isomers **114**, from which the phosphono analog of *N*-acetyl-α-D-mannosamine 1-phosphate (**115**) was obtained. It is worthwhile to note that *N*-acetyl-α-D-mannosamine 1-phosphate is a key intermediate in the biosynthesis of *N*-acetylneuraminic acid (Neu5Ac), and its phosphono analog **115** is a potential inhibitor of this process. Also the catalytic reduction of the oxime **111** with $Pt(OH)_2$ afforded the manno isomer, in this case in the debenzylated form [41].

The stereochemical outcome of the catalytic reductions of the oxime **111** is in contrast with that of the oximes of O-glycosides. It has been shown in fact that in O-glycosides, the reduction of similar oximes affords the gluco-isomer when the anomeric substituent is α-oriented, and the manno-isomer in the case of β-glycosides [42]. This inversion of the expected stereochemistry could be explained in the light of a possible coordination of the phosphonate with the metal catalyst, which favors the attack from the α-face. This hypothesis is supported by the fact that the reduction of the oxime **111** with the non-coordinating diborane affords the gluco derivative **112**. Acetylation of **112**, followed by treatment with iodotrimethylsilane, afforded **113**, the desired phosphono analog of *N*-acetyl-α-D-glucosamine 1-phosphate.

3.5 Chemoenzymatic C-Glycosylation

The syntheses of C-glycosides described to date require the protection of the hydroxyl groups of the sugar, a proper functionalization of the anomeric center, and then the deprotection of the product obtained. An interesting goal would be to obtain C-glycosides without effecting any protection-deprotection step. We studied this possibility, exploiting the ability of aldolases to build-up stereoselectively the skeleton of deprotected ketoses [43], and the possibility to reduce the hemiketal form of a deprotected ketose into a C-glycoside [44].

Reaction of ribose 5-phosphate **116** with dihydroxyacetone phosphate, catalyzed by fructose 1,6-diphosphate aldolase from rabbit muscle (RAMA) affords the ketose diphosphate **117**. Dihydroxyacetone phosphate was formed in situ from fructose 1,6-diphosphate by action of RAMA and triose phosphate isomerase (TPI). The diphosphate **117** was dephosphorylated enzymatically using acid phosphatase, and the ketose **118** was reduced directly into the α-C-mannoside **119** by treatment with bistrimethylsilyltrifluoroacetamide, trimethylsilyltriflate and triethylsilane (Scheme 28) [45].

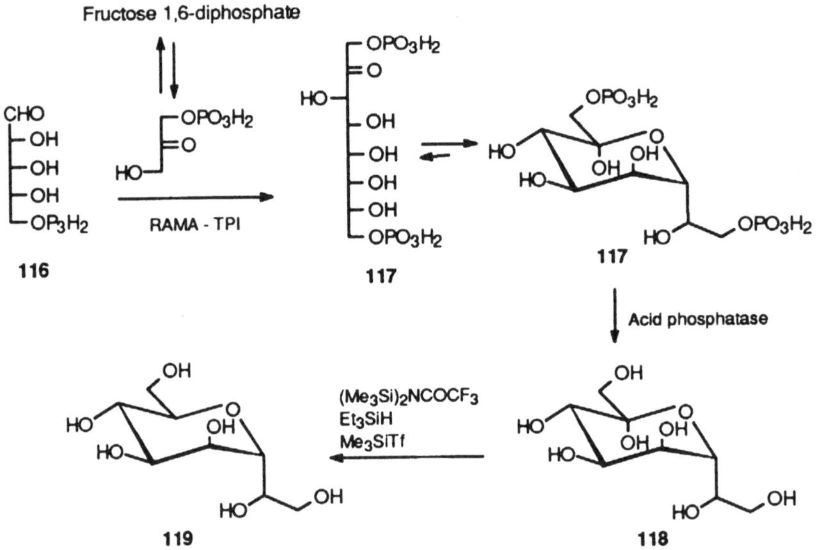

Scheme 28. RAMA = Rabbit muscle aldolase; TPI = Triose phosphate isomerase

4
Conclusions

The interest in C-glycosides is growing with the knowledge of the various roles that carbohydrates play in physiological and pathological events. C-glycosides are in fact potential inhibitors of carbohydrate processing enzymes and can act as stable analogs of physiologically active glycoforms. Different C-glycosylation procedures have been proposed, and some of them are highly stereoselective. The synthesis of C-glycosides of specific biological interest (such as the phosphono analog of *N*-acetyl-α-D-glucosamine 1-phosphate or some C-disaccharides) can however be very difficult. This article has presented our efforts in the synthesis of C-glycosides of biological interest, giving en route an overview of some of the main methods of C-glycosylation.

5
References

1. Hurd CD, Bonner WA (1945) J Am Chem Soc 67:1972
2. Hurd CD, Holysz RP (1950) J Am Chem Soc 72:1732, 1735, 2005
3. Helferich B, Bettin L (1961) Chem Ber 1158
4. Suhadolnik RJ (1970) Nucleoside antibiotics. Wiley-Interscience, New York
5. De Clercq E (1989) J Antimicrob Chemotherapy 23:35
6. Kobata A (1987) Malignant transformational changes of the sugar chain and their clinical applications. In: Greene MI, Hamaoca T (Eds) Development and recognition of the transformed cells. Plenum New York, p 385
7. Bernacki RI, Niedbala MJ, Korytnyk W (1985) Cancer Metastasis Rev 4:81
8. Pili R, Chang J, Patris RA, Mueller RA, Chrest FJ, Passaniti A (1995) Cancer Res 55:2920

9. Hers HG, (1984) Biochem Soc Trans 12:729
10. Hart GW (1992) Curr Op Cell Biol 4:1017
11. Tschesche R, Widera W (1982) Liebigs Ann 902
12. Nicotra F, Ronchetti F, Russo G (1982) J Org Chem 47:4459
13. Chem Abstr 65:790h
14. Lewis MD, Cha JK, Kishi Y (1982) J Am Chem Soc 104:4976
15. Nicotra F, Panza L, Russo G (1987) J Org Chem 52:5627
16. Nicotra F, Panza L, Russo G, Senaldi A, Burlini N, Tortora P (1990) J Chem Soc Chem Commun 1396
17. Dyer UC, Kishi Y (1988) J Org Chem 53: 3383; O'Leary DJ, Kishi Y (1993) J Org Chem 58:308
18. Carcano M, Nicotra F, Panza L, Russo G (1989) J Chem Soc Chem Commun 642; Lay L, Nicotra F, Pangrazio C, Panza L, Russo G (1994) J Chem Soc Perkin Trans 1 333
19. Boschetti A, Nicotra F, Panza L, Russo G, Zucchelli L (1989) J Chem Soc Chem Commun 1085
20. Lay L, Nicotra F, Russo G, Caneva E (1992) J Org Chem 57:1304
21. Cipolla L, Lay L, Nicotra F (1996) Carbohydr Lett 2:131
22. Wilcox CS, Long GH, Suh H (1984) Tetrahedron Lett 25:395; RajanBabu TU, Reddy GS (1986) J Org Chem 51:5458
23. Herpin TF, Motherwell WB, Tozer MJ (1994) Tetrahedron: Asymm 5:2269
24. Chapleur Y (1984) J Chem Soc Chem Commun 449
25. Cipolla L, Liguori L, Nicotra F, Torri G, Vismara E (1996) J Chem Soc Chem Commun, 1253
26. Shirahashi H, Murakami N, Watanabe M, Nagatsu A, Sakakibara J, Tokuda H, Nishino H, Iwashima A (1993) Chem Pharm Bull 41:1664
27. Ohrui H, Jones HJ, Moffatt JG, Maddox ML, Christensen AT, Bryan SK (1975) J Am Chem Soc 97:4602
28. Hanessian S, Ogawa T, Guindon Y (1974) Carbohydr Res 34:C12
29. Nicotra F, Ronchetti F, Russo G (1982) J Org Chem 47:5381
30. Nicotra F, Russo G, Ronchetti F, Toma L (1983) Carbohydr Res 124:C5
31. Fraser-Reid B, Dawe RD, Tulshian DH (1979) Can J Chem 57:1746
32. Pougny J-R, Nassr MAM, Sinaÿ P (1881) J Chem Soc Chem Commun 375
33. Nicotra F, Perego R, Ronchetti F, Russo G, Toma L (1984) Gazz Chim Ital 114:193
34. Reitz AB, Nortey SO, Maryanoff BE, Liotta D, Monahan R (1987) J Org Chem 52:4191
35. Nicotra F, Panza L, Ronchetti F, Russo G, Toma L (1987) Carbohydr Res 171:49
36. Nicotra F, Perego R, Ronchetti F, Russo G, Toma L (1984) Carbohydr Res 131:180
37. Nicotra F, Panza L, Ronchetti F, Toma L (1984) Tetrahedron Lett 25:5937
38. Boschetti A, Nicotra F, Panza L, Russo G (1988) J Org Chem 53:4181
39. Carcano M, Nicotra F, Panza L, Russo G (1989) J Chem Soc Chem Commun 297
40. Lay L, Nicotra F, Panza L, Verani A (1992) Gazz Chim Ital 122:345
41. Casero F, Cipolla L, Nicotra F, Panza L, Russo G (1996) J Org Chem, 61:3428
42. Lichtentaler FW, Kaji E (1985) Liebigs Ann Chem 1659 and references cited therein
43. Wong C-H, Halcomb RL, Ichikawa Y, Kajimoto T (1995) Angew Chem Int Ed Engl 34:412
44. Bennek JA, Gray GIR (1987) J Org Chem 52:892
45. Nicotra F, Panza L, Russo G, Verani A (1993) Tetrahedron: Asymm 4:1203

Thiooligosaccharides in Glycobiology

Hugues Driguez

Centre de Recherches sur les Macromolécules Végétales, (CERMAV-CNRS), affiliated with University Joseph Fourier, BP 53, 38041 Grenoble cedex 9, France

The chemical syntheses of thiooligosaccharides which were reported up to the beginning of 1996 are reviewed. Therefore, this article brings up to date previous reviews on thiosugars: (1963) Adv Carbohydr Chem 18:123, (1991) Studies in Natural Products Chemistry, Attar-ur-Rahman (ed), Elsevier, Amsterdam, 8E, p 315 and (1995) Progress in Biotechnology, Petersen SB, Svensson B, Pedersen S (eds) Elsevier, Amsterdam, 10, p 113. Because of the large amount of work published in this area, this paper mainly focuses on the synthesis of thiooligosaccharides which were shown to be recognized by proteins.

Table of Contens

Topics in Current Chemistry, Vol. 187

1 Introduction

1.1 Introductory Remarks

Carbohydrate-binding proteins are generally classified into two categories: immunoglobulins and lectins on the one hand and enzymes on the other, and interactions with their specific ligands or substrates have been recognized as the basis of glycobiology. Carbohydrate-protein recognition has been known for a long time to be important in the biotransformation of natural polysaccharides, and has been demonstrated in recent years as the essential process for biological transfer of information in living organisms.

Investigation of carbohydrate-protein interactions is then necessary to promote, for instance, new concepts in glycotherapy and for the engineering of enzymes. This study can be conducted using either natural oligosaccharides and modified proteins obtained by site-directed mutagenesis, or native proteins and non-natural oligosaccharides. This second approach, which depends heavily on synthetic carbohydrates, was first performed by Lemieux' group [1] and then by several laboratories [2].

In general, enzyme-substrate interactions are more difficult to analyse by introducing specific modifications in the substrate since these interactions involve recognition, binding and then the catalytic process. In some cases it is, however, possible to obtain some information about the geometry of the catalytic site or determine the mechanism of action of these enzymes [3]. A more versatile approach to map the active site of glycosyl-hydrolases is to utilize enzyme-resistant substrates which are competitive inhibitors. In these oligosaccharides the oxygen(s) of the bond(s) to be cleaved must be replaced with methylene group(s), nitrogen or sulfur atom(s). Since several contributions have dealt with *C*- and *N*-glycosides [4], the aim of this article is to review the synthesis and use of thiooligosaccharides: oligosaccharides in which at least one interosidic oxygen atom is substituted by a sulfur atom.

1.2 Abbreviations

Only those not found in Dodd JS (1986) the ACS Style Guide edn. Amer Chem Soc Publications, NW are given:

- cyst: cysteamine
- DMPU: 1,3-dimethyl-2-oxohexahydropyridine
- DMAP: 4-dimethylaminopyridine
- TESOTf: triethylsilyltriflate

2 Synthesis of Reducing Thiooligosaccharides

Several methods have been published for the synthesis of thiooligosaccharides. Apart from recent work using traditional glycosylation methodologies, most of the methods involved S_N2-type substitutions of thiolate anions on glycosyl halides, or of 1-thio-donors on acceptors bearing good leaving groups.

2.1 Traditional Glycosylation Methodologies

2.1.1 *1,6-Anhydroglycoses as Donors*

1,6-Anhydro derivatives of mono- and disaccharides have been particularly useful as glycosyl donors and acceptors for the production of dextran-type oligosaccharides [5].

Under acidic conditions, benzylated 1,6-anhydro-β-D-glucose (**1**) was also used as a donor with 4-thio-glucose (**2a**) or 4′-thiomaltose derivatives (**2b**) and the condensation afforded α-linked thio-di (**3a**) or trisaccharides (**3b**) in fair yields [6] (Scheme 1). It is noteworthy that 1,6-anhydro-di- and trisaccharide are unreactive under these conditions.

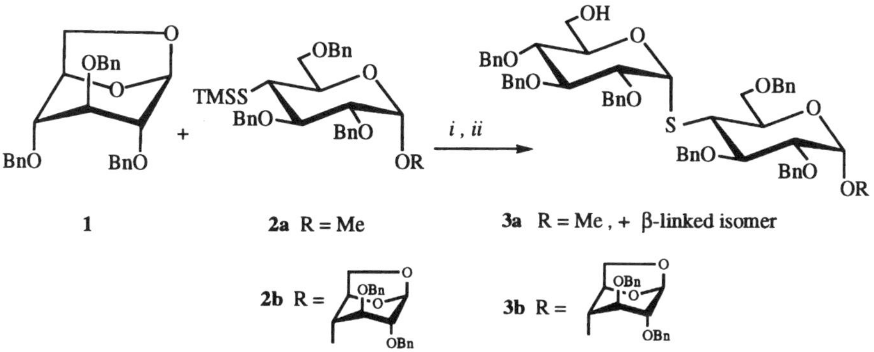

Scheme 1. *i*, ZnI_2 $(CH_2Cl)_2$; *ii*, K_2CO_3, MeOH; (**3a**) + β-linked isomer, 66%; (**3b**), 48%

2.1.2
The Trichloroacetimidate Method

This method has been extensively used in the establishment of *O*-glycosidic linkages [7]. Recently, a thioanalogue of kojibioside was synthesized using this coupling method [8a]. The trichloroacetimidate of benzylated β-D-glucopyranose (**5a**) as donor and the 2-thiol glycopyranoside (**4**) as acceptor afforded the expected disaccharide (**6a**) in a highly stereoselective manner (Scheme 2). However, glycosylation of the thiol (**4**) with the acetylated α-D-glucopyranose trichloroacetimidate (**5b**) gave a 1:2.3 mixture of α- and β- thiodisaccharides (**6b**). The rearrangement of an orthoester intermediate was responsible for this lack of stereoselectivity.

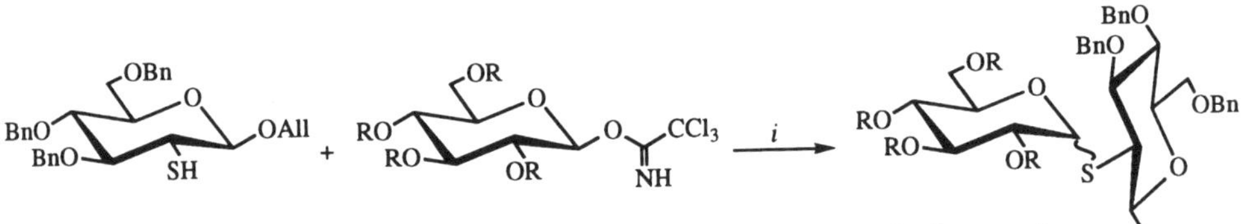

4

5a R = Bn
5b R = Ac

6a R = Bn α - only
6b R = Ac

Scheme 2. *i*, TESOTf (CH_2Cl_2); (**6a**) α-only, 70%; (**6b**) $\alpha + \beta$- isomers (1/2.3), 59%

2.1.3
Concluding Remarks

Both methods present some drawbacks. The most important is the use of benzylated synthons. Free disaccharides were obtained by catalytic hydrogenation using a large amount of catalyst or by Birch's reduction which implies two more steps (acetylation for work-up and deacetylation) [8b].

2.2
S_N2-Displacement Using 1-Thioglycoses as Donors

The nucleophilic attack of an anomeric alkoxide on primary triflates afforded disaccharides in fair to good yields (50–80%) but the displacement of secondary triflate was not as effective (around 20%) [7]. Furthermore, due to the basicity of the alkoxide, only ethers or ketal protecting groups have to be used in these coupling reactions. Since sulfur is less basic than oxygen and its nucleophilicity much greater, 1-thiolate can be generated and used in the presence of base-sensitive protecting groups.

2.2.1
Synthesis of 1-Thioglycoses

The most utilized synthons for the synthesis of thiooligosaccharides have been the per-acetylated 1-thioglycoses; they have to be selectively deacetylated and activated in situ, or activated in a two-step procedure by total deacetylation. *O*-Acetylated 1-thioglycoses may also be used after their selective transformation into 1-thiolates.

2.2.1.1 Synthesis of 1,2-trans-Thioglycoses

Glycosyl halides (**7a–e**) were stereoselectively transformed into 1,2-*trans*-thioglycoses by: i) (**8a–d, 8j**) a two-step procedure via the pseudothiourea derivatives [9, 10a]; the substitution of halide by thiourea is mostly a S_N1-type reaction since acetylated 1-thi*o*-α-D-mannose (**8b**) was obtained from acetobromomannose (**7b**) [9c]; ii) (**8e–i**) using thiolates in protic and aprotic solvents [10], or under phase transfer catalysis conditions [11]. Another approach involved the reaction of thioacetic acid with 1,2-*trans*-per-*O*-acetylated glycoses catalyzed with zirconium chloride [12]. The 1,2-*trans*-peracetylated 1-thioglycoses (**8e–h**) were obtained in high yield. No anomerized products could be detected in these reactions (Fig. 1).

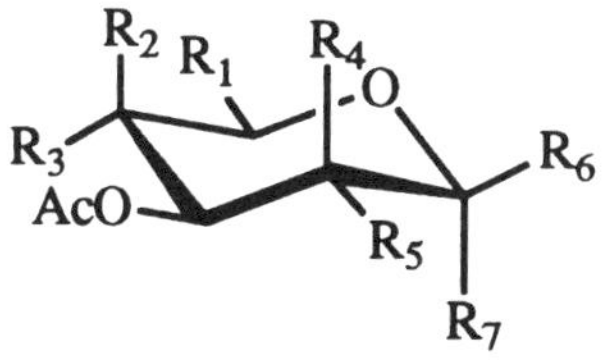

7a R_1=CH_2OAc, R_2=R_4=H, R_3=R_5=OAc, R_6=Br
7b R_1=CH_2OAc, R_2=R_5=H, R_3=R_4=OAc, R_6=Br
7c R_1=CH_2OAc, R_3=R_4=H, R_2=R_5=OAc, R_6=Br
7d R_1=H, R_2=R_4=H, R_3=R_5=OAc, R_6=Br
7e R_1=CH_2OAc, R_2=R_4=H, R_3=OAc, R_5=NHAc, R_6=Cl

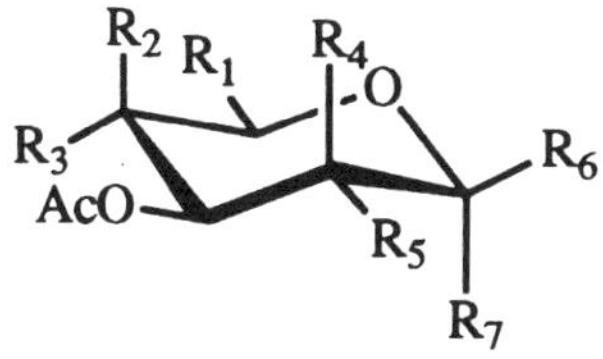

8a R_1=CH_2OAc, R_2=R_4=R_7=H, R_3=R_5=OAc, R_6=SH
8b R_1=CH_2OAc, R_2=R_5=R_6=H, R_3=R_4=OAc, R_7=SH
8c R_1=CH_2OAc, R_3=R_4=R_7=H, R_2=R_5=OAc, R_6=SH
8d R_1=R_2=R_4=R_7=H, R_3=R_5=OAc, R_6=SH
8e R_1=CH_2OAc, R_2=R_4=R_7=H, R_3=R_5=OAc, R_6=SAc
8f R_1=CH_2OAc, R_2=R_5=R_6=H, R_3=R_4=OAc, R_7=SAc
8g R_1=CH_2OAc, R_3=R_4=R_7=H, R_2=R_5=OAc, R_6=SAc
8h R_1=R_2=R_4=R_7=H, R_3=R_5=OAc, R_6=SAc
8i R_1=CH_2OAc, R_2=R_4=R_7=H, R_3=OAc, R_5=NHAc, R_6=SAc
8j R_1=CH_2OAc, R_2=R_4=R_7=H, R_3=OAc, R_5=NHAc, R_6=SH

Fig. 1. Synthesis of 1,2-trans-Thioglycoses

2.2.1.2
Synthesis of 1,2-cis-Thioglycoses

The first syntheses of 1,2-*cis*-thioglycoses (α-D-gluco- and β-D-manno- derivatives) have been achieved by the reaction in acetone of alkyl or benzyl xanthate or potassium thioacetate with the corresponding 1,2-*trans*-glycosyl halides [13]. More recently, tetra-*O*-acetyl-1-*S*-acetyl-1-thio-α-D-glucopyranose (**10a**) (Scheme 3) has been obtained i) by reaction of β-acetochloroglucose (**9a**) with either potassium thioacetate in HMPA or the tetrabutylammonium salt of thioacetic acid in toluene [14]; ii) by peroxide-induced addition of thioacetic acid to the pseudo-glucal (**11**) [15].

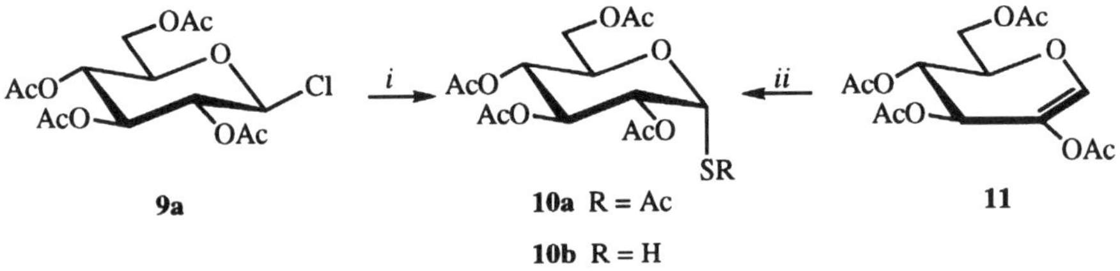

9a **10a** R = Ac **10b** R = H **11**

Scheme 3. *i*, KSAc (HMPA, 36%) or $(Bu)_4$ NSAc (toluene, 46%); *ii*, HSAc (acetone, 66%)

2.2.1.3
Selective S-Deacetylation of 1-Thioglycoses

Several methods were described for the selective de-*S*-acetylation of *O*-acetyl protected 1-thioglycoses. Sodium methoxide in methanol at low temperature (below –20 °C) was known to afford mainly the de-*S*-acetylated compound [16] or exclusively this compound when the reaction was quenched at low temperature by adding H+ resin [17]. Demercuration of tetra-*O*-acetyl-1-phenylmercury(II)-thio-β-D-glucopyranose (**12**) (Scheme 4) obtained by treatment of (**8e**) with phenylmercury(II)acetate afforded a convenient synthesis of tetra-*O*-acetyl-1-thio-β-D-glucose (**8a**) [18]. This sequence applied to the α-anomer (**10a**) (Scheme 3) led to the expected de-*S*-acetylated compound (**10b**) [19]. Chemoselective deprotection of thioacetate at the anomeric position of peracetylated 1-thioglycoses was also achieved in good yield by action of cysteamine in acetonitrile or hydrazinium acetate in DMF [20, 11].

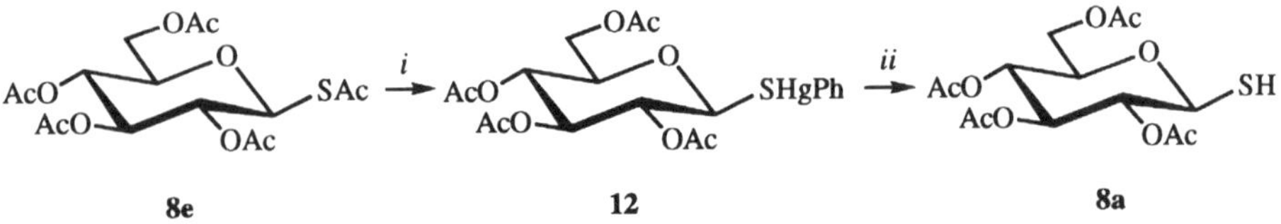

8e **12** **8a**

Scheme 4. *i*, PhHgOAc (EtOH, 75%); *ii*, H_2S (EtOH, 81%)

2.2.2
Establishment of 1,6-Thiolinkages

2.2.2.1
Synthesis of 1,2-trans-Thioglycosides

The first examples of reducing disaccharides formed via a thioglycosidic linkage was reported in 1967. Hutson obtained in low yield (19%) the thioglycosidic analogue of gentiobioside (**14a**) using this procedure (Scheme 5) [21]. This low yield may be explained by the poor leaving group properties of the tosyl group especially in acetone. 6-Thiogentiobiose (**14b**) and 6-thioallolactose (**14c**) were obtained in 73 and 59% yield, respectively, by halogen substitution of the 6-deoxy-6-iodo-glucose (**13b**) in DMF and acetone, respectively (Scheme 5) [22]. The synthesis of dithiogentiotriose (**14e**) was accomplished in 50% yield by sodium methoxide treatment of (**14d**) and then coupling with (**13b**) in DMPU [**22b**]. So the yield of a thiooligosaccharide synthesis under these conditions is essentially influenced by the right balance between the reactivity of the donor and the acceptor, and that can be achieved by a combination of various factors: the protecting groups of the donor, the leaving group on the acceptor and the solvent.

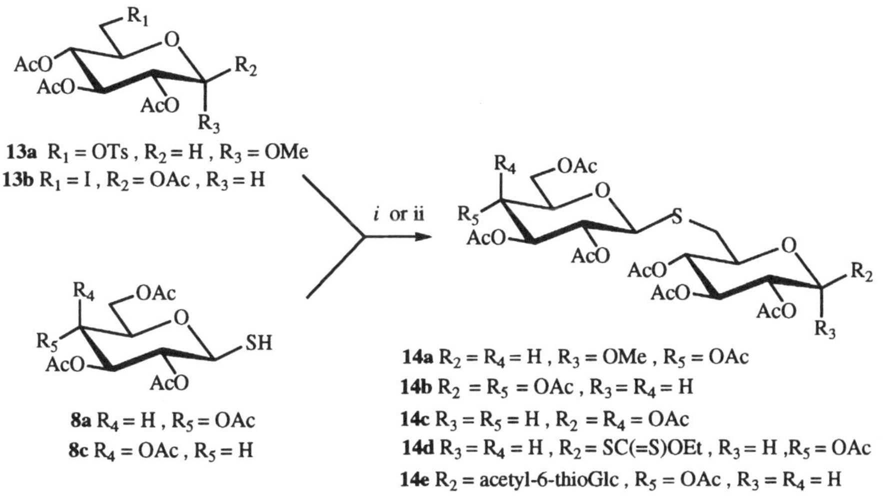

13a R_1 = OTs , R_2 = H , R_3 = OMe
13b R_1 = I , R_2 = OAc , R_3 = H

8a R_4 = H , R_5 = OAc
8c R_4 = OAc , R_5 = H

i or ii

14a R_2 = R_4 = H , R_3 = OMe , R_5 = OAc
14b R_2 = R_5 = OAc , R_3 = R_4 = H
14c R_3 = R_5 = H , R_2 = R_4 = OAc
14d R_3 = R_4 = H , R_2 = SC(=S)OEt , R_3 = H ,R_5 = OAc
14e R_2 = acetyl-6-thioGlc , R_5 = OAc , R_3 = R_4 = H

Scheme 5. (**14a, 14c**): *i*, K_2CO_3 (acetone-water); (**14b, 14d, 14c**): *ii*, NaH (THF); (DMF)

2.2.2.2
Synthesis of Branched-Cyclodextrins.

6-*O*-Tosylcyclomaltoheptaose (**15a**) treated with 2 eq of the sodium salt of either 1-thi*o*-*α*- (**16a**) or 1-thi*o*-*β*-D-glucopyranose (**16b**) in DMPU at 70 °C for 5 h afforded the expected branched cyclodextrins (**17a, 17b**) in 66 and 60% yield respectively [23] (Scheme 6).

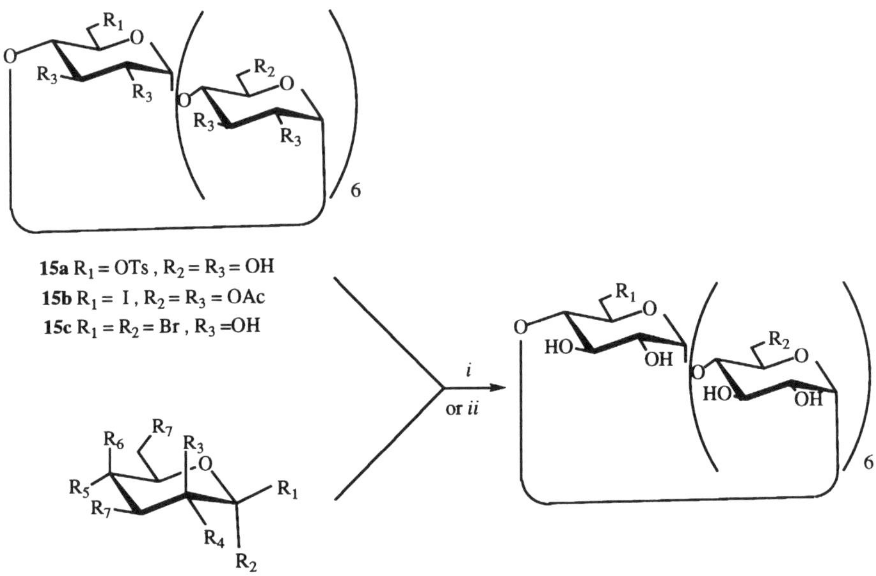

15a R_1 = OTs , R_2 = R_3 = OH
15b R_1 = I , R_2 = R_3 = OAc
15c R_1 = R_2 = Br , R_3 =OH

8b R_1 = R_4 = R_6 = H , R_2 = SH , R_3 = R_5 = R_7 = OAc
8e R_1 = SAc , R_2 = R_3 = R_6 = H , R_4 = R_5 = R_7 = OAc
8f R_1 = R_4 = R_6 = H , R_2 = SAc , R_3 = R_5 = R_7 = OAc
8g R_1 = SAc , R_2 = R_3 =R_5 = H , R_4 = R_6 = R_7 = OAc

10a R_1 = R_3 = R_6 = H , R_2 = SAc , R_4 = R_5 = R_7 = OAc

16a R_1 = R_3 = R_6 = H , R_2 = SNa , R_4 = R_5 = R_7 = OH
16b R_1 = SNa , R_2 = R_3 = R_6 = H , R_4 = R_5 = R_7 = OH
16c R_1 = R_3 = R_5 = H , R_2 = SNa , R_4 = R_6 = R_7 = OH
16d R_1 = SNa , R_2 = R_3 = R_5 = H , R_4 = R_6 = R_7 = OH

i or *ii*

17a R_1 = α-1-thioGlc , R_2 = OH
17b R_1 = β-1-thioGlc , R_2 = OH
17c R_1 = β-1-thioGal , R_2 = OH
17d R_1 = α-1-thioMan , R_2 = OH
17e R_1 = R_2 = α-1-thioGal
17f R_1 = R_2 = β-1-thioGal

Scheme 6. (**17a**–**17d**): *i*, DTE, cyst. (HMPA, RT); (**17e**, **17f**): *ii*, (DMPU, 70 °C)

The branched cyclodextrins (CDs, **17a**, **17b**) and their analogues with *β*-D-galactosyl and *α*-D-mannosyl residues (**17c**, **17d**) have also been prepared under mild conditions by the approach depicted in Scheme 6 [24,25]. Selective in situ *S*-deacetylation and activation was obtained by treatment of peracetylated 1-thioglycoses (**10a**, **8e**, **8g**) by cysteamine in the presence of dithioerythritol in HMPA [26]. This method was very efficient for the synthesis of branched CDs (**17a**) (80%), (**17b**) (60%), and (**17c**) (85%) when the acceptor molecule (**15b**) bearing primary iodide was used. However, peracetylated 1-thio-*α*-D-mannose (**8f**) failed as a donor under these conditions, but tetra-*O*-acetyl-1-thio-*α*-mannose (**8b**) afforded the expected CD (**17d**) in high yield (83%).

Since tosyl group and bromide have the same efficiency as leaving groups, the synthesis in good yield of heptakis-6-thioglycosyl-*β*-cyclodextrin (**17e**, **17f**) from the heptakis(6-bromo-6-deoxy)-*β*-cyclodextrin (**15c**) needs, as expected, the sodium salts of 1-thiogalactose (**16c**, **16d**) and high temperature (Scheme 6) [27].

2.2.2.3
Synthesis of 1,2-cis-Thioglycosides

Reaction of peracetylated 1-thio-α-D-glucopyranose (**10a**) with peracetylated 6-deoxy-6-iodomaltose (**18**), 6′-deoxy-6′-iodomaltose (**21a**) and 6^{ω}-deoxy-6^{ω}-iodomalto-oligosaccharides (**21b, 21c**) gave the corresponding *S*-α-glucosylthiomaltose derivatives (**20, 23a–c**) in excellent yields (Scheme 7) [24, 28, 29a].

Methyl 6-thioisomaltoside (**23d**) was also prepared by condensation in HMPA of the 6-iodo-glucoside (**21d**) and the thiolate (**16a**) (Scheme 7) [29b].

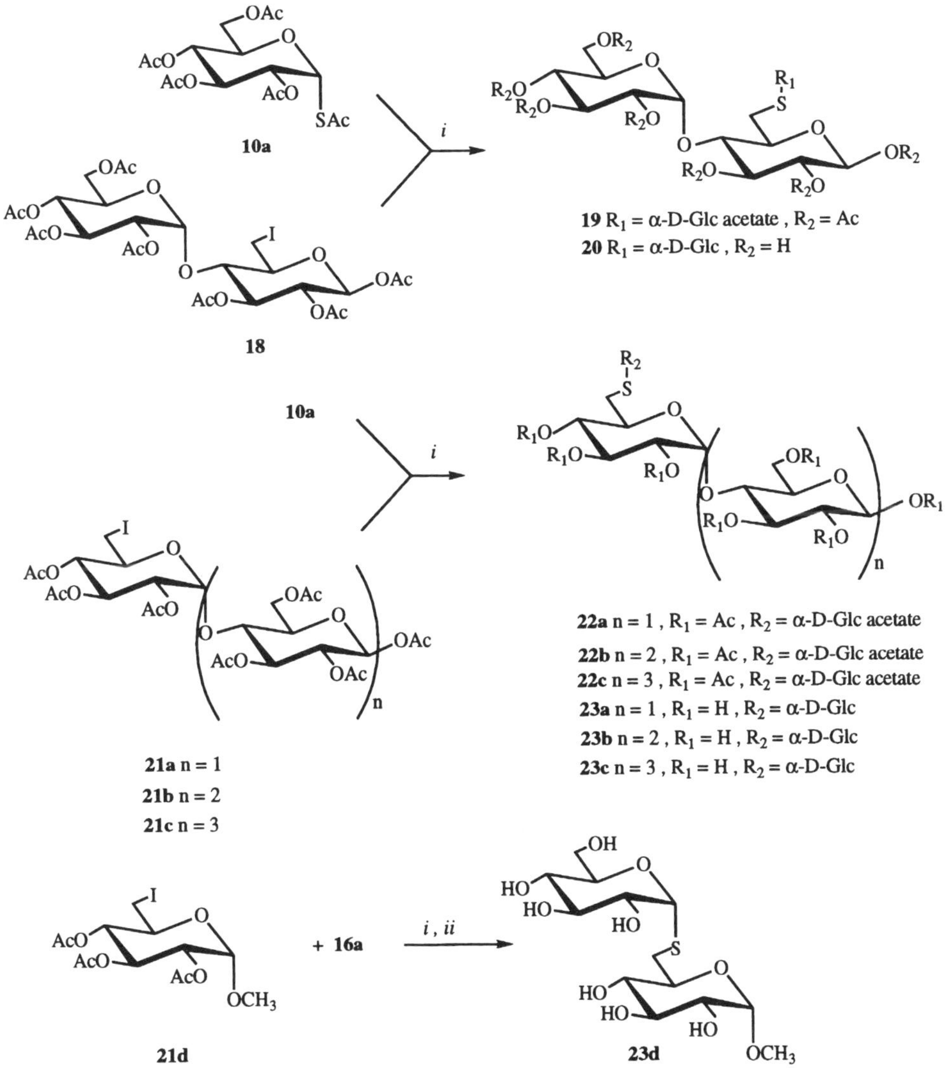

10a

18

19 R_1 = α-D-Glc acetate, R_2 = Ac
20 R_1 = α-D-Glc, R_2 = H

21a n = 1
21b n = 2
21c n = 3

22a n = 1, R_1 = Ac, R_2 = α-D-Glc acetate
22b n = 2, R_1 = Ac, R_2 = α-D-Glc acetate
22c n = 3, R_1 = Ac, R_2 = α-D-Glc acetate
23a n = 1, R_1 = H, R_2 = α-D-Glc
23b n = 2, R_1 = H, R_2 = α-D-Glc
23c n = 3, R_1 = H, R_2 = α-D-Glc

21d + **16a** → (*i, ii*) **23d**

Scheme 7. *i*, DTE, cyst (HMPA, RT, 80–86%); *ii*, NaOMe–MeOH; *iii*, (HMPA, RT)

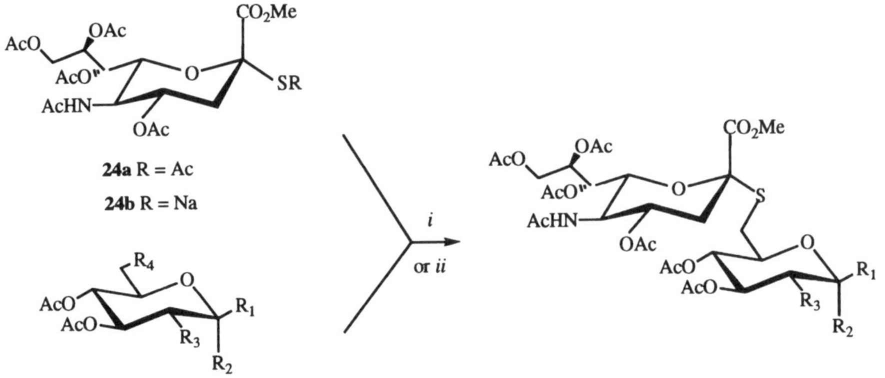

24a R = Ac
24b R = Na

28a $R_1 = H$, $R_2 = OMe$, $R_3 = NHAc$, $R_4 = Br$
28b $R_1 = H$, $R_2 = OMe$, $R_3 = OAc$, $R_4 = Br$
28c $R_1 = O(CH_2)_2N_3$, $R_2 = H$, $R_3 = NHAc$, $R_4 = OTs$

29a $R_1 = H$, $R_2 = OMe$, $R_3 = NHAc$
29b $R_1 = H$, $R_2 = OMe$, $R_3 = OAc$
29c $R_1 = O(CH_2)_2N_3$, $R_2 = H$, $R_3 = NHAc$
29d $R_1 = SAc$, $R_2 = H$, $R_3 = OAc$
29e $R_1 = S$-ceramide, $R_2 = H$, $R_3 = OAc$

Scheme 8. (**29a**, **29b**): *i*, Et_2NH (DMF, RT); (**29c**), *ii*, (DMF, 45 °C)

However, two recent papers suggest that DMF is also a good solvent for this type of condensation under appropriate activation conditions of the donor. A simple and efficient method for the preparation of thioglycosides of *N*-acetylneuraminic acid has been developed [30]. This procedure involves the selective in situ *S*-deacetylation and activation of the 2-*S*-acetyl Neu5Ac (**24a**) and the displacement of primary bromide of methyl glycosides (**28a**, **28b**). The desired $\alpha(2\rightarrow6)$ thioglycosides (**29a**, **29b**) were obtained in 75 and 82% yield. Condensation of the sodium salt (**24b**), freshly derived from (**24a**) by selective *S*-deacetylation with sodium methoxide [16], with (**28c**) in DMF at 45 °C also gave the expected compound (**29c**) in 76% yield (Scheme 8) [31].

The $\alpha(1\rightarrow6)$ thiofucosyl linkage was obtained in exceptionally high yield (99%) by the substitution of the tosyl group of (**28d**) with the sodium salt of 1-thio α-l-fucopyranose tetraacetate (**30c**) generated from the free thiol (**30b**) by NaH treatment in DMF (Scheme 9) [32].

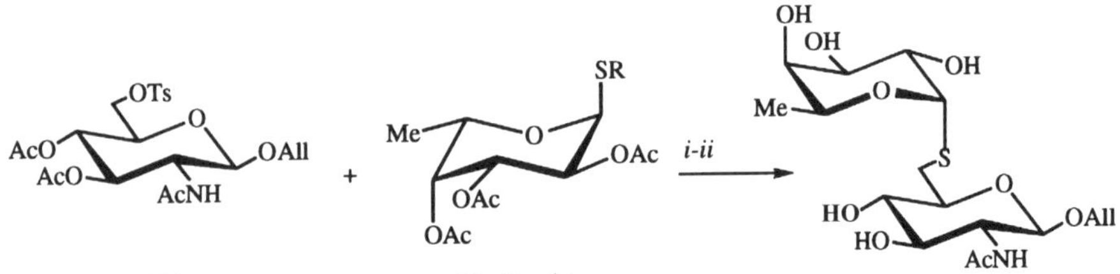

28d

30a R = Ac
30b R = H
30c R = Na

31

Scheme 9. *i*, NaH (DMF); *ii*, Ac_2O–DMAP (pyridine), *iii*, NaOMe–MeOH

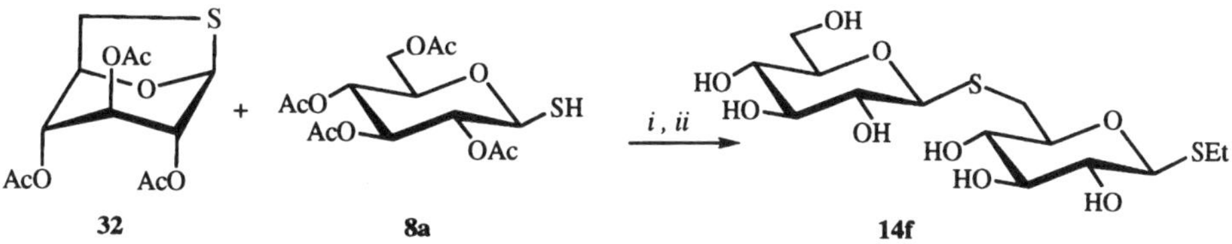

32 + 8a → (i, ii) 14f

Scheme 10. *i*, Et_3OBF4 (CH_2Cl_2); *ii*, NaOMe–MeOH

A recent improved preparation of 1,6-epithio-*β*-D-glucose (**32**) [33] and its selective opening at C-6 with a thioglucose such as (**8a**) [34] could be an interesting alternative method for the production of complex thiooligosaccharides since the resulting *β*-thiodisaccharide (**14f**) can act as a donor in *O*-glycoside synthesis (Scheme 10).

2.2.3
Establishment of 1,4-Thiolinkages

The first synthesis of 4-thiomaltosyl, 4-thiocellobiosyl and 4-thio-digalactobiosyl derivatives was reported in 1978 [14a] (see Sect. 2.3).

However the most efficient and reliable procedure involved the S_N2 displacement of a triflate group at C-4 of the acceptor with activated 1-thioglycoses.

2.2.3.1
Synthesis of 4-Thiodisaccharides

4-Thiomaltose (**37**) (Scheme 11) and 4-thiocellobiose (**40a**) (Scheme 12) were obtained in HMPA in 50–70% yield from the reaction of either acetylated 4-*O*-triflyl-1,6-anhydrogalactopyranose (**33**) or benzoylated 4-*O*-triflyl galactoside (**34a**) with the sodium salt of 1-thio-*α*- of *β*-D-glucopyranose (**16a**) and (**16b**) respectively [14a, 35].

Under the same conditions, 4-thioxylobiose (**40b**) was also easily obtained by the coupling of compounds (**16e**) with (**34b**) (Scheme 12) [36].

In the presence of a crown ether in THF, the nucleophilic substitution of the 4-triflate of (**34a**) by the sodium salt derivative of (**8a**) led to 4-thiocellobiose (**40a**) in good yield [37].

The lactosyl ceramide analogue (**39c**) was also obtained by condensation, in acetone-methanol, of 4-*O*-triflyl galactosyl sphingosine (**34c**) with thiogalactose (**8c**) generated in situ from the corresponding pseudothiourea derivative [38].

Thiodisaccharides in the GlcNAc series were recently synthesized. The substitution reaction of the triflate at C-4 of 2-acetamido- and 2-azido-D- galactosides (**34d**) and (**34e**) in DMF with thiolate nucleophiles (**8i**) or (**8j**) and (**30b**) afforded the expected disaccharides (around 50 and 68%), from which the deprotected (**41**) and (**42**) were obtained [**39a, 32**] (Scheme 13). The coupling of the triflate (**34f**) with the thiol (**8j**) in DMF in the presence of cysteamine gave a 63% yield of the expected thiodisaccharide which was converted into (**41**) in high yield [39b].

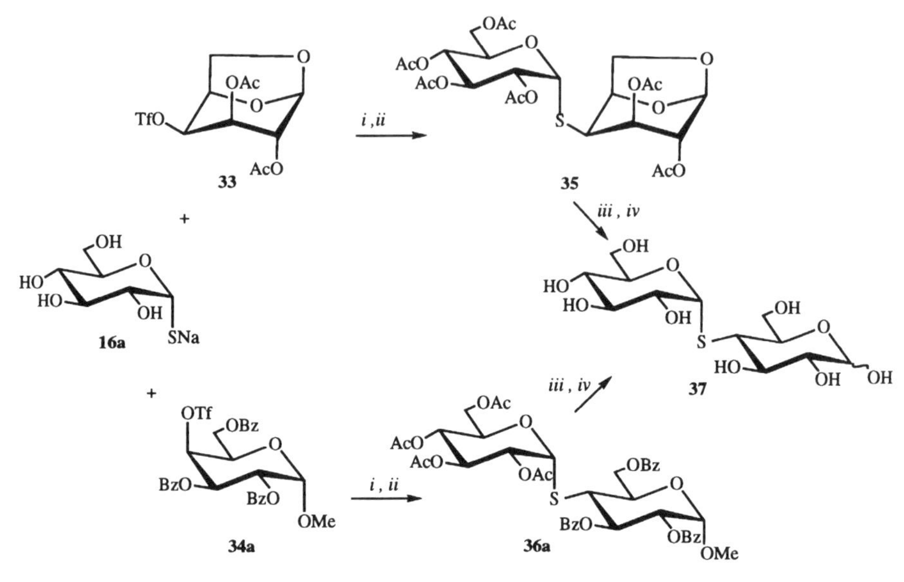

Scheme 11. *i*, (HMPA); *ii*, Ac_2O–DMAP (pyridine); *iii*, H_2SO4, HOAc, Ac_2O; *iv*, NaOMe–MeOH

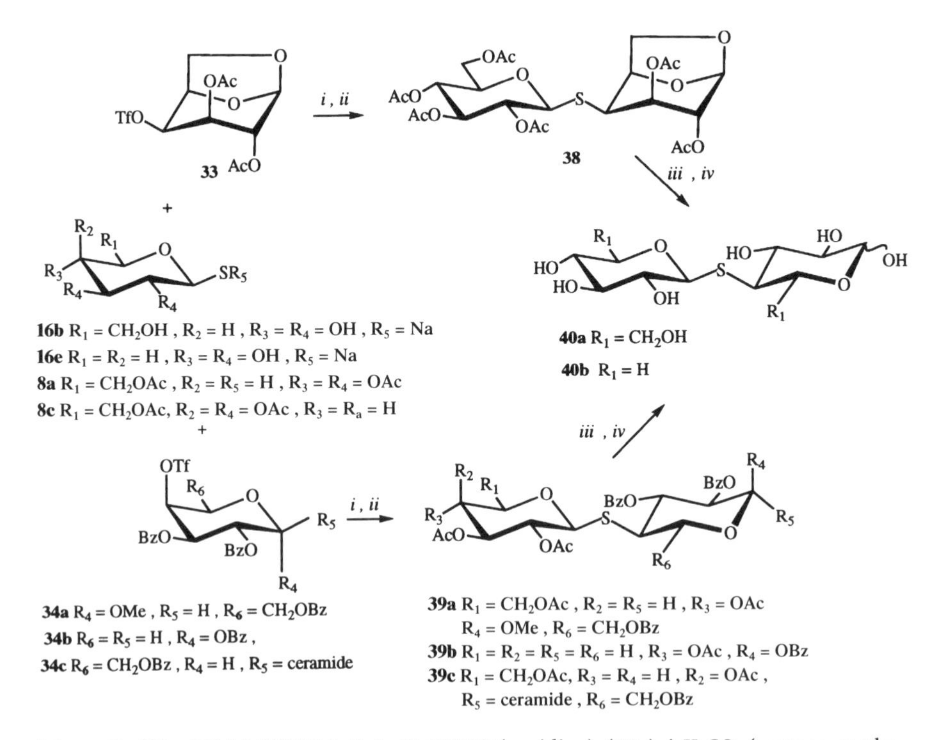

Scheme 12. (**39a**, **39b**) *i*, (HMPA); *ii*, Ac_2O–DMAP (pyridine); (**39c**), *i*, K_2CO_3 (acetone–methanol), *iii*, H_2SO_4, HOAc, Ac_2O; *iv*, NaOMe–MeOH

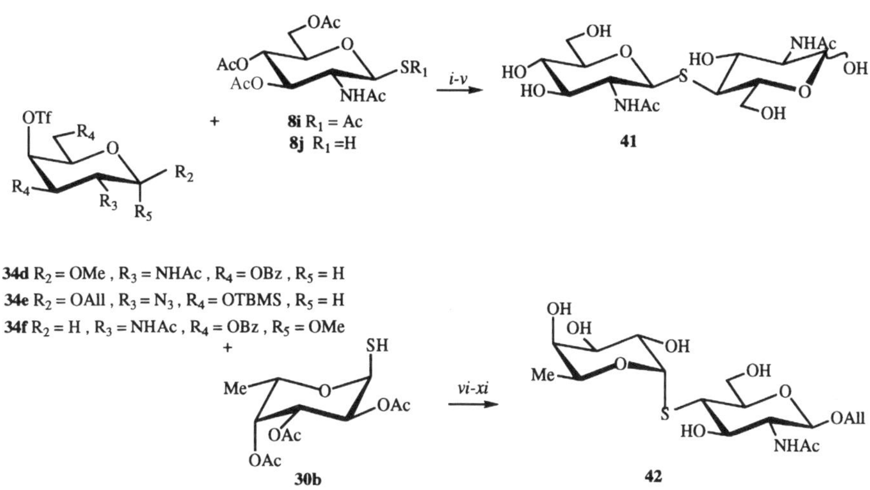

8i R_1 = Ac
8j R_1 = H

34d R_2 = OMe , R_3 = NHAc , R_4 = OBz , R_5 = H
34e R_2 = OAll , R_3 = N_3 , R_4 = OTBMS , R_5 = H
34f R_2 = H , R_3 = NHAc , R_4 = OBz , R_5 = OMe

Scheme 13. (**41**) *i*, NaH (DMF) or cyst (DMF); *ii*, NaOMe–MeOH; *iii*, Ac_2O–DMAP (pyridine); *iv*, H_2SO_4, HOAc, Ac_2O; *v*, NaOMe–MeOH; (**42**) *vi*, NaH (DMF), *vii*, Ac_2O–DMAP (pyridine); *viii* Bu_4NF (THF); *ix*, Ac_2O–DMAP (pyridine); *x*, H_2S (pyridine–H_2O); *xi*, NaOMe–MeOH

2.2.3.2
Synthesis of 4-Thiooligosaccharides.

The key compounds in the synthesis of 4,4′-dithiocello- and -xylotrioses are the per-acetylated 1,4-dithiodisaccharides (**43a, 43b**) (Scheme 14) which are obtained from acylated (**39a, 39b**) by hydrogen bromide treatment and displacement of the halide by potassium or tetrabutylammonium thioacetate. Total deacetylation of the donors and reaction with triflated acceptors (**34a**) or (**34b**) afforded the expected trisaccharides which can be converted, after conventional treatment, into free trisaccharides (**44d, 44e, 44g and 44h**) [40]. Using the same stepwise procedure, but with activation of the donor by selective in situ *S*-deacetylation and activation, the cellotetraoside (**44i**) and cellopentaoside (**44j**) were obtained in good yield [41].

Very recently using the same strategy, the condensation of (**43c**) and (**34f**) afforded the thiochitooligosaccharides (**44f**) and (**44k**) in moderate yield [39b] (Scheme 14).

A large number of hemithiocellodextrins were also synthesized. Hemithiocellodextrins can be considered as 4-thiocellobiosyl repeating units linked by $\beta(1\rightarrow4)$ oxygen linkages or as cellobiosyl repeating units linked by sulfur bridges. For instance, hemithiocellodextrins belonging to the former family were obtained by enzymic oligomerization of 4-thio-β-cellobiosyl fluoride (**39e**), derived from (**40a**), using cellulases in a buffer/organic solvent system [42a]. Water-soluble hemithiocellodextrins (**40c–40g**) were formed and isolated in 4.5, 7.5, 5.7, 5.0 and 20% yield respectively (Scheme 15).

The tetrasaccharide (**44l**) and hemithiocellotetraoside (**44m**) were also obtained from the laminaribiosyl (**43d**) or the cellobiosyl donor (**43a**) and the lactosyl acceptor (**39f**) (Scheme 15) [42b, c].

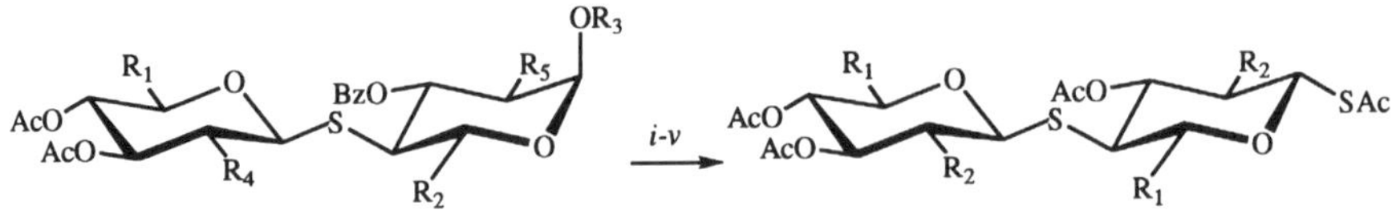

39a $R_1 = CH_2OAc$, $R_2 = CH_2OBz$, $R_3 = Me$, $R_4 = OAc$, $R_5 = OBz$
39b $R_1 = R_2 = H$, $R_3 = Bz$, $R_4 = OAc$, $R_5 = OBz$
39d $R_1 = CH_2OAc$, $R_2 = CH_2OBz$, $R_3 = Me$, $R_4 = R_5 = NHAc$

i-v →

43a $R_1 = CH_2OAc$, $R_2 = OAc$
43b $R_1 = H$, $R_2 = OAc$
43c $R_1 = CH_2OAc$, $R_2 = NHAc$

43a + 34a
43b + 34b —vi→
43c + 34f

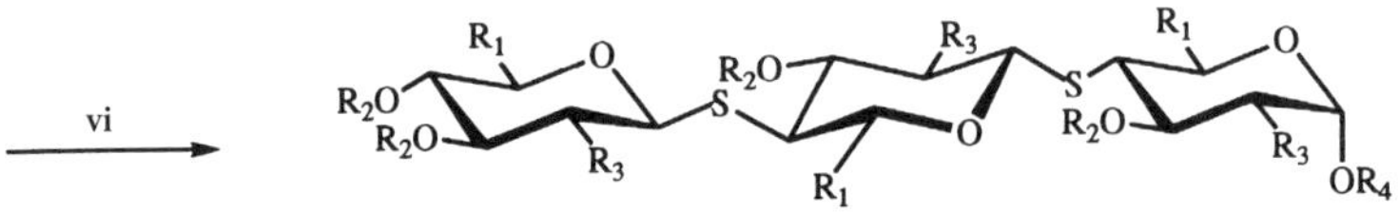

44a $R_1 = CH_2OAc$, $R_2 = Ac$, $R_3 = OAc$, $R_4 = Me$
44b $R_1 = H$, $R_2 = R_4 = Ac$, $R_3 = OAc$
44c $R_1 = CH_2OAc$, $R_2 = Ac$, $R_3 = NHAc$, $R_4 = Me$
44d $R_1 = CH_2OH$, $R_2 = H$, $R_3 = OH$, $R_4 = Me$
44e $R_1 = R_2 = R_4 = H$, $R_3 = OH$
44f $R_1 = CH_2OH$, $R_2 = H$, $R_3 = NHAc$, $R_4 = Me$

44a, 44b, 44f —vi→

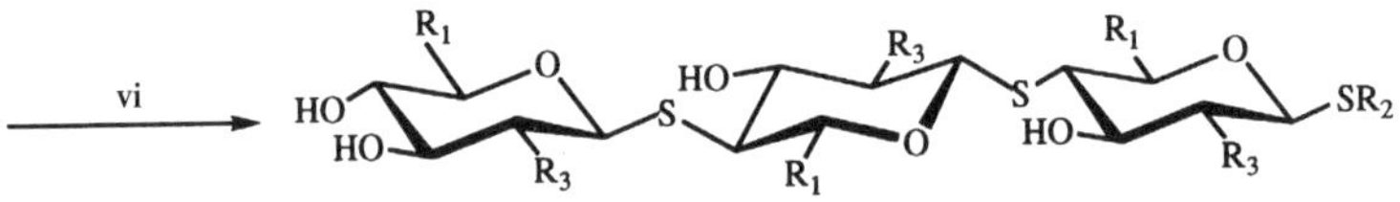

44g $R_1 = CH_2OH$, $R_2 = C_6H_4NH_2$ or C_6H_4NHAc, $R_3 = OH$
44h $R_1 = H$, $R_2 = C_6H_4NH_2$ or C_6H_4NHAc, $R_3 = OH$
44i $R_1 = CH_2OH$, $R_2 =$ [structure], $R_3 = OH$
44j $R_1 = CH_2OH$, $R_2 =$ [structure], $R_3 = OH$
44k $R_1 = CH_2OH$, $R_2 =$ [structure], $R_3 = OH$

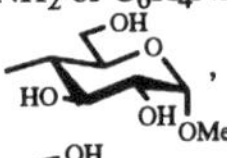

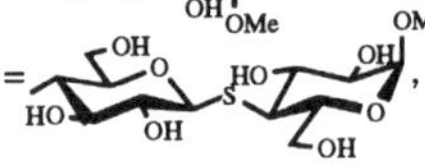

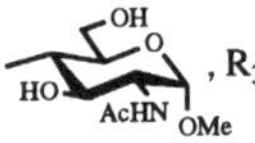

Scheme 14. *i,* NaOMe-MeOH; *ii,* Ac_2O-DMAP (pyridine); *iii,* H_2SO_4, HOAc-Ac_2O (in gluco series); *iv* HBr-HOAc or HCl-CH_2Cl_2 (**for 39d**); *v* KSAc (HMPA); *vi,* cyst., DTE (HMPA) or (DMF for **43c+34f**).

In the thiomaltose series the same approach was used for the synthesis of 4,4′-dithiomaltotriosides (**48a, 48b**) (Scheme 16) [43]. However the sequence of reactions is much longer since the *β*-chloride intermediate (**46**) was formed in 83% yield only when the peracetylated *β*-4-thiomaltose (**45**) was used, and the 1,4-dithiomaltose peracetate (**47b**) was obtained via the corresponding *S*-benzoyl analogue (**47a**). Under these conditions the 4,4′-dithiomaltotriosides were obtained in ~20% overall yield from the 4-thi*o*-*β*-maltose peracetates (**45**).

For these reasons a more versatile route has been designed based on a new acceptor molecule which already possesses at C-1 an *α*-thiol function able to react under the condensation conditions. To be effective, this protecting group

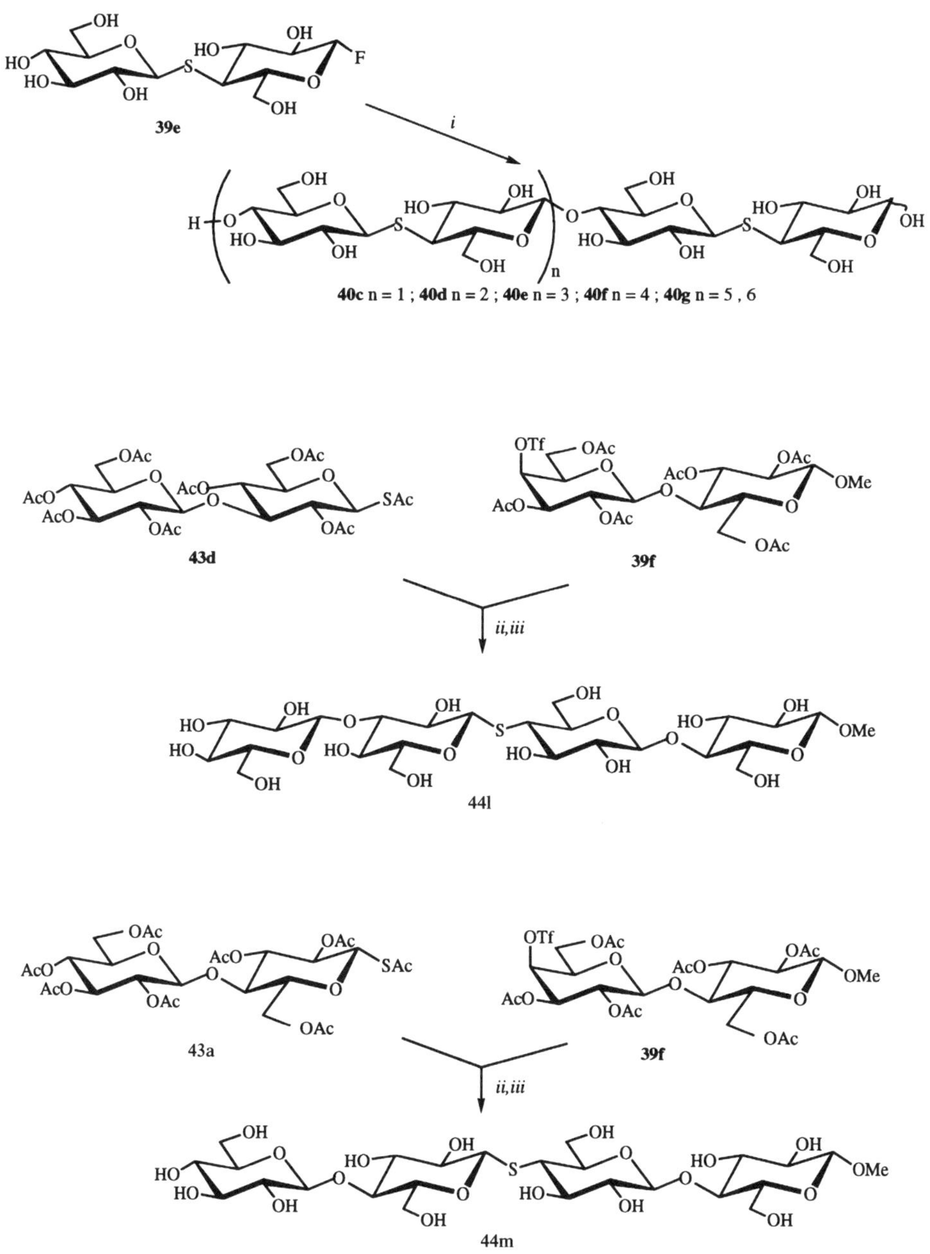

39e

40c n = 1 ; **40d** n = 2 ; **40e** n = 3 ; **40f** n = 4 ; **40g** n = 5 , 6

43d

39f

44l

43a

39f

44m

Scheme 15. *i,* cellulases (CH_3CN-buffer); *ii,* Et_2NH (DMF); *iii,* NaOMe–MeOH

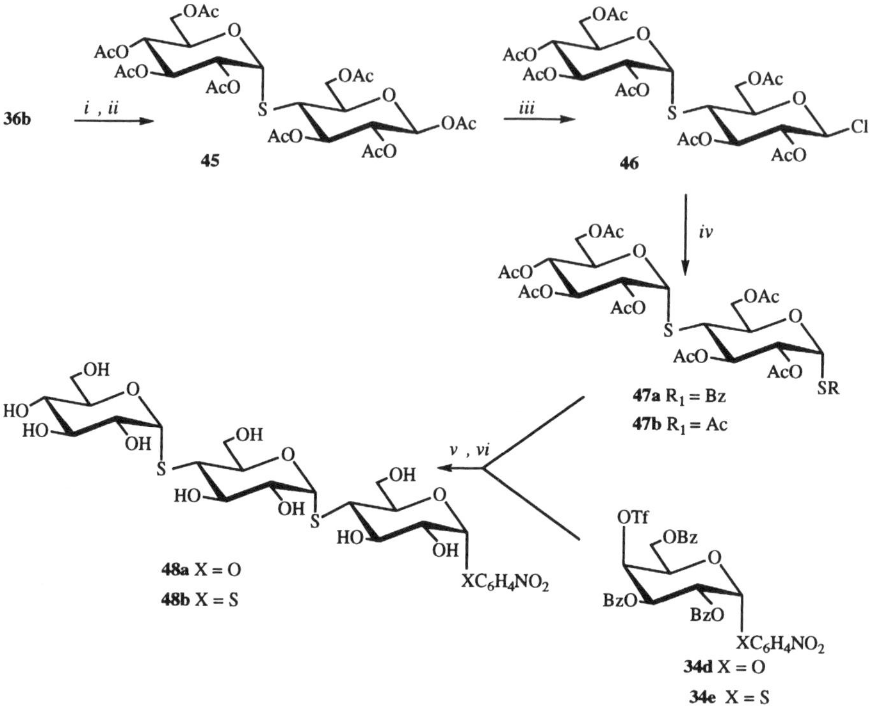

Scheme 16. *i,* HBr–AcOH (CH_2Cl_2); *ii,* AgOAc, HOAc, Ac_2O; *iii,* Cl_2CHOCH_3, $BF_3 \cdot Et_2O$ (CH_2Cl_2); *iv* BzSH, NaH (HMPA), *v* NaOMe–MeOH, *vi* Ac_2O-pyridine; *vii* NaOMe–MeOH

has to be easily and selectively removed or transformed into an *S*-acetyl group. The trityl groups fulfilled all these conditions. Trityl α-thioglucoside (**49a**) and galactoside (**49b**) were easily obtained by S_N2 displacement of the β-acetochloroglycoses (**9a or 9b**) using the tetrabutylammonium salt of triphenylmethanethiol. Typical experiments showing the versatility of (**49a**) and (**49b**) are reported in Schemes 17 and 18 [44].

A very efficient synthesis of methyl 4,4′-dithiomaltotrioside (**52**) was achieved based on: i) the selective in situ *S*-deacetylation and activation [26] of the

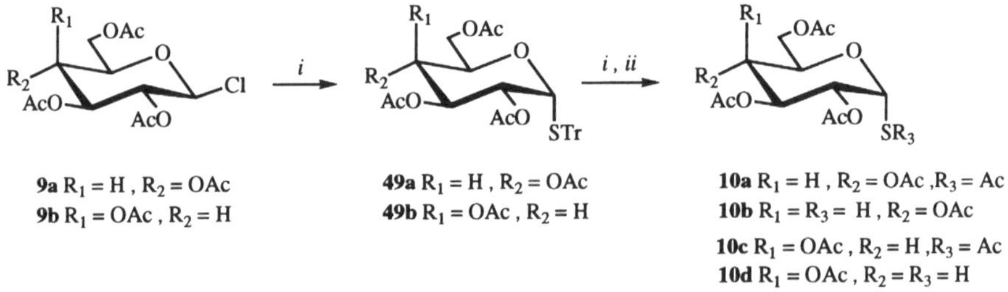

Scheme 17. *i,* (toluene); *ii,* PhHgOAc (MeOH-CH_2Cl_2); *iii,* H_2S (CH_2Cl_2-pyridine-Ac_2O)

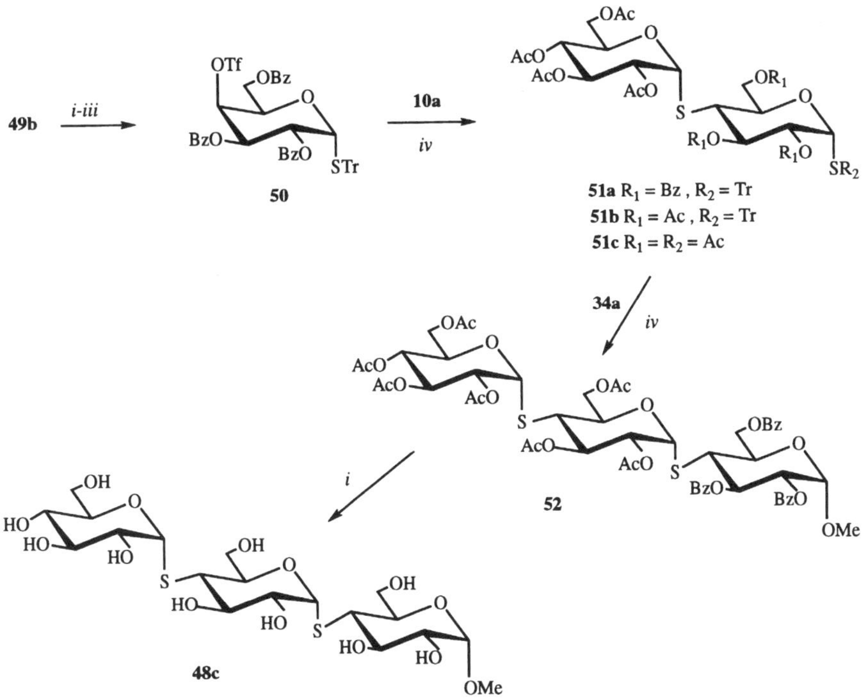

Scheme 18. *i*, NaOMe–MeOH; *ii*, BzCl (pyridine); *iii*, Tf_2O (CH_2Cl_2–pyridine); *iv*, cyst. DTE (HMPA)

donor (**10a**); ii) *S*-tritylgalactoside (**49b**) as acceptor and then the transformation of the disaccharide (**51a**) into the *S*-acetyl derivative (**51c**) in 95% yield (Scheme 18) [45].

The same methodology extended to maltose afforded the bifunctional synthons (**55b, 55d**) which led to the hemithiomaltooctaoses (**56d–g**) in good yield. However the reaction of (**56g**) led to the synthesis of the hemithiocyclomaltooctaose (**57a**) in a disappointing yield of 3% (Scheme 19) [46].

A more attractive chemo-enzymatic approach for the production of hemithiocyclodextrins has been developed. More than ten years ago, cyclodextrin glucosyltransferase (CGTase) was shown to use α-maltosyl fluoride for the synthesis of α-, β-, and γ-cyclodextrins [47] and has been used for the chemoenzymatic synthesis of regioselectively substituted cyclodextrins [48]. For this purpose, 4-thio-α-maltosyl fluoride (**58**) was easily prepared from (**45**). Incubation of (**58**) in the presence of pure CGTases afforded hemithiocyclodextrins (**57a**), (**57b**) and (**57c**) isolated in 14, 16 and 15% yields respectively [49].

2.2.4
Establishment of 1,3-Thiolinkages

The preparation of $\alpha(1\rightarrow3)$ and $\beta(1\rightarrow3)$-thiodisaccharides in the gluco and xylo series was readily achieved in high yield by S_N2 displacement of a triflate

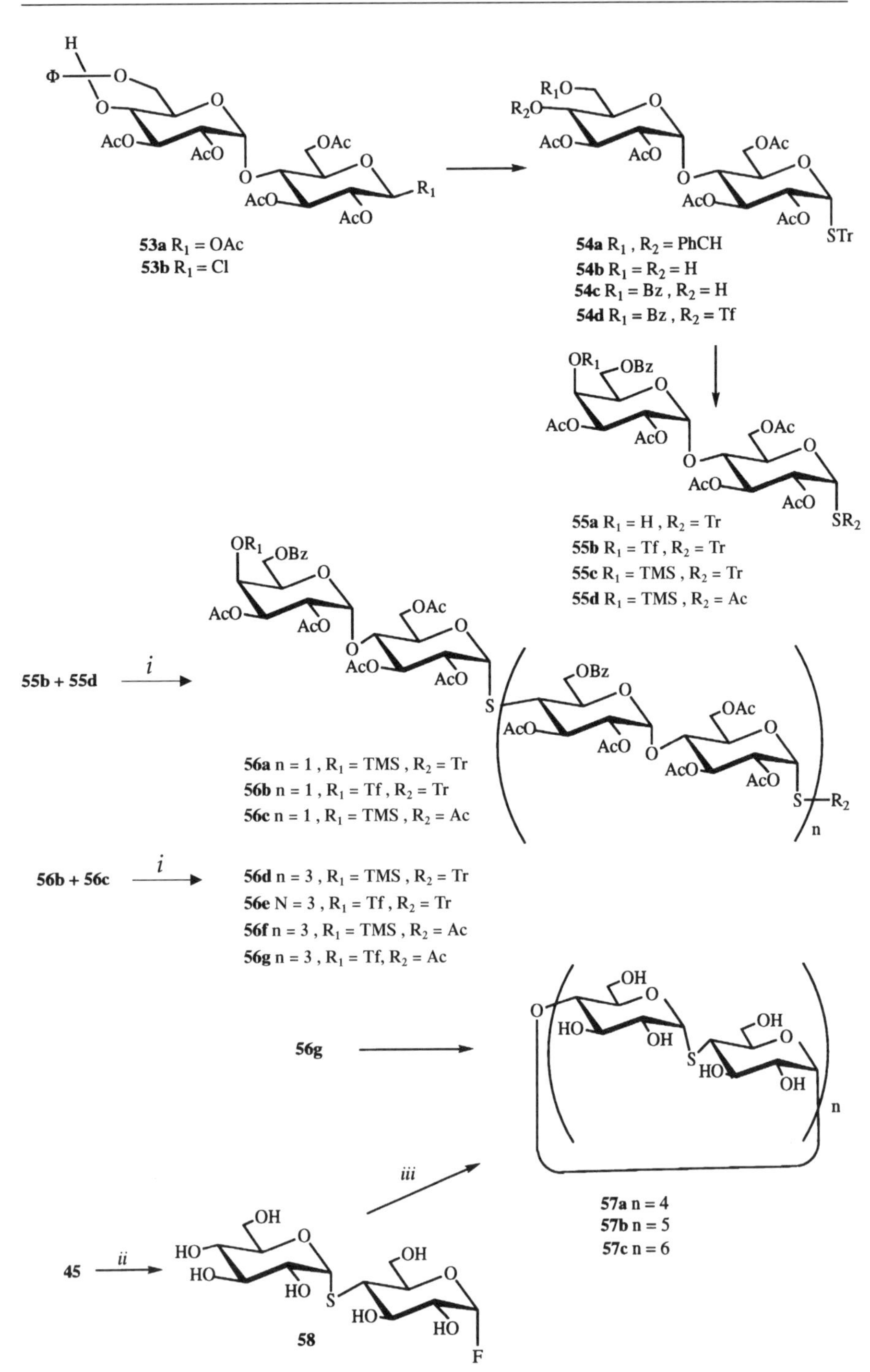

Scheme 19. *i*, cyst, DTE (HMPA); *ii*, HF(pyridine); *iii*, CGTase (0.2 M phosphate buffer pH 6.5)

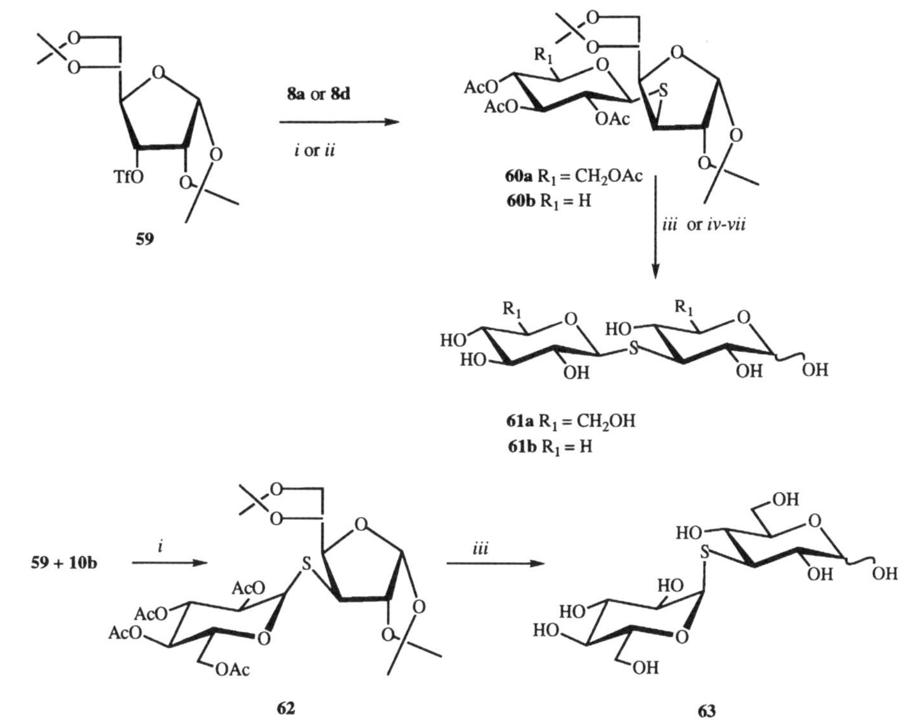

8a or 8d

i or *ii*

59

60a $R_1 = CH_2OAc$
60b $R_1 = H$

iii or *iv-vii*

61a $R_1 = CH_2OH$
61b $R_1 = H$

59 + 10b

62

63

Scheme 20. *i*, NaH (THF, DMF); *ii*, NaH, Kriptofix 21 (THF), *iii*, TFA (H_2O); *iv*, AcOH-H_2O; *v*, $NaIO_4$ (H_2O-EtOH), $NaBH_4$ (i-propanol); *vi*, ACOH-H_2O; *vii*, NaOMe-MeOH

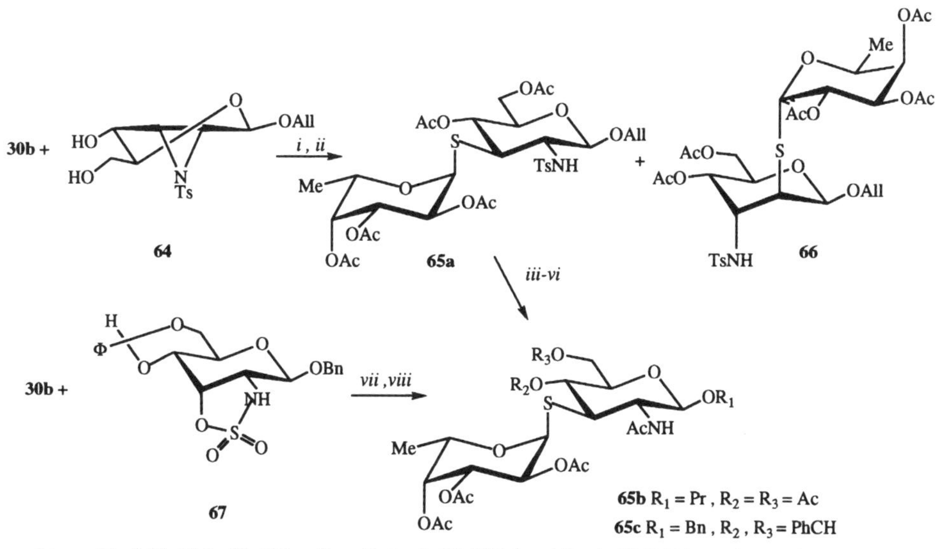

Scheme 21. *i*, NaOMe-MeOH, reflux; *ii*, Ac_2O-DMAP (pyridine); *iii*, Pd/C, H_2 (MeOH); *iv*, NaOMe-MeOH; *v*, Na/NH_3 (THF); *vi*, Ac_2O-DMAP (pyridine); *vii*, NaH (DMF); *viii*, H_2SO_4 (H_2O-THF)

group in the allofuranose (**59**) with the sodium salt of compounds (**8a, 8d, 10b**) in DMF, HMPA or THF in the presence of a crown ether. A conventional procedure was then used for the production of 3-thioxylobiose (**61b**) from (**60b**) (Scheme 20) [22b, 29b, 50].

The ring opening of the tosyl-aziridine (**64**) with 1-thiofucose (**30b**) in the presence of sodium methoxide in methanol under reflux gave a mixture of thiodisaccharide (**65a**) and (**66**) in a 2:1 ratio in very good yield. Compound (**65a**) was then converted into (**65b**) (Scheme 21) [32].

An improved method for the synthesis of α-fucosyl $(1\rightarrow 3)$-3-thio-glycoside (**65c**) involved the nucleophilic displacement of a cyclic sulfamidate derived from allosamine (**67**). The regioselective opening of (**67**) with (**30b**) led to (**65c**) in good yield (Scheme 21) [51].

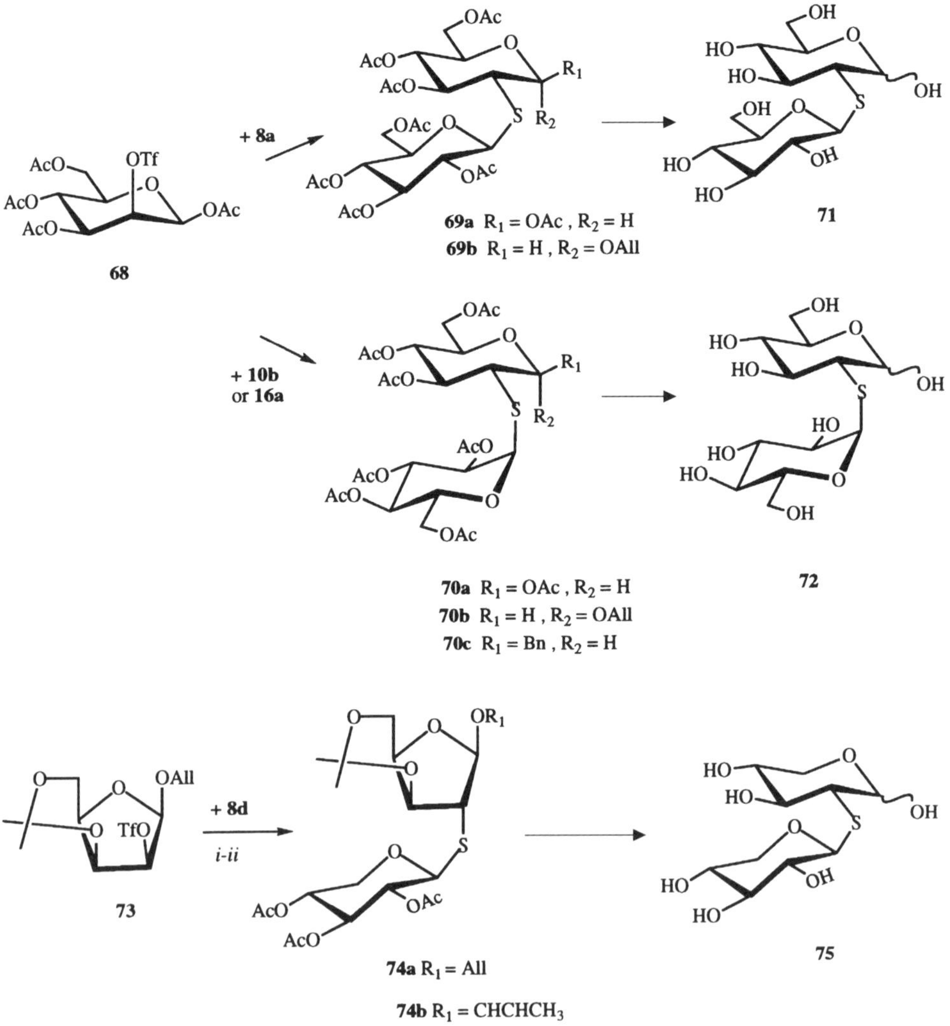

68
+ 8a
+ 10b
or 16a
69a R1 = OAc, R2 = H
69b R1 = H, R2 = OAll
71
70a R1 = OAc, R2 = H
70b R1 = H, R2 = OAll
70c R1 = Bn, R2 = H
72
73
+ 8d
i-ii
74a R1 = All
74b R1 = CHCHCH3
75

Scheme 22. *i*, NaH (THF); *ii*, (DMF)

2.2.5
Establishment of 1,2-Thiolinkages

2-Thioglucobioses (**69a**) and (**70a**) were easily prepared from 2-*O*-triflyl-mannose tetraacetate (**68**) and 1-thio-glucoses (**8a**) or (**10b**). However, Zemplen *O*-deacetylation led to an intractable mixture of compounds. Thus, prior to methoxide treatment, the readily cleavable allyl group was introduced at the anomeric position (Scheme 22) [20] of the acetylated disaccharides (**69a**) and (**70a**). Benzyl β-thiokojibioside (**70c**) was also prepared by condensation of (**68**) and (**16a**) [29b].

The allyl group was also used as the aglycon of the acceptor molecule (**73**) during the synthesis of 2-thioxylobiose (**75**) (Scheme 22) [50].

Oxirane-ring opening of talopyranose (**76**) with (**30a**) afforded a mixture of thiodisaccharides (**77**) and (**78**) in very good yield. Conventional treatment of (**77**) led to the allyl-2-thio-fucosyl-galactoside (**79**) (Scheme 23) [32].

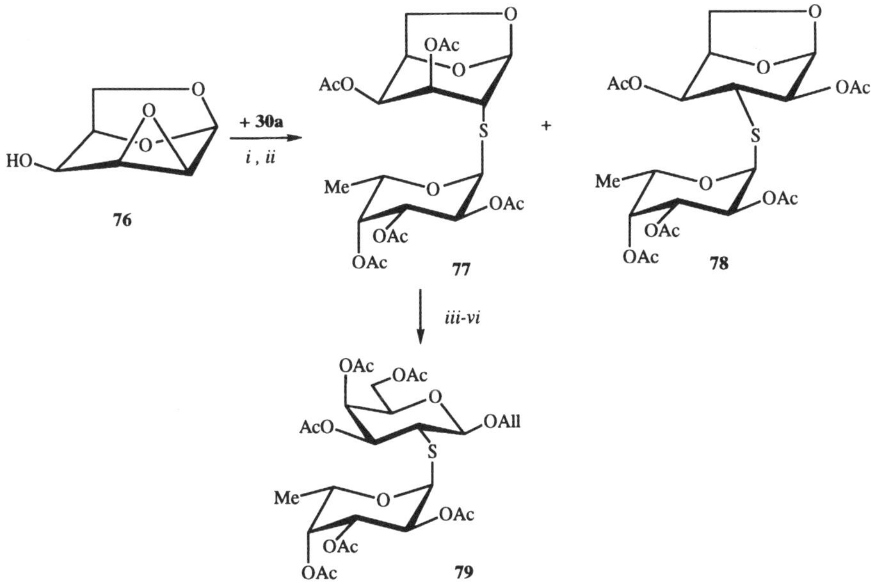

Scheme 23. *i*, NaOMe-MeOH (reflux); *ii*, Ac_2O-DMAP (pyridine); *iii*, H_2SO_4-Ac_2O-HOAc; *iv* NH_2NH_2-AcOH (DMF); *iv* CCl_3CN, DBU (CH_2Cl_2); *vi*, All-OH, $BF_3 \cdot OEt_2$ (CH_2Cl_2)

2.2.6
Establishment of 1,3- and 1,6-Thiolinkages

Preparation of the branched trisaccharide (**80**) was readily achieved in good yield by nucleophilic substitution of the iodo derivative (**60d**), derived from (**60a**), with (**8a**) in DMF (Scheme 24) [22b]. The introduction of a thiol functionality at the anomeric position of (**80**) using a known procedure and the condensation of this donor with the two acceptors (**59**)

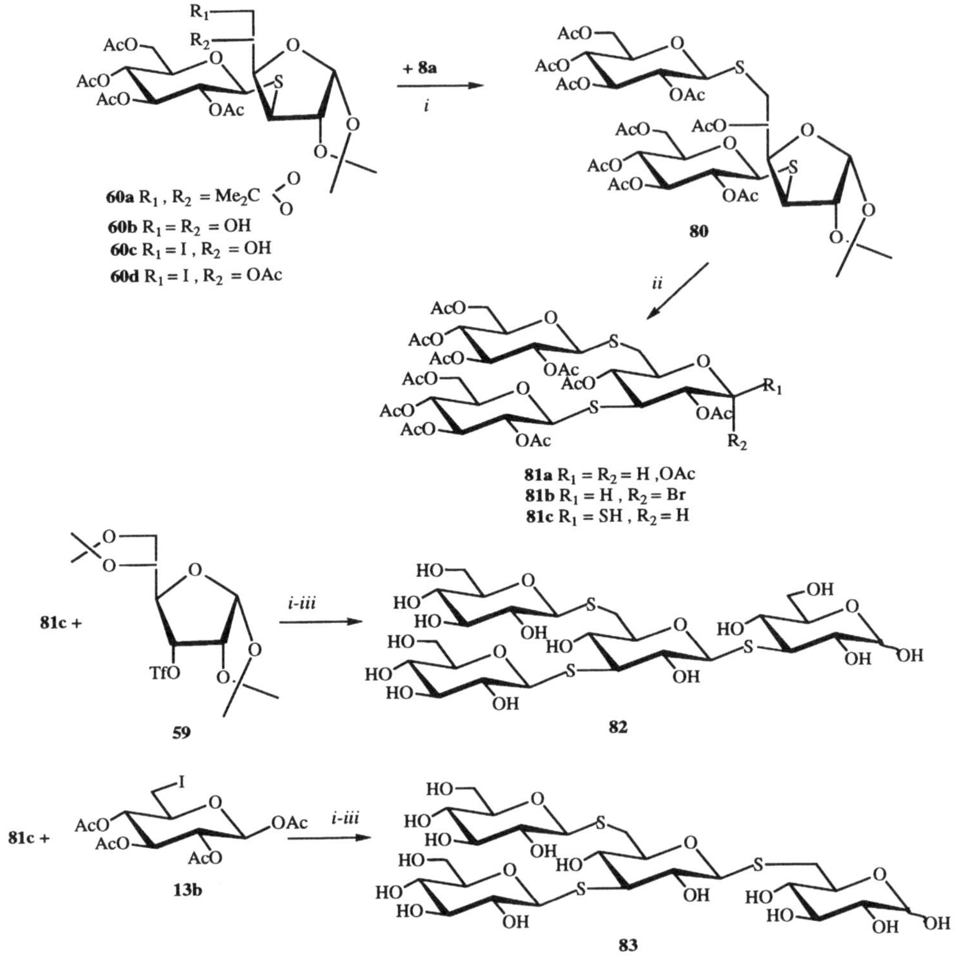

60a $R_1, R_2 = Me_2C$
60b $R_1 = R_2 = OH$
60c $R_1 = I, R_2 = OH$
60d $R_1 = I, R_2 = OAc$

81a $R_1 = R_2 = H, OAc$
81b $R_1 = H, R_2 = Br$
81c $R_1 = SH, R_2 = H$

Scheme 24. *i*, NaH (THF/DMF); *ii*, NaOMe–MeOH 90%, TFA–H_2O, Ac_2O-pyridine; *iii*, (90% TFA–H_2O)

and (**13b**) afforded the two tetrasaccharides (**82**) and (**83**) in excellent yields [52].

2.3
S_N2-Displacement Using Thiolate on Glycoside Acceptors

This alternative method for the synthesis of thiolinkages involved the displacement of an electrophilic group at C-1 of donor molecules with a thiolate at the primary or secondary position of acceptors.

This coupling strategy was used for the synthesis of methyl 6-thiogentiobioside (**14a**) and gave, in fact, better yield (55% vs 9%) than the method described in Sect. 2.2.2 [21].

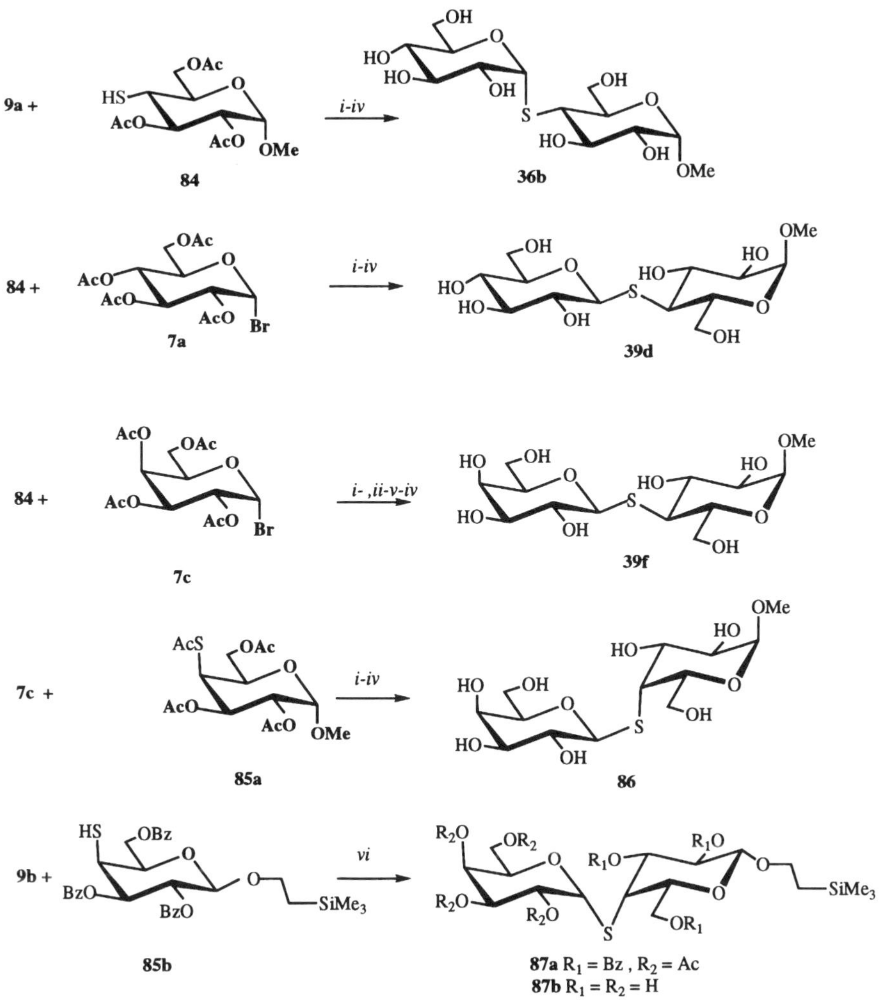

Scheme 25. *i*, NaOMe-MeOH; *ii*, (HMPA); *iii*, Ac_2O (pyridine); *iv* NaOMe-MeOH; *v*, BzCl (pyridine), *vi*, CS_2CO_3 (DMF)

The first synthesis of methyl 4-thiomaltoside (**36b**) (**34%**), 4-thiocellobioside (**39d**) (52%) and 4-thiodigalactobioside (**86**) (56%) was reported 18 years ago by using the S_N2 displacement of 1,2-*cis*- or 1,2-*trans*- glycosyl halides (**9a**), (**7a**) and (**7c**) with methyl 4-thi*o*-*α*-D-glycopyranoside sodium salts generated from the acetylated methyl glycosides (**84**) and (**85a**) (Scheme 25) [14a].

Surprisingly 4-thiolactoside (**39f**) was obtained only by this method, the reverse coupling failed to give the expected thiodisaccharide (Scheme 25) [53].

Using the same procedure, the synthesis of the thiogalabioside (**87b**) was reported very recently [54]. Also, this year an orthogonal synthesis of a thio-linked analogue of the sialyl Lewis X (**93**) has been described using known coupling reactions (Scheme 26) [55].

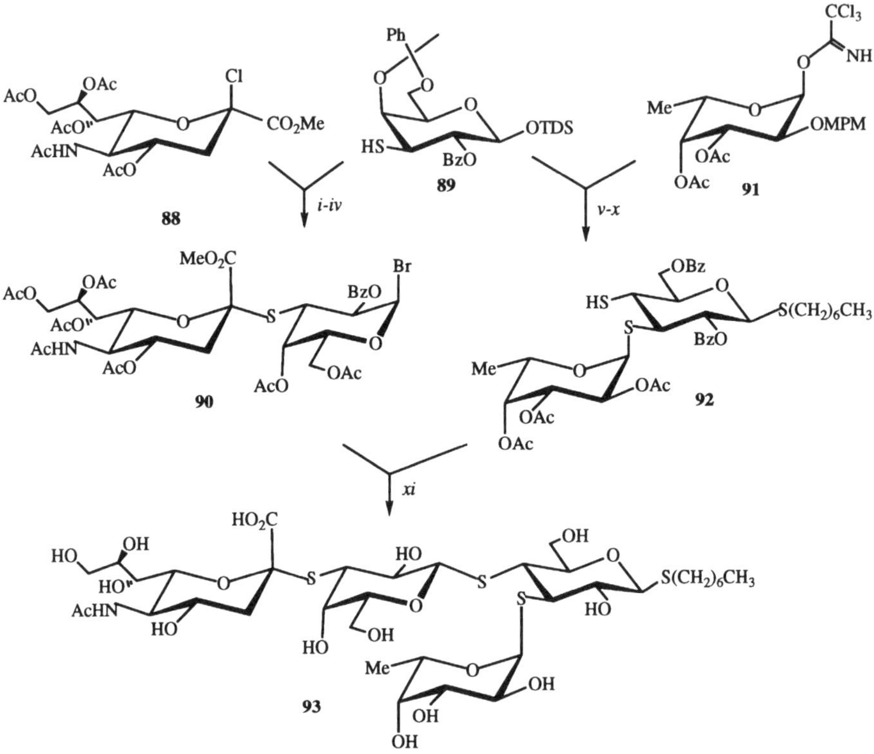

Scheme 26. *i*, NaH, Kryptofix 21 (THF); *ii*, TFA (H_2O), Ac_2O (pyridine); *iii*, TBAF (THF), Ac_2O (pyridine); *iv*, HBr-AcOH; *v*, TMSOTf (CH_2Cl_2); *vi*, CAN (CH_3CN-H_2O); Ac_2O (pyridine); *vii*, EtSH, p-TsOH, BzCN, NEt_3; *viii*, Tf_2O (pyridine), KSAc (THF); *ix*, TBAF (THF), CCl_3CN, DBU, HSC_7H_{15}; x, $N_2H_4 \cdot$ HOAc (DMF); x*i*, NaH (DMF), NaOMe–MeOH, 0.2 N KOH

2.4 Michael Addition to Unsaturated Acceptors

A new method of stereoselective synthesis of thiodisaccharides by conjugate addition of 1-thi*o*-*β*-D-glucose (**8a**) or 1-thi*o*-*α*-L-fucose (**30b**) to an unsaturated carbohydrate molecule was also used for the preparation of D-glucopyranosyl- and of L-fucopyranosyl-thiodisaccharides [56].

3 Synthesis of Non-Reducing Thiodisaccharides

3.1 Synthesis of (1→1) Thioglycosyl-glycosides

β, *β*-Thiotrehalose was synthesized at the beginning of this century and, since then, its analogues in the *xylo*, *galacto* and *N*-acetyl-*glucosamino* series have been

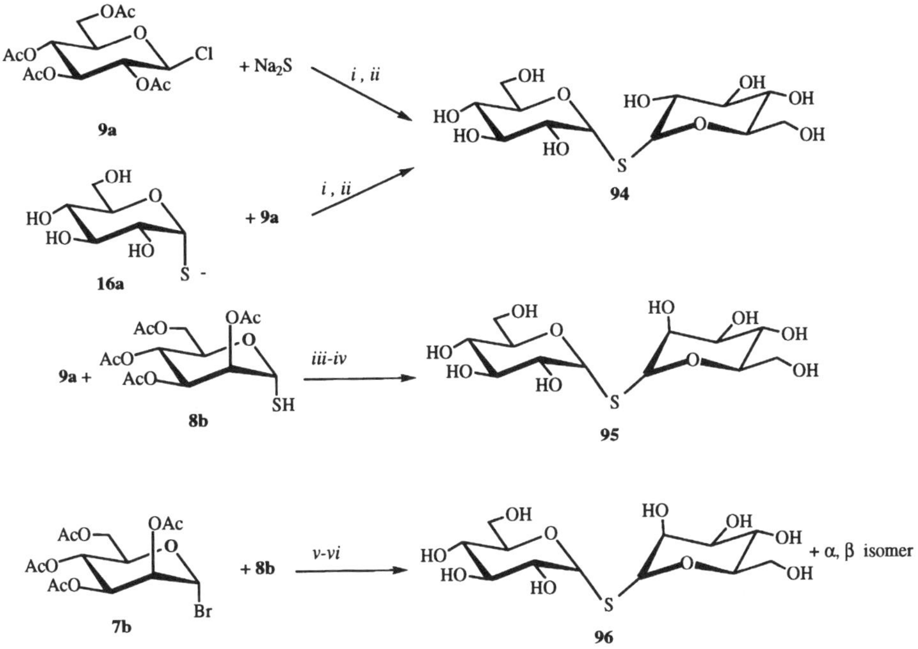

Scheme 27. *i*, (HMPA) Ac_2O (pyridine); *ii*, NaOMe–MeOH; *iii*, K_2CO_3 (HMPA–H_2O); *iv*, NaO-Me–MeOH; *v*, K_2CO_3 (acetone–H_2O); *vi*, NaOMe–MeOH

obtained under various conditions [57]. However the importance of such compounds is rather limited as substrate analogues for any enzyme since the natural (1→1) glucosyl-glucoside is α,α-trehalose. The first thio-analogues of α,α-trehalose (**94–96**) were obtained in 1978 using a one- or two-step procedure (Scheme 27) [14a]. These results demonstrate that the reaction in HMPA proceeds without any neighbouring group participation. It has also been shown that hydrogen sulfide and D-glucose dissolved in hydrogen fluoride react and yield a mixture of 1-thiotrehaloses (**94**), and its α,β- and β,β-isomers [58].

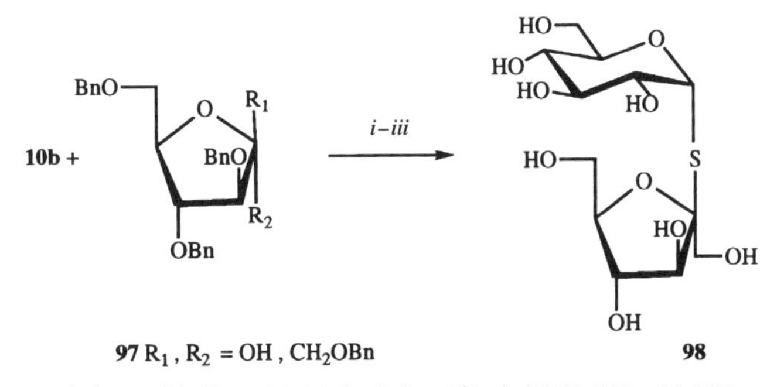

Scheme 28. *i*, $ZrCl_4$ (CH_2Cl_2); *ii*, Na (NH_3) Ac_2O (pyridine); *iii*, NaOMe–MeOH

3.2
Synthesis of (1→2) Thioglycosyl-ketosides

The true analogue of sucrose, 1-thiosucrose (**98**) was obtained using an acid-catalysed condensation involving the two synthons (**10b**) and (**97**) (Scheme 28) [19]. It is worthy of note that the classical method failed to give any condensation products.

4
Thiooligosaccharides and Proteins

4.1
Conformational Equilibria of Thiooligosaccharides

The recognition step, the first event in carbohydrate protein interactions, is dependent on the overall conformation of the oligosaccharide in aqueous solution. Therefore, it is important to determine whether the conformational characteristics of the natural compounds are reflected in the substrate analogues. The most usual methods to achieve this goal is the combination of NMR spectroscopy, molecular mechanics calculations and X-ray data. On this basis, only few conformational studies of thiodisaccharides were reported.

4.1.1
4-Thiomaltosyl Residue

For methyl 4-thio-α-maltoside (**36b**) (Scheme 25), the crystal structure has been solved [59]. Investigations of its conformations in different solvents using quantum-mechanical PCILO energy minimization have been published [60]. Its conformation in water was also studied by NMR and Monte-Carlo simulation [61]. From the X-ray data it has been found that the C–S bond is longer than the corresponding C–O bond in maltose by around 0.4 Å but that the τ angle is smaller; so the net effect is an increase of 0.35 Å for the non-bonded C(1′)...C(4), and the absence of intramolecular hydrogen-bond between O-2′ and O-3 in thiomaltose. However, all the studies in solution clearly demonstrate that the overall conformation of (**36b**) is similar to that of the parent maltosyl residue, and that the thioglycosidic linkage presents a high degree of flexibility.

4.1.2
4-Thiogalabioside

The conformation of (**87b**) (Scheme 25) was investigated by NMR and computational methods. It has also been shown that its overall conformation is similar to that of the natural galabioside. The 4-thiocompound is more flexible, but its O2′-O6 distance is more than 0.8 Å larger than in the parent compound, which precludes an intramolecular hydrogen-bond between these two positions [54].

4.1.3
1,1-Thiodisaccharides

A comparative study between the conformations of α,α-1-thiotrehalose (**94**), its epimer (**95**) and the parent disaccharides revealed that the non-bonded H(1)...H(1′) is longer (3.0 vs 2.83 Å) in the thioseries [62].

4.1.4
Concluding Remarks

No data were reported on 1,2-*trans*-thioglycosides. However, in accordance with a weaker *exo*-anomeric effect in thiooligosaccharides than in the *O*-glycosides, these compounds should also be more flexible.

The variance in binding between *O*- and *S*-oligosaccharides is expected to be small, provided that the differences in bond lengths and ϕ/ψ angles are compensated by the flexibility of the molecule.

4.2
Recognition of (1→6) Thiooligosaccharides

The galactose-specific cell-wall lectin of *Kluyveromyces bulgaricus* mediates the flocculation process of this yeast. The abilities of 6-thio-galactosyl-cyclodextrins (**17c, 17e and 17f**) (Scheme 6) to inhibit in vitro this flocculation were evaluated. The minimum inhibitory concentrations which totally inhibit the flocculation process for these compounds and for the best known ligands (*p*-nitrophenyl galactosides) have the same order of magnitude [27]. The complexing properties of such cyclodextrins may be used to target active molecules to specific cells.

Glucoamylase is one of the most important industrial enzymes which is used to produce glucose from starch. Most fungal enzymes possess a raw starch binding domain (SBD) which is distinct from the catalytic domain. To get a better understanding of the role of this additional domain during the enzymatic process, the 6^{ω}-S-α-D-glucopyranosyl-6^{ω}-thiomalto-oligosaccharides (**23a–23c**) (Scheme 7) were prepared. These compounds were not hydrolyzed by glucoamylases. It has been shown that (**23a**) exerts mild stimulation of the activity of the isoform G1, in which the starch binding domain is present, towards soluble starch and weak inhibition during hydrolysis of panose. However, these properties were not found in G2, a truncated-form of G1, in which the starch binding domain is absent [28]. The dissociation constants of (**23a–23c**) were determined by UV difference spectroscopy [29] and the thermodynamics of ligand binding to the starch-binding domain of glucoamylase, both in the SBD fragment and in glucoamylase G1, were also described [63]. All these results show that these compounds bind only to the starch binding domain.

Several ganglioside analogs containing an α-thioglycoside of sialic acid, derived for example from (**29c**) (Scheme 8), have been found to be potent inhibitors of sialidase activities of different subtypes of influenza virus [64].

α-L-Fucosidases are important enzymes in the metabolism of biological glycoconjugates. The disaccharide (**31**, Scheme 9) showed mixed-type inhibition during the hydrolysis of p-nitrophenyl α-L-fucopyranoside by commercially available α-L-fucosidases from bovine kidney and epididymis [32].

4.3
Recognition of (1→4) Thiooligosaccharides

With the aim of studying the glycosphingolipid metabolism, thiolactosylceramide (**39c**) (Scheme 12) was obtained. The corresponding free disaccharide has been shown to be totally resistant to the action of a mixture of GM_1-β-galactosidase and glucocerebrosidase [38].

The inhibitory activity of compound (**42**) (Scheme 13) was examined during the hydrolysis of *p*-nitrophenyl α-L-fucopyranoside by α-L-fucosidases. As reported above, this compound showed mixed inhibition [32].

The most complete studies of (1→4) thiooligosaccharides were achieved in the cellobiose and maltose series with cellulases and α-amylases of various origins.

Because of the complex structure and the insolubility of cellulose, the precise mechanism of recognition and action of cellulases have been examined by the use of inhibitors and artificial substrates. Cellulases have been found, so far, in 11 families of glycosyl hydrolases [65].

Compound (**44 g**, NHAc form) (Scheme 14) was found to be a competitive inhibitor for CBHI cellulase (family 7) from *Trichoderma reesei*, when 4-methylumbelliferyl β-lactoside was used as substrate. Therefore (**44 g**, NH_2 form) was coupled to CH-Sepharose 4B, and the affinity gel was very effective for the purification of cellobiohydrolases from a crude commercial cellulolytic extract of *T. reesei* [40c]. Using the same approach aryl 1,4-dithioxylobioside and 1,4,4′-trithioxylotrioside (**44 h**, NH_2 form) were coupled to CH-Sepharose 4B to give affinity gels which were used for the purification of xylanases [40a,b].

It has been shown that methyl thiocellotrioside (**44d**), tetraoside (**44i**) and pentaoside (**44j**) are potent inhibitors of endoglucanase I (EGI, family 7) and cellobiohydrolase II (CBHII, family 6) from *Humicola insolens* [41]. Furthermore, a crystal of EGI of *Fusarium oxysporum* was soaked in a solution of compound (**44j**), and a clear density for three sugar rings was seen in subsites −2, −1 and +1. It has been found that the pyranose ring in subsite -1 adopts a boat conformation leading to a pseudo-axial orientation for the scissile bond (Fig. 2) [66].

The use of methyl glycoside of hemithiocellodextrin (**40c**) and tetrasaccharide (**44j**) as inhibitors of cellulases belonging to various families has also been investigated. These compounds were not split under the experimental conditions used, and were effective inhibitors (K_I in µmol/l range) [67].

Kinetic investigations were performed using the tetrasaccharide (**44i**), its higher homologs of DP 5 and DP 6, and 1,3:1,4-β-D-glucanase isolated from *Bacillus licheniformis* [42b]. As expected, all these compounds were resistant to enzymatic cleavage and have been shown to act as competitive inhibitors (K_I in mmol/l range).

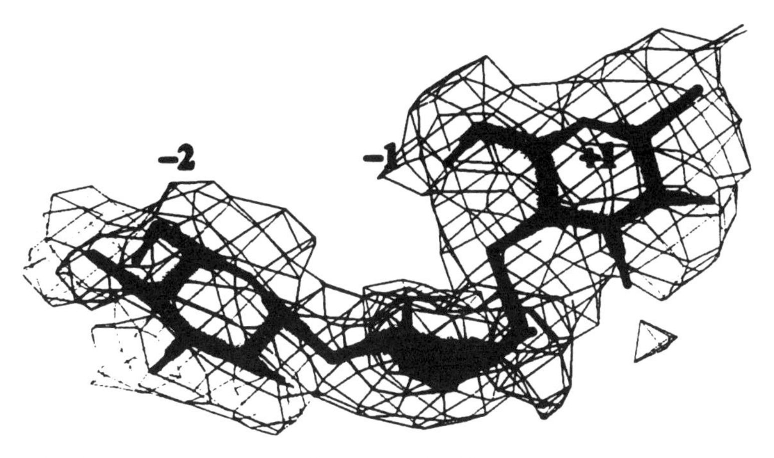

Fig. 2. *Fusarium oxysporum* EG1: 4-thiooligosaccharide complex

The interactions of α-amylases, mainly porcine pancreatic α-amylase, and thiomaltodextrins have been investigated. In 1980, the 3-D structure of porcine pancreatic α-amylase was reported [68], and an analogue of the thiomaltotrioside (**48b**), prepared by standard condensation between (**34e**) and (**51c**), was effective to label the active site and to identify a second binding site on the surface of the protein molecule.

A few years later, the trisaccharide (**48a**) was shown to be a chromogenic substrate for human and porcine pancreatic α-amylase [43a]. More recently, the use of methyl 4,4′-dithio-maltotrioside (**48c**) in further defining the mechanism of action of porcine α-amylase has been investigated [45].

Lectin-like receptors of various bacteria and viruses recognize the galabiose unit of glycolipids as the smallest ligand. The 4-thiogalabioside (**87b**) has been shown to be ten times less recognized than its oxygen analogues by a bacterial pilus adhesion [54].

4.4 Recognition of (1→3) Thiooligosaccharides

In the series of α-L-fucopyranosyl-thio-glycosides, the most powerful inhibitor of α-L-fucosidases was the $\alpha(1\rightarrow3)$ linked isomer obtained by deprotection of the disaccharide (**65b**). It has been shown to be a competitive inhibitor (K_I= 0.65 mmol/l) [32].

4.5 Recognition of Non-Reducing Thiodisaccharides

It has been found that both 1-thio-α,α-trehalose (**94**) and α-D-glucopyranosyl 1-thio-α-D-mannopyranoside (**95**) are inhibitors for cockchafer trehalase [69].

Thiosucrose (**98**) was also a good inhibitor for the two enzymes which act on sucrose, the levansucrase from *Bacillus subtilis* and the yeast invertase [19].

5
Conclusion

Up to the beginning of the 1990s most of the work on thiooligosaccharides was devoted to their synthesis. During the last few years, tailor-made complex or sophisticated thiooligosaccharides were obtained and used both in biochemical and X-ray studies for getting a better understanding of the mechanism of action of glycanases.

In the future, the ease of access of these non-natural oligosaccharides should also demonstrate potential in vivo, in glycotherapy and drug-targeting for instance.

Acknowledgements. I would like to express my sincere gratitude to all the colleagues who have been involved in this work. In particular, I would like to dedicate this paper to Dr. M. Blanc-Muesser who spent the last months of her life improving the methodology for the formation of thiolinkages in carbohydrates. This work was supported by the CNRS and in part by several E.C. programmes.

6
References

1. Lemieux RU, Du MH, Spohr U, Acharya S, Surolia A (1994) Can J Chem 72:158 and references cited therein
2. Lowary TL, Eichler E, Bundle DR (1995) J Org Chem 60:7316 and references cited therein
3. (a) Knowles JKC, Lentovaara P, Murray M, Sinnot ML (1988) J Chem Soc Chem Commun 1401 (b) Bock K, Sigurskjold BW (1990) In: Attar-ur-Rahman (ed) Studies in natural products chemistry, vol 7. Elsevier, Amsterdam, p 29
4. Beau JM, Gallagher T, Nicotra F, de Raadt A, Ekhart CW, Ebner M, Stütz AE, Stick RV: see this volume
5. (a) Uryu T, Libert H, Zachoval J, Schuerch C (1970) Macromolecules, 3:345 (b) Schuerch C (1973) Acc Chem Res 6:184 (c) Kobayashi K, Ichikawa H, Sumitomo H (1990) Macromolecules 23:3708
6. Wang LX, Sakairi N, Kuzuhara H (1990) J Chem Soc Perkin Trans 1:1677
7. Schmidt RR (1986) Angew Chem Int Ed Engl 25:212 and references cited therein
8. (a) Andrews JS, Pinto BM (1995) Carbohydr Res 270:51 (b) Andrews JS, Pinto BM (1994) Tetrahedron: Asymm 5:2367
9. (a) Cerny M, Vrkoc J, Stanek J (1959) Collect Czech Chem Commun 24:64 (b) Cerny M, Pacak T (1959) Collect Czech Chem Commun 24:2566 (c) Matta KL, Girotra RN, Barlow JJ (1975) Carbohydr Res 43:101
10. (a) Horton D, Wolfrom ML (1962) J Org Chem 27:1794 (b) Horton D (1963) Methods Carbohydr Chem 433 and references cited therein
11. Park WKC, Meunier SJ, Zanini D, Roy R (1995) Carbohydr Lett 1:179
12. Defaye J, Driguez H, Ohleyer E, Orgeret C, Viet C (1984) Carbohydr Res 130: 317
13. (a) Sakata M, Haga M, Tejima S, Akagi M (1964) Chem Pharm Bull 12:652 (b) Tejima S, Maki T, Akagi M (1964) Chem Pharm Bull 12:528
14. (a) Blanc-Muesser M, Defaye J, Driguez H (1978) Carbohydr Res 67:305 (b) Blanc-Muesser M, Defaye J, Driguez H (1982) J Chem Soc Perkin Trans 1:15
15. Gadelle A, Defaye J, Pedersen C (1990) Carbohydr Res 200:497
16. Hasegawa A, Nakamura J, Kiso M (1986) Carbohydr Chem 5:11
17. (a) Roy R, Zanini D, Meunier SJ, Romanowska A (1994) ACS Symp Ser 560:104 (b) Cottaz S, Driguez H (unpublished results)

18. Ferrier RJ, Furneaux RH (1977) Carbohydr Res 57:73
19. Defaye J, Driguez H, Poncet S, Chambert R, Petit-Glatron MF (1984) Carbohydr Res 130:299
20. Defaye J, Guillot JM (1994) Carbohydr Res 253:185
21. Hutson DH (1967) J Chem Soc (C) 442
22. (a) Boos W, Schaedel P, Wallenfels K (1967) Eur J Biochem 1:382 (b) Contour-Galcera MO, Guillot JM, Ortiz-Mellet C, Pflieger-Carrara F, Defaye J, Gelas J (1996) Carbohydr Res 281:99
23. Defaye J, Gadelle A, Guiller A, Darcy R, O'Sullivan T (1989) Carbohydr Res 192:251
24. Cottaz S, Driguez H (1989) Synthesis 755
25. Lancelon-Pin C, Driguez H (1992) Tetrahedron Lett 33:3125
26. Blanc-Muesser M, Driguez H (1988) J Chem Soc Perkin Trans 1:3345
27. de Robertis L, Lancelon-Pin C, Driguez H, Attioui F, Bonaly R, Marsura A (1994) Biorg Med Chem Lett 4:1127
28. Cottaz S, Driguez H, Svensson B (1992) Carbohydr Res 228:299
29. (a) Apparu C, Driguez H, Williamson G, Svensson B (1995) Carbohydr Res 277:313 (b) Comber RN, Friedrich JD, Dunshee DA, Petty SL, Secrist III JA (1994) Carbohydr Res 262:245
30. Bennett S, von Itzstein M, Kiefel MJ (1994) Carbohydr Res 259:293
31. Hasegawa A, Terada T, Ogawa H, Kiso M (1992) J Carbohydr Chem 11:319
32. Hashimoto H, Shimada K, Horito S (1994) Tetrahedron: Asymm 34:4953
33. Driguez H, McAuliffe JC, Stick RV, Tilbrook MG, Williams SJ (1996) Austr J Chem 49:343
34. Lundt I, Skelbaek-Pedersen B (1981) Acta Chem Scand B 35:637
35. Rho D, Desrochers M, Jurasek L, Driguez H, Defaye J (1982) J Bacteriol 149:47
36. Defaye J, Driguez H, John M, Schmidt J, Ohleyer E (1985) Carbohydr Res 139:123
37. Hamacher K (1984) Carbohydr Res 128:291
38. Albrecht B, Pütz U, Schwarzmann G (1995) Carbohydr Res 276:289
39. (a) Wang LX, Lee YC (1995) Carbohydr Lett 1:185 (b) Wang LX, Lee YC (1996) J Chem Soc Perkin Trans 1:581
40. (a) Ohleyer E (1982) PhD Thesis Grenoble (b) Comtat J, Defaye J, Driguez H, Ohleyer E (1985) Carbohydr Res 144:33 (c) Orgeret C, Seillier E, Gautier C, Defaye J, Driguez H (1992) Carbohydr Res 224:29
41. Schou C, Rasmussen G, Schulein M, Henrissat B, Driguez H (1993) J Carbohydr Chem 12:743
42. (a) Moreau V, Driguez H (1996) J Chem Soc Perkin Trans 1:525 (b) Moreau V, Viladot JL, Samain E, Planas A, Driguez H, Biorg Med Chem in press (c) Moreau V, Driguez H, to be submitted for publication
43. (a) Blanc-Muesser M, Defaye J, Driguez H, Marchis-Mouren G, Seigner C (1984) J Chem Soc Perkin Trans 1:1885 (b) Blanc-Muesser M, Driguez H, unpublished results
44. (a) Blanc-Muesser M, Vigne L, Driguez H (1990) Tetrahedron Lett 31:3869 (b) Blanc-Muesser M, Driguez H, Joseph B, Viaud MC, Rollin P (1990) Tetrahedron Lett 31:3867
45. Blanc-Muesser M, Vigne L, Driguez H, Lehmann J, Steck J, Urbahns K (1992) Carbohydr Res 224:59
46. Blanc-Muesser M, Driguez H, unpublished results
47. Hehre EJ, Mizokami K, Kitahata S (1983) Denpun Kagaku 30:70
48. (a) Cottaz S, Apparu C, Driguez H (1991) J Chem Soc Perkin Trans 1:2235 (b) Apparu C, Cottaz S, Bosso C, Driguez H (1995) Carbohydr Lett 1:349
49. Bornaghi L, Utille JP, Driguez H, to be submitted for publication
50. Defaye J, Guillot JM, Biely P, Vrsanska M (1992) Carbohydr Res 228:47
51. Aguilera B, Fernandez-Mayoralas A (1996) J Chem Soc Chem Commun 127
52. Contour-Galcera MO, Ding Y, Ortiz-Mellet C, Defaye J (1996) Carbohydr Res 282: 119
53. Reed LA, Goodman L (1981) Carbohydr Res 94:91
54. Nilsson U, Johansson R, Magnusson G (1996) Chem Eur J 2:295
55. Eisele T, Toepfer A, Kretzschmar G, Schmidt RR (1996) Tetrahedron Lett 37:1389

56. (a) Witczak ZJ, Sun J, Mielguj R (1995) Biorg Med Chem Lett 5:2169 (b) Becker B, Thimm J, Thiem J (1996) J Carbohydr Chem, in press
57. (a) Schneider W, Wrede F (1917) Ber Dtsch Chem Ges 50:793 (b) Schneider W, Beuther A (1919) Ber Dtsch Chem Ges 52:2135 (c) Akagi M, Tejima S, Haga M (1961) Chem Pharm Bull 9:360 (d) Stanek J, Sindlerova M, Hanova O, Cerny M, Pacak J (1965) Collect Czech Chem Commun 30:2494 (e) Chrétien F, Di Cesare P, Gross B (1988) J Chem Soc Perkin Trans 1:3297
58. Defaye J, Gadelle A, Pedersen C (1991) Carbohydr Res 217:51
59. Pérez S, Vergelati C (1984) Acta Cryst B40:294
60. Mazeau K, Tvaroska I (1992) Carbohydr Res 225:27
61. Bock K, Duus JO, Refn S (1994) Carbohydr Res 253:51
62. Bock K, Defaye J, Driguez H, Bar-Guilloux E (1983) Eur J Biochem 131:595
63. Sigurskjold BW, Svensson B, Williamson G, Driguez H (1994) Eur J Biochem 225:133
64. Suzuki Y, Sato K, Kiso M, Hasegawa A (1990) Glycoconjugate J 7:349
65. Henrissat B, Bairoch A (1993) Biochem J 293:781
66. Sulzenbacher G, Schulein M, Driguez H, Henrissat B, Davies G (1996) Biochem 35:15280
67. Armand S, Moreau V, Henrissat B, Driguez H, to be published
68. Payan F, Haser R, Pierrot M, Frey M, Astier JP (1980) Acta Cryst B36:416
69. Defaye J, Driguez H, Henrissat B, Bar-Guilloux E (1980) Some recent aspects of the specificity and mechanisms of action of trehalases. In: Marshall JJ (ed) Mechanisms of saccharide polymerization and depolymerization. Academic Press, New York, p 331

Aldonolactones as Chiral Synthons

Preparation of Highly Functionalized, Optically Active Pyrrolidines, Piperidines, Carbocycles and Tetrahydrofurans

Inge Lundt

Department of Organic Chemistry, The Technical University of Denmark, DK-2800 Lyngby, Denmark

This article focusses on the synthetic potential of selectively activated aldonolactones. These are obtained without using any protection group strategy, thus avoiding one of the less attractive features usually associated with polyhydroxylated compounds. The selectively brominated as well as the sulfonylated lactones are suitable starting materials for the preparation of hydroxylated and amino substituted pyrrolidines and piperidines, obtained by simple treatment with ammonia. Boiling in water transformed the activated lactones into hydroxylated tetrahydrofurans. Radical-initiated internal ring closure of brominated aldonolactones with unsaturation yields functionalized cyclopentanes regio- and stereospecifically. This results in the synthesis of optically pure carbasugars. Easy manipulation of the hydroxy groups in the cyclopentane-lactones obtained gives access to other hydroxy/amino substituted cyclopentanes.

The type of compounds synthesized are structures of biological importance, and the use of aldonolactones as optically active synthons yields optically pure products.

Table of Contents

1
Introduction

The aim of this article is to focus on the diversity of aldonolactones as chiral synthons. The chemistry of aldonolactones was an almost unexplored area when, in 1979, we started our investigations on the reaction of aldonolactones with hydrogen bromide in acetic acid thereby obtaining bromodeoxyaldonolactones [1, 2]. These compounds have over the years proven to be very versatile compounds for stereoselective synthesis, both in the carbohydrate field, giving access to otherwise less readily obtainable sugars, and as chiral, optically pure synthons in a broader sense within organic chemistry.

Since these first investigations many other research groups have recognized the synthetic prospects of aldonolactones as compounds having different functionalities: the lactone function and hydroxyl groups with different reactivities, namely the one α to the lactone carbon, which is of similar or enhanced reactivity as compared to a primary hydroxy group, followed by several secondary hydroxy groups. Aldonolactones are thus compounds ready for selective transformations. A recent review [3] concentrates on such reactions, while two previous reviews concentrate on the chemistry of D- and L-gulono- [4] and of D-ribono-1,4-lactone [5], respectively. There is also a review by Fleet describing the synthetic prospects of protection group chemistry of aldonolactones and the use of these derivatives in the preparation of biological active compounds [6]. Work by Chittenden and coworkers concerning the chemistry of vitamin C [7] and D-glucono-1,5-lactone [8] as chirons should also be mentioned here.

In the following, the emphasis has mainly been put on transformations of aldonolactones without any use of protection group chemistry. After selective conversion of one or two hydroxy groups into leaving groups, these chiral synthons can be transformed stereoselectively. The bromo- or sulfonate groups introduced at the α and/or the ω positions exhibit different reactivity, the α substituent being the more reactive. Thus regioselective reduction, epoxidation or nucleophilic substitutions can be performed as well as reactions at both centers. By treatment with strong bases, stereoselective rearrangements via epoxides can take place giving rise to new aldonic acid derivatives. Having this variety of functionalized aldonolactones at hand, their use as synthons for the preparation of highly functionalized pyrrolidines, piperidines, carbocyclic compounds or tetrahydrofurans has been studied. Such compounds can be visualized as carbohydrates in which the ring oxygen has been replaced with either a nitrogen or a carbon atom, or, in the case of tetrahydrofurans, as sugars lacking the anomeric oxygen. The lack of the acetal function renders the compounds stable towards hydrolysis.

During the last ten years or more there has been a growing interest in compounds which could mimic sugars, since they have been shown to interfere with biological systems, and many groups have dedicated their research to the synthesis of such compounds.

Our work has mostly been focussed on the potential of selective transformation of aldonolactones, as will be described in this article, rather than on the synthesis of defined target molecules. Thus, the purpose of this paper is also to inspire researchers from the non-carbohydrate field to recognize these chiral synthons for further use in the synthesis of *e.g.* complex natural products.

A broader use of carbohydrates in this context has been treated by Hanessian in "The Chiron Approach" [9] and a recent book entitled "Carbohydrate Building Blocks" has just appeared [10].

2
Preparation of Selectively Functionalized Aldonolactones

2.1
Bromodeoxyaldonolactones

2.1.1
Reaction of Aldonolactones with Hydrogen Bromide in Acetic Acid

Many aldonolactones or salts of aldonic acids are easily available, either commercially, or by oxidation of the parent sugar with bromine. Reaction of these compounds with hydrogen bromide in acetic acid yields bromodeoxyaldonolactones. The bromine is either introduced α to the lactone function, or at the primary position, or in both positions. In the strongly acidic medium, formation of acetoxonium ions may take place. A subsequent opening of these ions with bromide ions yields acetylated bromohydrins. Thus, introduction of bromine into the lactones is dependent on the ability to form such ions. Lactones having the hydroxy groups *cis* within the lactone ring form an acetoxonium ion, while *trans* oriented hydroxy groups do not. Opening of a 2,3-acetoxonium ion has been found to take place exclusively at C-2. Thus, bromine is introduced at this

Table 1. Bromodeoxyaldonolactones from reaction of aldonolactones with HBr/HOAc

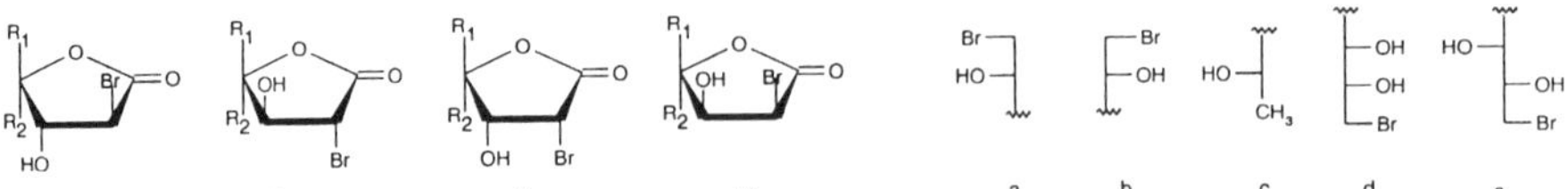

Name of aldono-1,4-lactone	Structure			Starting Material	Ref.
	formula	R_1	R_2		
2-bromo-2-deoxy-D-erythono-	III	H	H	Ca D-threonate	see text
2-bromo-2-deoxy-L-erythono-	IV	H	H	Ca L-threonate	11
2-bromo-2-deoxy-D-threono-	I	H	H	K D-erythronate	see text
2-bromo-2-deoxy-L-threono-	II	H	H	K L-erythronate	11
2-bromo-2-deoxy-D-arabinono-	I	CH_2OH	H	D-ribono-1,4-lactone	14
2,5-dibromo-2,5-dideoxy-D-arabinono-	I	CH_2Br	H	D-ribono-1,4-lactone	14
5-bromo-5-deoxy-D-arabinono-				K D-arabonat	15
2,5-dibromo-2,5-dideoxy-D-lyxono-	IV	CH_2Br	H	NH_4 D-xylonate	15
2-bromo-2-deoxy-D-xylono-	II	CH_2OH	H	D-lyxono-1,4-lactone	14
2,5-dibromo-2,5-dideoxy-D-xylono-	II	CH_2Br	H	K D-lyxonate	14
6-bromo-6-deoxy-D-altrono-				Ca D-altronate	16
2,6-dibromo-2,6-dideoxy-D-altrono-	I	a	H	D-allono-1,4-lactone	17
6-bromo-6-deoxy-D-galactono-, tri-*O*-acetat				D-galactono-1,4-lactone	2
2,6-dibromo-2,6-dideoxy-D-galctono-	II	H	a	D-talono-1,4-lactone	17
2,6-dibromo-2,6-dideoxy-L-galctono-	I	b	H	K L-talonate	18
2,6-dibromo-2,6-dideoxy-D-glucono-	II	a	H	D-mannono-1,4-lacton	19
2,6-dibromo-2,6-dideoxy-L-glucono-	I	H	b	ethyl L-mannonate	20
2-bromo-2,6-dideoxy-L-glucono-	I	H	c	L-rhamnono-1,4-lactone	21
6-bromo-6-deoxy-D-idono-				D-idonic acid	16
2,6-dibromo-2,6-dideoxy-D-idono	I	H	a	D-gulono-1,4-lactone	17
2,6-dibromo-2,6-dideoxy-L-idono-	II	b	H	L-gulono-1,4-lactone	7
2,6-dibromo-2,6-dideoxy-D-mannono-	IV	a	H	D-glucono-1,5-lactone	19
2,6-dibromo-2,6-dideoxy-L-mannono-	III	H	b	Ca L-gluconate	20
2,7-dibromo-2,7-dideoxy-D-*glycero*-D-*ido*-heptono-	I	H	d	D-*glycero*-D-*gulo*-heptono-1,4-lactone	22
2,7-dibromo-2,7-dideoxy-D-*glycero*-D-*galacto*-heptono-	II	H	d	Ba D-*glycero*-D-*talo*-heptonate	23
2,7-dibromo-2,7-dideoxy-D-*glycero*-L-*gluco*-heptono-	I	H	e	D-*glycero*-L-*manno*-heptonamide	23

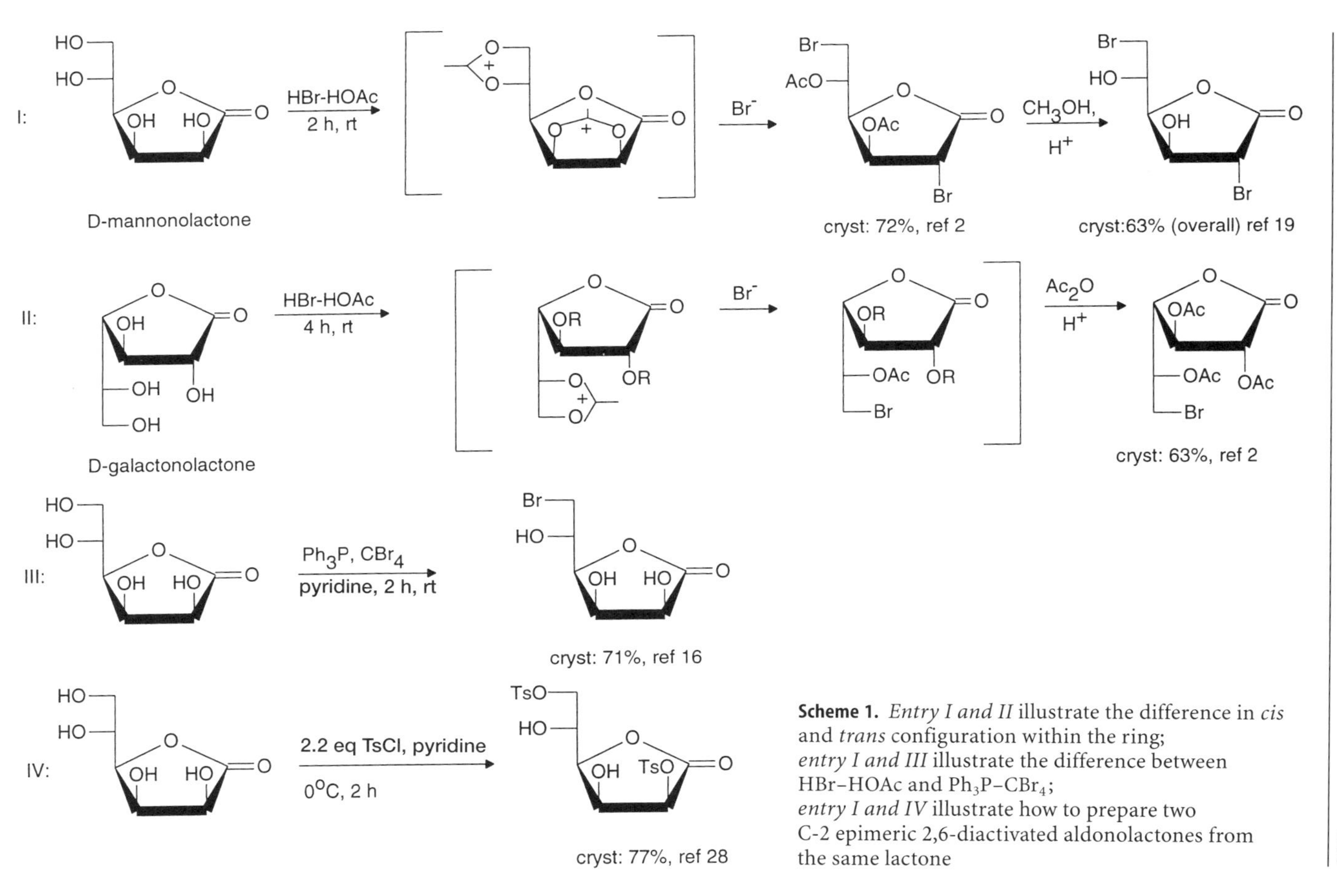

Scheme 1. *Entry I and II* illustrate the difference in *cis* and *trans* configuration within the ring; *entry I and III* illustrate the difference between HBr–HOAc and $Ph_3P–CBr_4$; *entry I and IV* illustrate how to prepare two C-2 epimeric 2,6-diactivated aldonolactones from the same lactone

position with inversion of the configuration as compared to the starting lactone (Scheme 1 entry I). This holds for the aldonolactones with from four to seven carbon atoms which we have investigated (Table 1).

Aldono-1,4-lactones, having six or more carbon atoms also form an exocyclic aceteoxonium ion between the primary and the neighbouring hydroxy group when treated with hydrogen bromide in acetic acid. By the opening of this ion, bromine is easily introduced at the primary carbon (Scheme 1, entry II). Pentonolactones also yield 5-bromo-5-deoxy-lactones when treated with HBr–HOAc, although the reaction requires 1–4 days. Since no exocyclic acetoxonium ion can be formed, a direct substitution probably takes place in these cases.

The principle of the reactions between aldonolactones and HBr–HOAc is illustrated in Scheme 1 (entries I and II), and the mono- and di-bromodeoxylactones prepared in this way are listed in Table 1. The 2-bromo-2-deoxy-D-erythrono- and D-threonolactone can be prepared analogously to the L-isomers [11] from the salts of D-threonic and D-erythronic acid [12], respectively. The former can be prepared by oxidative degradation of D-xylose following an analogous procedure described by Humphlett [13].

2.1.2
Reaction of Aldonolactones with Triphenylphosphine/Carbontetrahalide

Triphenylphosphine in combination with carbontetrabromide [24, 25] or with iodine [26] is known to convert primary hydroxy groups into the corresponding halide. This method is complementary to the HBr–HOAc reaction, since the halide is introduced only at the primary carbon, leaving the secondary hydroxy groups untouched. The method is thus useful for the preparation of ω-halogenated aldonolactones, having the 2- and the 3-hydroxy groups *cis* oriented. In this way 5-bromo-5-deoxy-D-ribono- [27], -D-lyxono- [27] as well as 6-bromo-6-deoxy-D-mannono-1,4-lactone [16] have been prepared. The reactions, which are performed in pyridine at around 60 °C, are sensitive to the reaction conditions and to the structure of the lactone. Thus, mannonolactone gave a high yield (71%) of the 6-bromo derivative [16] (Scheme 1, entry III), whereas ribonolactone was converted into the 5-bromoderivative in a 50% yield [27]; lyxonolactone gave only a modest yield [27]. We recently improved the yield by prolonging the reaction time to 4 days.

2.2
Selectively Tosylated/Mesylated Aldonolactones

Selective tosylation or mesylation of unprotected aldonolactones, to give the primary sulfonate only, seemed not to be possible since the 2-hydroxy group has about the same reactivity in these reactions as the primary one [28, 29]. Selective di-*O*-tosylation was possible by treatment of aldonolactones with 2.2 mole equivalents of tosyl chloride to give 2,5-di-*O*-tosylated pentono- or 2,6-di-*O*-tosylated hexono-1,4-lactones [28]. The yields of crystalline ditosylates were high, when the OH-2 and OH-3 groups were *cis*- oriented and also *cis* to the side chain, and somewhat lower when they were both *trans* to the side chain. By this method [28] the following selective *O*-tosylated aldono-1,4-lactones were prepared and isolated by direct crystallization: 2-*O*-tosyl-L-erythrono-, 2,5-di-*O*-tosyl-D-ribono- and D-lyxono-, 2,6-di-*O*-tosyl-D-mannono- and D-talono-, 2-*O*-tosyl-L-rhamnono-, and 2,7-di-*O*-tosyl-D-*glycero*-D-*gulo*-heptono-1,4-lactone. The 2,6-di-*O*-tosylated D-allono- and D-gulonolactones were obtained as syrups, whereas the corresponding di-*O*-mesylates could be prepared in a crystalline state [28]. If the OH-2 and the OH-3 groups were *trans* oriented it was only possible to obtain a crystalline 2,5-di-*O*-tosylate from D-arabinonolactone, and only in a modest yield (24%) [28].

Selective tosylation introduces leaving groups in the α and ω positions in the aldonolactones, the same positions which were substituted with bromine if the lactones were treated with HBr–HOAc. In the latter case, bromine was introduced with inversion of the configuration at C-2, whereas the tosylation does of course not alter the configuration at this center. Thus, the two methods give activated aldonolactones with different stereochemistry at C-2 (Scheme 1, entries I and IV).

The aim of the methods just discussed was to introduce leaving groups selectively in aldonolactones *without any use of protection groups.* However, if this is not possible, it is nessessary to protect hydroxy groups selectively. This can most simply be achieved by using ketal protection prior to tosylation/mesylation or other manipulations. This simple method often yields selective protections of hydroxy groups in polyols. The protecting groups are afterwards removed by simple acid treatment. Preparation and synthetic uses of acetonides of aldonolactones has been extensively elaborated by Fleet and coworkers [30]. Another simple transacetalization of D-gluconolactone to give 3,4; 5,6-di-*O*-isopropylidene-D-gluconolactone methyl ester, leaving the OH-2 group free for transformations, has also been described [31].

3 From Bromodeoxy/Tosyloxy Aldonolactones to Other Lactone Synthons

3.1 Deoxyaldonolactones by Selective Reduction

The dibromodideoxylactones shown in Table 1 as well as the α,ω-di-*O*-tosylated aldonolactones discussed in Sect 2.2 possess leaving groups of different reactivities, the bromine/tosyloxy group α to the lactone function being the more reactive. It was found that debromination of the C-2 bromine could be performed selectively by treatment with aqueous hydrazine [32]. Similarly, we have also used this reagent to remove *O*-tosyl groups [33] α to the lactone function. The method was previously described for deoxygenation of an *O*-mesyl group α to a lactone function [34]. Using this procedure we have obtained 2-deoxylactones functionalized at the primary position. The reaction of dibromomannonolactone **2** with hydrazine to give the corresponding 2-deoxylactone **3** is shown in Scheme 2.

Selective hydrogenolysis of bromodeoxylactones can also be achieved by catalytic reduction in the presence of an acid acceptor (triethylamine). Thus, reaction of 2,6-dibromo-2,6-dideoxy-D-mannonolactone (**2**) (Scheme 2) under

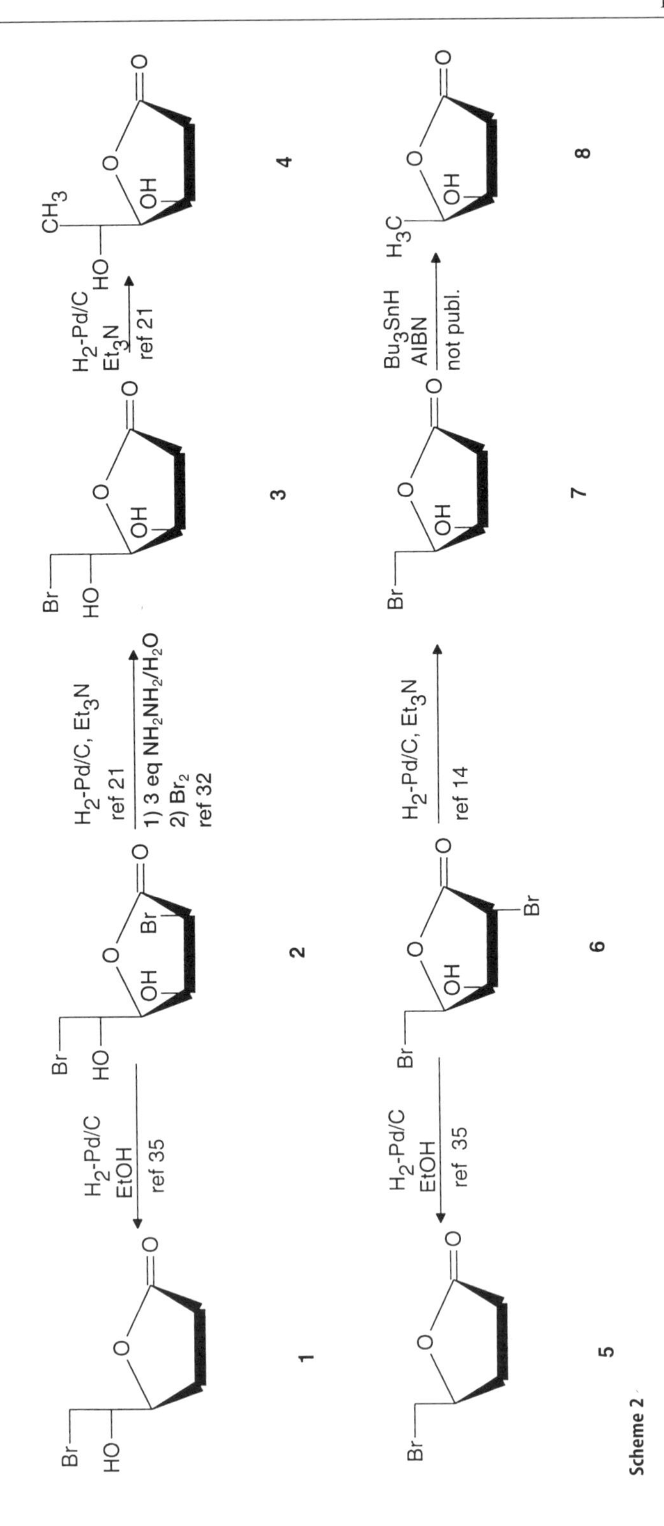
H2-Pd/C
EtOH
ref 35
H2-Pd/C, Et3N
ref 21
1) 3 eq NH2NH2/H2O
2) Br2
ref 32
H2-Pd/C
Et3N
ref 21
H2-Pd/C
EtOH
ref 35
H2-Pd/C, Et3N
ref 14
Bu3SnH
AIBN
not publ.
1
2
3
4
5
6
7
8

Scheme 2

these conditions gave within 1 h the 6-bromo-2,6-dideoxy-D-*arabino*-hexono-1,4-lactone (**3**), which by prolonged reaction gave the corresponding 2,6-dideoxylactone **4** [21]. Other dibromohexonolactones [17] and heptonolactones [22] behave similarly.

The 2,5-dibromo-2,5-dideoxy-D-xylono-1,4-lactone (**6**) (Scheme 2) can also be selectively reduced to give 5-bromo-2,5-dideoxy-D-*threo*-pentonolactone (**7**) by reaction with either aqueous hydrazine as mentioned above, or by consumption of 1 mole of hydrogen [14]. Further reaction with hydrogen and Pd/C leads to 3-hydroxypentanoic acid by an initial reductive elimination [15]. Reduction of the primary bromine in these cases can be obtained by means of tributyltin hydride [25] to give **8**.

If, however, the catalytic hydrogenolysis of a 2,6-dibromohexonolactone was performed with no acid acceptor present, an unusual reduction took place to give a 6-bromo-2,3,6-trideoxyhexonolactone. Thus, **2** gave 6-bromo-2,3,6-trideoxy-D-*erythro*-hexonolactone (**1**) (Scheme 2) [35]. Similar reductions took place with other hexono- and pentonolactones independent of the relative stereochemistry of the C-2 and C-3 substituents [35]. A 7-bromo-2,3,7-trideoxyheptonolactone has also been synthesized in this way [22]. The initial step in the reactions was shown to be a reductive elimination to give an intermediate 2,3-unsaturated lactone which subsequently was saturated [35].

3.2 Epoxyaldonolactones

Optically pure chiral epoxides are valuable synthons, since reaction with nucleophiles gives products of defined stereochemistry. Chirality can be introduced into an achiral molecule by means of a chiral auxillary, as in the enantioselective epoxidation of allylic alcohols introduced by Katsuki and Sharpless [36]. The extension of this "reagent controlled strategy" has been manifold and may be exemplified here by the total synthesis of the L-hexoses [37]. Introducing chirality in a molecule using asymmetric reagents or catalysts generally yields products with high optical purity, but modification of compounds from the "chiral pool" might often be an attractive alternative. Chiral epoxides can thus easily be prepared starting from a chiral bromohydrin using basic conditions. From the selectively brominated aldonolactones described in Sect. 2.1 and collected in Table 1, a number of optically pure monoepoxy- and diepoxy-aldonolactones have been synthesized by treatment of the bromolactones with a base under non-aqueous conditions [19]. In unpublished work, we have also used the selectively tosylated aldonolactones for the preparation of epoxylactones, or *trans*-epoxides of open aldonic acids/esters. The bases used were potassium or cesium fluoride [19], potassium carbonate [19] or sodium hydride [7] used in an organic solvent.

Brief treatment of 2,6-dibromo-2,6-dideoxy-D-mannono- (**2**) or -D-gluconolactone (**9**) (Scheme 3) with potassium fluoride or potassium carbonate in acetone gave the 6-bromo-2,3-monoepoxymannonolactone (**10**), whereas after a more prolonged reaction time the diepoxylactone **11** was formed [19]. This selectivity reflects the difference in reactivity of the two bromine atoms, as also discussed for the selective reductions above. An interesting observation was that

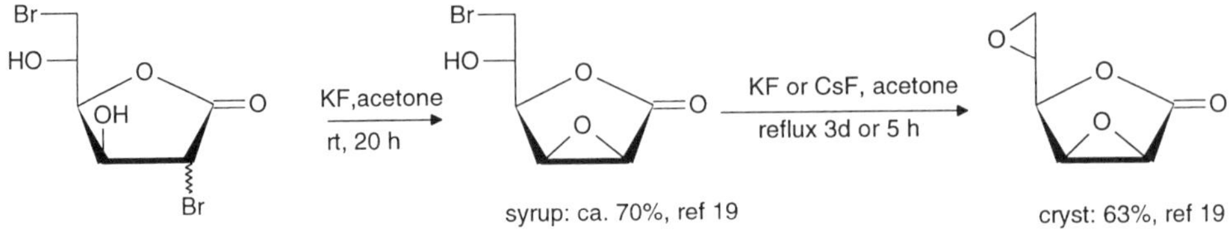

KF,acetone rt, 20 h → syrup: ca. 70%, ref 19 **10** → KF or CsF, acetone reflux 3d or 5 h → cryst: 63%, ref 19 **11**

2 (manno)
9 (gluco)

Scheme 3

both the *trans*- isomer (gluco-, **9**) and the *cis*-isomer (manno-, **2**) yielded the manno-epoxide **10**. This is due to the base lability of the proton α to the lactone, causing an epimerization at C-2 under these conditions. Only the *trans*-bromohydrin gives the *cis*-epoxide within the lactone ring.

A number of mono- and di-epoxyaldonolactones with four and five carbons [38] as well as with six and seven carbon atoms [19] have been described.

3.3 New Aldonolactones by Stereoselective Isomerization at One or More Carbon Atoms

When aldonolactones are dissolved in strong aqueous base the lactone ring is immediately opened to give the acid salt. If bromodeoxylactones are treated similarly, besides opening of the lactone ring, epoxides are also formed. Depending on the pH and the structure of the substrate different reactions may occur.

3.3.1 *Inversion at One Carbon Atom*

If 5-bromo-2,5-dideoxy-D-*threo*-pentono-1,4-lactone (**7**) was treated with aqueous potassium hydroxide and subsequently acidified, 2-deoxy-L-*erythro*-pentono-1,4-lactone (2-deoxy-L-ribonolactone) (**12**) (Scheme 4) was formed, and isolated as the crystalline di-*O*-benzoate [14]. Similarly, when 5-*O*-mesyl-2,3-*O*-isopropylidene-D-lyxonolactone (**13**) (Scheme 4) was treated with 3 mol equivalents of potassium hydroxide followed by acidification, L-ribonolactone (**14**) was formed as the only product. The lactone was isolated as the crystalline 3,4-*O*-benzylidene-L-ribono-1,5-lactone [27]. Thus, in both examples a stereoselective inversion at C-4 had taken place.

By monitoring the reaction of **13** with base by ^{13}CNMR spectroscopy, the 4,5-epoxy carboxylate **C** was observed after 5 min together with a small amount of the the final product **D**. The reaction was complete within about 4 h. Opening of the epoxide **C** to give **D** was most likely preceded by an intramolecular substitution by the carboxylate at C-4. This resulted in an inversion at C-4 and formation of a lactone, which was immediately opened in the basic medium. The intramolecular *exo*-opening rather than an *endo*-opening

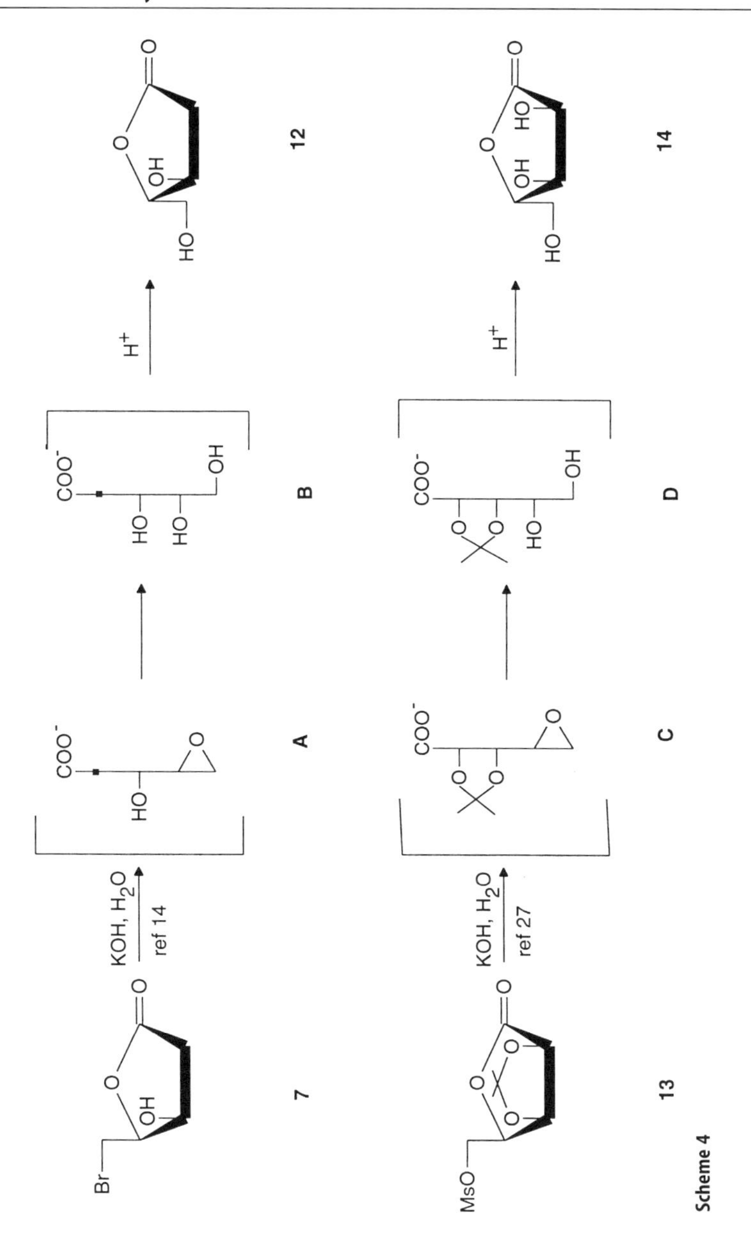

Scheme 4

of the epoxide followed Baldwin's rules [39]. In this example a more favored five-membered transition state, rather than a six-membered one, was thus involved.

Similarly, 5-bromo-5-deoxy-D-lyxonolactone gave the potassium L-ribonate, exclusively, when treated with a strong solution of potassium hydroxide [27]. The method was also used to prepare L-lyxonolactone from D-ribonolactone [27].

3.3.2
Inversion at Two or More Carbon Atoms: Epoxide Rearrangements

If a polyol with a primary epoxide is treated with strong base, a rearrangement to the more stable secondary epoxide takes place, as observed by Payne [40]. If an unprotected bromodeoxyaldonolactone is treated with strong base a similar rearrangement may be expected. We found that if 6-bromo-2,6-dideoxy-D-*arabino*-hexonolactone (**3**) (Scheme 5) was treated with 4 molar equivalents of potassium hydroxide, 2-deoxy-L-*ribo*-hexono-1,4-lactone (**15**) was isolated as the only product after work up [32].

By monitoring the reaction using ^{13}C NMR spectroscopy the primary epoxy carboxylate **A** (Scheme 5) was observed immediately, together with a secondary epoxide, probably **B**. Within 30 min the only spectroscopically detectable product was the L-*ribo*-hexonate **C**. Thus, the 5,6-epoxide had rearranged to the 4,5-epoxide, causing inversion at C-5. Intramolecular substitution by the carboxylate at C-4 in an *exo*-mode, as described above, resulted in inversion at this center. Stereoselective inversion at two carbon atoms has thus taken place. The product was isolated as the tri-*O*-acetyl-2-deoxy-L-*ribo*-hexono-1,4-lactone [32]. Similar stereoselective inversions at C-4 and C-5 in 6-bromo-3,6-dideoxy-hexono-1,4-lactones take place by treatment with strong aqueous base [41].

Stereoselective inversion at three carbon atoms was found when 7-bromo-2,3,7-trideoxy-D-*arabino*-heptono-1,4-lactone (**16**) (Scheme 5) was treated with a strong aqueous base. Similar epoxide rearrangements as described above took place to give only one compound **G**, which by acidification yielded 2,3-dideoxy-L-*arabino*-heptono-1,4-lactone (**17**) [42]. The intermediate epoxides **D** and **E** were observed in the ^{13}C NMR spectra, whereas the epoxide **F** was not present in detectable amounts, although it is the opening of this epoxide at C-4 by the carboxylate which yields the final product (Scheme 5). Presumably, **F** has a less stable *cis*-configuration compared to the *trans*-epoxide **E**.

The general feature in these reactions in strong potassium hydroxide is the Payne rearrangement of primary epoxides to secondary epoxides, an equilibrium between several epoxides being established. The intramolecular opening of the epoxide by the carboxylate always takes place in the *exo*-mode, preferentially via a five-membered transition state. The epoxide to be opened in this way is not always present in detectable amounts according to the NMR spectra [42], as discussed above, probably due to different stability of *cis* and *trans* epoxides. In the stereoselective inversions at several carbon atoms just discussed, it has, however, to be emphasized that the pH must be at least 14, using an excess of potassium hydroxide. Using the less basic potassium carbonate, epoxides are opened by simple hydrolysis without any previous rearrangements [32, 41, 42].

Finally, it should be pointed out that in some cases epoxides are opened by a remote hydroxy group within the molecule to give the five-membered ethers [41]. We are currently investigating how the conformation of an open chain epoxy polyol might be responsible for cyclic ether formation [43].

The access to new stereoisomeric aldonolactones expands the pool of activated synthons which can be used, as described in the following paragraphs.

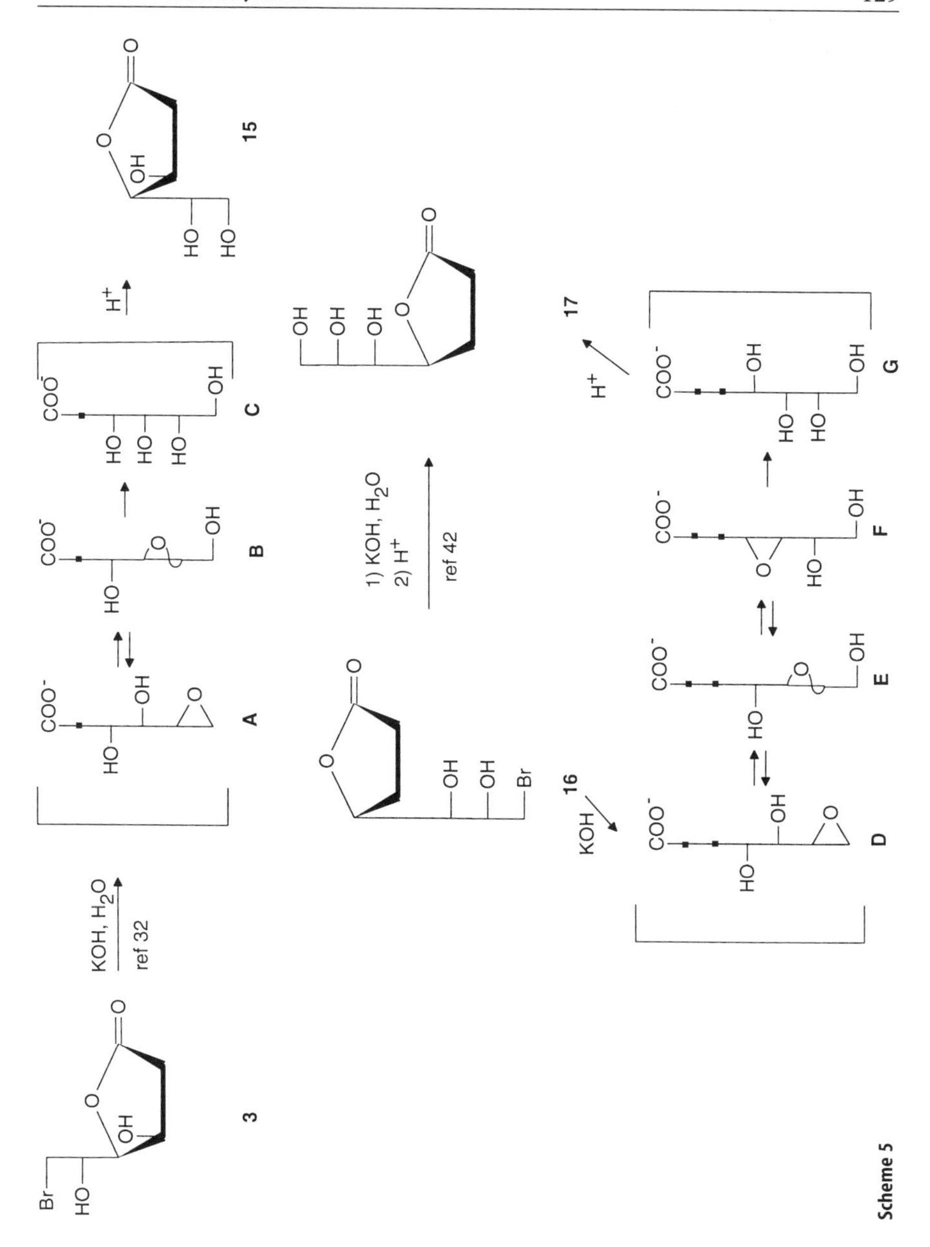

Scheme 5

3.4
2,3-Unsaturated Aldono-1,4-Lactones

The last type of useful synthons prepared from bromodeoxyaldonolactones which should be discussed here is the 2,3-unsaturated aldono-1,4-lactones, easily obtained from 2-bromo-2-deoxyaldonolactones. The method involves a

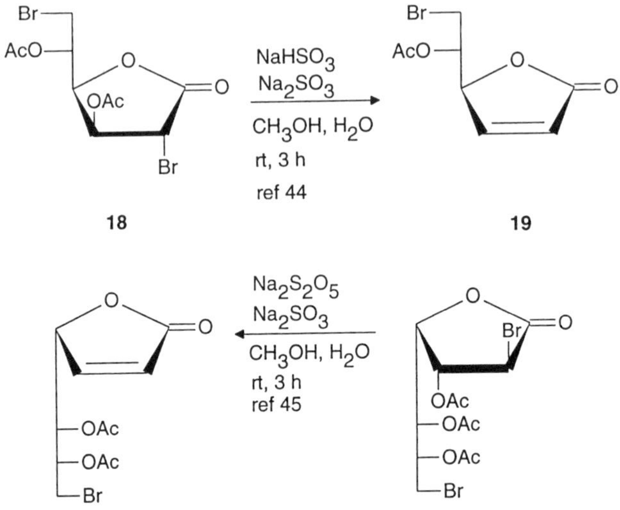

$NaHSO_3$, Na_2SO_3, CH_3OH, H_2O, rt, 3 h, ref 44: **18** → **19**

$Na_2S_2O_5$, Na_2SO_3, CH_3OH, H_2O, rt, 3 h, ref 45: **21** → **20**

Scheme 6

very smooth reductive elimination of the C-2 bromine and a *trans*-3-acetoxy group by use of sodium sulfite in methanol [44]. This is shown in Scheme 6 for the reaction of tri-*O*-acetyl-2,6-dibromo-2,6-dideoxy-D-glucono-1,4-lactone (**18**) to give **19** [44]. The method has been used for the preparation of 2,3-unsaturated- heptono-(**20**) [45] and pentono-1,4-lactones [46].

4 Activated Aldonolactones as Chiral Synthons

4.1 General

The functionalized lactone synthons described in the preceding section are prone to nucleophilic substitutions. Under *non-aqueous* conditions the bromo- and sulfonyloxygroups give substitution at the same carbon atoms bearing the leaving group. This has been used to introduce nitrogen via azide displacement to give mono- and di-azidodeoxyaldonolactones, which can be converted into aminodeoxyaldonic acids/lactones [11, 47, 48].

Under non-aqueous conditions the epoxy function in the epoxylactones can be opened only by Lewis acid assistance. Thus, 2-fluoro-2-deoxy-lactones have been prepared from 2,3-epoxylactones by treatment with HF-amine complexes [38, 49, 50], while 5,6-epoxylactones yield 6-deoxy-6-fluoro-lactones by this treatment [49, 50]. Likewise, BF_3-assisted opening of a 2,3- epoxy function with $TMSN_3$ [51] gave a 2-azido-2-deoxy-lactone [52]. In all cases the opening of the epoxide is a *trans*-opening, and it is noteworthy that under acidic conditions the nucleophile attacks at C-2 or at the primary position, similar to the opening of acetoxonium ions by bromide ions in the preparation of the bromodeoxylactones, as discussed in the preceding paragraph.

In the following paragraph the nucleophilic substitution of some of the activated lactones under *aqueous* conditions will be discussed in detail, namely the reactions with aqueous ammonia, thereby yielding polyhydroxylated pyrrolidines (azafuranoses) and piperidines (azapyranoses).

4.2 Hydroxylated Pyrrolidines and Piperidines – Glycosidase Inhibitors

Polyhydroxylated pyrrolidines and piperidines, which are sugar analogues having the ring oxygen replaced by a basic nitrogen, have been found to inhibit glycosidases, a diverse group of enzymes that catalyze the hydrolysis of glycosidic linkages. Specific inhibitors of glycosidases are of interest in a number of quite different applications. These include the mechanistic study of the enzymes themselves [53, 54] as well as approaches to identify the active site in the glycosidases [55]. Compounds inhibiting the oligosaccharide-processing enzymes thus exhibit properties of potential medicinal interest, as for example in the study of diabetes, cancer and viral deseases [56]. Azasugars act as glycosidase inhibitors, presumably by mimicking the transition state of the protonated pyranose-ring, developed during the hydrolysis of a glycosidic linkages. Both five- and six-membered azasugars fit into this model, which perhaps cannot be used precisely, since the conformation of the transition state might be unknown. These many questions to be answered have stimulated the interests in the development of effective procedures for the synthesis of azasugars for investigation of glycosidase reactions and for developement of effective and specific glycosidase inhibitors. The results of interaction of potential glycosidase inhibitors with various enzymes are useful in further developing this area [57].

In addition to the glycosidase inhibitors of azasugars with a nitrogen replacing the ring oxygen, another type, having a nitrogen at the place of the anomeric position, has recently been synthesized [58, 59] and proven to be a very potent inhibitor.

The synthetic approaches to obtain azasugars have been numerous. Examples may be given from the extensive work by Fleet and coworkers, in which sugars have been used originally as starting materials in multistep synthesis, using protection/deprotection chemistry [60], and later also aldonolactones [61], to reach the target molecules. Other approaches using a combination of chemical and enzymatic reactions have been developped by Wong and coworkers [57a, b, 62], whereas Stütz and coworkers use enzymatic isomerization of defined substrates to attain the targets [63]. The stereocontrol in the different approaches is either due to the chirality of the starting material, or is determined by a chiral catalyst, the enzyme.

The strategy for our syntheses was to use the reaction between the selectively activated aldonolactones and ammonia and, in few steps, using cheap materials without protection group chemistry, to obtain 5- or 6-membered azasugars.

4.2.1
Synthesis of Hydroxylated Pyrrolidines

Preparation of alicyclic compounds having a heteroatom within the ring can be performed by ring closure between an appropiate nucleophile and an acyclic compound having leaving groups at the correct positions [64]. Inspection of the dibromolactones in Table 1 reveals that a number of stereoisomeric substrates with four and five carbon atoms between the leaving groups are available. Thus, a simple reaction with ammonia might lead to formation of five-membered pyrrolidines and six-membered piperidines. However, in non-protected bromohydrins epoxides might be formed as the primary reaction products when exposed to base, as discussed above.

4.2.1.1
Reaction of Dibromodideoxyhexonolactones with Ammonia

When the dibromodideoxy-D-mannono-1,4-lactone (**2**) (Scheme 7) was treated with aqueous ammonia the 3,6-dideoxy-3,6-imino-D-allonamide (**22**) was formed as the only product [20]. Conversion to the ethyl ester and subsequent reduction of the ester with sodium borohydride transformed **22** into the known 1,4-dideoxy-1,4-imino-L-allitol (**23**) [20].

Formation of the pyrrolidine ring revealed that the nucleophile had been attatched to C-6 and C-3 of the bromolactone **2**, and not to the carbons having the leaving groups originally. The mechanism, as revealed by monitoring the reaction by ^{13}CNMR spectroscopy, involved rapid conversion of the lactone into the 2,3;5,6-diepoxy carboxamide (**25a**) (Scheme 7). Opening at the primary carbon by NH_3 gave the primary amine **26**, which by attack at C-3 in an *exo*-opening of the 2,3-epoxide, yielded the 5-membered ring **22**. Thus, two simple reactions, namely treatment of the bromolactone with NH_3 and reduc-

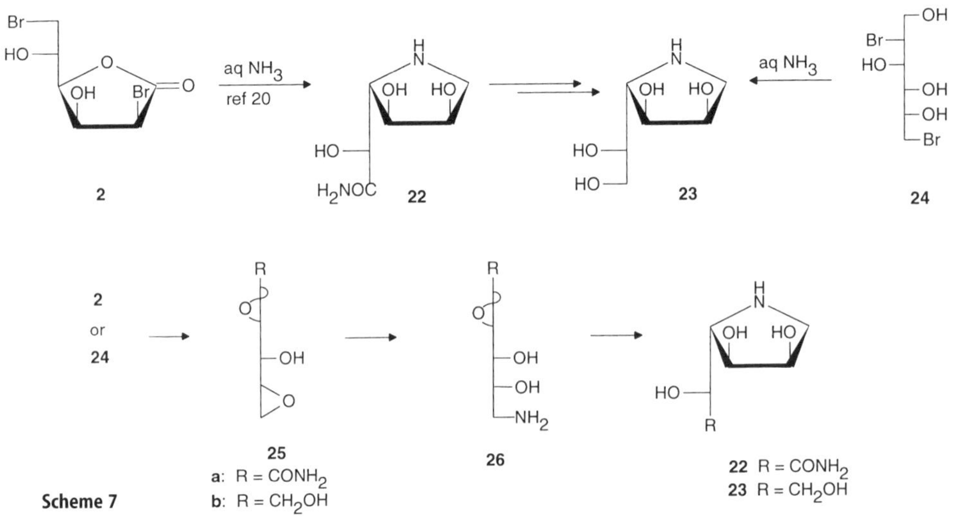

Scheme 7

tion of the carboxylic function with $NaBH_4$, gave the pyrrolidine **23**, isolated as the crystalline hydrochloride, without any need for chromatographic purification.

Alternatively, reaction of the dibromomannitol **24**, which was easily obtained by reduction of the bromolactone **2** with $NaBH_4$, also gave the pyrrolidine **23** if reacted with aqueous ammonia. The mechanism involves the formation of a similar 2,3;5,6-diepoxide (**25b**) (Scheme 7) as discussed for the carboxamide above. It is worth noting that a secondary epoxide is formed and not the primary 1,2-epoxide. This is general for all the 2-bromo-2-deoxy-alditols investigated. In some cases, however, the bromine might be substituted by a remote hydroxy group, forming a five-membered ether instead. We are currently interested in the question of a relation between the conformation of the bromopolyol and the formation of an epoxide vs. a tetrahydrofuran.

By using the concept of reacting a dibromohexitol with ammonia the four enantiomeric pairs of 1,4-dideoxy-1,4-imino-hexitols with *allo-*, *talo-*, *galacto-*, and *ido-* configurations, have been synthesized from the 2,6-dibromo-2,6-dideoxyhexonolactones with D- and L-*manno-*, D- and L-*gluco-* [20], D- and L-*galacto*, and D- and L-*ido*-configurations [18], respectively.

Finally, the 2,6-dibromo-2,6-dideoxy-D-altrono-1,4-lactone (**27**) (Scheme 8), the last mentioned isomer in Table 1 with six carbon atoms, gave, if reduced with sodium borohydride, the corresponding dibromohexitol **28**. This was however not stable, since it was converted slowly into a 3,6-anhydride **29** (Scheme 8) [65]. If the dibromolactone **27** was treated with ammonia a mixture containing anhydrides was obtained. Knowing that the reactions with ammonia proceed via epoxides, the 2,3-epoxylactone was preformed from the dibromolactone using potassium fluoride in acetone. Thus the OH-3 was blocked towards anhydride formation. The lactone **30** was then treated with NH_3 to give the pyrrolidine carboxylate **31**, which was subsequently converted into 1,4-dideoxy-1,4-imino-L-gulitol (**32**) (Scheme 8) [65].

To summarize, crystalline 1,4-dideoxy-1,4-imino-hexitols are readily prepared from 2,6-dibromo-2,6-dideoxy-hexonolactones in gram quantities without any use of chromatographic separations. The bromolactones may, prior to treatment with aqueous ammonia, either be converted into the 2,3-epoxides by treatment with KF or K_2CO_3 in acetone, or may be reduced with sodium borohydride in water to give the bromodeoxyhexitols.

4.2.1.2
Reaction of 2-Amino-6-bromo-2,6-dideoxyaldonolactones with Base

Since reaction of dibromohexonolactones with ammonia resulted in ring closure between C-3 and C-6, as described in the preceding paragraph, selective introduction of an amino group at C-2 should cause ring closure from this carbon, to either C-6 or C-5.

Previously, we have shown that selective substitution of the C-2 bromine could be performed by reaction of the dibromolactone with sodium azide in acetonitrile [47]. Using DMF as a solvent the C-6 bromine was also substituted [47]. In the case of dibromomannonolactone **2** (Scheme 9) the azido group was

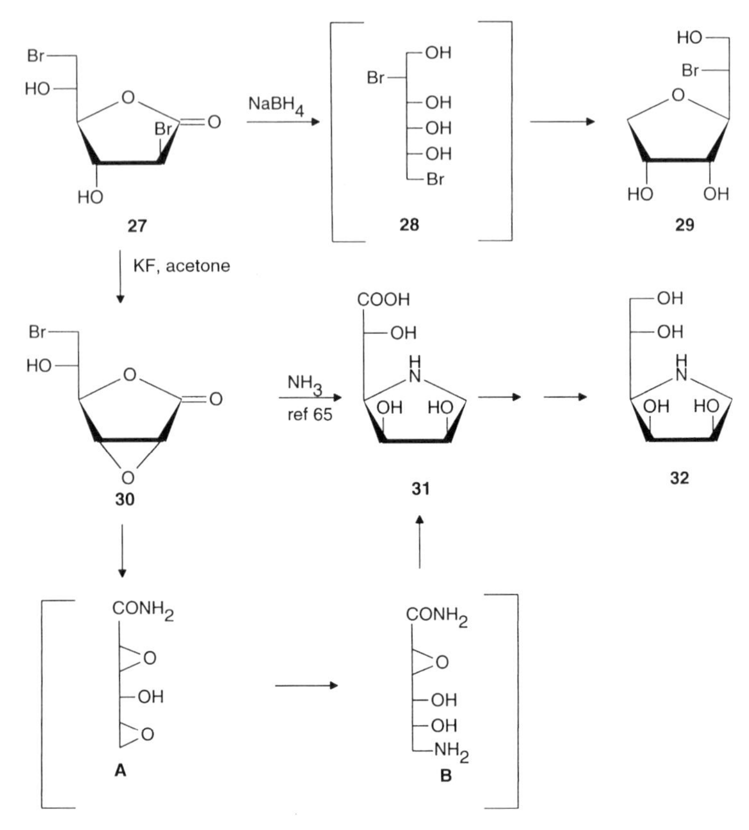

Scheme 8

introduced with retention of configuration. Starting with the C-2 epimeric dibromogluconolactone the same 2-azido-6-bromo-2,6-dideoxy-D-mannonolactone (**33**) was obtained. This is due to the acidity of the C-2 hydrogen, both in the bromo- as well as in the azidolactones, since they epimerize in the presence of base (NaN_3). In the tetrono- and pentonolactone series we obtained both C-2 epimeric azides in different ratios [11]. The crude 2-azido-6-bromo-mannonolactone was reduced catalytically in the presence of HCl to prevent reduction of the bromine [35] and to protect the amino group formed. The crystalline aminobromolactone **34** was obtained in about 40% yield from **2**.

The C-2 epimeric 2-azidolactone with gluco-configuration **38** (Scheme 9) was synthesized by opening of the 2,3-epoxy-6-bromolactone **10**. This in turn was prepared from the dibromolactone **2** by treatment with K_2CO_3 in acetone (Scheme 3). The cheap D-gluconolactone was thus the precursor of **10**. Opening of the epoxide with $TMSN_3$ in the presence of a Lewis acid gave the 2-azido-6-bromolactone **38** in good yield [52]. Catalytic reduction in the presence of HCl gave the corresponding aminolactone **39**.

Ring closure of the amino-bromolactones **34** and **39** was now performed. Based on our experience with the reaction of bromolactones with different

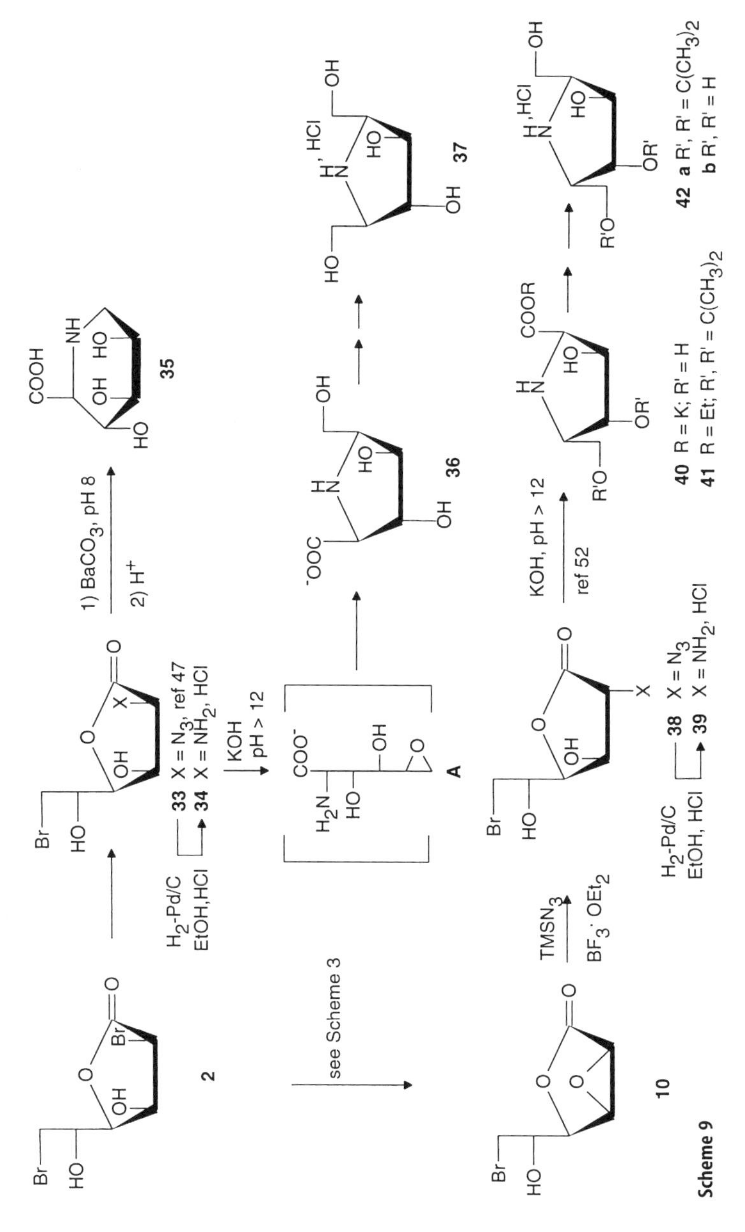

Scheme 9

bases, we expected rapid formation of a primary epoxide. Keeping the pH around 8, epoxide formation may be avoided. Thus by titration of **34** with $BaCO_3$, we were able to obtain a six-membered ring as a result of a direct attack of the C-2 amino group at C-6. It took, however, 3 days for complete conversion, since it was difficult to keep the pH from rising higher than 8. The product obtained had the D-manno-configuration, and could be converted into the pipecolic acid **35** [60a].

If, on the other hand, **34** was treated with an excess of aqueous potassium hydroxide the pyrrolidine derivative **36** (Scheme 9) was formed immediately. Conversion of the carboxylic group to the primary alcohol gave **37**, the structure of which was finally proven to be 2,5-dideoxy-2,5-imino-D-glucitol (**37**) by comparison with literature data. Crystalline **37** can be obtained in gram amounts from dibromomannolactone using cheap reagents and no protection groups, as shown in Scheme 9. The reaction proceeds via the 2-amino-5,6-epoxide **A** which, by an internal nucleophilic *exo*-opening of the epoxide by the amino group, yields a 5-membered pyrrolidine ring. The presence of the strongly nucleophilic NH_2-group causes opening of the epoxide ring prior to any epoxide migration as discussed for epoxy aldonic acids in Sect. 3.3.

The 2,5-iminoglucitol **37** was recently prepared both by a chemoenzymatic synthesis [66] and by chemical means [67], but the synthesis just discussed is an attractive alternative for preparation of this strong α-D-glucosidase inhibitor.

The C-2 epimeric 2-amino-6-bromo-2,6-dideoxy-D-gluconolactone (**39**) was treated similarly with strong aqueous potassium hydroxide to give the C-2 epimeric pyrrolidine carboxylic acid **40**, which was converted into the 2,5-dideoxy-2,5-imino-iditol (**42b**). This compound can be obtained from the cheap D-mannitol [68, 69]. Our approach gives the intermediate carboxy amino acid **40**, which can be protected easily as the mono-isopropylidene ethyl ester **41** [52]. Reduction of the ester gives the selectively protected iminoiditol **42a**, not otherwise obtainable from this C-2 symmetric molecule. Thus the differentiation of the two primary hydroxy groups is possible. Further studies in this context are in progress. The iminoiditol **42b** having C-2 symmetry [70] has also attracted interest as a chiral auxiliary in asymmetric synthesis [69, 71].

4.2.2
Synthesis of Hydroxy/Amino Substituted Piperidines

4.2.2.1
Reaction of Dibromodideoxyheptonolactones/Heptitols with Ammonia

2,7-Dibromo-2,7-dideoxy-heptonolactones are reognized as synthons for the preparation of hydroxylated piperidines, since they have one more carbon between the two leaving groups than the dibromohexonolactones, which yielded pyrrolidines when reacted with ammonia. Thus, if these lactones (Table 1), or their corresponding dibromoheptitols, were treated with either aqueous or pure liquid ammonia, hydroxylated piperidines were obtained. The reaction between the bromoheptonolactone **43** having D-*glycero*-D-*ido*-configuration, and ammonia is shown in Scheme 10. The reaction proceeds, as for the bromohexonolactones, via a diepoxy carboxamide, which is opened at the primary position by NH_3 to give the intermediate **A**. An intramolecular *exo* opening of the 2,3-epoxide by the primary amino group yielded the piperidine **44** [23]. The other two isomeric dibromoheptonolactones mentioned in Table 1 were reduced with sodium borohydride to the 2,7-dibromo-2,7-dideoxyheptitol, prior to treatment with aqueous ammonia. Thereby the crystalline piperidines **46** and **47** were obtained [23].

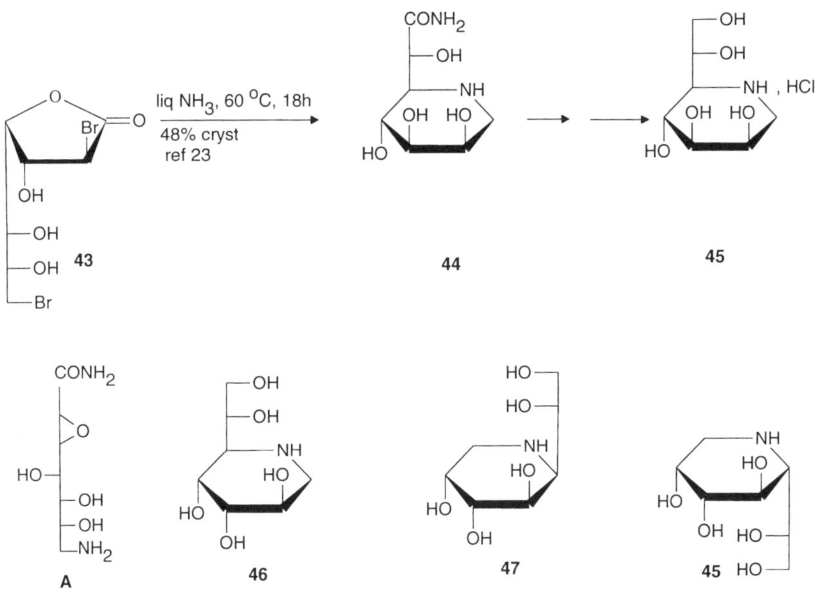

Scheme 10

4.2.2.2
Reaction of C-5 Substituted Pentonolactones with Ammonia

Piperidines might also be easily obtained by reduction of δ-lactams. Aldonolactams having ring sizes of five, six, or seven atoms have previously been prepared by displacement of ω-mesyloxy aldonolactones with sodium azide. Hydrogenation of the ω-azido group to the ω-amino derivatives was followed by ring enlargement of to give the sugar lactams [72].

We found that 5-bromo-5-deoxy- or 5-*O*-tosylated/mesylated pentonolactones gave 5-amino-5-deoxy-pentonolactams when treated with aqueous ammonia. As an example, the reaction of 5-bromo-5-deoxy-D-arabinonolactone with ammonia to give the lactam **48a**, isolated as the crystalline isopropylidene derivative **48b**, is shown in Scheme 11. The reaction proceeds via a 5,6-epoxy-carboxamide **A**, which is observed in ^{13}C NMR spectra of the reaction mixture. Opening of the epoxide to give the 5-amino carboxamide **B**, followed by internal lactamization, gave the product **48**. Reduction of the lactam function using sodium borohydride/trifluoro acetic acid yielded the trihydroxylated piperidine **49** with *arabino*-configuration [73]. The 1,5 dideoxy-1,5-imino-D-arabinitol (**49**) has recently been synthesized using 5-azido-5-deoxy-D-arabinofuranose as the key intermediate [74].

Similarly, we have found that treatment of 5-bromo-2,5-dideoxy- and 5-bromo-2,3,5-trideoxy-pentono-1,4-lactones with aqueous ammonia yielded lactams. As an example, **7** and **5**, readily available from dibromolactones as shown in Scheme 2, gave the lactams **50** and **51**, respectively (Scheme 11). These might also be reduced to the corresponding hydroxylated piperidines.

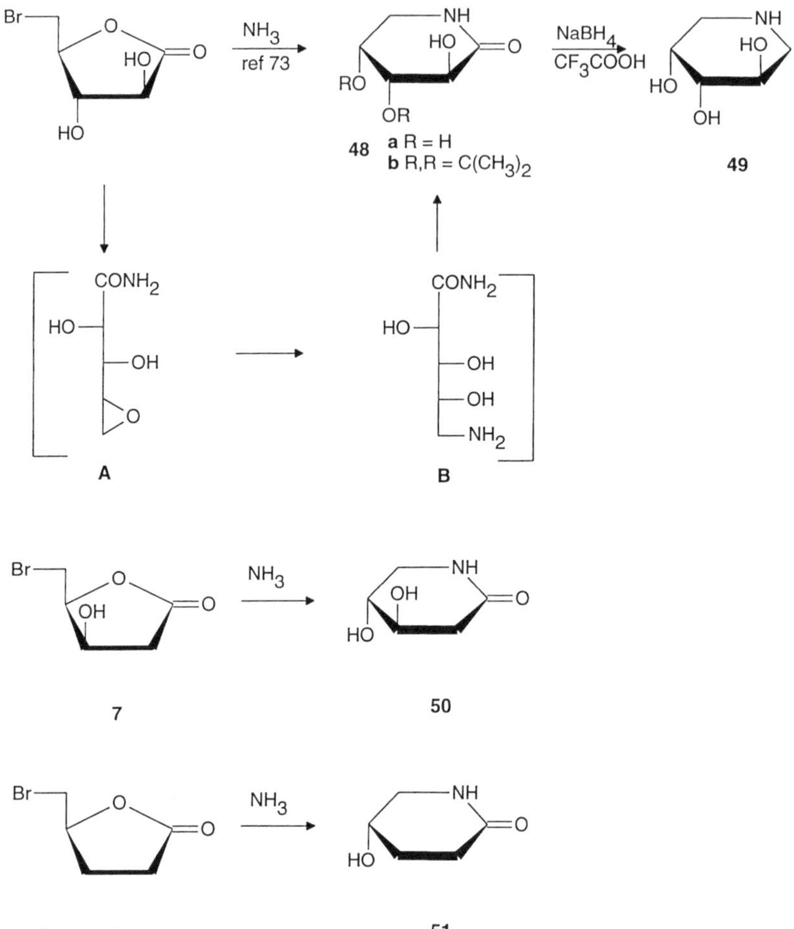

Scheme 11

This very simple procedure for synthesizing 1,5-dideoxy-1,5-iminoalditols [73] is further illustrated in Scheme 12 for the preparation of the iminoribitol **54** and the imino-L-lyxitol **58** [75]. They were both synthesized from the same starting lactone, the 5-*O*-mesyl-2,3-*O*-isopropylidene-D-ribonolactone (**52**). Reaction of **52** with ammonia gave the lactam **53**, which was reduced as discussed above and worked up further to give the ribitol derivative **54** [73]. Stereoselective inversion at C-4 of the mesylated ribonolactone **52** was achieved by treatment with strong base, as discussed in Sect. 3.3. By keeping the pH not lower than 3 during the work up the 2,3-*O*-isopropylidene-L-lyxono-1,4-lactone (**55**) could be isolated. Mesylation to give **56**, followed by treatment with ammonia gave the lactam **57**, which was elaborated further to give the 1,5-imino-L-lyxitol (**58**) [73]. The 1,5-imino-xylitol [76], the last of the four stereoisomeric 1,5-iminopentitols, was prepared similarly [73].

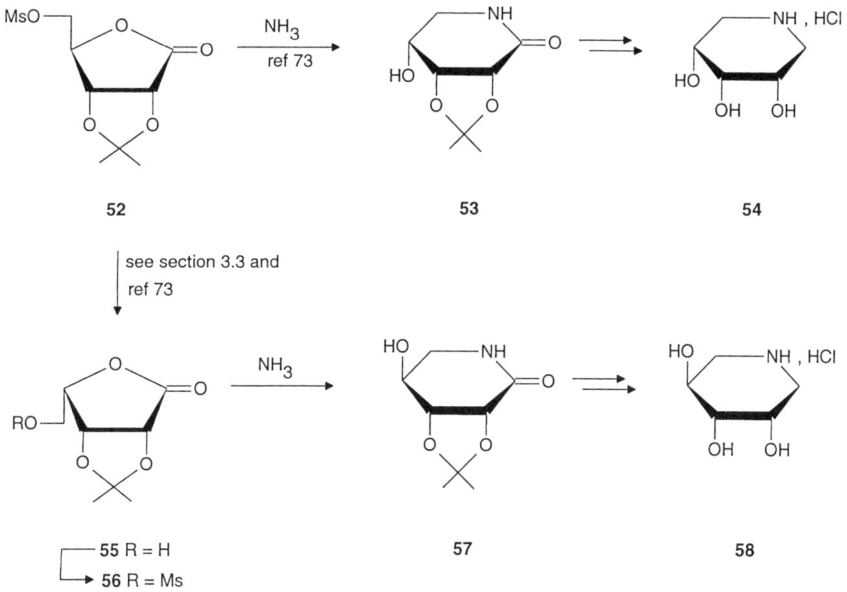

MsO O O NH3 ref 73 NH O HO O O NH , HCl HO OH OH

52 53 54

see section 3.3 and ref 73

O O RO O O NH3 HO NH O O O HO NH , HCl OH OH

55 R = H
56 R = Ms
57
58

Scheme 12

4.2.2.3
Reaction of 2,5-Disubstituted Pentonolactones with Ammonia

Following the idea to synthesize δ-lactams from pentonolactones activated at C-5, we undertook the investigation of the reaction between ammonia and 2,5-dibromo-2,5-dideoxy-pentono-1,4-lactones. It was found that 2-amino-2-deoxy-1,5-lactams were formed. Thus, the reaction of α,ω-dibromo-α,ω-dideoxy-pentono-1,4-lactones with ammonia took another pathway compared to the corresponding hexono-1,4-lactones. In the latter case only one nitrogen atom was incorporated in the products giving pyrrolidines, as discussed in Sect. 4.2.1.1, whereas the dibromopentonolactones were substituted with nitrogen at the carbon atoms originally having the leaving groups. Thus, the 2,5-dibromo-2,5-dideoxy-D-xylono-1,4-lactone (**6**) gave 2,5-diamino-2,5-dideoxy-D-xylonolactam (**60**) (Scheme 13) [77]. Nitrogen had been introduced at C-5 and C-2, the latter with retention of configuration. Again the ^{13}C NMR spectra of the reaction mixture revealed the mechanism, showing that a 2,3;4,5-diepoxide was an intermediate. Opening at the primary carbon to give the primary amine, followed by lactamization gave the 2,3-epoxylactam **59** in the course of around 1 h. Opening of this *cis* epoxide yielded the 2-aminolactam **60** [77]. This mechanism was confirmed by treatment of the crystalline 2,3-epoxide derived from **6** [38] with ammonia to give the epoxylactam **59**. The 2,3-epoxy-lactam **59** has now been isolated in a crystalline state [58b].

Opening of the cyclic 2,3-*cis*-epoxy carboxamide with ammonia at C-2 was in contradiction to opening of acyclic 2,3-epoxy carboxamides, since 3-amino-3-deoxy-aldonic acids were obtained in the latter cases [77]. Attack at C-3 would be in accordance with the findings in simple 2,3-epoxy carboxamides [78].

The question then arises if a regioselective opening of a 2,3-*trans*-epoxy carboxamide derived from aldonic acids would occur. The 2,5-di-*O*-tosyl-D-ribono-1,4-lactone (**62**) (Scheme 13) was used to find an answer to this question. If treated with ammonia, the 2,5-diamino-2,5-dideoxy-D-ribono-1,5-lactam (**63**) was obtained as the only product [79]. The ^{13}C NMR spectra of the reaction mixture showed the formation of the diepoxy amide **A** which was opened at C-5 by ammonia. In this case no internal lactamization could occur, due to the *trans* 2,3-epoxy group in **B** (Scheme 13). Thus, a regioselective opening of an acyclic 2,3-epoxy carboxamide took place at C-2. The reaction was complete within 6 days.

By studying the nucleophilic opening of 2,3-epoxy carboxylic acid derivatives it has been found that the regioselectivity was dependent on the acid function (ester, amide, carboxylate), as well as on the nucleophile [78]. In the examples

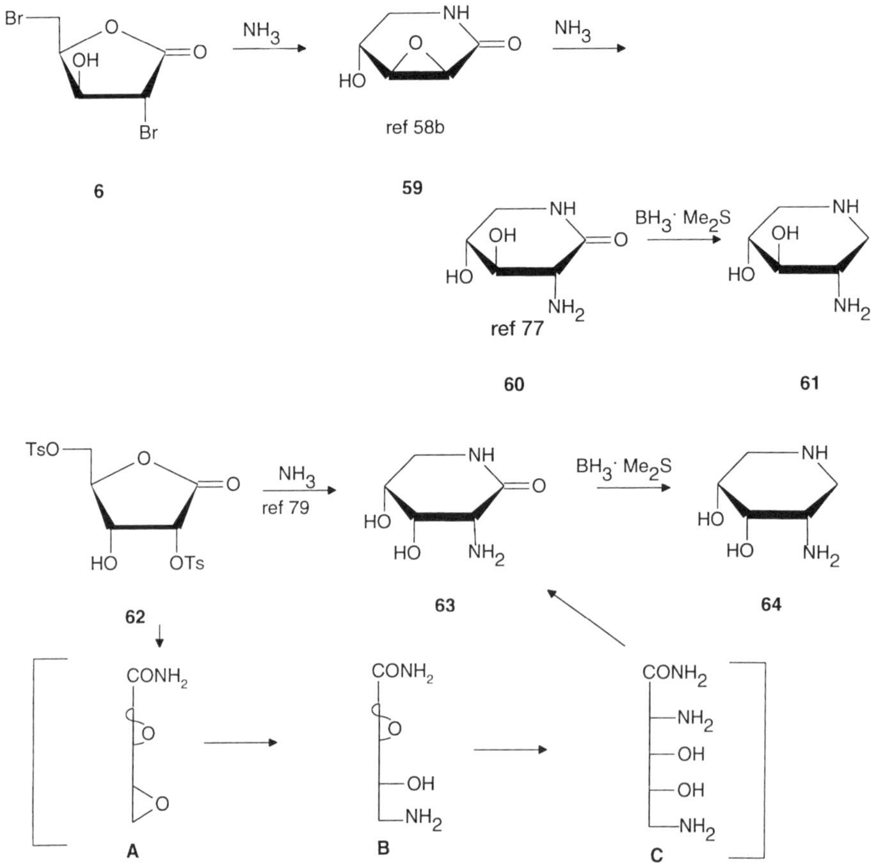

Br OH O O Br **6** NH_3 → NH O O HO ref 58b **59** NH_3 →

NH OH O HO NH_2 ref 77 **60** $BH_3 \cdot Me_2S$ → NH OH HO NH_2 **61**

TsO O O HO OTs **62** NH_3 ref 79 → NH O HO HO NH_2 **63** $BH_3 \cdot Me_2S$ → NH HO HO NH_2 **64**

$CONH_2$ O O **A** → $CONH_2$ O OH NH_2 **B** → $CONH_2$ NH_2 OH OH NH_2 **C**

Scheme 13

just discussed, the differences of the two substrates are the configurations around the epoxide and, perhaps more importantly, cyclic contra acyclic amide structures. Thus, steric and electronic effects may also be taken into consideration in opening of the epoxy function.

The two lactams obtained, **60** and **63**, were silylated in situ followed by borane reduction to give the 2-amino-1,2,5-trideoxy-1,5-imino-D-xylitol (**61**), and 2-amino-1,2,5-trideoxy-1,5-imino-D-ribitol (**64**), respectively [79].

4.2.3
Biochemical Results

Investigation of the inhibitory effects on glycosidases has been carried out for the hydroxylated pyrrolidines [65] and piperidines [73]. Among the pyrrolidines prepared only the 1,4-dideoxy-1,4-imino-L-allitol (**23**) (Scheme 7) and its C-5 isomer showed any remarkable effects. They inhibited lysosomal α-mannosidase rather than the processing α-mannosidases I and II [65], and their specificity is in accord with the structural requirements of azafuranose analogues of mannose for inhibiting mammalian α-D-mannosidases [80].

Among the azapyranoses prepared, the 1,5-dideoxy-1,5-imino-xylitol exhibit strong inhibition towards β-glucosidase whereas the corresponding D-arabino isomer (**49**) (Scheme 11) was a strong α-L-fucosidase inhibitor [73] as also shown recently [74]. Interestingly, the three iminoheptitols **45**–**47** (Scheme 10) also strongly inhibited α-L-fucosidase from human liver [73]. These compounds all have the same absolute stereochemistry at the hydroxy bearing carbon atoms within the ring and in this respect are identical to L-fucose. They thus fulfil the minimal structural motif necessary for the inhibition of α-L-fucosidase [73,74,81].

Preliminary investigations show that the 2-amino-1,2,5-trideoxy-D-ribitol (**64**) (Scheme 13) is a strong β-D-glucosidase inhibitor [79]. These results, together with the observation that iminosugars, in which the nitrogen atom displaces the oxygen in anomeric position [58, 59], were strong β-D-glucosidase inhibitors show that further synthetic work is needed in order to obtain more specific and powerful glycosidase inhibitors.

4.3
Hydroxylated Cyclopentanes

4.3.1
General

Highly functionalized cyclopentanes as well as cyclohexanes are structural features found in biologically interesting compounds. The terms "pseudosugar" or "carbasugar" have often been used to describe these structures, since they belong to the class of compounds in which the ring oxygen in either a furanose or a pyranose has been replaced by a methylene group. The lack of the acetal moiety in these compounds preserves them from hydrolysis. This is the important feature regarding their resistance in biological systems, as it was for the hydroxylated pyrrolidines and piperidines discussed above. Like those, the carbasugars and related compounds may act as, for example, enzyme inhibitors or antibiotics. Considerable efforts have been dedicated to development of efficient syntheses of these compounds [82]. The starting materials may be noncarbohydrate precursors such as symmetric bornene derivatives, among which the 7-oxanorbornenic system has been the most popular, thereby yielding racemic carbapyranoses [83–86]. By resolving either the starting materials [83, 86], or at an early stage in the synthesis [85], chiral compounds can be obtained. Stereoselective transformations of such "naked sugars" have proven them to be versatile chiral synthons [85]. From optically active norbornene derivatives, enantiopure carbapentofuranoses have recently been prepared [86]. By use of carbohydrate precursors optically pure compounds are obtained, as shown in the synthesis of carba-α-D-arabinofuranose from D-arabinose [87].

The importance of the carbafuranoses stems from the interest in having access to carbocyclic analogues of nucleosides [88], which have attracted particular attention as anti-tumor and anti-viral agents, (–)-carbovir being one of the more popular synthetic target molecules [89]. (Fig. 1)

Since the isolation of the chitinase inhibitor allosamidin from *Streptomyces* [90], the aminohydroxy-substituted cyclopentanes have been recognized as powerful and specific inhibitors of glycosidases [91]. The synthesis of (+) [92], as well as of racemic [93] mannostatin, which is a strong mannosidase inhibitor, should be mentioned here (Fig. 1).

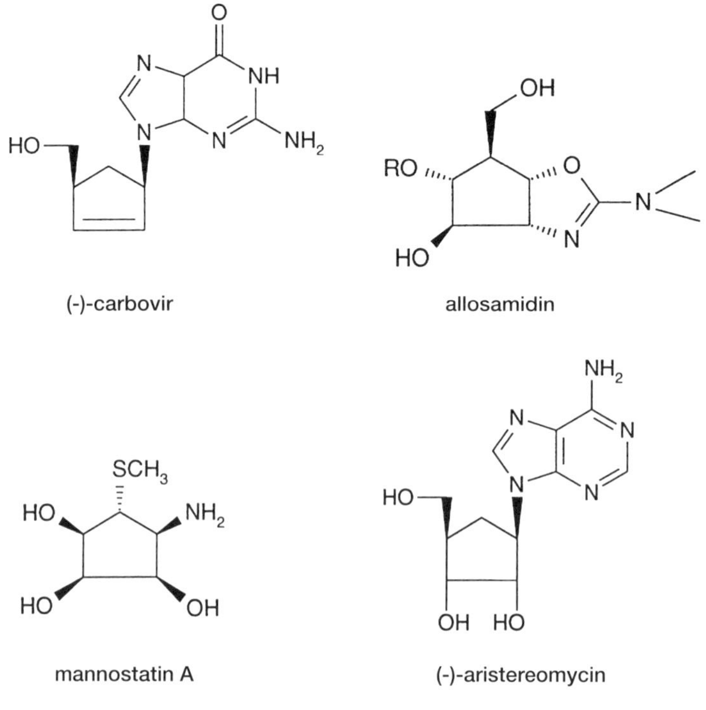

(-)-carbovir

allosamidin

mannostatin A

(-)-aristereomycin

Fig. 1

Stereocontrol in the synthesis of such compounds is quite challenging and the use of synthons from the chiral pool in stereocontrolled transformations might be an attractive alternative to be considered, as shown recently [94].

The easy access to activated aldonolactones, as described above, led us to envisage the possibility of using ω-bromo-ω-deoxy-2,3-unsaturated aldono-1,4-lactones (Sect. 3.4) as precursors for highly functionalized cyclopentanes.

4.3.2
Radical Induced Cyclization

The conversion of carbohydrate derivatives into functionalized cyclohexanes and cyclopentanes has recently been reviewed [95]. The key step is the formation of carbon-carbon bonds, and different approaches have been used for this purpose. Radical reactions have in the last decade been recognized as valuable in this context [96] since the regio- and stereocontrol may frequently be predictable [97].

Our approach was to use the unsaturated bromodeoxylactones in an intramolecular radical reaction, since these compounds possess both the radical precursor and the radical trap within the same molecule. Thus, reacting the unsaturated bromodeoxyheptonolactone **20** (Scheme 14) with tributyltin hydride and a radical initiator, the bicyclic lactone **65a** was obtained in a quantitative yield within 1 h. The stereocontrol in the reaction was determined by the structure of the product, since the compound obtained has two fused cyclopentane rings which can only be *cis* anellated. The radical **A**, which is the intermediate, was trapped by the tin hydride. The stereochemistry of the newly formed chiral center is determined by the configuration at C-4 in the educt **20** [45].

As discussed in Sect. 3.4, the synthon **20** was prepared from the dibromoheptonolactone, which in turn was obtained from the cheap commercially available D-*glycero*-D-*gulo*-heptono-1,4-lactone (Table 1). The other isomeric dibromoheptonolactones shown in Table 1, which were prepared from the heptonates, obtained from chain extension of D-mannose and D-galactose, respectively, were also converted into unsaturated bromodeoxyheptonolactones. Finally, we obtained 2-*O*-acetyl-7-bromo-3,7-dideoxy-D-*xylo*-hept-2-enono-1,4-lactone and the corresponding D-*lyxo*-isomer by the Kiliani extension of D-gulose. These substrates were all cyclized to cyclopentane lactones, stereoisomers of **65** [98].

The ring closure of the 2-acetoxy-substituted 2,3-unsaturated lactone **66** (Scheme 14) proceeded as smoothly as for the lactones having an unsubstituted double bond, giving also only one product. In this case, however, *two* new chiral centers have been generated. The stereochemistry at the first center is obvious as being determined by the configuration at C-4 of **66**. Upon formation of the bicyclic structure, the molecule adopts a cupped shape and the radical will be trapped from the less hindered side, which is *anti* to the newly formed cyclopentane ring, to give **67** only [45].

The radical cyclization of an unsaturated lactone having a nitrogen substituent at C-2 was investigated. This study revealed that the radical cyclization proceeded smothly only when the amino group was converted into a trifluoracetamido group. The rationale behind this conversion was to make the nitrogen

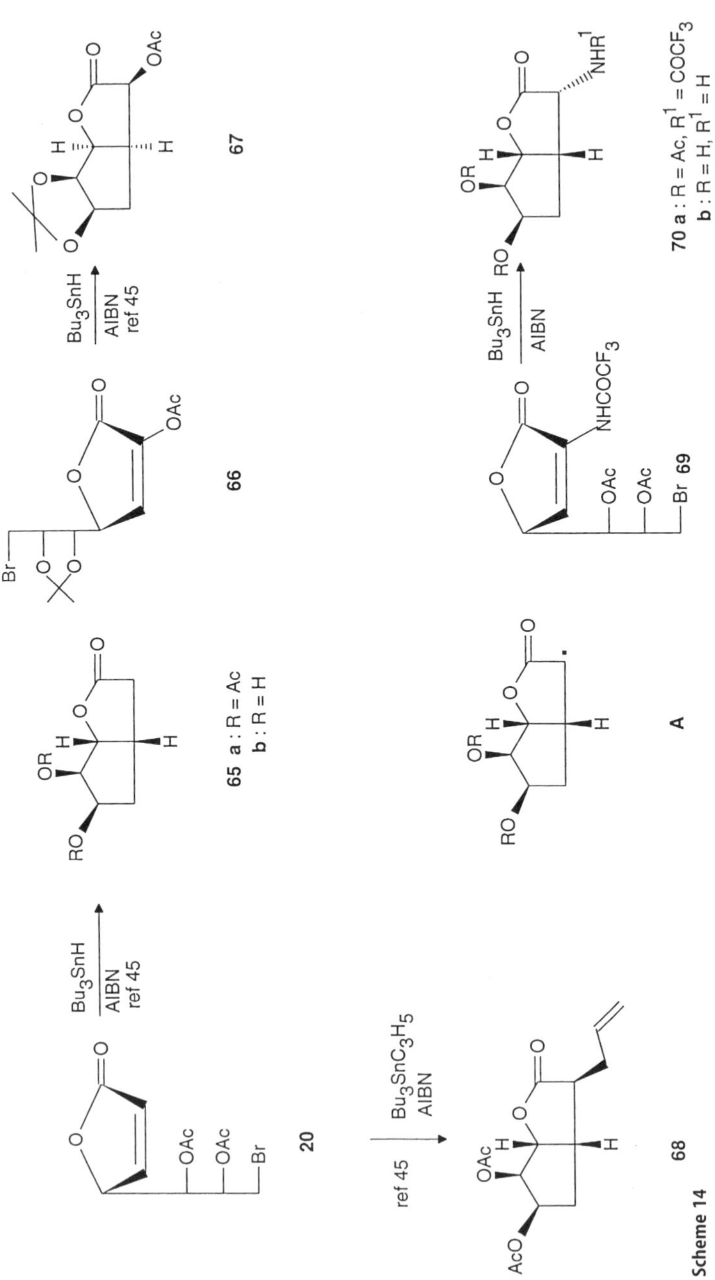

20 → (Bu_3SnH, AIBN, ref 45) → 65 a : R = Ac; b : R = H

66 → (Bu_3SnH, AIBN, ref 45) → 67

20 → ($Bu_3SnC_3H_5$, AIBN, ref 45) → 68

A

69 → (Bu_3SnH, AIBN) → 70 a : R = Ac, $R^1 = COCF_3$; b : R = H, R^1 = H

Scheme 14

substituent less electron donating, thereby changing the electron density of the double bond in order to accelerate the cyclization. Thus, the trifluoracetamide **69** (Scheme 14) gave **70a** with a small amount (6%) of the C-2 epimer. The main product could be crystallized directly in 76% yield [98a].

In a study of C-allylation of 2-bromo-2-deoxy-pentonolactones using allyl-tributylstannane and a radical initiator, mixtures of C-2-allylated pentono-1,4-lactones were obtained [99]. This may indicate that the selectivity in the reactions discussed above is determined by the cyclopentane ring.

We also tried to trap the presumed intermediate radical **A** using acrylonitrile, butyl vinyl ether or allyltributylstannane as the trapping reagents. In the first two cases mixtures of the 2-deoxy-lactone **65** and the C-2-alkylated product were obtained. Treatment of **20** with allyltributylstannane gave **68** as a homogeneous product (Scheme 14) in a fully stereocontrolled reaction [45, 98b].

The potential of using free radical reactions as synthetic methods will be increased when the origin of selectivity, especially in acyclic systems, is better understood [100, 101].

4.3.3
Carbasugars

The bicyclic lactones produced by the radical cyclizations discussed above were transformed into a series of carbafuranoses by reduction of the lactone moiety to the corresponding alcohol [98]. Sodium borohydride and borane-dimethyl sulfide were used as complementary reducing reagents for this purpose. Sodium borohydride is a widely used reagent for the reduction of aldonolactones to alditols in aqueous solution, as also discussed in Sect. 4.2 [20]. A prerequisite for this reduction is the activation of the carbonyl group by an electron withdrawing substituent in the α-position. Reductions using borane-dimethyl sulfide are normally performed in tetrahydrofuran or dioxan, thus making the solubility of the lactone a limiting factor. The lactone **65b** (Scheme 14) and isomers thereof, possessing no α-substituent to the lactone function, were reduced with the borane complex. Thus, **65b** gave the polyhydroxylated cyclopentane **71** (1*R*,2*R*,3*R*,4*R*)-4-(2-hydroxyethyl)-cyclopentane-1,2,3-triol (Scheme 15), which may be regarded as a carbocyclic analogue of 5-deoxy-α-L-*xylo*-hexofuranose [45, 98b]. This is the C-3 epimer of an already known cyclopentanetetraol, prepared in connection with studies on the biosynthesis of aristereomycin (see Fig. 1). The cyclopentanetetraol was the major product formed in a radical cyclization of an acyclic unsaturated bromoester [102], thus showing, that acyclic precursors do not cyclize with full stereocontrol [95, 102].

The 2-oxygenated lactone **67** was reduced by means of sodium borohydride to give the protected carbasugar **72**, which by hydrolysis yielded carba-β-D-mannofuranose (**73**). Similarly, the 2-amino substituted lactone **70b** was reduced to carba-5-amino-5-deoxy-α-L-glucofuranose (**74**) [98a] (Scheme 15).

Carbapentofuranoses could be obtained by periodate cleavage of an exocyclic diol in the isopropylidene protected carba hexofuranoses, as illustrated for **72**. The protected carba pentofuranose **75** was obtained and deprotected to give carba-β-D-lyxofuranose (**76**).

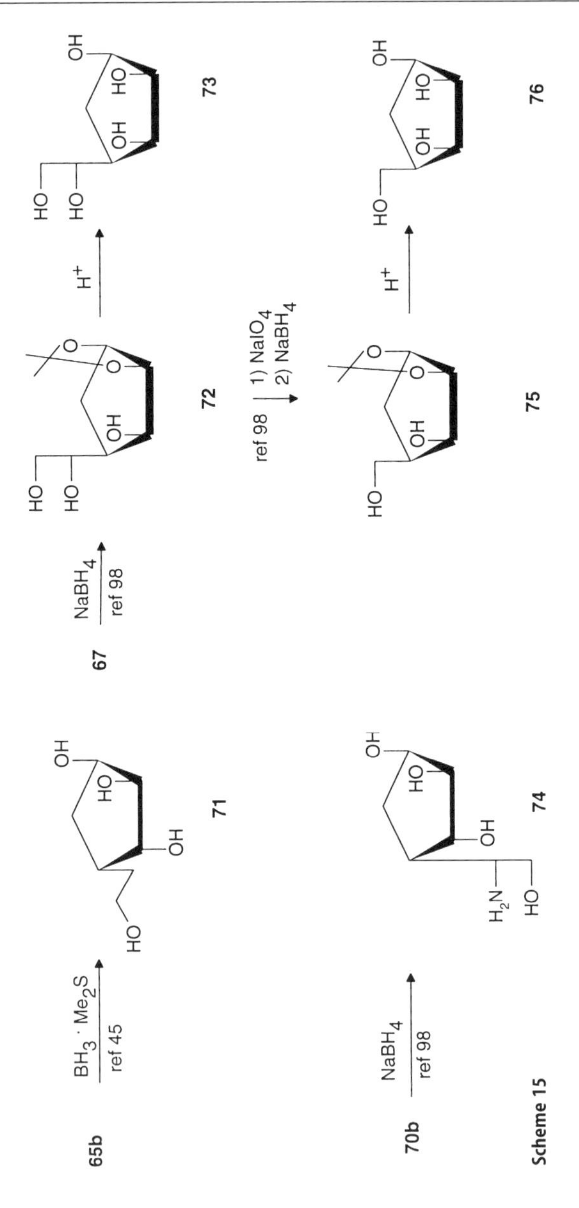

Scheme 15

The strategy for the synthesis of carba-hexofuranoses and -pentofuranoses is summarized in Scheme 16. Five new carbaanalogues of hexofuranoses were synthesized: **74, 73** and the enantiomer α-L-mannofuranose, and the α-L and β-D-glucofuranoses. The carba-analogues of the 5-deoxyhexofuranoses with α-L- and β-D-*lyxo*-, and α-L-*xylo*-configurations, which we have also prepared [98b], have recently been described, either with the same configuration, or as the enantiomeric compound, or as a racemic mixture [86]. Several of the hitherto known compounds were now obtained in a crystalline state

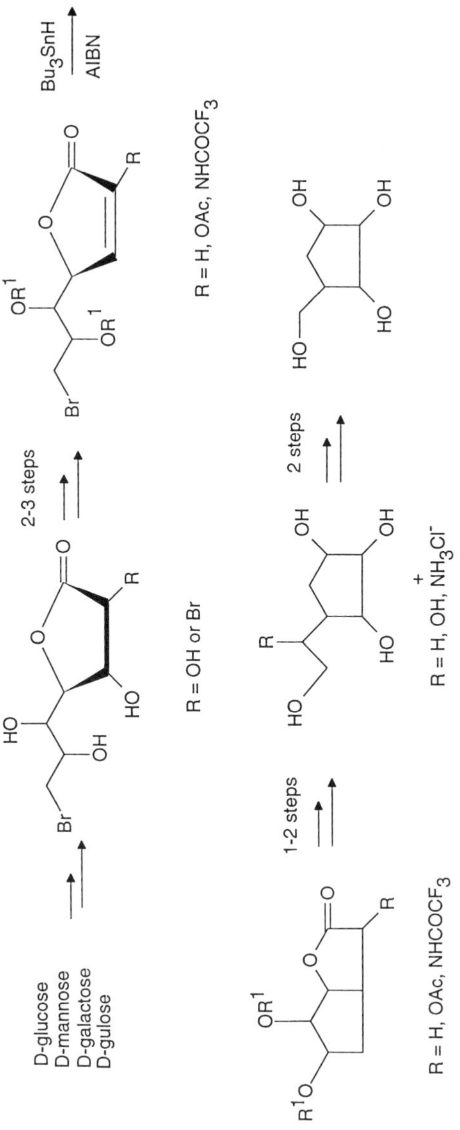

Scheme 16

and with higher numerical values for the optical rotations [98b], as compared to the reported non-crystalline compounds [86]. An explanation for the difference in optically purity is found in the starting material for the synthesis. (+)-Norborn-5-en-2-on, obtained from enzymatic resolution of *endo*-norborneyl acetate, with an e.e. of 86% was used [86] whereas we have used optically pure carbohydrate precursors. The limitation of our method is, however, the presently limited availability of stereoisomeric heptonolactone precursors.

4.3.4
Hydroxy/Amino Substituted Cyclopentanes

The easily available bicyclic cyclopentane **65** prepared by a free radical 5-*exo*-trig-C–C cyclization of the unsaturated bromoheptonolactone **20**, as outlined above, is envisioned as a synthon for stereoselective transformations of the substitution pattern in the cyclopentane ring. Treatment of **65b** with hydrogen bromide in acetic acid gave the bromoacetate **77a** as the only product (Scheme 17). The crystalline bromohydrin **77b** was treated with potassium fluoride in acetone, following the procedures discussed in Sect. 3.2 [19] to give the crystalline epoxide **78**. We have studied opening of this epoxide with various nitrogen

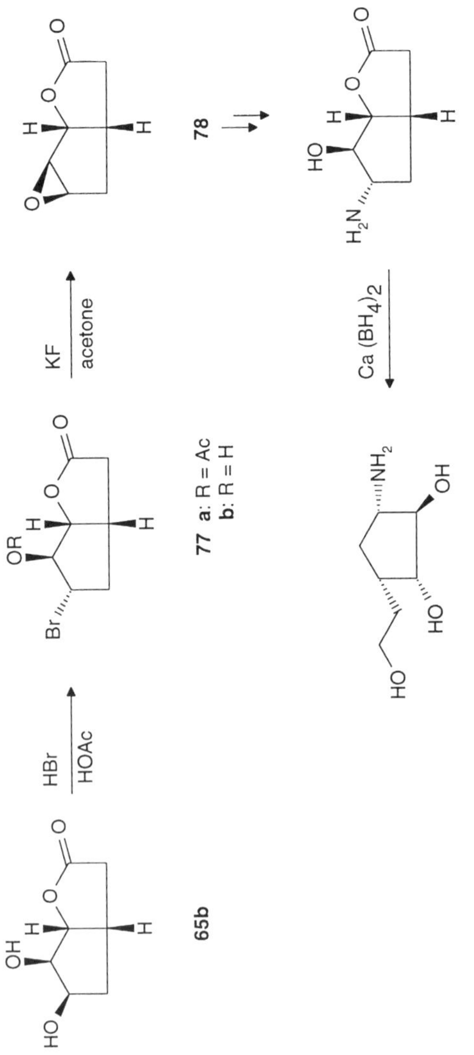

Scheme 17

nucleophiles to give the amino hydroxy cyclopentane **79**. The nitrogen was always introduced stereospecifically as shown in **79**. Thus, the regioselectivity observed in the reactions leading to both **77** and **79** may be induced by the anellated lactone ring. Reduction of the lactone, as discussed for preparation of the carbasugars above, gave the amino hydroxy substituted cyclopentane **80**. This may be regarded as the carba-analogue of 1-amino-1,5-dideoxy-β-L-*xylo*-hexofuranose. Further work along this line is in progress.

Other approaches to synthesize highly substituted cyclopentanes, including amino and hydroxy groups, from γ- and δ-lactones, has been radical and anionic Michael cyclizations of the α-iodo-γ- and -δ-lactones [94]. Likewise, methods using radical cyclization to oxime ethers have been reported to give amino substituted cyclopentanes [95, 103, 104]. It should be noted that although only one isomer is often obtained [103], such cyclizations generating a secondary radical may not be stereospecific [95, 100, 101].

4.4 Hydroxylated Tetrahydrofurans

4.4.1 *General*

Highly substituted tetrahydrofurans are structures found in biologically active compounds in a wide range of complexity, and a variety of approaches have been developed to synthesize such structures [105–107]. Tetrahydrofurans having alkyl substituents at both C-2 and C-5 are important compounds as precursors for C-nucleoside antibiotics and more complex natural products [85]. One of the most applied methods to prepare the furan moity is the ring closure by attack of an OH- or OR-group on an electrophilically activated double bond [108], a method which has been used in only few cases within the carbohydrates [109]. Recently the successful ring closure via a radical-initiated cyclization of β-alkoxyacrylates, in which the functionalized alkoxy group was prepared from tartaric acids, was reported to give stereoselectively 2,5-*cis*-substituted tetrahydrofurans [110], which may be regarded as C-furanosides.

Intramolecular displacement by a benzyloxy group to form a tetrahydrofuran ring with concomitant debenzylation has been recognized as an undesired side reaction [111, 112] but is also useful in stereocontrolled formation of poly-substituted tetrahydrofurans [111, 113].

Ring contraction of readily available sugar derivatives has been used in synthetic approaches towards C-nucleosides, and the first ring contraction under acidic conditions of benzylated 2-bromo-2-deoxy-1,5-lactones to give 2,5-disubstituted tetrahydrofurans with complete invertion at C-2, has been reported [114]. The two separable bromolactones used were prepared from tri-*O*-acetyl-D-glucal.

A wide range of acetal-protected aldono-1,5-lactones ready for preparing C-2 activated aldonolactones has been published [115] and used for preparation of highly functionalized tetrahydrofurans. The ring contractions were initiated either by methoxide [116] or by acid [117], and clean inversion at the carbon bearing the leaving group was observed. Thus, these methods using simple protecting group strategies provide access to substituted tetrahydrofurans.

4.4.2
Tetrahydrofurans from Bromodeoxy/Tosyloxy Substituted Aldonolactones

The selectively activated aldonolactones (Sect. 2) which were prepared without any use of protecting groups, have been used as synthons for simple transformation into chiral, hydroxylated tetrahydrofurans, as exemplified in Scheme 18. The 6-bromo-6-deoxyhexonolactones with D-galacto- (**77**) and D-manno- (**80**) configurations, gave the 3,6-anhydrides **78** and **81**, respectively, when boiled in water for a couple of hours. In the case of the galacto-isomer the lactone ring must open prior to formation of the ether **78**, whereas the manno-isomer, having the OH-3 *cis* to the side chain, can form an anhydride with preservation of the lactone ring. This gave the bicyclic compound **81** [16]. Conversion of the carboxyl group of **78** to a primary alcohol gave the optically active 1,4-anhydro-D-galactitol, a compound not otherwise obtainable by simple acid treatment of galactitol, since it is a meso form [118]. The 6-bromo-6-deoxy hexonolactones were reduced to the corresponding 6-bromo-6-deoxy-hexitols which, by boiling in water, similarly gave 3,6-anhydrides [16]. In this way a number of 2-substituted hydroxylated tetrahydrofurans were prepared in high yields [16]. Similar 3,6-anhydro-2-bromo-2-deoxy-hexono-1,4-lactones have been prepared from 2,6-dibromo-2,6-dideoxy-hexonolactones by boiling in *tert*-amylalcohol and simultaneously distilling off volatile materials [7].

Along the same line, we have investigated the behaviour of the di-*O*-tosylated-aldonolactones, discussed in Sect. 2.2. Due to lower solubility compared to the bromolactones, a water-dioxane mixture was used. When boiled in this medium, the ditosylated mannonolactone **84** (Scheme 18) gave the 2-*O*-tosylated-3,6-anhydride **85** [119]. Similarly, the di-*O*-tosylated-D-talonolactone (**82**) gave the 3,6-anhydride **83** when boiled in water-dioxane. It is worth noting that a 3,6-anhydride, and not a 2,5-anhydride, was formed. In both cases substitution at the primary carbon occurred, and no further reaction to dianhydrides was observed. The two new compounds may be interesting chirons since, besides the tetrahydrofuran ring and a carboxylic acid function, they also have a leaving group α to the lactone/acid function.

If the 2-*O*-tosylated L-rhamnono-1,4-lactone (**88**) was boiled in water-dioxane the 2,5-anhydride (**89**) was obtained in good yield. The reaction time was 16 h in this case, compared to 2–4 h in the examples in which a primary tosyloxy group was substituted. The 2,5-anhydro-carboxylic acid **89** was isolated as the methyl ester, and was shown to have the L-*gluco*-configuration [117]. Thus, a clean inversion at C-2 had taken place.

From the commercially available D-*glycero*-D-*gulo*-heptono-1,4-lactone the 2,7-di-*O*-tosyl-D-*glycero*-D-*gulo*-heptono-1,4-lactone can be prepared by selective tosylation (Sect. 2.2). By reaction of the lactone with HBr–HOAc, the

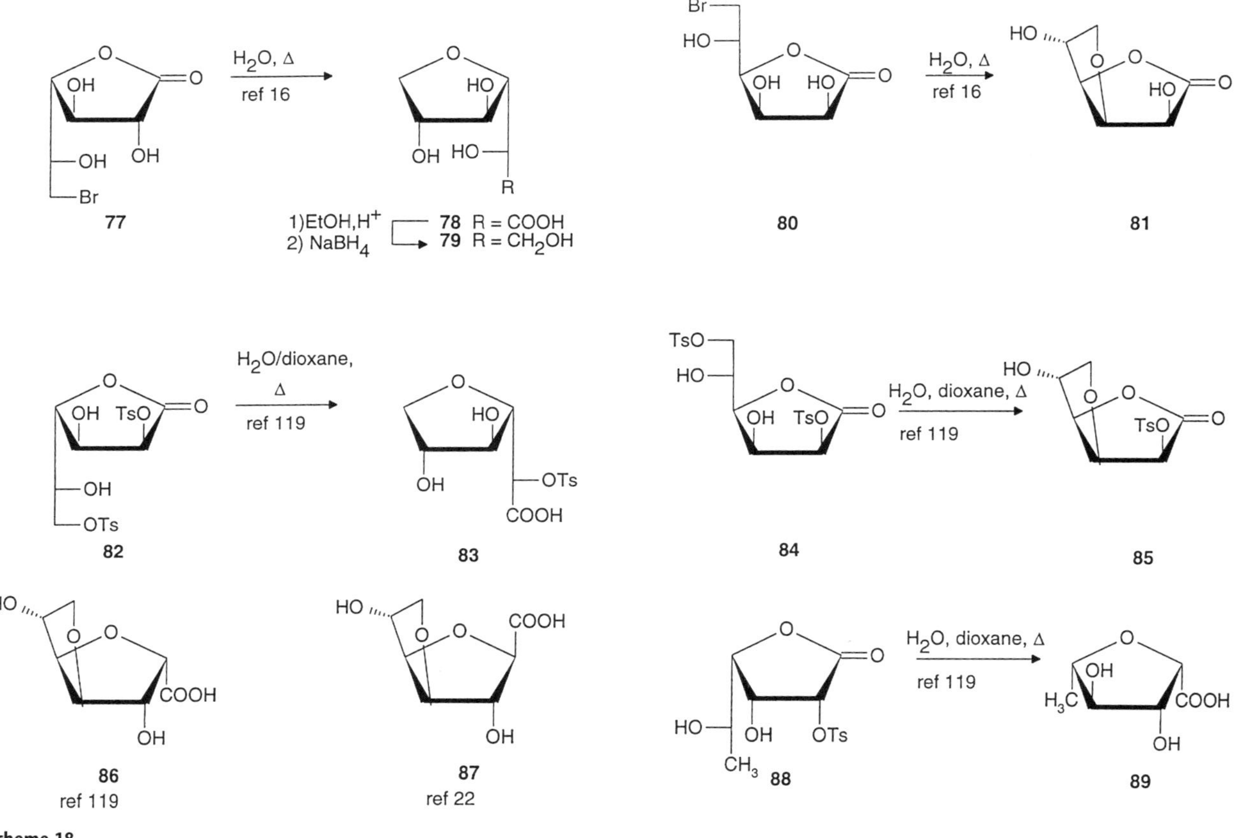
77
H_2O, Δ
ref 16
1)EtOH,H+
2) $NaBH_4$
78 R = COOH
79 R = CH_2OH
80
H_2O, Δ
ref 16
81
82
H_2O/dioxane,
Δ
ref 119
83
84
H_2O, dioxane, Δ
ref 119
85
86
ref 119
87
ref 22
88
H_2O, dioxane, Δ
ref 119
89

Scheme 18

2,7-dibromo-2,7-dideoxy-D-*glycero*-D-*ido*-heptono-1,4-lactone can be prepared (Sect. 2.1.1). Thus two 2,7-diactivated heptonolactones, epimeric at C-2 were available. If the ditosylate was boiled in water-dioxane for 1 h, the 2,5;4,7-dianhydride **86** with D-*glycero*-D-*ido*-configuration was formed [119]. If the dibromolactone was boiled in water for the same time the dianhydride **87** was formed [22]. This dianhydride was isomeric with **86**, and thus both reactions proceeded with inversion at C-2. Thus, internal substitution of a secondary tosyloxy group to form a 2,5-anhydride will be more dependent on the conformation of the substrate giving a favorable transition state, rather than on secondary vs. primary leaving groups.

2-Substituted tetrahydrofurans can be prepared readily by boiling bromodeoxypentonolactones in water. A few exampes are shown in Scheme 19. 5-Bromo-5-deoxy-D-arabinonolactone (**90**) gave exclusively the 2,5-anhydro-D-arabonic acid (**91**), proven by conversion to the known 1,4-anhydro-D-lyxitol **92**. The 2,5-anhydro-D-ribonic acid (**94**) was formed from 2-bromo-2-deoxy-D-arabinonolactone (**93**) (Scheme 19), as well as from 5-bromo-5-deoxy-D-ribonolactone (**95**) if boiled for a few hours in water. This clearly shows again that these reactions give products with inversion at the carbon bearing the leaving group [120].

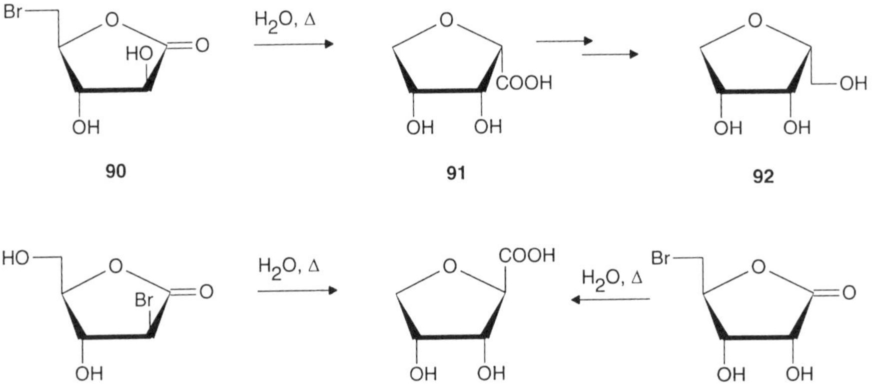

Scheme 19

In summary, we have shown that bromo- and tosyloxydeoxy-aldono-1,4-lactones can be transformed into tetrahydrofuran derivatives by boiling in water. The mechanism involves opening of the lactone ring and nucleophilic displacement by the hydroxy group in question. The reactions proceed stereospecifically with inversion at secondary carbon atoms. If 3,6-anhydrides are formed between an OH-3 and the C-6 in a lactone having these groups in a *cis* relationship, the lactone ring may be preserved. We have also shown that 6-bromo-6-deoxy-hexitols similarly give 3,6-anhydrohexitols [119].

5
Conclusion

Carbohydrate lactones have proven to be versatile starting materials for a range of complex target structures, including nitrogen heterocycles (azasugars), carbocycles (carbasugars) and oxygen heterocycles (tetrahydrofurans, C-furanosides). The selective introduction of leaving groups, followed by ring closure, induced by an appropiate nucleophile, leads to the above-mentioned compounds without any use of protective groups. The easily accessible α- and/or ω-substituted bromodeoxyaldonolactones are especially powerful synthons. Besides being substrates for preparation of the type of structures mentioned above, they can be isomerized stereoselectively at one or more carbon atoms, giving rise to otherwise less readily available polyhydroxylated chiral compounds. On the other hand, they can also be substrates for the preparation of chiral compounds having only one chiral center, since methods for selective deoxygenation are available. Thus, the possibilities seem to be manyfold.

In planning the synthesis of biologically active compounds, strategies using aldonolactones or other compounds from the "chiral pool" should therefore continue to be considered, since they can provide attractive routes in comparison with alternative methods by asymmetric synthesis.

Acknowledgements. The work described in this article has its origin in the discovery of the reaction of aldonolactones with hydrogen bromide in acetic acid to give bromodeoxyaldonolactones. This work, which is treated mainly in Sects. 2 and 3, was a result of a very fruitful and inspiring collaboration with my colleagues Christian Pedersen and Klaus Bock (now at the Carlsberg Laboratory). I acknowledge them for many stimulating discussions over the years, which resulted in a number of co-authored publications, many of which are cited in this article.

References

1. Pedersen C, Bock K, Lundt I (1978) Pure Appl Chem 50:1385
2. Bock K, Lundt I, Pedersen C (1979) Carbohydr Res 68:313
3. de Lederkremer RM, Varela O (1994) Adv Carbohydr Chem Biochem 50:125
4. Crawford TC (1981) Adv Carbohydr Chem Biochem 38:287
5. Bhat KL, Chen S-Y, Joullié MM (1985) Heterocycles 23:691
6. Fleet GWJ (1993) Sugar lactones as useful starting materials. In: Krohn K, Kirst HA, Maas H (eds) Antibiotics and antiviral compounds: chemical synthesis and modification. VCH Verlagsgesellschaft, Weinheim, p 333
7. Vekemans JAJM, Dapperens CWM, Claessen R, Koten AMJ, Godefroi EF, Chittenden GJF (1990) J Org Chem 55:5336 and references cited therein
8. Regelin H, Chittenden GJF (1989) Recl Trav Chim Pays-Bas 108:330 and references cited therein
9. Hanessian S (1983) Total synthesis of natural products. The chiron approach. Pergamon, Oxford
10. Bols M (1996) Carbohydrate Building Blocks. Wiley, New York
11. Bols M, Lundt I (1988) Acta Chem Scand B42:67
12. Bock K, Lundt I, Pedersen C (1983) Acta Chem Scand B37:341
13. Humphlett WJ (1967) Carbohydr Res 4:157
14. Bock K, Lundt I, Pedersen C (1981) Carbohydr Res 90:17
15. Bock K, Lundt I, Pedersen C (1982) Carbohydr Research 104:79
16. Lundt I, Frank H (1994) Tetrahedron 50:13285

17. Bock K, Lundt I, Pedersen C, Refn S (1986) Acta Chem Scand B40:740
18. Lundt I, Madsen R (1993) Synthesis 720
19. Lundt I, Pedersen C (1992) Synthesis 669
20. Lundt I, Madsen R (1993) Synthesis 714
21. Bock K, Lundt I, Pedersen C (1981) Carbohydr Res 90:7
22. Bock K, Lundt I, Pedersen C, Sonnichsen R (1988) Carbohydr Res 174:331
23. Lundt I, Madsen R (1995) Synthesis 787
24. Anisuzzaman AKM, Whistler RL (1978) Carbohydr Res 61:511
25. Chen S-Y, Joullié MM (1984) J Org Chem 49:2168
26. (a) Garegg PJ, Samuelsson B (1980) J Chem Soc Perkin 1:2866; (b) Papageorgiou C, Benezra C (1984) Tetrahedron Lett 25:6041
27. Kold H, Lundt I, Pedersen C (1994) Acta Chem Scand 48:675
28. Lundt I, Madsen R (1992) Synthesis 1129
29. Pfaendler HR, Maier FK (1989) Liebigs Ann Chem 691
30. Bell AA, Nash RJ, Fleet GWJ (1996) Tetrahedron: Asymm 7:595 and references cited therein
31. Regeling H, de Rouville E, Chittenden GJF (1987) Recl Trav Chim Pays-Bas 106:461
32. Bock K, Lundt I, Pedersen C (1984) Acta Chem Scand B38:555
33. Godskesen M, Lundt I, unpublished results
34. Paulsen H, Stoye D (1966) Chem Ber 99:908
35. Lundt I, Pedersen C (1986) Synthesis 1052
36. Katsuki T, Sharpless KB (1980) J Am Chem Soc 102:5974
37. Ko SY, Lee AWM, Masamune S, Reed LA, Sharpless KB, Walker FJ (1990) Tetrahedron 46:245
38. Bols M, Lundt I (1990) Acta Chem Scand 44:252
39. (a) Baldwin JE (1976) J Chem Soc Chem Commun 734; (b) Baldwin JE, Cutting J, Dupont W, Kruse L, Silberman L, Thomas RC (1976) J Chem Soc Chem Commun 736
40. Payne GB (1962) J Org Chem 27:3819
41. Bock K, Lundt I, Pedersen C (1986) Acta Chem Scand B40:163
42. Bock K, Lundt I, Pedersen C (1988) Carbohydr Res 179:87
43. Frank H, Lundt I, to be published
44. Vekemans JAJN, Franken GAM, Dapperns CWM, Godefroi EF, Chittenden GJF (1988) J Org Chem 53:627
45. Horneman AM, Lundt I, Søtofte I (1995) Synlett 918
46. Lundt I, unpublished results
47. Bock K, Lundt I, Pedersen C (1987) Acta Chem Scand B41:435
48. Bols M, Lundt I (1992) Acta Chem Scand 46:298
49. Lundt I, Albanese D, Landini D, Penso M (1993) Tetrahedron 49:7295
50. Jünnemann J, Lundt I, Thiem J (1994) Acta Chem Scand 48:265
51. Janairo G, Kowollik N, Voelter W (1987) Liebigs Ann Chem 165
52. Mikkelsen G, Christensen T, Bols M, Lundt I (1995) Tetrahedron 36:6541
53. Sinnot ML (1990) Chem Rev 90:1171
54. Legler G (1990) Adv Carbohydr Chem Biochem 48:319
55. Withers SG, Aebersold R (1995) Protein Science 4:361
56. Winchester B, Fleet GWJ (1992) Glycobiology 2:199
57. (a) Wong C-H, Provencher L, Porco JA, Jung S-H, Wang Y-F, Chen L, Wang R, Steensma DH (1995) J Org Chem 60:1492; (b) Kajimoto T, Liu KK-C, Pederson RL, Zhong Z, Ichikawa Y, Porco JA, Wong C-H (1991) J Am Chem Soc 113:6187; (c) van den Broek LAGM, Vermaas DJ, Heskamp BM, Boeckel CAA, Tan MCAA, Bolscher JGM, Ploegh HL, van Kemenade FJ, de Goede REY, Miedema F (1993) Recl Trav Chim Pays-Bas 112:82
58. (a) Jespersen TM, Dong W, Sierks MR, Skrydstrup T, Lundt I, Bols M (1994) Angew Chem Int Ed Engl 33:1778; (b) Jespersen TM, Bols M, Sierks MR, Skrydstrup T (1994) Tetrahedron 50:13449
59. Ischikawa M, Igarashi Y, Ischikawa Y (1995) Tetrahedron Lett 36:1767
60. (a) Fleet GWJ, Ramsden NG, Witty DR (1989) Tetrahedron 45:327; (b) Fleet GWJ, Smith PW (1987) Tetrahedron 43:971; (c) Fleet GWJ, Smith PW (1986) Tetrahedron 42:5685

61. (a) Collin WF, Fleet GWJ, Haraldsson M (1990) Carbohydr Res 202:105; (b) Behling JR, Campbell AL, Babiak KA, Ng JS, Medich J, Farid P, Fleet GWJ (1993) Tetrahedron 49:3359
62. Look GC, Fotsch CH, Wong C-H (1993) Acc Chem Res 26:182
63. (a) de Raadt A, Ebner M, Ekhardt CW, Fechter M, Lechner A, Strobel M, Stütz AE (1994) Catalysis Today 22:549; (b) de Raadt A, Stütz AE (1992) Tetrahedron Lett 33:189
64. Furneaux RH, Lynch GP, Way G, Winchester B (1993) Tetrahedron Lett 34:3477
65. Lundt I, Madsen R, Al Daher S, Winchester B (1994) Tetrahedron 50:7513–7520
66. (a) Kajimoto T, Chen L, Liu KK-C,Wong C-H (1991) J Am Chem Soc 113:6678, 9009; (b) Liu KK-C, Kajimoto T, Chen L, Zhong Z, Ichikawa Y, Wong C-H (1991) J Org Chem 56:6280; (c) Legler G, Korth A, Berger A, Ekhart C, Gradnig G, Stütz AE (1993) Carbohydr Res 250:67
67. Reitz AB, Baxter E (1990) Tetrahedron Lett 31:6777
68. Duréault A, Portal M, Depezay JC (1991) Synlett 225
69. Shing TKM (1988) Tetrahedron 44:7261
70. Martin M-T, Morrin C (1986) Heterocycles 24:901
71. (a) Masaki Y, Oda H, Kazuta K, Usui A, Itoh A, Xu F (1992) Tetrahedron Lett 33:5089; (b) Shi M, Satoh Y, Makihara T, Masaki Y (1995) Tetrahedron: Asymm 6:2109
72. Hanessian S (1969) J Org Chem 34:675
73. Godskesen M, Lundt I, Madsen R, Winchester B (1996) Bioorg Med Chem 4:1857
74. Legler G, Stütz AE, Immich H (1995) Carbohydr Res 272:17
75. Bernotas RC, Papandreou G, Urbach J, Ganem B (1990) Tetrahedron Lett 31:3393
76. Norris P, Horton D, Levine BR (1995) Tetrahedron Lett 36:7811
77. Bols M, Lundt I (1991) Acta Chem Scand 45:280
78. Chong JM, Sharpless KB (1985) J Org Chem 50:1563
79. Godskesen M, Lundt I (1996) unpublished results
80. Winchester B, Al Daher S, Carpenter NC, Cenci di Bello I, Choi SS, Fairbanks AJ, Fleet GWJ (1993) Biochem J 290:743
81. Winchester B, Barker C, Baines S, Jacob GS, Namgoong SK, Fleet G (1990) Biochem J 265:277
82. Suami T, Ogawa S (1990) Adv Carbohydr Chem Biochem 48:21
83. Suami T (1990) Top Curr Chem 154:257
84. Aceña JL, Arjona O, de la Pradilla RF, Plumet J, Viso A (1994) J Org Chem 59:6419
85. Vogel P, Fattori D, Gasparoni F, Le Drian C (1990) Synlett 173
86. Marschner C, Baumgartner J, Griengl H (1995) J Org Chem 60:5224
87. Yoshikawa M, Yokokawa Y, Inoue Y, Yamaguchi S, Murakami N (1994) Tetrahedron 50:9961
88. Agrofoglio L, Suhas E, Farese A, Condom R, Challand SR, Earl RA, Guedj R (1994) Tetrahedron 50:10611
89. (a) Taylor SJC, Sutherland AG, Lee C, Wisdom R, Thomas S, Roberts SM (1990) J Chem Soc Chem Commun 1120; (b) Trost BM, Li L, Guile SD (1992) J Am Chem Soc 114:8745; (c) Hildbrand S, Troxler T, Scheffold R (1994) Helv Chim Acta 77:1236
90. Sakuda S, Isogai A, Matsumoto S, Suzuki A, Koseki K (1986) Tetrahedron Lett 27:2475
91. Kobayashi Y, Miyazaki H, Shiozaki M (1994) J Org Chem 59: 813 and references cited therein
92. King SB, Ganem B (1994) J Am Chem Soc 116:562
93. Trost BM, van Vranken DL (1993) J Am Chem Soc 115:444
94. Elliott RP, Hui A, Fairbanks AJ, Nash RJ, Winchester BG, Way G, Smith C, Lamont B, Storer R, Fleet GWJ (1993) Tetrahedron Lett 34:7949
95. Ferrier RJ, Middelton S (1993) Chem Rev 93:2779
96. Giese B (1986) Radicals in organic synthesis: formation of carbon-carbon bonds, 1st edn. Pergamon, Oxford
97. Curran DP, Porter NA, Giese B (1996) Stereochemistry of radical reactions, 1st edn. VCH Verlagsgesellschaft, Weinheim
98. (a) Horneman AM (1996) Ph D Thesis; (b) Horneman AM, Lundt I (1997) Tetrahedron (to be published)
99. Hanessian S, Léger R, Alpegiani M (1992) Carbohydr Res 228:145

100. Rajanbabu TV (1991) Acc Chem Res 24:139
101. Lesueur C, Nouguier R, Bertrand MP, Hoffmann P, De Mesmacher A (1994) Tetrahedron 50:5369 and the extensive references cited therein
102. Roberts SM, Shoberu KA (1992) J Chem Soc Perkin Trans 1:2625
103. Ingall AH, Moore PR, Roberts SM (1994) J Chem Soc Chem Commun 93
104. Kiguchi T, Tajiri K, Ninomiya I, Naito T, Hiramatsu H (1995) Tetrahedron Lett 36:253
105. Bovin TL (1987) Tetrahedron 43:3309
106. Cardillo G, Orena M (1990) Tetrahedron 46:3321
107. Makabe H, Tanaka A, Oritani T (1994) J Chem Soc Perkin Trans 1:1975
108. Rychnovsky SD, Bartlett PA (1981) J Am Chem Soc 103:3963
109. Wilson P, Shan W, Mootoo DR (1994) J Carbohydr Chem 13:133
110. Lee E, Park CM (1994) J Chem Soc Chem Commun 293
111. Dehmlow H, Mulzer J, Seilz C, Strecker AR, Kohlmann A (1992) Tetrahedron Lett 33:3607
112. Barbaud C, Bols M, Lundt I, Sierks MR (1995) Tetrahedron 51:9063
113. Martin OR, Yang F, Fang X (1995) Tetrahedron Lett 36:47
114. Toril S, Okumoto H, Hikasa (1989) Chemistry Express 4:535
115. (a) Beacham AR, Bruce I, Choi S, Doherty O, Fairbanks AJ, Fleet GWJ, Skead BM, Peach JM, Saunders J, Watkin DJ (1991) Tetrahedron: Asymm 2:883; (b) Bichard CJF, Fairbanks AJ, Fleet GWJ, Ramsden NG, Vogt K, Doherty O, Pearce L, Watkin DJ (1991) Tetrahedron: Asymm 2:901
116. Choi SS, Myerscough PM, Fairbanks AJ, Skead BM, Bichard CJF, Mantell SJ, Son JC, Fleet GWJ, Saunders J, Brown D (1992) J Chem Soc Chem Commun 1605
117. Wheatley JR, Bichard CJF, Mantell SJ, Son JC, Hughes DJ, Fleet GWJ, Brown D (1993) J Chem Soc Chem Commun 1065
118. Bock K, Pedersen C, Tøgersen H (1981) Acta Chem Scand B 35:441
119. Frank H, Lundt I (1995) Tetrahedron 51:5397
120. Kold H, Lundt I, Pedersen C (unpublished)

Chemical and Chemo-Enzymatic Approaches to Glycosidase Inhibitors with Basic Nitrogen in the Sugar Ring

Anna de Raadt · Christian W. Ekhart · Michael Ebner · Arnold E. Stütz

Institut für Organische Chemie, Technische Universität Graz, Stremayrgasse 16, A-8010 Graz, Austria

Due to the vital functions of enzymes with glycosidase and glycosyl transferase activities in living systems, inhibitors of such biocatalysts have attracted considerable interest in the recent past. Low molecular weight glycosidase inhibitors are structurally frequently related to the natural substrates or, more importantly, to the oxocarbonium ion like transition state of glycoside hydrolysis. Sugar shaped inhibitors with a basic nitrogen instead of oxygen in the ring, so-called imino sugars, by the combination of their interesting biological properties and their challenging chemical structures, are certainly one of the most attractive classes of glycosidase inhibitors known to date. While inhibitors of D-glucosidases as well as D-mannosidases have found widespread interest and have been extensively reviewed, quite a few other structurally related compounds with marked effects on other glycosidases have been discovered as natural products or synthesized. This account will attempt to survey some of these classes of compounds and cover the literature to April 1996 as available in Graz.

Table of Contents

Topics in Current Chemistry, Vol. 187

1 Introduction

Glycoside hydrolases (glycosidases) are enzymes that degrade poly- and oligosaccharides into monomers or cleave bonds between sugars and non-carbohydrate aglycons. From the degradation of starch to the highly organized and complex process of glycoprotein trimming, these enzymes have many functions essential for the existence and survival of living systems. Inhibitors of glycosidases have served as important tools in investigations carried out to understand the catalytic action of the active site and to differentiate between various types of glycoside hydrolases and their modes of action [1]. Furthermore, quite a range of low molecular weight reversible inhibitors exhibit interesting biological properties associated with the inhibition of key-steps in carbohydrate and glycoprotein metabolism such as anti-diabetic [2], anti-cancer [3], anti-retroviral [4], anti-malarial [5], and insect antifeedant [6] activities. Amongst these compounds, sugar analogues with basic nitrogen instead of oxygen in the ring (frequently and according to carbohydrate nomenclature incorrectly addressed as "aza-sugars") have attracted considerable and widespread attention. Since the first synthesis of such a sugar derivative conducted by Paulsen et al. [7] and the discovery of the first natural representative [8] of this class of compounds at about the same time, a vast number of publications concerned with their chemistry and biochemistry has appeared. Supported by tremendous interest, the research area under consideration has emerged over the past decade as one of the most thriving and prosperous in carbohydrate and organic natural products chemistry. Renowned scientists such as Sharpless, Whitesides, and Effenberger, just to mention a few, have made valuable contributions probing with their methods and approaches the challenging chemical structures under consideration. Not surprisingly, the high profile of the field, especially in the first "gold-rush"-like period of development, has also attracted less than desirable contributions [9]. In the case of the example referred to, the high activity of the scientific community in this area was quickly self-policing with such a development [10, 11].

Due to the abundance and biological importance of the above mentioned enzymes, the areas of D-glucosidase as well as D-mannosidase inhibitors such as

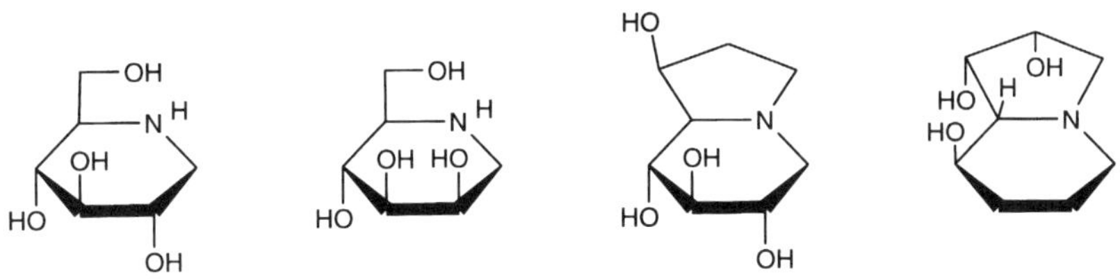

Fig. 1

1-deoxynojirimycin (**1**) [12, 13], 1-deoxymannojirimycin (**2**) [14], as well as their bicyclic counterparts castanospermine (**3**) [15, 16] and swainsonine (**4**) [16], just to mention a few, have been covered by excellent recent reviews concerning their syntheses [17, 18] , biological activities [12, 19–21]] as well as structure-activity relationships [12, 20–23]. This account aims to survey the literature on some less mainstream, but in terms of synthetic demands as well as biological significance, equally attractive classes of compounds such as D-galactosidase, D-hexosaminidase, α-L-fucosidase, and pyranosiduronidase inhibitors. In addition, a few selected examples of related substances with interesting or unusual structures and biological properties will also be highlighted. Depending on the class of enzymes in question, different levels of inhibition are considered to be efficient or not. For example, a good inhibitor of α-L-fucosidases or α-D-galactosidases has a K_i in the nanomolar range while an inhibitor for D-mannosidases with a K_i in the 10 micromolar range can already be regarded as highly active [1]. In addition, the way inhibitory data are reported can vary from K_i to IC_{50} to the degree of inhibition of an enzyme at a given inhibitor concentration (such as 1 mmol/l). Furthermore, the individual experimental conditions and, most importantly, the p*H*-values at which the inhibition experiments are conducted have great influence on the inhibitory activities. Therefore, it can be quite difficult for the average organic chemist to extract and compare inhibition data from different literature sources.

2 D-Galactosidase Inhibitors

2.1 Piperidine Derivatives

Several routes to 1-deoxygalactonojirimycin (**5**) have been published to date. 1,5-Dideoxy-1,5-imino-D-galactitol (**5**), the epimer of 1-deoxynojirimycin (1) at C-4, was first synthesized by Paulsen and co-workers in 1980 [24]. These workers employed 1,6-anhydro- α-D-galactofuranose (**6**) as the starting material.

This compound is easily available by the pyrolysis of D-galactose [25] or, on the small scale, by Angyal and Beveridge's method [26]. Regioselective catalytic oxidation led to the corresponding 5-ulose which, via the oxime and subsequent reduction, could be converted into inhibitor (**5**) in approximately 6% overall

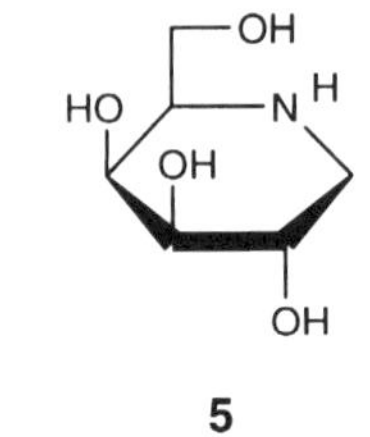

Fig. 2

yield. This relatively low isolated yield was due to the isolation of the chemically labile intermediate *N*-protected 5-amino-5-deoxy-D-galactose. Six years later, Legler devised a route previously employed for the synthesis of nojirimycin [27] from 1,2;5,6-di-*O*-isopropylidene- α-D-glucofuranose. Here a highly regioselective configurational inversion at C-4 introduced by Meyer zu Reckendorf [28]

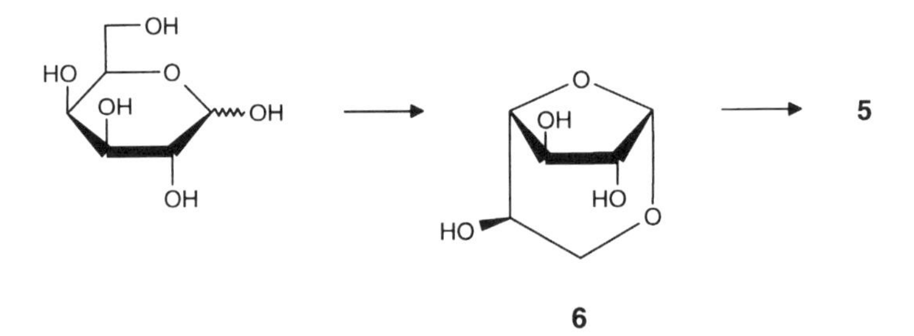

Fig. 3

was the key step. In 12 steps and 20% overall yield the desired compound was furnished and could be shown to be a very potent inhibitor of a range of α- as well as β-galactosidases from various sources [29]. Following their synthesis of 1-deoxynojirimycin from D-glucose [30], Ganem and co-workers synthesized compound (**5**) from methyl α-D-galactopyranoside via an intramolecular aminomercuration method [31] which took advantage of the reductive ring-opening procedure of 6-bromodeoxy pyranosides (**7**) first introduced by Bernet and Vasella [32]. From a suitably protected intermediate of their synthesis, they also produced a related suicide inhibitor (**8**) containing an aziridine ring formed by C-5, C-6, and the ring-nitrogen. This compound was demonstrated to be an extremely potent irreversible inhibitor of α-galactosidase from green coffee beans [33].

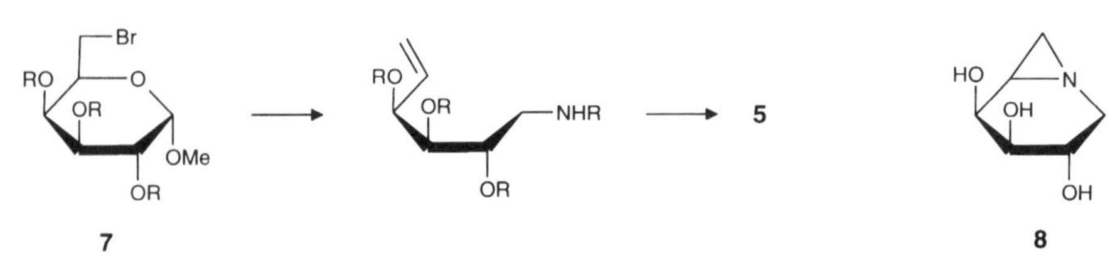

Fig. 4

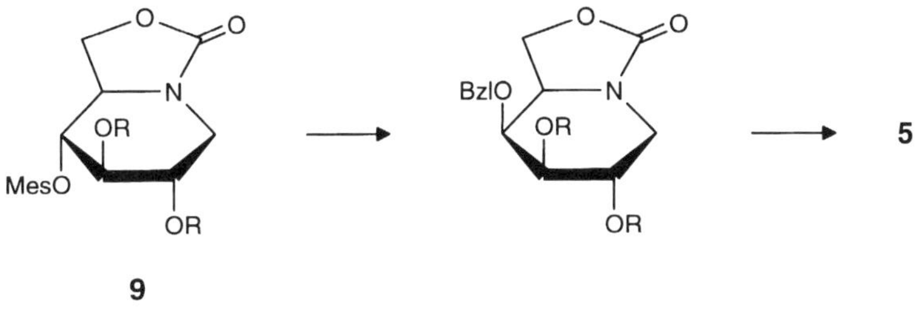

Fig. 5

Taking advantage of their easy access to preparative amounts of 1-deoxynojirimycin [34], a BAYER group used the suitably protected compound (**9**) to prepare 1-deoxygalactonojirimycin. Employing essentially standard carbohydrate procedures [35], the 4-mesylate was nucleophilically displaced by a suitable oxygen nucleophile.

In 1987, a compound named (+)-galactostatin by the investigators was found in the culture broth of a *Streptomyces* species [36] and could be identified as 5-amino-5-deoxy-D-galactose [37]. The corresponding product of reduction at C-1, identical with 1-deoxygalactonojirimycin, was named 1-deoxygalactostatin [38] leading to an unfortunate accumulation of generic names for this compound. The identity of 1-deoxygalactostatin and 1-deoxygalactonojirimycin was established by an L-tartaric acid based synthesis [39]. Another non-carbohydrate-based approach was chosen by Dondoni and co-workers by the chain-extension of L-serine for the synthesis of the enantiomer of (**5**) [40]. The preparation of the natural enantiomer was not described but was expected to follow the same route starting from D-serine. Several bio-organic syntheses have also been published. Syntheses based on a fuculose-1-phosphate aldolase mediated addition of dihydroxyacetone phosphate to 3-azido-2-hydroxy-propanal [41] (Scheme 1) also led to compound (**5**) [42, 43].

The precursor to (**5**) was made diastereoselectively by Fessner and co-workers [44]. When the intermediate ketose 1-phosphate was immediately submitted to hydrogenation conditions, 1,5,6-trideoxy-1,5-imino-D-galactitol (1,6-dideoxygalactonojirimycin, **10**) was obtained in fair yield [42]. The latter derivative was also obtained by a hetero Diels-Alder cycloaddition of a benzyloxycarbonyl nitroso dienophile to (*E*,*E*)-sorbaldehyde dimethylacetal [45].

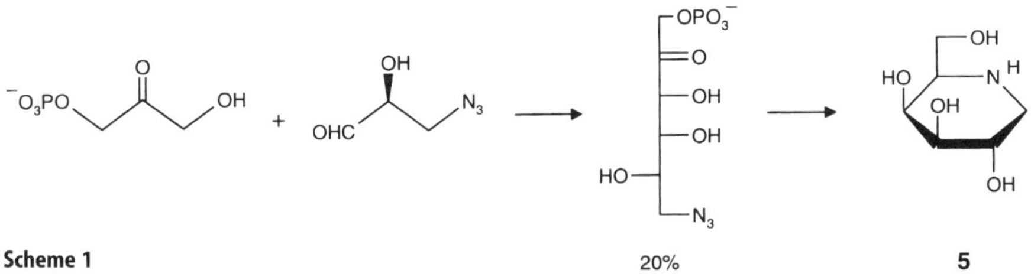

Scheme 1

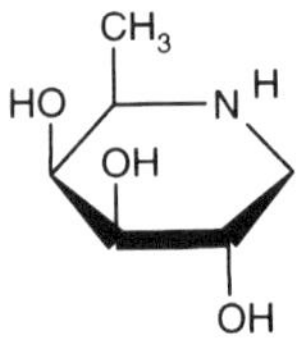

10

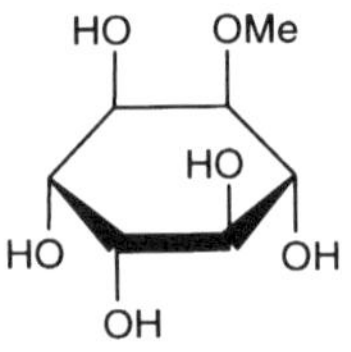

11

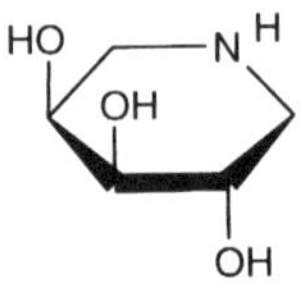

13

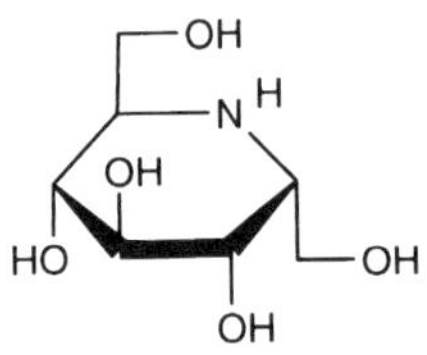

14

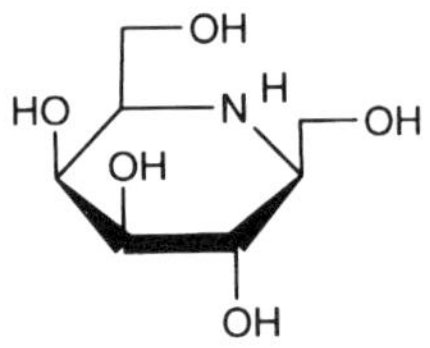

15

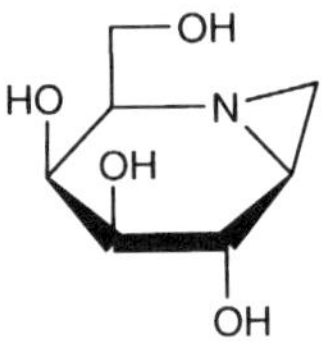

16

Fig. 6

Not able to reproduce a previously reported approach [30] to (**5**) from D-galactose, Tyler and co-workers designed an efficient sequence starting from L-sorbose [46], via a partially protected 6-azido-6-deoxy-L-tagatose derivative, and obtained an overall yield of 20%. Ogawa and co-workers [47] prepared D-galactonojirimycin as well as 1-deoxygalactonojirimycin from L-quebrachitol (**11**), a natural product found in the serum of the rubber tree.

A de novo approach featuring a biocatalytic functionalization step of benzene or bromobenzene (**12**, x = Br) was reported by Johnson and his group [48].

The synthesis of 1,5-dideoxy-1,5-imino-L-arabinitol (**13**), a derivative of (**5**) lacking the hydroxymethyl side chain at C-5 was reported [49]. This compound did not exhibit any appreciable activity with unspecified galactosidases.

In 1987 2,6-dideoxy-2,6-imino-D-*glycero*-L-*gulo*-heptitol (**14**), a chain-extended derivative of 1-deoxynojirimycin, was synthesized as a potential glucosidase inhibitor [50] and soon thereafter extracted from the leaves of the Panamesian plant *Omphalea diandra* L [51]. Coined "α-homonojirimycin" by the discoverers, this novel type of inhibitor exhibited pronounced activity against gluco-

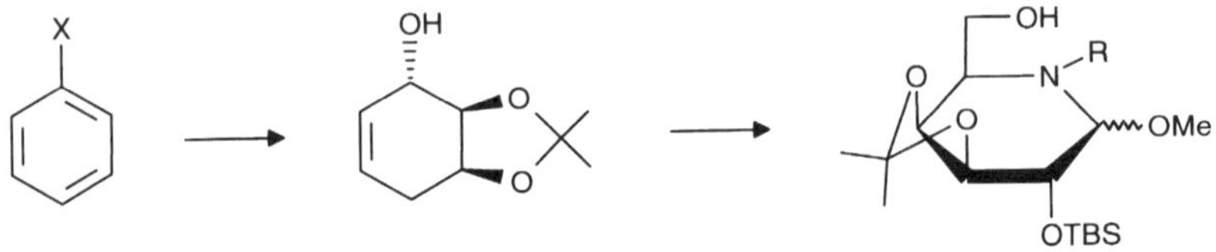

12

Scheme 2

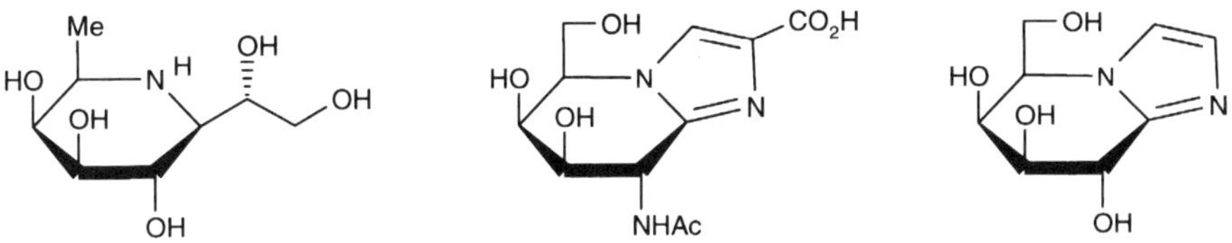

17 18 19

Fig. 7

20

R = NH_2, NMe_2, $NH\text{-}NH_2$, NH-OH

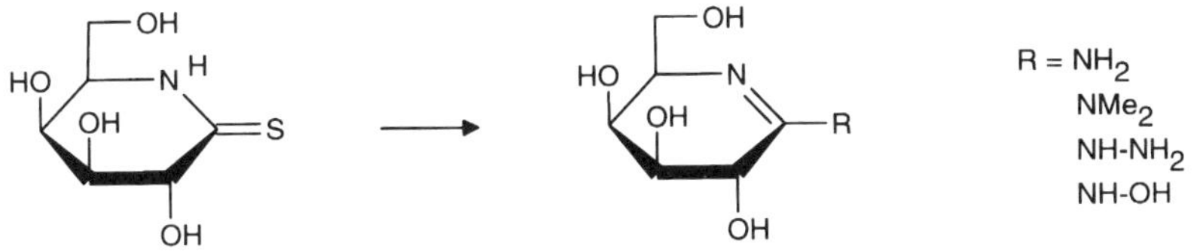

Fig. 8

sidases from mouse gut. The first synthesis of the related "homo"-analogue (**15**) of 1-deoxygalactonojirimycin was recently reported by Martin and co-workers [52], who took advantage of a sophisticated variation of an approach previously designed by Bernotas and Ganem [30]. In the course of their investigation, an interesting 1,*N*-aziridine derivative (**16**) of compound (**9**) was also synthesized. No biological data have been available from the literature for either compound. The synthesis of a protected chain-extended analogue of the 6-deoxy derivative of compound (**9**) was recently reported [53] as an intermediate in a C-disaccharide synthesis. Deprotected 3,7,8-trideoxy-3,7-imino-D-*threo*-L-*galacto*-octitol (**17**) was prepared by the same group and, despite the close resemblance of this compound to (**5**), was found to be rather an inhibitor of β-D-glucosidases than D-galactosidases [54].

Based on the unusual structure of the recently discovered, very powerful hexosaminidase inhibitor nagstatin (**18**) [55], analogue (**19**) was synthesized [56] and shown to be a very powerful inhibitor of β-D-galactosidase from *E. coli* [57]. Amidine, amidrazone, and amidoxime derivatives (**20**) of D-glucono- [58] as well as D-galactono-1,5-lactam were synthesized via the corresponding thionolactams and found to inhibit α- as well as β-D-galactosidases in the micromolar range [59]. A significantly lower degree of inhibitory power was found for a guanidine derivative (**21**) structurally related to D-galactose which was subsequently synthesized [60]. Low inhibition was also found for functionalized guanidinium derivatives (**22**) synthesized by Lehmann and his group.

Bols and co-workers [61] synthesized isofagomine (**23**), a compound in which the basic hetero-atom replaces the anomeric carbon, and showed that this new type of synthetic sugar analogue is one of the strongest inhibitors of β-glucosidase from almonds known to date. Following this innovative work, Ichikawa and Ichikawa prepared (3*R*,4*R*,5*R*)-3(*R*)-hydroxymethylpiperidine (1,5-dideoxy-2-C-hydroxymethyl-1,5-imino-D-arabinitol, **24**) [62] from D-lyxose. This was

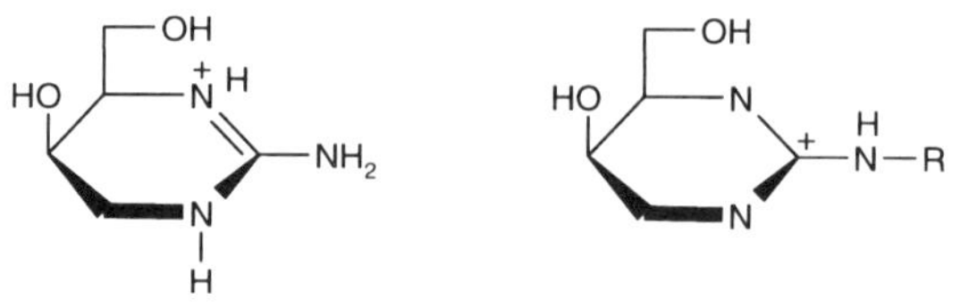

Fig. 9

21 22

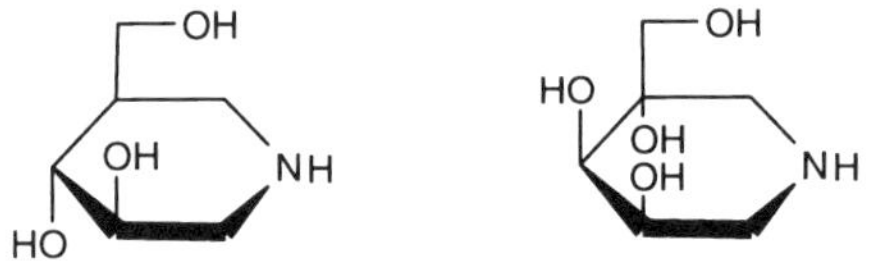

Fig. 10

23 24

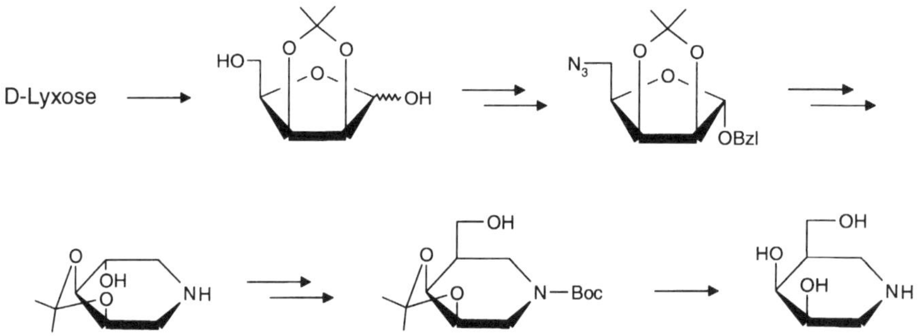

Scheme 3

25

demonstrated to be a good inhibitor of β-galactosidase from *Aspergillus oryzae*. Extending this line of research these workers synthesized piperidine (**25**) that inhibits the same enzyme with a K_i-value in the nanomolar range [63] (Scheme 3).

2.2 Pyrrolidine Derivatives

Quite a few of the five-membered ring sugar analogues with basic nitrogen in the ring have been found to be at least as potent inhibitors of glycosidases as their piperidine counterparts. For example, 2,5-dideoxy-2,5-imino-D-mannitol (**26**), a virtually isosteric analogue of 1-deoxynojirimycin (**1**), inhibits a range of α- and β-glucosidases distinctly more pronounced than the latter [1]. This fact was attributed to the flatter ring system of the pyrrolidine derivative which is believed to resemble the oxocarbonium ion-like transition state of the glycoside hydrolysis more closely than the piperidine chair.

Analogously, in the field of D-galactosidase inhibitors, five-membered ring 1,4-dideoxy-1,4-imino-D-lyxitol (**27**), first synthesized by Fleet and co-workers,

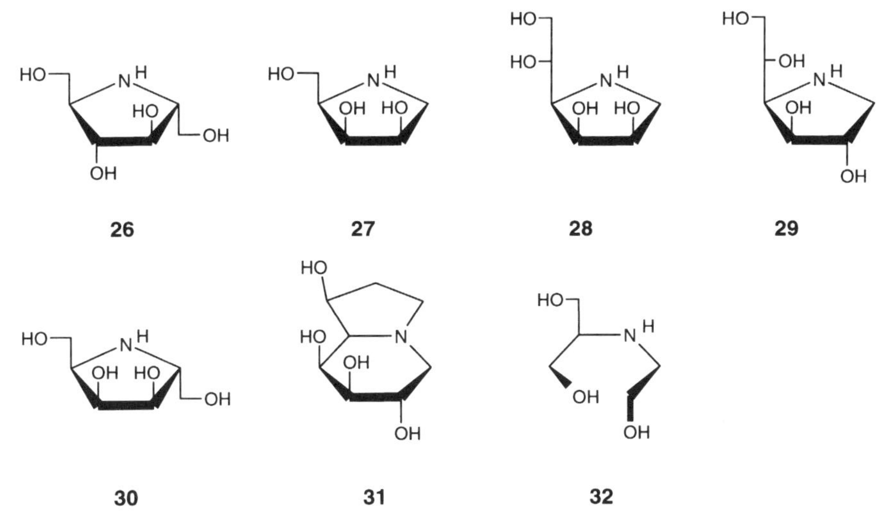

Fig. 11

was indeed found to be a surprisingly potent inhibitor of α-D-galactosidase with a K_i of 100 nmol/l [64]. Two additional syntheses of this compound were published by the same group [65, 66]. Other synthetic approaches to this substance were reported starting from 2-*O*-benzyl-D-glycerinaldehyde [67] via a Henry reaction (8 steps, 17% overall yield) and (*R*)-glutamic acid [68]. Structure (**27**) was also obtained as an undesired by-product in a (2*S*)-3,4-dehydroproline based synthesis [69] of 1,4-dideoxy-1,4-imino-D-ribitol. A synthesis from D-glucose was reported in1988 [70]. Interestingly, the homologous 1,4-dideoxy-1,4-imino-D-mannitol (**28**) was found not to be an inhibitor of galactosidases [66]. In contrast, Lundt, Winchester and their co-workers could show that structurally related 1,4-dideoxy-1,4-imino-L-iditol (**29**) is a potent inhibitor of α-D-galactosidase [71].

2,5-Dideoxy-2,5-imino-D-talitol (**30**), synthesized by Wong and co-workers, was demonstrated to strongly inhibit an α-galactosidase with a K_i-value of 50 nmol/l which is still about 30 times less potent than 1-deoxygalactonojirimycin itself [72].

2.3 Bicyclic Systems

The synthesis of 8-epi-castanospermine (**31**), a bicyclic analogue of (**5**), was reported [48, 73], but no biological data were given.

2.4 Others

Various open-chain analogues [74] of 1-deoxynojirimycin, such as (**32**), were synthesized as potential glucosidase inhibitors and, surprisingly, exhibited good inhibitory activities with K_i-values around 10 µmol/l. Due to the nature of their structures some of the compounds reported could also interact with galactosidases. Further investigations into this interesting area will be necessary to appreciate the scope and limitations of such a series of compounds.

3 D-Hexosaminidase Inhibitors

N-Acetyl-*β*-D-glucosaminidases are widespread in plants as well as animals. They are also associated with the intracellular degradation of glycolipids as well as glycoproteins and genetically caused deficiencies which result in various lipid storage disorders such as GM_2-gangliosidosis and Sandhoff's disease [75]. Because of their poor ability to discriminate between *N*-acetyl-*β*-D-glucosaminides and *N*-acetyl-*β*-D-galactosaminides, these enzymes are also called hexosaminidases.

Inhibitors of *β*-D-glucosaminidases and *β*-D-galactosaminidases, commonly also called *β*-D-hexosaminidase inhibitors, have attracted much interest as useful tools for glycobiologists.

3.1 Piperidine Derivatives

The paradigmatic hexosaminidase inhibitor with basic nitrogen in the ring, 2-acetamido-1,2-dideoxynojirimycin (2-acetamido-1,2,5-trideoxy-1,5-imino-D-glucitol, **33**) has not been found naturally to date. This inhibitor was first prepared by Fleet and co-workers [76, 77] in a 17-step synthesis from D-glucose via methyl 3-*O*-benzyl-2,6-dideoxy-2,6-imino-*α*-D-mannopyranoside together

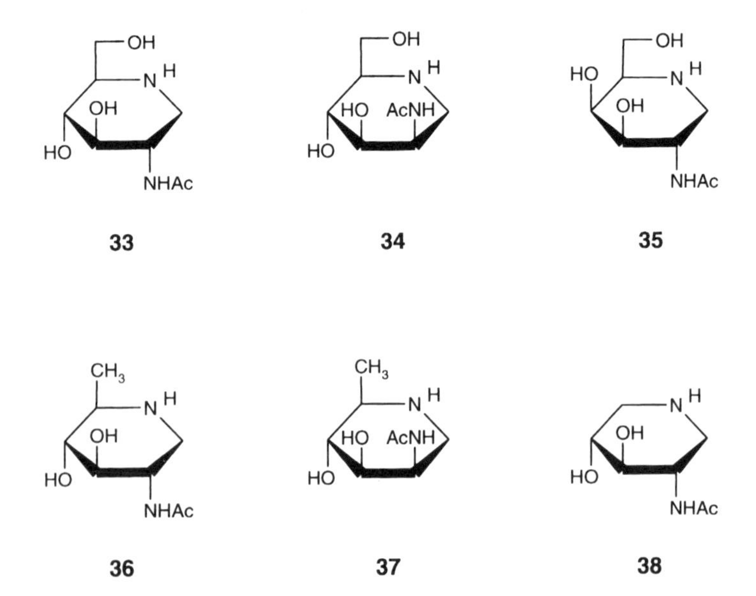

Fig. 12

with the corresponding D-*manno* epimer (**34**). Compound (**33**) turned out to be an extremely powerful inhibitor of *N*-acetyl-*β*-D-glucosaminidases from plants and mammalian sources with K_i-values of less than 1 μmol/l. Interestingly, only weak inhibition was found with the *N*-acetyl-*β*-D-glucosaminidase from *Aspergillus niger*. Inhibitor (**33**) was also synthesized by a BAYER group in 10 steps from 1-deoxynojirimycin (**1**) as the starting material [78]. Kappes and Legler [79] achieved a synthesis in 11 steps from *N*-acetyl-D-glucosamine. They also found comparatively poor inhibition of *N*-acetyl-D-glucosaminidases from *Asp. niger* and *Helix pomatia*. The D-*galacto* epimer (**35**) was prepared by a BAYER group in 25 steps from 1-deoxynojirimycin (**1**) and exhibited excellent inhibitory power with *N*-acetyl-*α*-D-galactosaminidase from *Charonia lampas* [80, 81]. Another synthetic approach based on the chemical modification of intact 1-deoxynojirimycin gave access to compound (**33**) as well as its 2-acetamido-D-*manno*- and D-*galacto* analogues (**34**) and (**35**), respectively [82]. Compound (**33**), its epimer at C-2 (**34**), as well as their 6-deoxy derivatives (**36**) and (**37**) were prepared by Wong and co-workers, a fructose 1,6-diphosphate aldolase (FDP-aldolase) catalyzed aldol reaction being a key step of their approach [83]. No inhibition constants were given for the two new 6-deoxy derivatives. A preparatively attractive synthesis from *N*-acetyl-D-glucosamine leading to compound (**33**) in six steps and with about 10% overall yield was devised by Furneaux and his group [84] (Scheme 4).

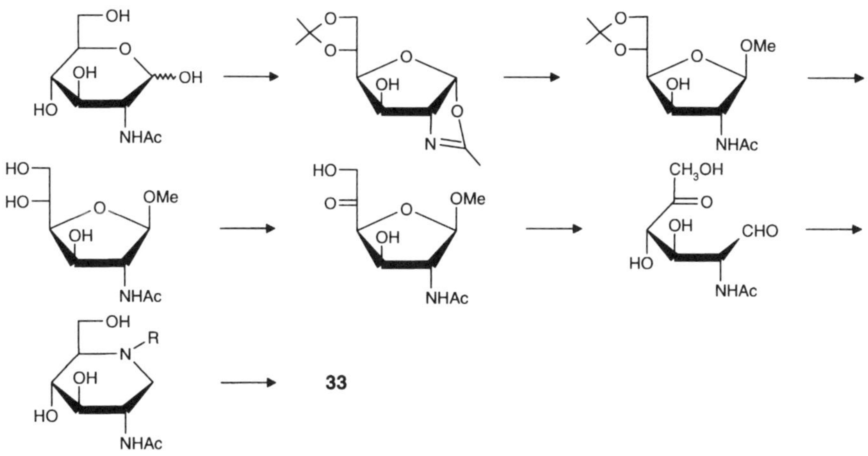

OH O OH OH HO NHAc → O O O OH O N → O O O OMe OH NHAc →
HO HO O OMe OH NHAc → HO O O OMe OH NHAc → CH_2OH O OH CHO HO NHAc →
OH N R OH HO NHAc → **33**

Scheme 4

The synthesis of *nor*-derivative (**38**), lacking the hydroxymethyl group at C-5, was reported by Bernotas and Ganem [85] who employed their previously established method [28]. Hasegawa and co-workers described the synthesis of an analogue of a bacterial peptidoglycan fragment (**39**) containing 2-acetamido-1,2-dideoxynojirimycin [86].

The same group recently reported the chemical synthesis of sialyl-Lewis X and A epitope analogues containing the *N*-methyl derivative of (**33**) [87].

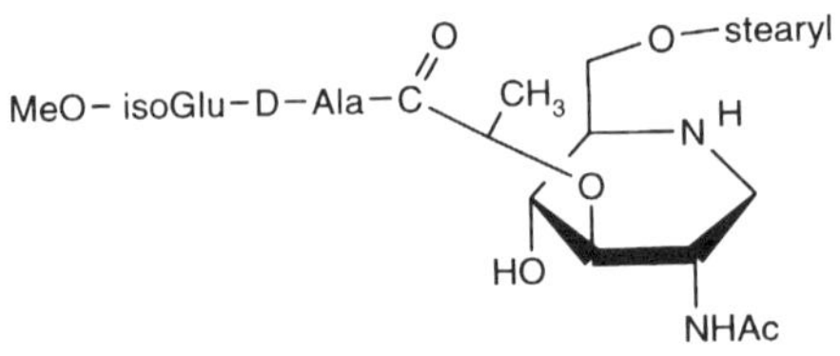

39

Fig. 13

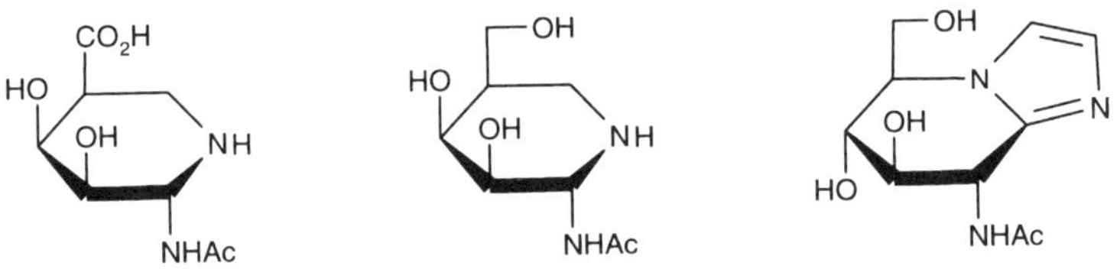

40 **41** **42**

Fig. 14

Some non-natural derivatives of the *β*-glucuronidase and *N*-acetylneuraminidase inhibitor siastatin B (**40**) [88], found in the culture broth of a *Streptomyces* species, were found to be potent inhibitors of chicken liver and bovine epidermis hexosaminidases with IC_{50} values of less than 1 µg per ml [89]. Inhibitors such as (**41**) can be regarded as relatives of isofagomine [61], mentioned previously, in which not the ring oxygen but the anomeric carbon has been formally replaced by a basic nitrogen.

The recent discovery of nagstatin (**18**) [85], a novel inhibitor of *N*- acetyl-*β*-D-glucosaminidases isolated from culture filtrates of *Streptomyces amakusaensis* in which the ring nitrogen is a constituent of an imidazole ring, has triggered interest in the synthesis of structurally simplified analogues lacking the acetic acid side chain [56]. Some of these compounds, such as (**42**), are remarkably powerful hexosaminidase inhibitors with K_i values in the nmol/l-range [57]. The total synthesis of nagstatin (**18**) by the same route was recently reported [90].

3.2 Pyrrolidine Derivatives

In search for hexosaminidase inhibitors exhibiting higher selectivity than (**33**) and in order to gain further insight into structure activity relationships in this family of compounds, several five-membered ring analogues of (**33**) have been recently synthesized. Low inhibitory power with various hexosaminidases was found for 2-acetamido-1,2,4-trideoxy-1,4-imino-D-galactitol (**43**), which was prepared by a New Zealand group in 12 steps from *N*-acetyl-D-glucosamine [91]. This compound was also prepared by Giannis and co-workers and found to inhibit human lysosomal *β*-hexosaminidase A [92].

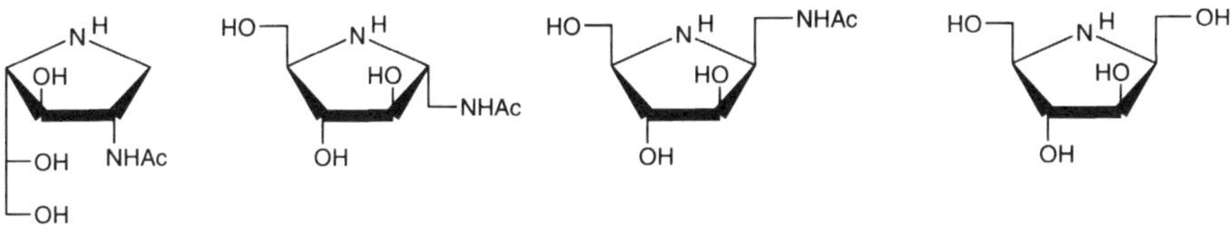

43 44 45 46

Fig. 15

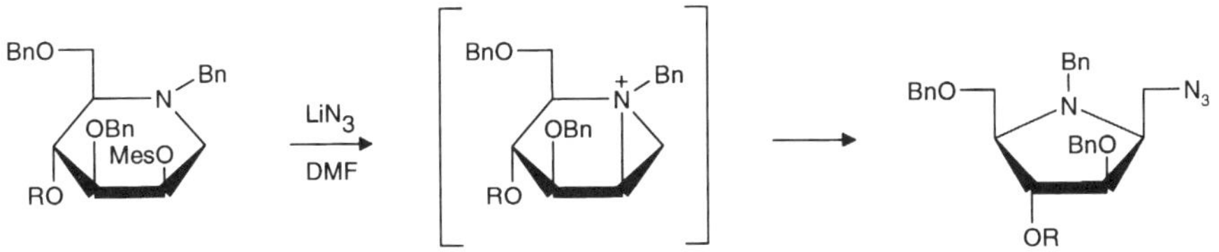

Fig. 16

Very interesting 5-membered ring hexosaminidase inhibitors were reported by Wong and co-workers prepared by the FDP-aldolase catalyzed condensation of (*S*)-3-azido-2-acetamidopropanal with dihydroxyacetone phosphate [93]. D-*Manno*- and D-*gluco*-configurated compounds (**44**) and (**45**), respectively, have the same structural relationships with 2-acetamido-1,2-dideoxynojirimycin (**33**) as have 2,5-dideoxy-2,5-imino-D-mannitol (**26**) and -glucitol (**46**) to 1-deoxynojirimycin (**1**). Compound (**44**) was found to be a very potent hexosaminidase inhibitor albeit not as powerful as the 6-membered ring analogue (**33**). In addition, the D-*gluco* epimer (**45**) appears to be about one magnitude less effective, an observation that parallels the comparison of compounds (**26**) and (**46**) as glucosidase inhibitors [94]. 1-Acetamido-1,2,5-trideoxy-2,5-imino-D-glucitol (**45**) was also prepared by Peter and his group via an interesting ring-contraction step [95].

These workers synthesized a suitably protected derivative of 1-deoxymannojirimycin, following Ganem's entry into this class of compounds, from 6-halogenodeoxy pyranosides [30, 31]. In particular, after the introduction of a leaving group at *O*-2 of the *N*-benzyl protected 1-deoxymannojirimycin derivative, the ring nitrogen competes with external nucleophiles such as azide ions for the nucleophilic displacement at C-2. The aziridinium ring system with D-*manno*-configuration is formed under inversion of configuration at C-2 and is preferentially opened by the external nucleophile at the primary position leading to ring contraction and giving rise to a direct precursor of (**45**). Asymmetrically substituted derivatives of the 2,5-dideoxy-2,5-imino-D-glucitol (**46**) have also recently been made available by the nucleophilic opening of D-mannitol derived bis-aziridines with various nucleophiles [96]. An alternative approach to the aldolase catalyzed synthesis of 2-keto sugars for the preparation of 2,5-dideoxy-2,5-imino-D-mannitol derivatives was found in the enzymatic isomerization of suitably substituted 5,6-bis-modified analogues of D-glucofuranose and L-ido-

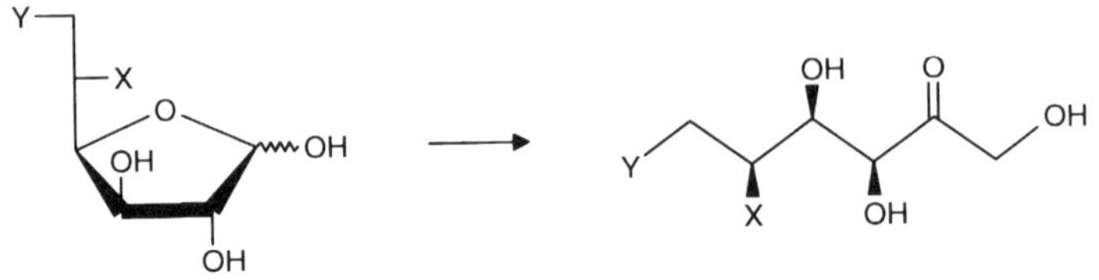

Fig. 17

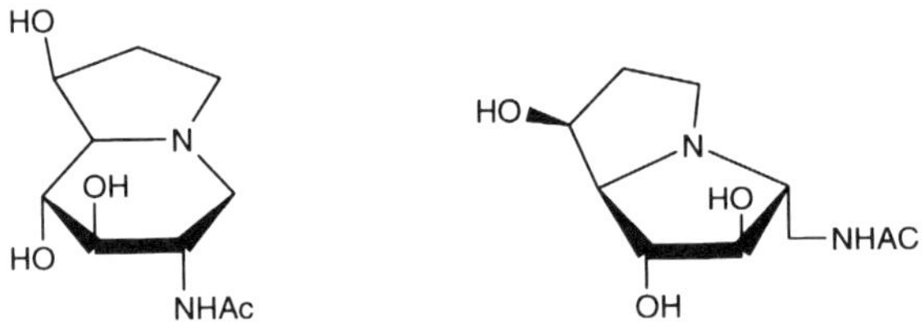

Fig. 18

furanose (**47**), respectively, into the corresponding open-chain D-fructoses and L-sorboses (**48**) with the aid of immobilized glucose isomerase [97, 98].

3.3 Bicyclic Systems

A very powerful inhibitor, which is a bicyclic analogue of (**33**), is 6-acetamido-6-deoxycastanospermine (**49**), a compound synthesized from castanospermine and tested by Liu and co-workers [99]. This compound, amongst many other 6-modified castanospermine derivatives, was also prepared by the Furneaux team [100]. Furthermore, by controlled ring-contraction of suitably substituted castanospermine derivatives, these workers gained access to 8-acetamido-8-deoxyaustraline (**50**), a byciclic analogue of 1-acetamido-1,2,5-trideoxy-2,5-imino-D-mannitol (**44**). This ring contraction reaction, which proceeds via a tricyclic aziridinium intermediate, is based on the same principle as the approach by Peter et al. to compound (**45**) [95].

4 α-L-Fucosidase Inhibitors

4.1 Piperidine Derivatives

The L-*fuco* analogue of 1-deoxynojirimycin, 1,5-dideoxy-1,5-imino-L-fucitol or 1-deoxy-L-fuconojirimycin (**51**), was first synthesized by Fleet and co-workers in 1985 by a multi-step sequence starting from methyl-α-D-glucopyranoside

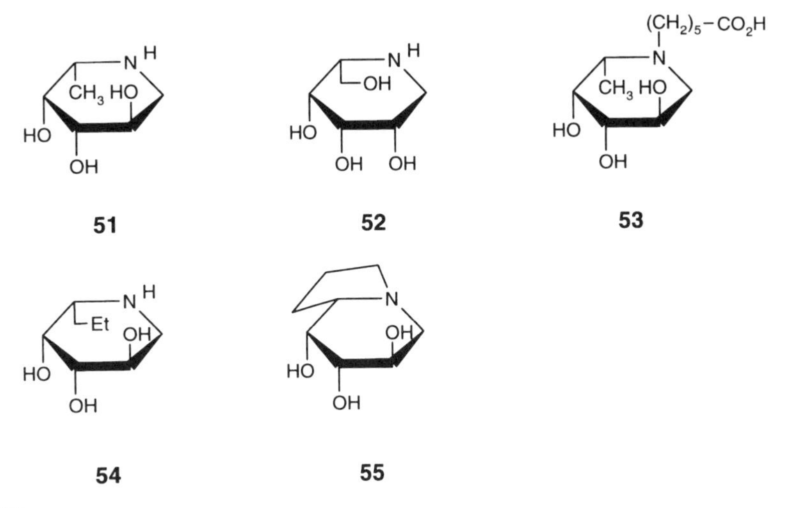

Fig. 19

[101]. Its remarkably low K_i-value of 5 nmol/l makes it one of the most potent inhibitors in the class of sugar analogues with basic nitrogen instead of oxygen in the ring.

A synthesis from 3,4,6-tri-*O*-acetyl-D-galactal (3,4,6-tri-*O*-acetyl-D-*lyxo*-hex-1-enitol) was devised by Paulsen and Matzke [102]. In addition, these workers studied methods of immobilizing the inhibitor on aminohexyl sepharose via spacers linked to the ring nitrogen. Such compounds in the D-*gluco* series had been shown by Legler and co-workers [103] to be very useful ligands for the isolation and purification of other glycosidases by affinity chromatography. Later, a five-step synthesis of deoxyfuconojirimycin from D-lyxonolactone was executed by Fleet and his group [104]. Tri-*O*-benzyl-β-D-ribofuranoside served as the starting material in a synthesis of 1,5-dideoxy-1,5-imino-L-talitol (**52**) [105] which is structurally related to (**51**). This compound was found to be a good inhibitor of α-L-fucosidase from bovine kidney with K_i 7 µmol/l. The immobilization of *N*-carboxypentyl-deoxyfuconojirimycin and the application of the affinity ligand was used for the isolation of an α-fucosidase from almond meal as described by Scudder, Fleet, and co-workers in 1990 [106]. This method gave an unprecedented degree of purification of this enzyme (163.000-fold). The efficient removal of the immobilized enzyme from highly active affinity ligands such as (**53**) can be a problem in terms of yields as well as possible denaturation of the enzyme due to the drastic conditions required.

Less active derivatives might therefore be desirable in some cases. In this context, Paulsen and co-workers synthesized a range of chain-extended as well as bicyclic analogues of 1-deoxyfuconojirimycin, amongst others the 6-C-ethyl (**54**) and 6-C-methyl derivatives as well as 1-deoxy-6,7,8-α-tri-epi-castanospermine (**55**). All of these derivatives exhibited considerable inhibitory activities towards α-L-fucosidase and were slightly to appreciably less active than the parent compound [107]. In an attempt to design a simple synthesis of 1-deoxyfuconojirimycin (**51**), Adelhorst and Whitesides prepared 2-acetamido-1,2-dideoxy-D-galactitol which was supposed to be oxidized at C-6 by galactose oxidase. However, this was not accepted as a substrate [108]. A synthesis of (**51**) from methyl tri-*O*-benzoyl-*α*-D-arabinofuranoside via the partially protected dialdose employing a one-carbon chain extension at C-5 as the key step leading to a D-*altro* configurated intermediate was reported by Takahashi and Kuzuhara [109].

This compound (**56**), by introduction of an azido group at C-5 under inversion of configuration at this carbon, gave the desired L-galactose derivative that was converted into (**51**) employing conventional carbohydrate methodology. Applying proven aldolase assisted chemistry, Wong and co-workers prepared a range of iminoalditols and tested their fucosidase inhibitory properties [110]. Amongst others, 1-deoxy-D-rhamnonojirimycin (1,5,6-trideoxy-1,5-imino-D-mannitol **57**) showed considerable activity albeit at least three orders of magnitude less than (**51**).

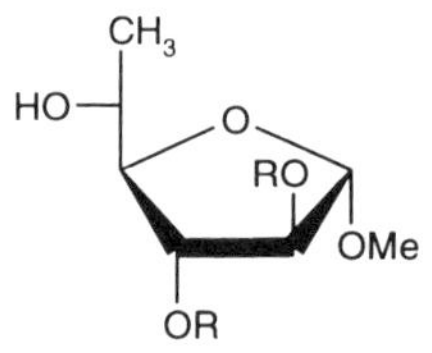

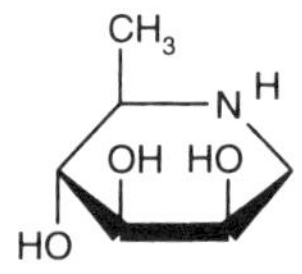

Fig. 20 **56** **57**

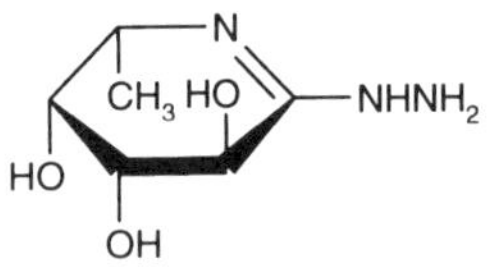

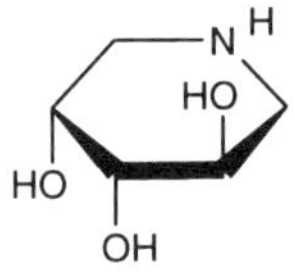

Fig. 21 **58** **59**

An amidrazone (**58**) derived from 5-amino-5-deoxy-L-fuconolactam was found to inhibit a recombinant human *α*-L-fucosidase with a K_i-value of 820 nmol/l [111]. A simple synthesis of 1,5-dideoxy-1,5-imino-D-arabinitol (**59**), previously prepared by Ganem et al. [49] as a potential mannosidase inhibitor, was applied to the affinity purification of *α*-L-fucosidase from bovine kidney by an improved method and the characterization of the enzyme thus obtained [112]. The relatively low affinity of this compound to the enzyme (K_i 2.2 µmol/l at pH 7) compared to 1-deoxyfuconojirimycin (**51**) turned out to be advantageous in terms of enzyme recovery and yield. Structurally related, suitably protected 5-amino-5-deoxy-D-arabinopyranose (**60**), was coupled with a *N*-acetyl-6-deoxy-6-thio-D-glucosaminide (**61**) to give a stable thioglycoside (**62**) [113].

The biological activity of the natural product *α*-homonojirimycin (2,6-dideoxy-2,6-imino-D-*glycero*-L-*gulo*-heptitol, **14**) as a glucosidase inhibitor [50]

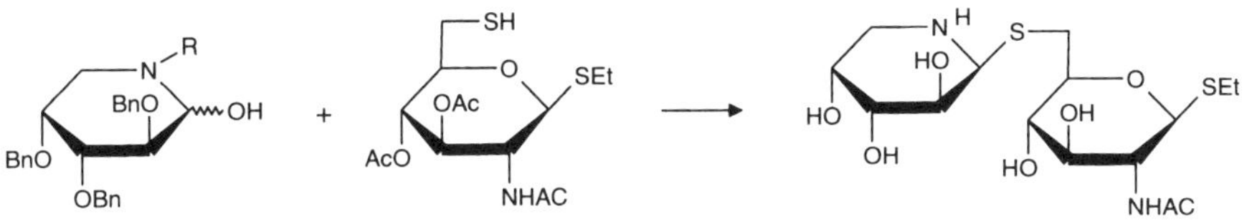

Fig. 22

triggered interest in other 2,6-dideoxy-2,6-iminoheptitols as inhibitors of various glycosidases. Analogues of 1-deoxyfuconojirimycin, such as (**63**), (**64**), and (**65**), were synthesized from a suitably protected 5-azido-5-deoxy-D-mannonolactone by addition of methyl lithium and vinyl as well as phenyl magnesium bromide to C-1. These 1-C-alkyl or 1-C-aryl derivatives of 1-deoxymannojirimycin were shown to be very powerful inhibitors of human liver fucosidase, 1,2,6-trideoxy-D-*glycero*-D-*galacto*-heptitol (**63**) being as potent as 1-deoxynojirimycin itself. Increasing chain length/bulkiness in the side chain led to activity loss of one order of magnitude with the C-ethyl (**64**) and two orders of magnitude with the C-phenyl derivative (**65**). Due to the inherent structural similarity of α-L-fucosidase and D-mannosidase inhibitors, many of the compounds under consideration inhibit both groups of enzymes and, therefore, due to this lack of selectivity, cannot be used for practical applications. Interestingly, it was found that the good mannosidase inhibitor α-homomannojirimycin (2,6-dideoxy-2,6-imino-D-*glycero*-D-*talo*-heptitol **66**), interferes to only a slight degree with fucosidases. The epimer of this compound at C-6, 2,6-dideoxy-2,6-imino-L-*glycero*-D-*galacto*-heptitol (**67**), a powerful fucosidase inhibitor with K_i 4.5 µmol/l, does not interact with either human lysosomal, Golgi II, or neutral α-mannosidases. Both of these compounds had been prepared from suitably protected heptono-1,5-lactones [115, 116]. 2,6,7-Trideoxy-2,6-imino-D-*glycero*-D-*gluco*-heptitol (**68**), previously prepared by Fleet et al. [114] from a carbohydrate starting material, was also synthesized by Effenberger and co-workers via a rabbit muscle aldolase catalyzed reaction as the key-step [117]. From the C-5 epimer of the previously used azidodeoxy lactone (**70**) [114], the desired [114] α-L-homofuconojirimycin (1,2,6-trideoxy-2,6-imino-L-*glycero*-D-*galacto*-heptitol **69**) was synthesized in 1993 by a Glaxo group [118] who employed a method based on Fleet chemistry. The inhibition of α-L-fucosidases by this compound was very similar to the effect exerted by the C-6-epimer [114]. As with the latter, no inhibition of α-D-mannosidase was observed. β-Homomannojirimycin (**67**) was also synthesized from 3-azido-3-deoxy-D/L-threose [119] and from the enriched L-enantiomer [120], which was obtained via a Sharpless epoxidation of (*Z*)-but-2-ene-1,4-diol, with the aid of D-fructose-1,6-diphospate aldolase.

A L-*fuco*-related analogue (**71**) of isofagomine [61], which was recently prepared from D-ribose by Ichikawa et al. following previously established chemistry [63, 121], was found to be an α-L-fucosidase inhibitor with K_i 8 µmol/l. No inhibition was found with tetrazole derivatives such as (**72**) [122] obtained from L-gulonolactone.

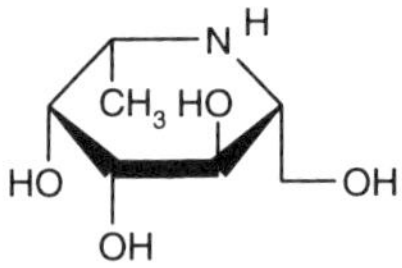

63 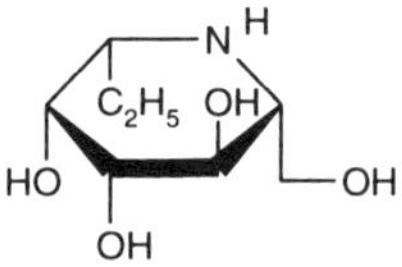 64 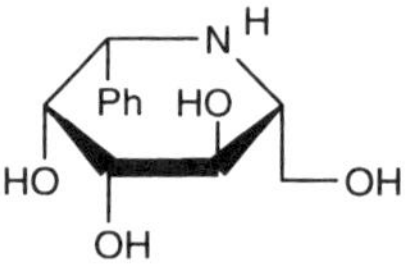 65

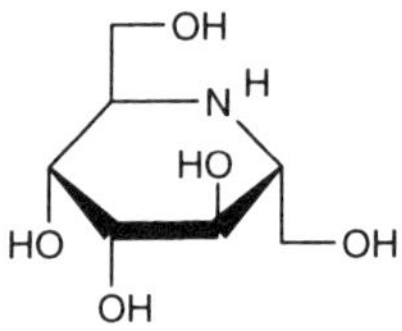

66 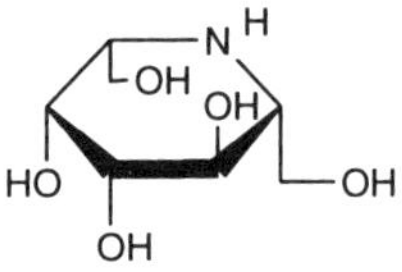 67 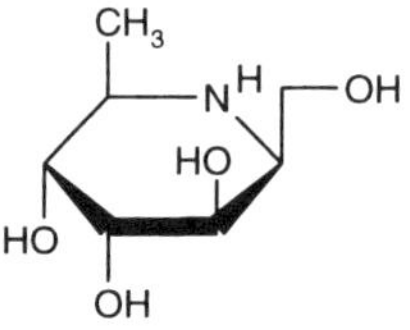 68

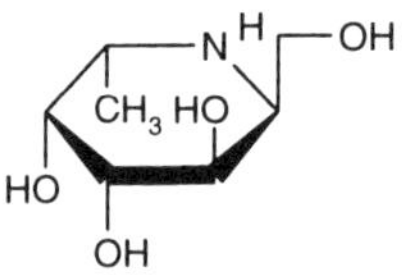

69 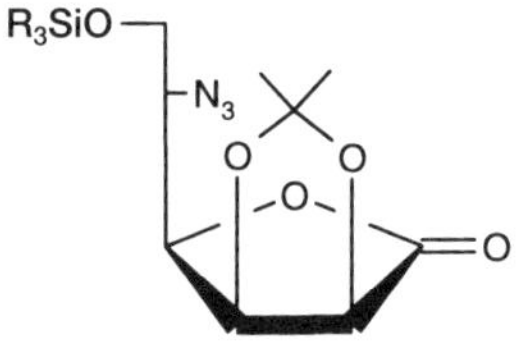 70

Fig. 23

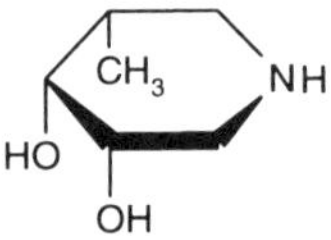

71 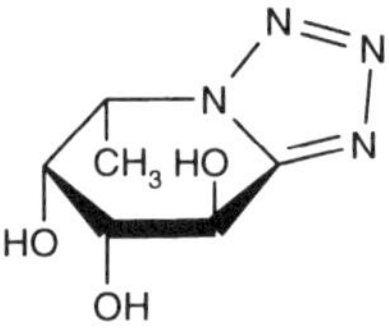 72

Fig. 24

4.2
Pyrrolidine Derivatives

1,2,5-Trideoxy-2,5-imino-L-iditol (**73**), synthesized with an aldolase assisted method by Wong and his group [110], although exhibiting considerable inhibitory activity, is still three orders of magnitude less efficient than 1-deoxyfuconojirimycin. Other five-membered ring analogues of 1-deoxyfuconojirimycin, also prepared by fucose 1-phosphate aldolase methodology and tested by these workers, were 1,2,5-trideoxy-2,5-imino-L-altritol (**74**), -D-glucitol (**75**), -D-mannitol (**76**), as well as -L-gulitol (**77**). All of these substances exhibited K_i-values in the micromolar range [123].

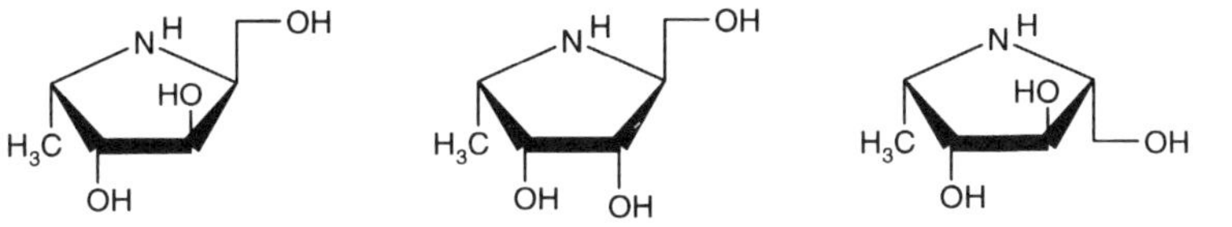

73 74 75

76 77

Fig. 25

The iminoaltritol (**74**) as well as the iminogulitol (**77**) could also be shown to be moderate inhibitors of a recombinant human α-1,3-fucosyltransferase.

Fucosidase inhibiting 1,4,5-trideoxy-1,4-imino-L-lyxitol (**78**) was prepared [124] from D-ribose via protected 5-amino-5-deoxy-L-lyxose (**79**) by a chemical route. This compound, as well as the 1-aminomethyl homologue (**80**), obtained by Kiliani-Fischer chain extension of (**79**), inhibited α-L-fucosidase with K_i around 2 μmol/l. 1,4-Dideoxy-1,4-imino-D-iditol (**81**) [71] was found to be a moderate inhibitor of the enzyme.

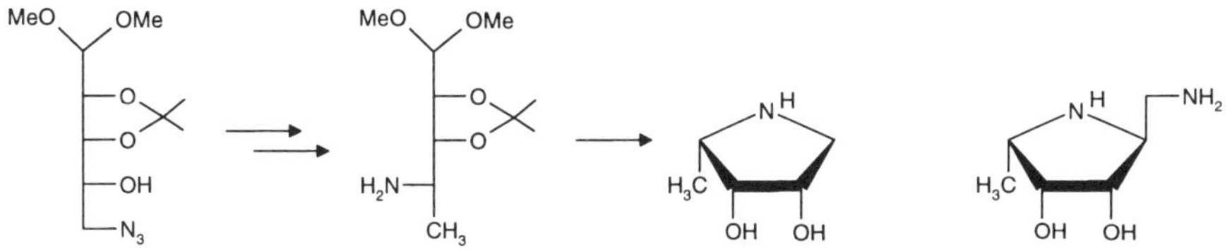

79 78 80

Fig. 26

5 Inhibitors of Other Glycosidases

5.1 Invertase Inhibitor

2,5-Dideoxy-2,5-imino-D-mannitol (**26**), a natural product found in *Derris eliptica* as well as *Lonchocarpus* sp. [126], was obtained from L-sorbose via 5-azido-5-deoxy-D-fructopyranose (**82**) by Card and Hitz [127] and characterized as a very potent inhibitor of β-D-fructofuranosidase (invertase).

This high activity was recently exploited by employing an immobilized compound (**83**), obtained in high yield by glucose isomerase catalyzed conversion of 5-azido-5-deoxy-D-glucofuranose into 5-azido-5-deoxy-fructopyranose and subsequent intramolecular reductive amination, as a affinity ligand for the isolation and charcterization of invertase from yeast [94]. Other approaches to 2,5-dideoxy-2,5-imino-D-mannitol were reported by Fleet and Smith [128] from D-glucose and by Reitz and Baxter as a by-product from the reductive cyclization of "5-keto-D-fructose" with benzhydrylamine [129]. In addition, Whitesides and co-workers reported a synthetic approach applying a rabbit muscle aldolase [130]. 2,5-Dideoxy-2,5-imino-D-mannitol has also been prepared from D-mannitol [131] and from a suitably protected D-arabinofuranose, the latter making use of an aminomercuration-type reaction [132].

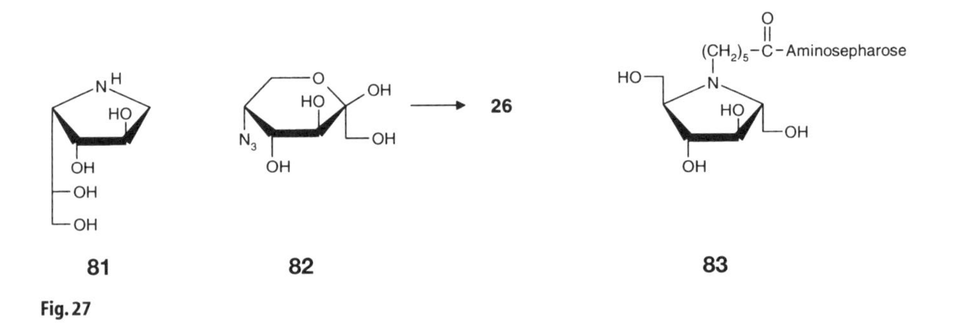

Fig. 27

5.2 D-Glucuronidase, L-Iduronidase, Neuraminidase, and Heparanase Inhibitors

The nitrogen-containing structural analogue (**84**) of D-glucuronic acid has been found occurring naturally [133] and was reported to be a D-glucuronidase as well as L-iduronidase inhibitor [134]. Total syntheses of this compound from D-glucofuranurono-6,3-lactone have been reported by Fleet and his group [135] in addition to others [136]. A potent β-glucuronidase inhibitor (K_i 79 nmol/l) with the basic nitrogen in the position of the anomeric carbon (**85**) was recently synthesized by Ichikawa and co-workers [137]. This compound being about 1000-fold as potent as the glucuronic acid analogue (**84**), is about half as active as D-glucaro-δ-lactam (**86**), which exhibits K_i 39 µmol/l.

The same workers found the corresponding L-iduronic acid congener (**87**) to be a moderate inhibitor of β-glucuronidase from human liver (K_i 1.3 µmol/l). The sodium salt of glucaro-δ-lactam and strucural analogues thereof were recently reported to be potent inhibitors of pulmonary metastases and tumour cell invasion [138].

Siastatin B (**40**), the first natural inhibitor of neuraminidase and a compound that also inhibits β-D-glucuronidase, was discovered in 1974 [139]. Two total syntheses of this compound have been published to date [140, 141]. Due to the fact that recent studies have shown that human tumours are associated with

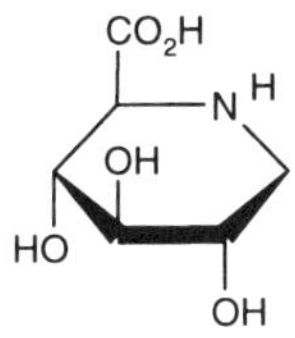

Fig. 28

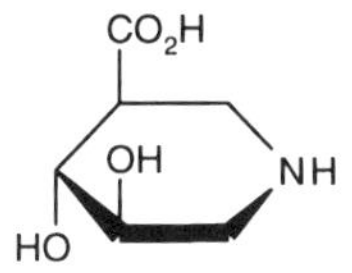

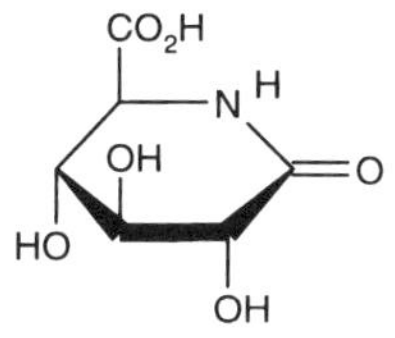

increased levels of β-D-glucuronidase activity and that this enzyme might play a role in the metastasis of tumour cells, much effort has been focused on the synthesis of siastatin-based inhibitors as potential antimetastatic drugs [142]. Natural diastereoisomers of siastatin B were recently found to exhibit interesting β-D-glucuronidase as well as tumour cell heparanase inhibiting effects [143]. Synthetic analogues (**88**) of siastatin related to L-iduronic acid were good inhibitors of experimentally induced metastatis in mice [144]. This fact was attributed to their inhibitory effects on α-L-iduronidase supposed to be excreted by human melanoma in the process of metastasis.

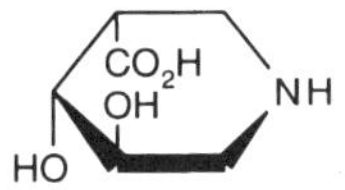

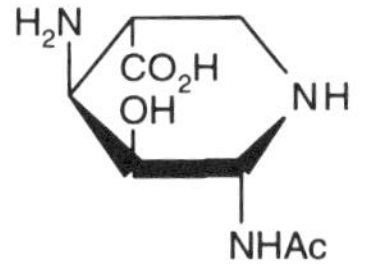

5.3 Rhamnosidase Inhibitors

Both enantiomers of rhamnose, 6-deoxy-mannose, occur naturally. L-Rhamnose is a constituent of glycosides such as naringin (**89**) as well as hesperidin, and a constituent in plant tissues as well as of bacterial cell walls. α-L-Rhamnosidases have been employed to debitter fruit juices [145].

Both enantiomers (**90**) and (**91**) of 1-deoxy-rhamnonojirimycin (1,5,6-dideoxy-1,5-imino-mannitol) were synthesized by Fleet and co-workers from D- and L-gulonolactones employing established methodology [146]. Interestingly, neither of the two was found to be an inhibitor of naringinase, a common α-L-rhamnosidase. This was attributed to the possibility that this enzyme depends on the natural aglycon of naringin for recognition/binding. 1-Deoxy-rhamnonojirimycins had previously been prepared by Wong and co-workers employing an enzymatic aldol reaction as the key-step [42, 147]. In contrast to previous findings [146], 1-deoxy-L-rhamnonojirimycin (**90**), prepared with the aid of *E. coli* L-rhamnulose-1-phosphate aldolase, was found to inhibit naringinase with K_i 34 µmol/l [148]. A more potent α-L-rhamnosidase inhibitor

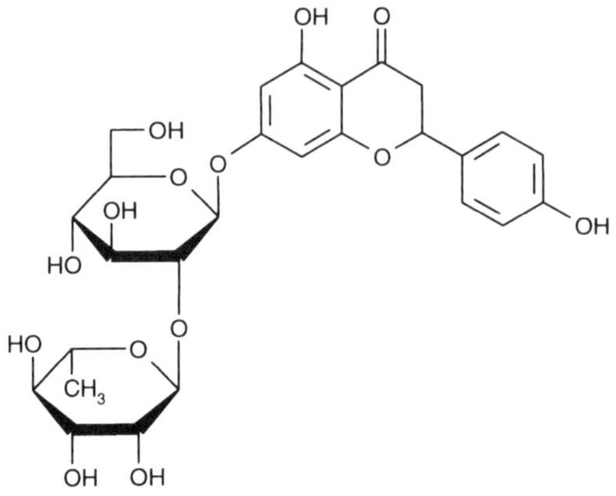

89

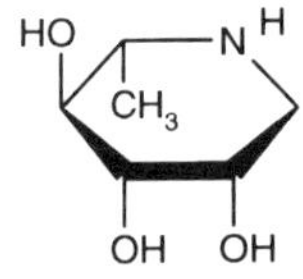

90 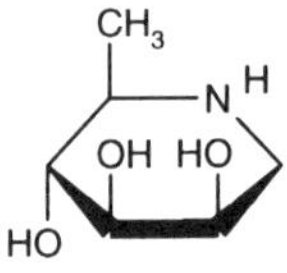 91 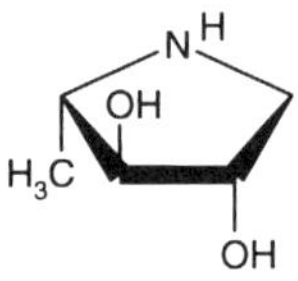 92

Fig. 29

than (**90**) was the 5-membered ring analogue (**92**) recently prepared by Wong and his group [149] inhibiting the enzyme with K_i 5.5 µmol/l [150].

5.4 Inhibitors of Soluble Lytic Transglycosylases

Growth and divison of most bacteria depends on the proper enlargement of their unique cell wall skeleton, the murein sacculus [151]. The bag-shaped macromolecule murein, a peptidoglycan, is made of glycan strands which are interlinked by peptides, thereby forming a covalently closed network that completely envelopes the cell [152]. Enzymes capable of cleaving bonds within the murein network are deemed to be essential for growth of the murein sacculus and splitting of the septum during cell division [153]. The action of such murein hydrolases has to be strictly coordinated with murein synthesizing enzymes, or autolysis of the cell would occur. Compounds known to interfere with this subtle balance of synthesis and lysis of the cell wall are, for example, the penicillins, by reaction with a class of enzymes called Penicillin Binding Proteins (PBP). Other agents behaving in a similar way could also act as antibacterial therapeutics. Bulgecins are a family of antibiotics also interfering with cell wall synthesis but they do not act on any PBP known. Rather, they inhibit bacterial α-specific enzymes acting on murein, called Soluble Lytic Transglycosylases (SLT).

These enzymes cleave the β-1,4-glycosidic bonds of peptidoglycan to produce small 1,6-anhydro-muropeptides (**93**) [154, 155]. Interestingly, an X-ray crystallography study of the structure of SLT has recently become available [156]. This enzyme was found to resemble closely the structure of lysozymes [157]. Remarkably, in this context, no significant sequence homology of amino acid residues could be detected. This successful structure elucidation should allow the design of non-natural specific inhibitors of this enzyme on the basis of structural information known from the bulgecins.

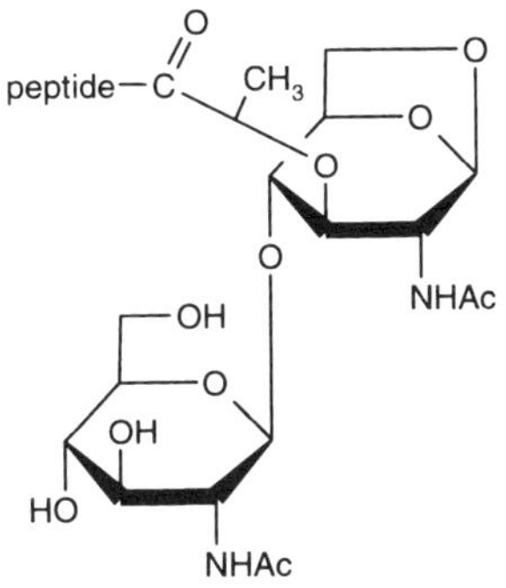

Fig. 30

Bulgecins were reported for the first time in 1982 [158]. Their structures, such as (**94**), were established soon afterwards [159, 160] and they were found to interfere with the cell-wall synthesis of gram-negative bacteria due to a unique mechanism inhibiting vital bacterial Soluble Lytic Transglycosylase (SLT) [161].

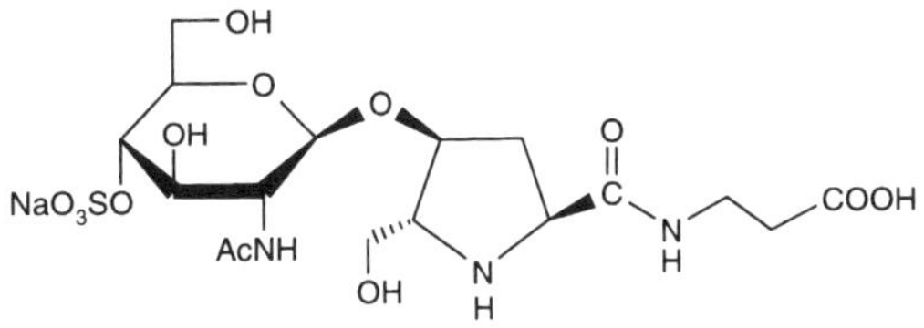

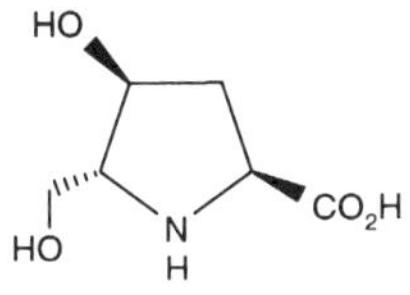

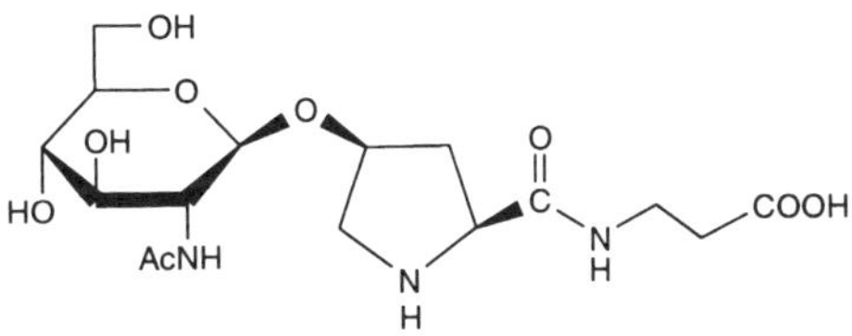

Fig. 31

As this enzyme does not exist in mammals and man, potential specific inhibitors could serve as non-toxic, no side effect antibiotics against gram-negative pathogenic organisms under consideration.

Several syntheses of bulgecinin (**95**) [135, 162–173] the five-membered partial ring structure of bulgecins, have been published. Only a few approaches to bulgecins [167, 174, 175] as well as non-natural derivatives thereof [176, 177] such as (**96**) have been reported to date.

5.5
Inhibitors of N-Ribohydrolases

N-Ribohydrolases have been found to be involved in novel pathways of purine salvage in protozoan parasites as well as in nucleic acid repair, and exhibit other interesting biological activities [178]. In order to investigate the molecular electrostatic potential surface of the enzyme from the trypanosome *Crithidia fasciculata*, several 1,4-dideoxy-1,4-imino-D-ribitol derivatives were synthesized as nucleoside analogues and their inhibitory powers were tested [179, 180]. In the course of this work, 1,4-dideoxy-1,4-imino-1(*S*)-phenyl-D-ribitol (**97**) was found to inhibit this enzyme with K_i 30 nmol/l.

6
Miscellaneous

Just to indicate the biological potential inherent in sugar analogues with basic nitrogen instead of oxygen in the ring, here are a few examples from the recent literature briefly outlined.

6.1
Antitumour Cisplatin Derivatives

Cisplatin is a widely used chemotherapeutic agent against various human cancers. Cisplatin-like complexes of free and partially protected 1-amino-1,2,5-trideoxy-2,5-imino-L-arabinitol (**98**) and (**99**) were recently synthesized and found to exhibit high cytotoxicities against a range of human cancer cell lines [181].

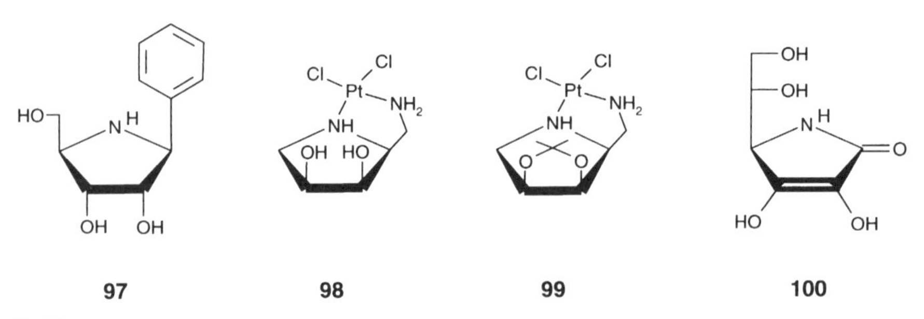

Fig. 32

6.2
A Vitamin C Analogue

The analogue (**100**) of vitamin C with nitrogen in the ring instead of oxygen was synthesized by a German group as the racemate. This was found to be a stronger reducing agent than the parent compound [182].

6.3
The Acaricidal Glycoside Gualamycin

The natural product gualamycin, recently obtained from the fermentation broth of a *Streptomyces* sp., was found to be a potent acaricide [183]. The disaccharide as well as the heterocyclic subunit (**101**) were synthesized to support the structure proof but the completion of the total synthesis was not reported [184].

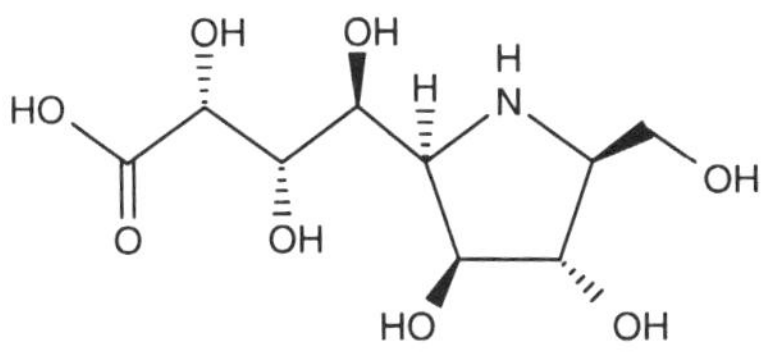

OH OH H H N HO OH O OH HO OH

Fig. 33 **101**

7
Perspectives

It has hopefully been shown in this account that the class of compounds under consideration is a fruitful research area with a wide range of already useful applications as well as a remarkable potential for future discoveries and innovations. The large number of non-natural derivatives of glycosidase inhibitors with nitrogen in the ring, such as fluorine-containing analogues and many others, should be an exciting playground for organic chemists and biochemists in the years to come. The same is true for the likely discoveries of related new natural products in plants, from marine sources, and in microorganisms that live under extreme environmental conditions, aided by the sophisticated state-of-the-art methods of extraction, detection, structure elucidation, synthesis, and derivatization.

References

1. Legler G (1990) Adv Carbohydr Chem Biochem 48:319
2. Rhinehart BL, Robinson KM, Payne AJ, Wheatley ME, Fisher JL, Liu PS, Cheng W (1987) Life Sci 41:2325
3. Ostrander GK, Scribner NK, Rohrschneider LR (1988) Cancer Res 48:1091
4. Fleet GWJ, Karpas A, Dwek RA, Fellows LE, Tyms AS, Peturrson S, Namgoong SK, Ramsden NG, Smith PW, Son JC, Wilson F, Witty DR, Jacob GS, Rademacher TW (1988) FEBS Lett 237:128
5. Bitonti AJ, McCann PP (1990) Eur Patent Appl EP423.728
6. Dreyer DL, Jones KC, Molyneux RJ (1985) J Chem Ecol 11:1045
7. Paulsen H, Sangster I, Heyns K (1967) Chem Ber 100:802
8. Inouye S, Tsuruoka T, Ito T, Niida N (1968) Tetrahedron 24:2125 and references cited therein
9. Ranjanikanth B, Seshadri R (1989) Tetrahedron Lett 30:755
10. Fleet GWJ, Ramsden NG, Carpenter NM, Petursson S, Aplin RT (1990) Tetrahedron Lett 31:405
11. Dax K, Gaigg B, Grassberger V, Kölblinger B, Stütz AE (1990) J Carbohydr Chem 9:479
12. van den Broek LAGM, Vermaas DJ, Heskamp BM, van Boeckel CAA, Tan MCAA, Bolscher JGM, Ploegh HL, van Kemenade FJ, de Goede REY, Miedema F (1993) Recl Trav Chim Pays-Bas 112:82
13. Hughes AB, Rudge AJ (1994) Nat Prod Rep 135
14. Winchester B, Al Daher S, Carpenter NC, Cenci di Bello I, Choi SS, Fairbanks AJ, Fleet GWJ (1993) Biochem J 290:743
15. Burgess K, Henderson I (1992) Tetrahedron 48:4045
16. Herczegh P, Kovacs I, Sztaricskai F (1993) In: Lukacs G (ed) Recent progress in the chemical synthesis of antibiotics and related microbial products, vol 2. Springer, Berlin Heidelberg New York, p 751
17. Look GC, Fotsch CH, Wong C-H (1993) Acc Chem Res 26:182
18. Wong C-H, Halcomb RL, Ichikawa Y, Kajimoto T (1995) Angew Chem Int Ed Engl 34:412
19. Truscheit E, Frommer W, Junge B, Müller L, Schmidt DD, Wingender W (1981) Angew Chem Int Ed Engl 20:744
20. Fuhrmann U, Bause E, Ploegh H (1985) Biochim Biophys Acta 825:95
21. Winchester B, Fleet GWJ (1992) Glycobiology 2:199
22. Tan A, van den Broek L, van Boeckel S, Ploegh H, Bolscher J (1991) J Biol Chem 266:14504
23. Winkler DA, Holan G (1989) J Med Chem 32:2084
24. Paulsen H, Hayauchi Y, Sinnwell V (1980) Chem Ber 113:2601
25. Köll P (1973) Chem Ber 106:3559
26. Angyal S J, Beveridge RJ (1979) Aust J Chem 31:1151
27. Legler G, Jülich E (1984) Carbohydr Res 128:61
28. Meyer zu Reckendorf W (1967) Angew Chem Int Ed Engl 6:177
29. Legler G, Pohl S (1986) Carbohydr Res 155:119
30. Bernotas RC, Ganem B (1985) Tetrahedron Lett 26:1129
31. Bernotas RC, Pezzone MA, Ganem B (1987) Carbohydr Res 167:305
32. Bernet B, Vasella A (1979) Helv Chim Acta 62:1990
33. Tong MK, Ganem B (1988) J Am Chem Soc 110:312
34. Kinast G, Schedel M (1981) Angew Chem Int Ed Engl 20:805
35. Heiker F-R, Schueller AM (1990) Carbohydr Res 203:314
36. Miyake Y, Ebata M (1987) J Antibiot 40:122
37. Miyake Y, Ebata M (1988) Agric Biol Chem 52:661
38. Miyake Y, Ebata M (1988) Agric Biol Chem 52:153
39. Aoyagi S, Fujimaki S, Yamazaki N, Kibayashi C (1991) J Org Chem 56:815
40. Dondoni A, Merino P, Perrone D (1991) J Chem Soc Chem Commun 1576
41. von der Osten CH, Sinskey AJ, Barbas CF, Pederson RL, Wang Y-F, Wong C-H (1989) J Am Chem Soc 111:3924

42. Kajimoto T, Chen L, Liu K-C, Wong C-H (1991) J Am Chem Soc 113:6678
43. Lees WJ, Whitesides GM (1992) Bioorg Chem 20:173
44. Fessner W-D, Badia J, Eyrisch O, Schneider A, Sinerius G (1992) Tetrahedron Lett 33:5231
45. Streith J, Defoin A (1996) Synlett 189
46. Furneaux RH, Tyler PC, Whitehouse LA (1993) Tetrahedron Lett 34:3609
47. Chida N, Tanikawa T, Tobe T, Ogawa S (1994) J Chem Soc Chem Commun 1247
48. Johnson CR, Golebiowski A, Sundram H, Miller MW, Dwaihy RL (1995) Tetrahedron Lett 36:653
49. Bernotas RC, Papandreou G, Urbach J, Ganem B (1990) Tetrahedron Lett 31:3393
50. Liu PS (1987) J Org Chem 42:4717
51. Kite GC, Fellows LE, Fleet GWJ, Liu PS, Scofield AM, Smith NG (1988) Tetrahedron Lett 29: 6483
52. Martin OR, Xie F, Liu L (1995) Tetrahedron Lett 36:4027
53. Baudat A, Vogel P (1996) Tetrahedron Lett 37:483
54. Baudat A, Picasso S, Vogel P (1996) Carbohydr Res 281:277
55. Aoyama T, Naganawa H, Suda H, Uotani K, Aoyagi T, Takeuchi T (1992) J Antibiot 45: 1557
56. Tatsuta K, Miura S, Ohta S, Gunji H (1995) Tetrahedron Lett 36:1085
57. Tatsuta K, Miura S, Ohta S, Gunji H (1995) J Antibiotics 48:286
58. Tong MK, Papandreou G, Ganem B (1990) J Am Chem Soc 112:6137
59. Papandreou G, Tong MK, Ganem B (1993) J Am Chem Soc 115:11682
60. Fotsch CH, Wong C-H (1994) Tetrahedron Lett 35:3481
61. Jespersen TM, Dong W, Sierks MR, Skrydstrup T, Lundt I, Bols M (1994) Angew Chem Int Ed Engl 33:1778
62. Ichikawa M, Ichikawa Y (1995) Bioorg Med Chem 3:161
63. Ichikawa Y, Igarashi Y (1995) Tetrahedron Lett 36:4585
64. Fleet GWJ, Nicholas SJ, Smith PW, Evans SV, Fellows LE, Nash RJ (1985) Tetrahedron Lett 26:3127
65. Austin GN, Baird PD, Fleet GWJ, Peach JM, Smith PW, Watkin DJ (1987) Tetrahedron 43: 3095
66. Bashyal BP, Fleet GWJ, Gough MJ, Smith PW (1987) Tetrahedron 43:3083
67. Wehner V, Jäger V (1990) Angew Chem Int Ed Engl 29:1169
68. Ikota N, Hanaki A (1987) Chem Pharm Bull 35:2140
69. Goli DM, Cheesman BV, Hassan ME, Lodaya R, Slama JT (1994) Carbohydr Res 259:219
70. Han S-Y, Liddell PA, Joullie M (1988) Synth Commun 18:275
71. Lundt I, Madsen R, Al Daher S, Wichester B (1994) Tetrahedron 50:7513
72. Wang Y-F, Takaoka Y, Wong C-H (1994) Angew Chem Int Ed Engl 33:1242
73. Hamana H, Ikota N, Ganem B (1987) J Org Chem 52:5494
74. Fowler PA, Haines AH, Taylor RJK, Chrystal EJT (1994) J Chem Soc Perkin Trans 1 2229
75. Legler G, Lüllau E, Kappes E, Kastenholz F (1991) Biochim Biophys Acta 1080:89
76. Fleet GWJ, Smith PW, Nash RJ, Fellows LE, Parekh RB, Rademacher TW (1986) Chem Lett 1051
77. Fleet GWJ, Fellows LE, Smith PW (1987) Tetrahedron 43:979
78. Böshagen H, Heiker F-R, Schüller AM (1987) Carbohydr Res 164:141
79. Kappes E, Legler G (1989) J Carbohydr Chem 8:371
80. Eur Pat Appl EP 298.350; (1989) Chem Abstr 111:58272
81. Schüller AM, Heiker F-R (1990) Carbohydr Res 203:308
82. Kiso M, Kitagawa M, Ishida H, Hasegawa A (1991) J Carbohydr Chem 10:25
83. Kajimoto T, Liu KK-C, Pederson RL, Zhong Z, Ichikawa Y, Porco JA, Wong C-H (1991) J Am Chem Soc 113:6187
84. Furneaux RH, Gainsford GJ, Lynch GP, Yorke SC (1993) Tetrahedron 49:9605
85. Bernotas RC, Ganem B (1987) Carbohydr Res 167:312
86. Ishida H, Kitagawa M, Kiso M, Hasegawa A, Azuma I (1990) Carbohydr Res 208:267
87. Kiso M, Furui H, Ishida H, Hasegawa A (1996) J Carbohydr Chem 15:1
88. Umezawa H, Aoyagi T, Komiyama T, Morishima H, Hamada M, Takeuchi T (1974) J Antibiot 27:963

89. Nishimura Y, Satoh T, Kudo T, Kondo S, Takeuchi T (1996) Bioorg Med Chem 4:91
90. Tatsuta K, Miura S (1995) Tetrahedron Lett 36:6721
91. Croucher PD, Furneaux RH, Lynch GP (1994) Tetrahedron 50:13299
92. Liessem B, Giannis A, Sandhoff K, Nieger M (1993) Carbohydr Res 250:19
93. Takaoka Y, Kajimoto T, Wong C-H (1993) J Org Chem 58:4809
94. Legler G, Korth A, Berger A, Ekhart C, Gradnig G, Stütz AE (1993) Carbohydr Res 250:67
95. Schumacher-Wandersleb MHMG, Petersen S, Peter MG (1994) Liebigs Ann Chem 555
96. Campanini L, Dureault A, Depezay J-C (1995) Tetrahedron Lett 36:8015
97. de Raadt A, Ebner M, Ekhart CW, Fechter M, Lechner A, Strobl M, Stütz AE (1994) Cat Today 22:549
98. Andersen SM, de Raadt A; Ebner M, Ekhart C, Gradnig G, Legler G, Lundt I, Schichl M, Stütz AE, Withers SG (1995) In: Rzepa HS, Goodman JG (eds) Electronic conference on trends in organic chemistry (CD-ROM). Royal Soc Chem Publications
99. Liu PS, Kang MS, Sunkara PS (1991) Tetrahedron Lett 32:719
100. Furneaux RH, Gainsford GJ, Mason JM, Tyler PC (1994) Tetrahedron 50:2131
101. Fleet GWJ, Shaw AN, Evans SV, Fellows LE (1985) J Chem Soc Chem Commun 841
102 Paulsen H, Matzke M (1988) Liebigs Ann Chem 1121
103. Hettkamp H, Legler G, Bause E (1984) Eur J Biochem 142:85
104. Fleet GWJ, Petursson S, Campbell AL, Mueller RA, Behling JR, Babiak KA, Ng JS, Scaros MG (1989) J Chem Soc Perkin Trans 1 665
105. Hashimoto H, Hayakawa M (1989) Chem Lett 1881
106. Scudder P, Neville DCA, Butters TD, Fleet GWJ, Dwek RA, Rademacher TW, Jacob GS (1990) J Biol Chem 265:16472
107. Paulsen H, Matzke M, Orthen B, Nuck R, Reutter W (1990) Liebigs Ann Chem 953
108. Adelhorst K, Whitesides GM (1992) Carbohydr Res 232:183
109. Takahashi S, Kuzuhara H (1992) Chem Lett 21
110. Dumas DP, Kajimoto T, Liu KK-C, Wong C-H, Berkowitz DB, Danishefsky SJ (1992) Bioorg Med Chem Lett 2:33
111. Schedler DJA, Bowen BR, Ganem B (1994) Tetrahedron Lett 35:3845
112. Legler G, Stütz AE, Immich H (1995) Carbohydr Res 272:17
113. Suzuki K, Hashimoto H (1994) Tetrahedron Lett 35:4119
114. Fleet GWJ, Namgoong SK, Barker C, Baines S, Jacob GS, Wichester B (1989) Tetrahedron Lett 30:4439
115. Bruce I, Fleet GWJ, Cenci di Bello I, Wichester B (1989) Tetrahedron Lett 30:7257
116. Bruce I, Fleet GWJ, Cenci di Bello I, Winchester B (1992) Tetrahedron 48:10191
117. Straub A, Effenberger F, Fischer P (1990) J Org Chem 55:3926
118. Andrews DM, Bird MI, Cunningham MM, Ward P (1993) Bioorg Med Chem Lett 3: 2533
119. Henderson I, Laslo K, Wong C-H (1994) Tetrahedron Lett 35:359
120. Holt KE, Leeper FJ, Handa S (1994) J Chem Soc Perkin Trans 1 231
121. Ichikawa M, Igarashi Y, Ichikawa Y (1995) Tetrahedron Lett 36:1767
122. Brandstetter TW, Davis B, Hyett D, Smith C, Hackett L, Winchester B, Fleet GWJ (1995) Tetrahedron Lett 36:7511
123. Wang Y-F, Dumas DP, Wong C-H (1993) Tetrahedron Lett 34:403
124. Wong C-H, Provencher L, Porco JA, Jung A-H, Wang Y-F, Chen L, Wang R, Steensma DH (1995) J Org Chem 60:1492
125. Welter A, Jadot J, Dardenne G, Marlier M, Casimir J (1976) Phytochem 15:747
126. Evans SV, Fellows LE, Shing T, Fleet GWJ (1985) Phytochem 24:1953
127. Card PJ, Hitz WD (1985) J Org Chem 50:891
128. Fleet GWJ, Smith PW (1987) Tetrahedron 43:971
129. Reitz AB, Baxter EW (1990) Tetrahedron Lett 31:6777
130. Hung RR, Straub JA, Whitesides GM (1991) J Org Chem 56:3849
131. Dureault A, Portal M, Depezay JC (1991) Synlett 225
132. Chorghade MS, Cseke CT, Liu PS (1994) Tetrahedron: Asymm 5:2251

133. Manning KS, Lynn DG, Shabanowitz J, Fellows LE, Singh M, Schrire BD (1985) J Chem Soc Chem Commun 127
134. Cenci Di Bello I, Dorling P, Fellows LE, Wichester B (1984) FEBS Lett 176:61
135. Bashyal BP, Chow H-F, Fleet GWJ (1986) Tetrahedron Lett 27:3205
136. Bernotas RC, Ganem B (1985) Tetrahedron Lett 26:4981
137. Igarashi Y, Ichikawa M, Ichikawa Y (1996) Tetrahedron Lett 37:2707
138. Tsuruoka T, Fukuyashi H, Azetaka M, Iizuka Y, Inouye S, Hosokawa M, Kobayashi H (1995) Jpn J Cancer Res 86:41
139. Umezawa H, Aoyagi T, Komiyama T, Morishima H, Hamada M, Takeuchi T (1974) J Antibiot 27:963
140. Nishimura Y, Wang W, Kondo S, Aoyagi T, Umezawa H (1988) J Am Chem Soc 110:7249
141. Nishimura Y, Wang W, Kudo T, Kondo S (1992) Bull Chem Soc Jpn 65:978
142. Nishimura Y, Kudo T, Kondo S, Takeuchi T (1994) J Antibiot 47:101 and references cited therein
143. Kawase Y, Takahashi M, Takatsu T, Arai M, Nakajima M, Tanzawa K (1996) J Antibiot 49:61
144. Nishimura Y, Satoh T, Adachi H, Kondo S, Takeuchi T, Azetaka M, Fukuyasu H, Iizuka Y (1996) J Am Chem Soc 118:3051
145. Chase T (1974) Adv Chem Ser 136:257
146. Fairbanks AJ, Carpenter NC, Fleet GWJ, Ramsden NG, Cenci de Bello I, Wichester BG, Al-Daher SS, Nagahashi G (1992) Tetrahedron 48:3365
147. Kajimoto T, Liu KK-C, Pederson RL, Zhong Z, Ichikawa Y, Porco JA, Wong C-H (1991) J Am Chem Soc 113:6187
148. Zhou P, Salleh HM, Chan PCM, Lajoie G, Honek J, Nambiar PTC, Ward OP (1993) Carbohydr Res 239:155
149. Provencher L, Steensma DH, Wong C-H (1994) Bioorg Med Chem 2:1179
150. Wong C-H, Provencher L, Porco JA, Jung S-H, Wang Y-F, Chen L, Wang R, Steensma DH (1995) J Org Chem 60:1492
151. Koch AL (1990) Res Microbiol 141:529
152. Glauner B, Höltje J-V, Schwarz U (1988) J Biol Chem 263:10088
153. Höltje J-V, Tuomanen EI (1991) J Gen Microbiol 137:441
154. Höltje J-V, Mirelman D, Sharon N, Schwarz U (1975) J Bact 124:1067
155. Engel H, Kazemir B, Keck W (1991) J Bact 173:6773 and references cited therein
156. Thunnissen A-MWH, Dijkstra AJ, Kalk KH, Rozeboom HJ, Engel H, Keck W, Dijkstra BW (1994) Nature 367:750
157. Hardy LW, Poteete AR (1991) Biochemistry 30:9457
158. Imada A, Kintaka K, Nakao M, Shinagawa S (1982) J Antibiot 35:1400
159. Shinagawa S, Kasahara F, Wada Y, Harada S, Asai M (1984) Tetrahedron 40:3465
160. Shinagawa S, Maki M, Kintaka K, Imada A, Asai M (1985) J Antibiot 38:17
161. Templin MF, Edwards DH, Höltje J-V (1992) J Biol Chem 267:20039
162. Bashyal, BP, Chow H-F, Fleet GWJ (1986) Tetrahedron Lett 27:3205
163. Wakayama T, Yamanoi K, Nishikawa M, Shiba T (1985) Tetrahedron Lett 26:4759
164. Bashyal BP, Chow HK, Hak F, Fleet GWJ (1987) Tetrahedron 43:423
165. Ofune Y, Hori K, Sakaitani M (1986) Tetrahedron Lett 27:6079
166. Ohta T, Hosoi A, Nozoe S (1988) Tetrahedron Lett 29:329
167. Barrett AGM, Pilipauskas D (1990) J Org Chem 55:5194
168. Hirai Y, Terada T, Amemiya Y, Momose T (1992) Tetrahedron Lett 33:7893
169. Jackson RFW, Rettie AB (1993) Tetrahedron Lett 34:2985
170. Yuasa Y, Ando J, Shibuya S (1994) J Chem Soc Chem Commun 1383
171. Jackson RFW; Rettie AB, Wood A, Wythes MJ (1994) J Chem Soc Perkin Trans 1 1719
172. Schmeck C, Hegedus LS (1994) J Am Chem Soc 116:9927
173. Oppolzer W, Moretti R, Zhou C (1994) Helv Chim Acta 77:2363
174. Barrett AGM, Pilipauskas D (1990) J Org Chem 56:2787 and references cited therein
175. Wakamiya T, Kimura Y, Takahashi S, Shimamoto K, Nishikawa M, Kanou K, Ohyama K, Shiba T (1991) Pept Chem 28:103

176. Wakamiya T, Yamanoi K, Kanou K, Shiba T (1987) Tetrahedron Lett 28:5887
177. Brown AG, Moss SF, Southgate R (1994) Tetrahedron Lett 35:451
178. Mazzella LJ, Parkin DW, Tyler PC, Furneaux RH, Schramm VL (1996) J Am Chem Soc 118:2111 and references cited therein
179. Horenstein BA, Zabinski RF, Schramm VL (1993) Tetrahedron Lett 34:7213
180. Horenstein B, Schramm VL (1993) Biochem 32:9917
181. Kim D-K, Kim Y-W, Kim H-T, Kim K H (1996) Bioorg Med Chem Lett 6:643
182. Stachel H-D, Zeitler K, Dick S (1996) Liebigs Ann Chem 103
183. Tsuchiya K, Kobayashi S, Harada T, Kurokawa T, Nakagawa T, Shimada N (1995) J Antibiot 48:626
184. Tatsuta K, Kitagawa M, Horiuchi T (1995) J Antibiot 48:741

The Synthesis of Novel Enzyme Inhibitors and Their Use in Defining the Active Sites of Glycan Hydrolases*,**

Robert V. Stick

Department of Chemistry, The University of Western Australia, Nedlands, Western Australia 6907

β-D-Glucosidases and β-D-glucan hydrolases are two important classes of enzymes that are responsible for various processes that occur in plants and bacteria. Here we describe our efforts, over the past decade, towards the design and synthesis of potential inhibitors of these enzymes. In particular, approaches are described to epoxyalkyl glycosides, deoxy halo epoxides, aziridines, thiirans and epoxyalkyl C-glycosides, all potentially irreversible, mechanism-based inhibitors of the above enzymes. Our efforts towards tetrahydro-1,2-oxazines are also included. Some time is devoted to the concept of "β-acarbose", a diastereoisomer of the naturally occurring acarbose, as a novel, competitive inhibitor of the above enzymes. Throughout this work, attention was continually directed to the synthesis of potential inhibitors in optically-pure form.

Table of Contents

---

* The original title to this article, "Inhibitions in Australia – Are There Any?" was, understandably, unacceptable to the editors.

** This article is dedicated to Professor FN (Norm) Lahey, natural product chemist and mentor.

1 Prologue

To most world travellers, Australia *is* a long way away, separated in the main by a good body of water from major land masses and deserving of the title "Downunder". This physical isolation has led to Australia having a unique collection of flora and fauna, both marine and terrestrial. It was not surprising, therefore, that early science in Australia focussed on local organisms from the sea and from the land and, in particular, chemists sought to extract from these sources the very substances that were responsible for some biological activity or therapeutic event associated with local folklore.

One particular example of the early extraction of a chemical from a plant, a small tree found in the rainforests of eastern Queensland, would be that of acronycine **1** [1, 2], an alkaloid currently undergoing clinical trials as an anti-cancer agent. More recently, a group in Perth at Murdoch University succeeded in extracting the alkaloid, swainsonine **2** from a desert shrub which causes poisoning in cattle [3]. Swainsonine has been found to exhibit an interesting spectrum of biological activity [4], especially as a potent, reversible inhibitor of various α-D-mannosidases [5].

It is regrettable that Australian chemists failed to discover another "local" alkaloid, namely castanospermine **3**. It was left to an overseas group to collect and extract the seeds from a Queensland legume to yield this valuable alkaloid [6]. Castanospermine seems to display a broad biological activity which surpasses even that of swainsonine, being a particularly powerful competitive inhibitor of both α- and β-D-glucosidases [7]. The alkaloid is plentiful enough in the

O OMe N Me O **1**

HO H OH N OH **2**

HO H OH HO N HO **3**

legume so that further collections by a New Zealand group have resulted in the commercial supply of castanospermine to Japan [8].

The extraction of another local plant indigenous to Western Australia, namely "smokebush" [*Conospermum* sp. (Proteaceae)], yielded several novel compounds [9] but failed to locate the traces of conocurvone **4** present, a chemical later identified after a biological screening program by an American group [10]. Conocurvone has promising activity against HIV.

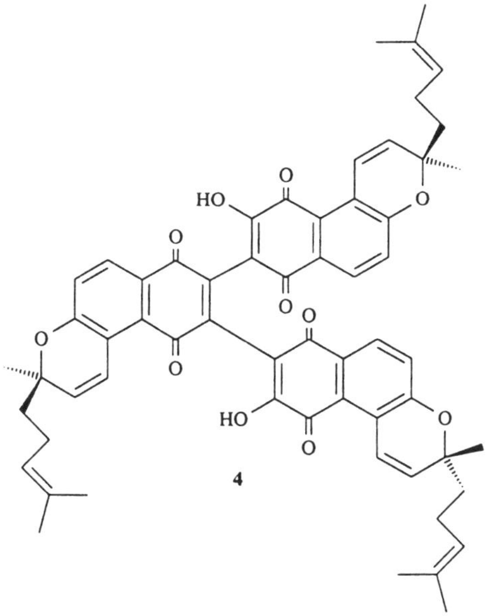

These few examples have shown the rich variety of chemical structures present in Australian plants. Although intensive efforts at the Roche Research Institute of Marine Pharmacology (Sydney) in the last decade failed to find any compounds in the local marine environment possessing significant biological activity, one feels that the time is ripe for a resurgence in natural product chemistry in Australia. This is probably reflected in the establishment of a huge extraction and screening program for local flora and fauna at Griffith University in Queensland, funded by Astra.

2 Introduction

The emergence of molecules such as swainsonine and castanospermine and their impact on the rapidly developing field of glycobiology saw significant gains in the understanding of the mechanism of action of enzyme inhibitors. In addition, chemists were able to design and synthesize "bigger and better" molecules for the rapid and complete inhibition of various target enzymes. Two recent examples are pertinent to carbohydrate chemistry in Australia.

Over the past two decades, research groups in Australia at the Commonwealth Scientific and Industrial Research Organization and several universities have worked together to isolate, purify, crystallize and structurally identify a neuraminidase from influenza virus. Biochemical investigations identified this enzyme as a key player in the cycle of infection known as "getting the flu". As a result, chemists were able to design a potent inhibitor of the viral neuraminidase, namely the alkene **5** and it is this molecule that is in advanced clinical trials as a true "cure" for the flu [11].

HO H O CO_2H HO HO AcHN NH H_2N NH

5

In the mid-1980s, my research group in Perth began a collaboration with a team of biochemists at La Trobe University in Melbourne to study the mechanism of action of a range of β-glycan hydrolases and cellulases found in various plants and bacteria. The approach was to involve synthetic enzyme inhibitors.

3
Epoxyalkyl Glycosides

In the classic studies by Sharon et al. [12] and Legler and Bause [13], epoxyalkyl glycosides were used to inhibit the action of various glycan hydrolases. In view of the almost certain presence of an acid residue at the active site and the two mechanisms which have been proposed for the action of glycan hydrolases, namely that involving hydrolysis with retention or inversion of configuration at the anomeric centre (Figs. 1 and 2) [14, 15], we decided to use epoxyalkyl

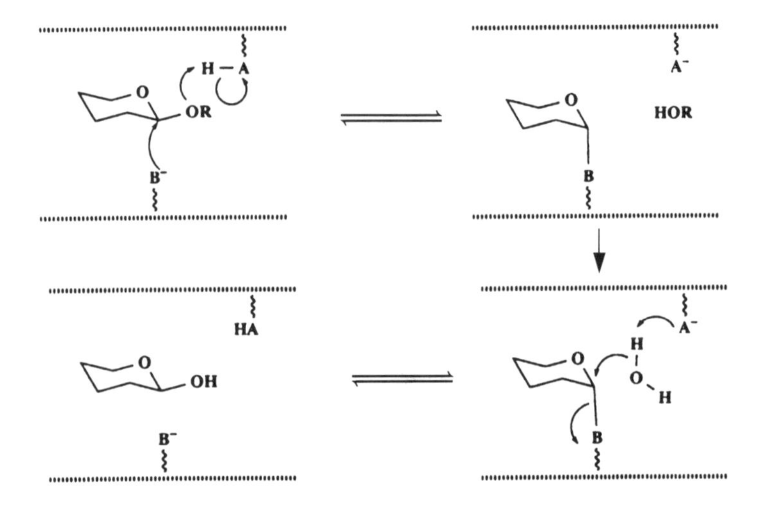

Fig. 1

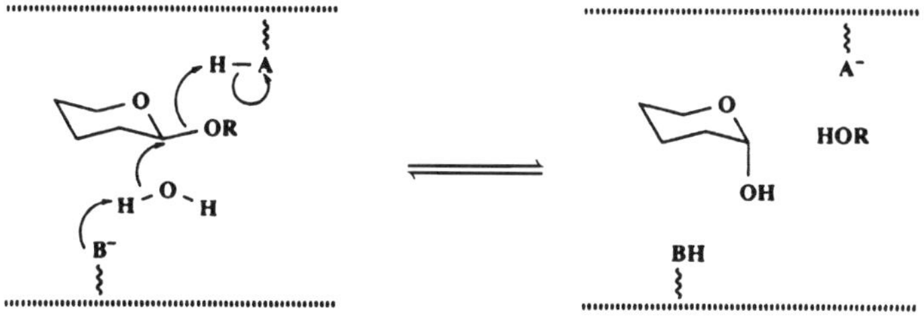

Fig. 2

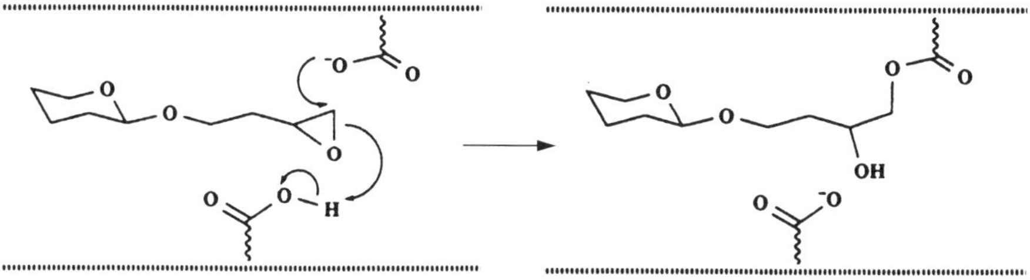

Fig. 3

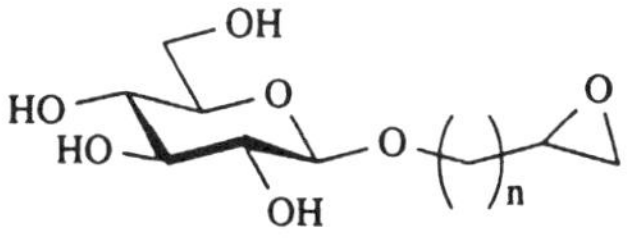

6 n = 1 - 3

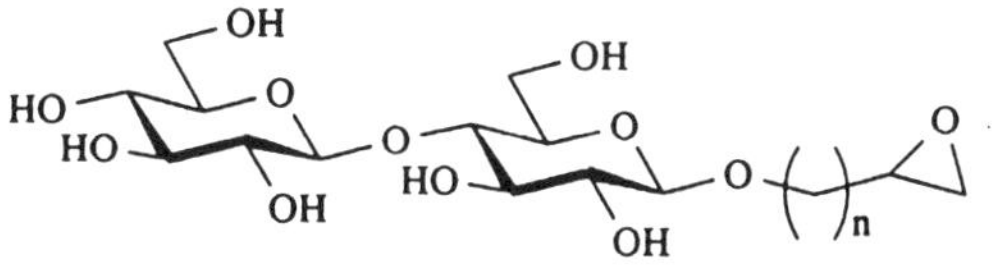

7 n = 1 - 3

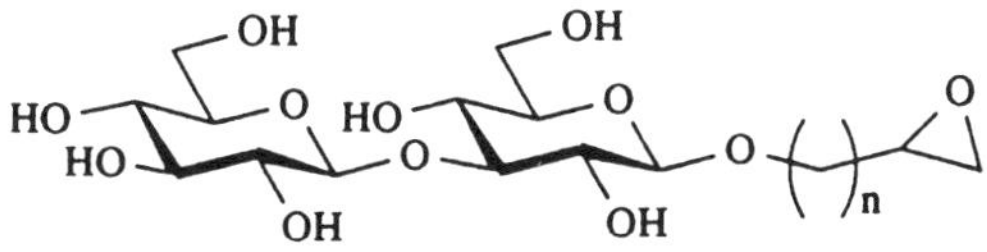

8 n = 1 - 3

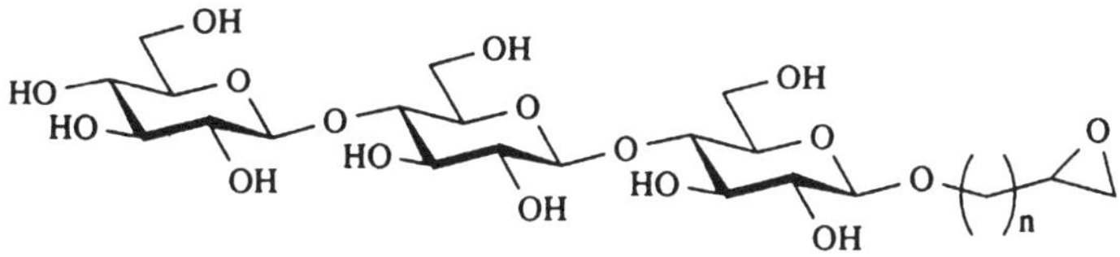

9 n = 3

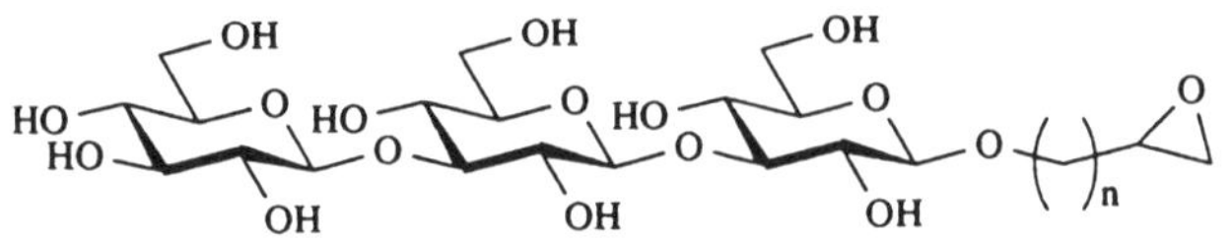

10 n = 3

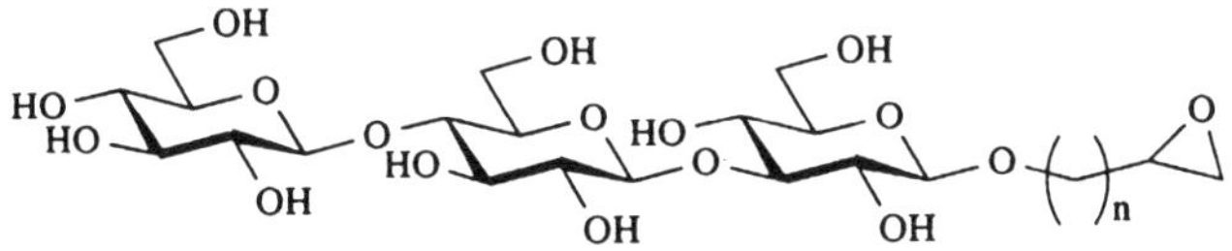

11 n = 3

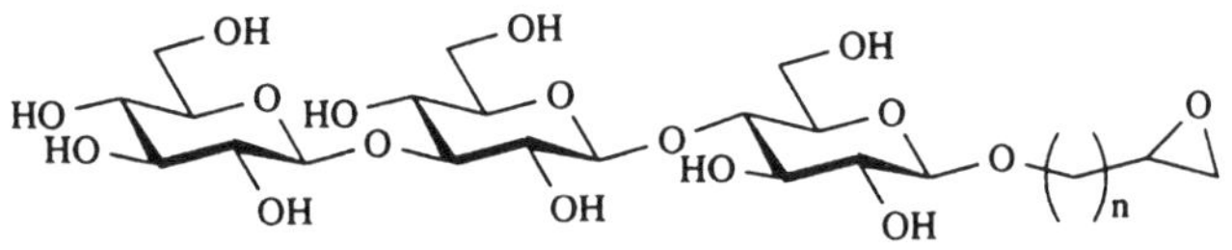

12 n = 3

β-glycosides as potential mechanism-based, irreversible, active-site directed inhibitors of our *β*-glucan hydrolases and cellulases (Fig. 3).

3.1 Synthesis

Early in our studies, therefore, we prepared a range of epoxyalkyl (propyl, butyl and pentyl) *β*-glycosides **6–12** of D-glucopyranose, cellobiose, laminaribiose, cellotriose, laminaritriose and two trisaccharides [16]. The syntheses utilized conventional Koenigs-Knorr methodology to prepare laminaribiose and the two trisaccharides and then the epoxyalkyl glycosides were constructed according to Scheme 1. The only points worthy of comment are that the intermediate pentenyl glycosides were best prepared using silver(I) rather than mercury(II) promoters (to minimize oxymercuration) and that the epoxidation step is now better performed using dimethyldioxirane [17] rather than the more conventional *meta*-chloroperbenzoic acid.

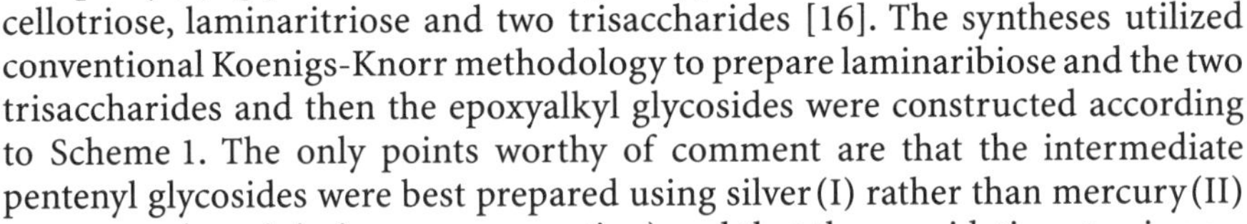

1. epoxidation
2. NaOMe
→ **6**

Scheme 1

3.2 Optically-Pure Epoxides

Although most of the glycosides **6–12** showed some inhibition towards the β-glycanases available to us [18], it was the epoxybutyl β-cellobioside **7** (n = 2) that was an efficient inhibitor of one particular enzyme. Also, in view of the chiral nature of an enzyme, it seemed an especially good idea to prepare molecules such as **7** in optically-pure form. We therefore began a program to synthesize the pure diastereoisomers **6a, 6b, 7a** and **7b**.

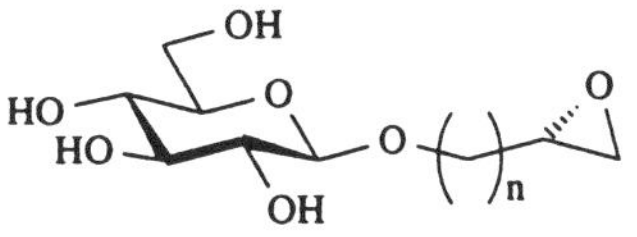

6a n = 1 - 3

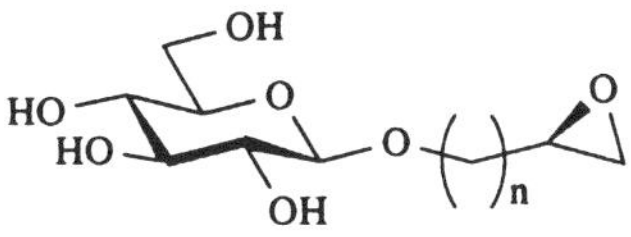

6b n = 1, 2

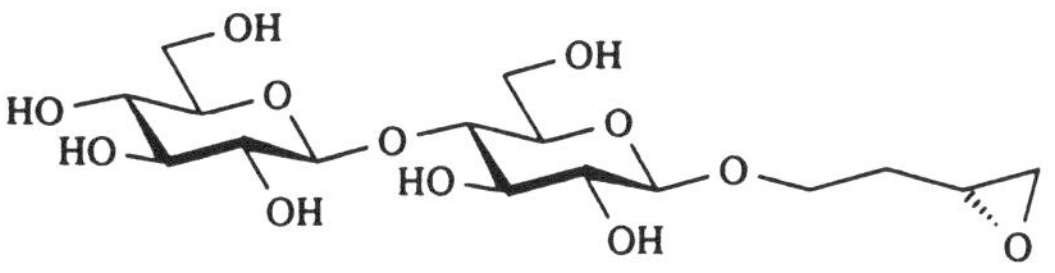

7a

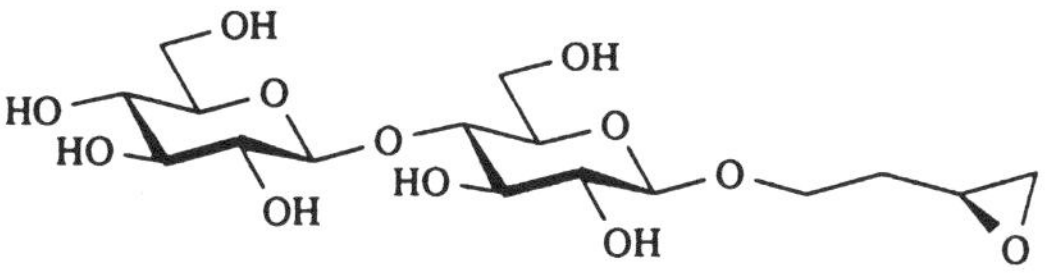

7b

The obvious route to these molecules involved glycosylation of the optically-pure alcohols **13a, 13b** followed by deprotection to give the diol, e.g. **14** which could be transformed into the desired epoxide **6a** (Scheme 2) [19]. It later

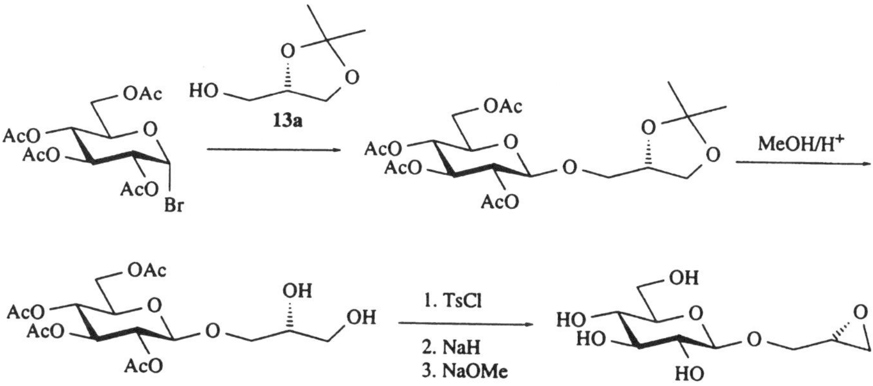

13a

MeOH/H+

1. TsCl
2. NaH
3. NaOMe

Scheme 2

14

6a

occurred to us that both of the diastereoisomers **6a, 6b** were available from the one hand of the alcohol, **13a** by an alternative manipulation of the diol **14** (Scheme 3) [19].

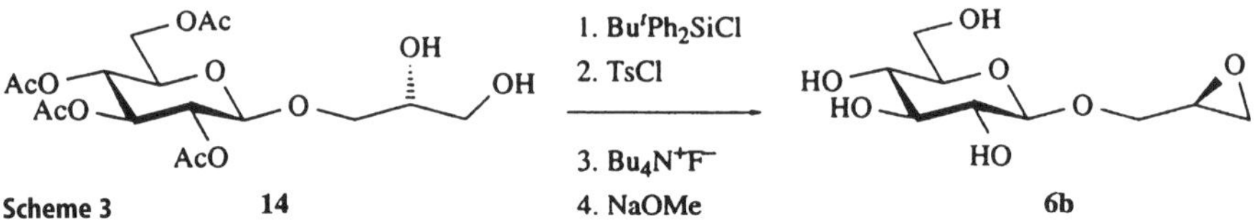

1. Bu^tPh_2SiCl
2. TsCl
3. $Bu_4N^+F^-$
4. NaOMe

Scheme 3 **14** **6b**

More recently, in light of the development of the Sharpless "asymmetric dihydroxylation" protocol [20], we have approached the synthesis of diols such as **14** (Scheme 2) from the alkene. Thus, treatment of the alkenyl β-D-glucosides **15** under the conditions of the Sharpless dihydroxylation gave a range of diols **16** with varying diastereoisomeric excesses (Table 1). One of these mixtures of diols, upon recrystallization, yielded the pure diastereoisomer, namely the diol **14**. This procedure now gives a very rapid and efficient entry into one of the precursor diols for the synthesis of the optically-pure epoxides [21].

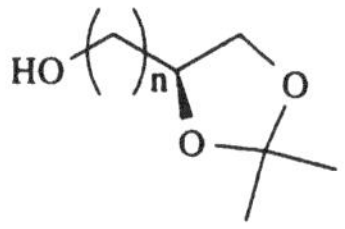

13a n = 1 - 3 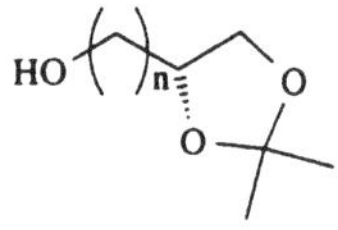 **13b** n = 1, 2

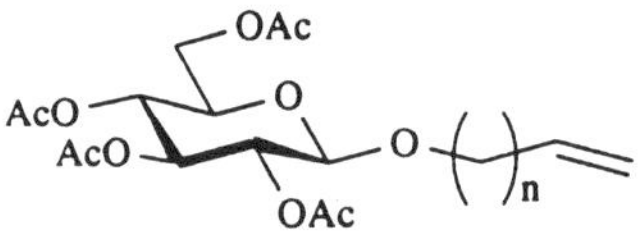

15 n = 1 - 4 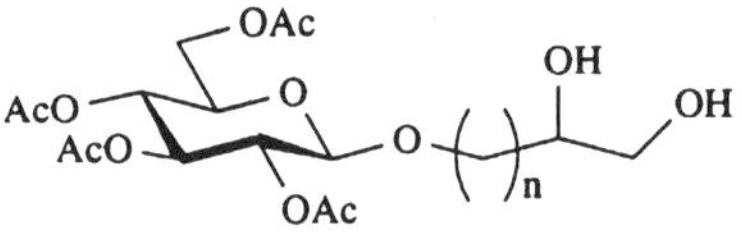 **16** n = 1 - 4

Earlier on in our work, we also used the other legendary Sharpless oxidation, the "asymmetric epoxidation" [22] for the transformation of the allylic alcohol **17** into the epoxide **18**. In the presence of (+)-diethyl tartrate, **17** was converted

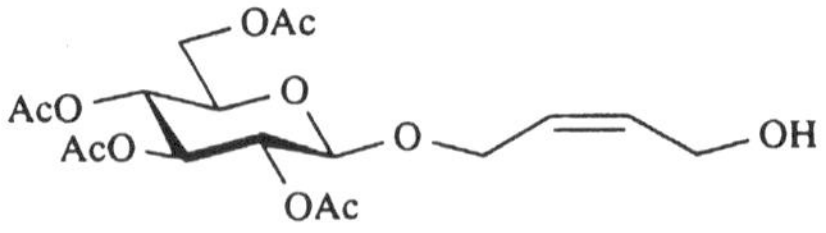

17 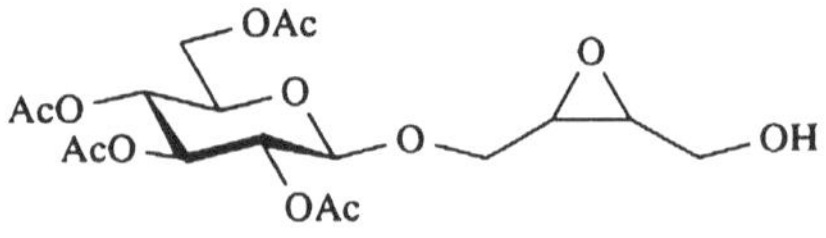 **18**

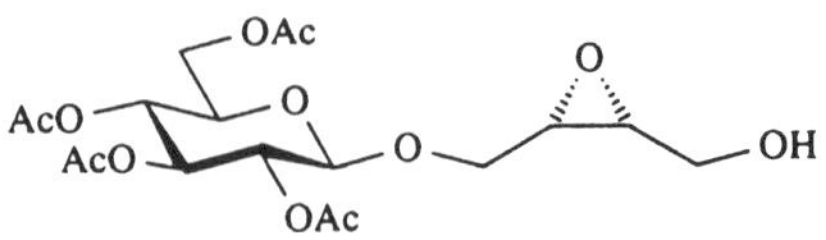

19 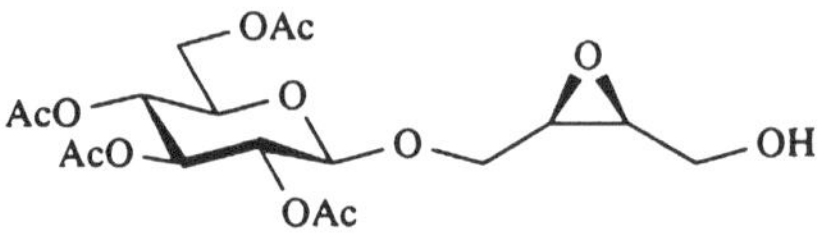 **20**

Table 1. Diols **16** with varying diastereoisomeric excesses obtained by treating alkenyl β-D-glucosides **15**

15	Ligand	**16** *R*:*S*
n = 1	Q[a]	3:2
	α[a]	1:1
	β[a]	19:1
2	Q	1:1
	α	2:3
	β	4:1
3	Q	1:1
	α	1:3
	β	7:3
4	Q	1:1
	α	1:3[b]
	β	17:3[b]

[a] (±)-3-Quinuclidinol, "α-AD mix", "β-AD mix".
[b] This ratio can be reversed.

almost exclusively into the epoxide **19**. Such was not the case when the enantiomeric (–)-diethyl tartrate was used; a mixture of **19** and **20** (1:9) resulted [19].

3.3 Deoxy Halo Epoxides

X-ray crystallography and NMR spectroscopy are both powerful tools for the investigation of protein structure and conformation. Vital information may be obtained about the binding sites and catalytic sites of enzymes, particularly if the enzyme can be crystallized with the natural substrate or some smaller, analogue molecule in place at the active site. It occurred to us that our epoxyalkyl glycosides, suitably modified to contain either an iodine (heavy) or fluorine (magnetically active) atom, would be ideal molecules to assist in X-ray crystallographic and NMR analyses, respectively.

In view of the efficient inhibition of a particular crystalline glycan hydrolase by the epoxylbutyl β-cellobioside **7** (n = 2), we decided to prepare the deoxy iodo derivative **21** to aid in the X-ray crystallographic analysis. We soon found that, although the alkene **22** was easily available as a direct precursor to our target, the epoxide functionality had to be introduced indirectly using bromohydrin **23** technology; any direct oxidation of **22** invariably led to some loss of the iodine atom [23].

The synthesis of the corresponding fluoride **24** posed different, but interesting, problems. The diol **25**, on treatment with diethylaminosulfur trifluoride (DAST), gave the required fluoride **26** and thence the epoxide **24** after introduction of the four-carbon aglycon. However, with an excess of DAST, the diol **25** was converted into the fluoro lactoside **27**, a rearrangement with some precedent [23,

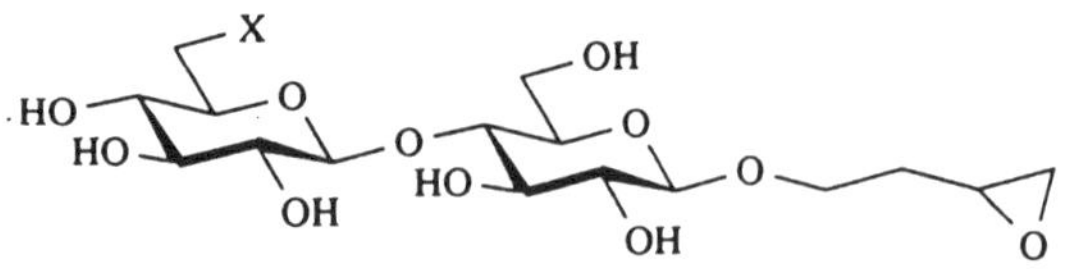

21 X = I
24 X = F

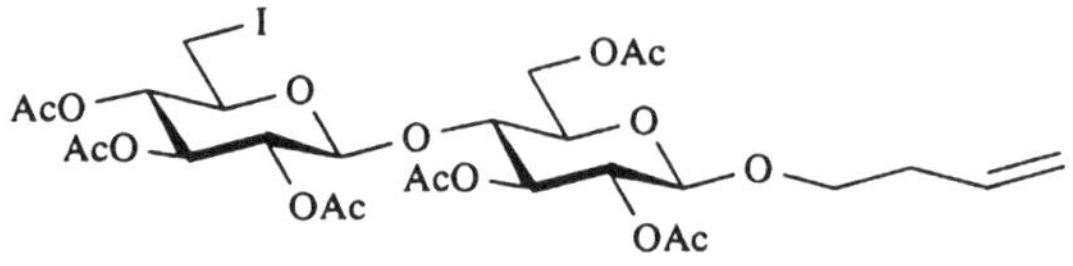

22

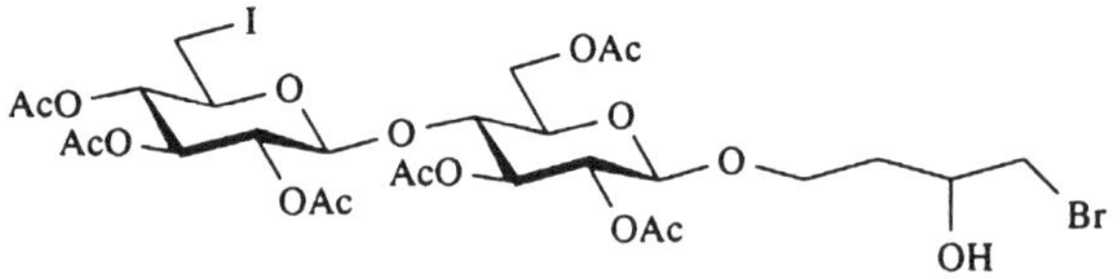

23

24]. Also prepared during this work were the previously unknown deoxy fluoro cellobioses **28** and **29**.

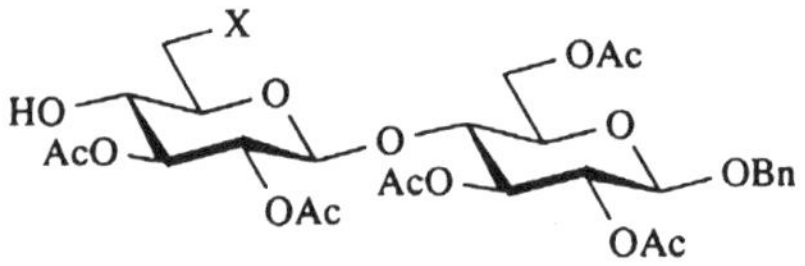

25 X = OH
26 X = F

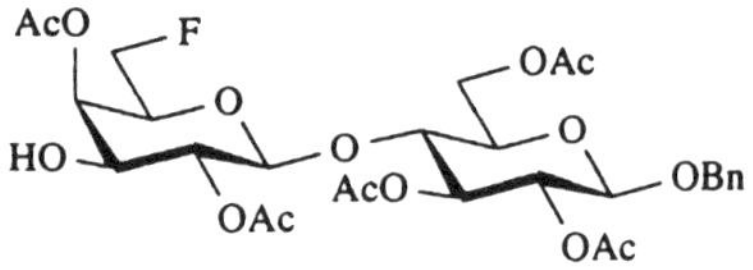

27

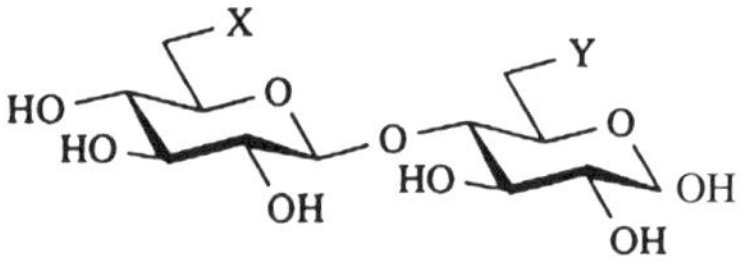

28 X = F, Y = OH
29 X = OH, Y = F

4 Aziridines and Thiirans

We had for some time considered that aziridines, because of their increased basicity and hence reactivity at the active site of a glycan hydrolase, would prove to be better inhibitors than the epoxides in many respects. Conversely, the less basic thiirans would be expected to be poorer inhibitors than the corresponding epoxides. We thus embarked on a synthesis of the aziridine **30** and the thiiran **31**.

Treatment of the epoxide **32** with sodium azide and ammonium chloride in dimethylformamide gave the azido alcohol **33** which, via the mesylate **34**, could be reduced to the aziridine **35**, a direct precursor of the aziridine **30** [21]. In a similar fashion, treatment of the epoxide **32** with thiourea [25] gave the thiiran **36** and thence the thiiran **31** directly.

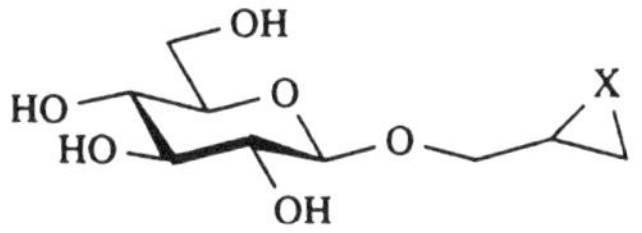

30 X = NH
31 X = S

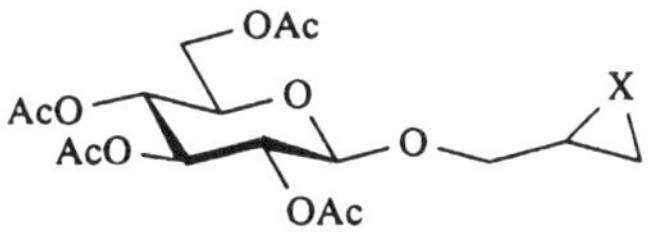

32 X = O
35 X = NH
36 X = S

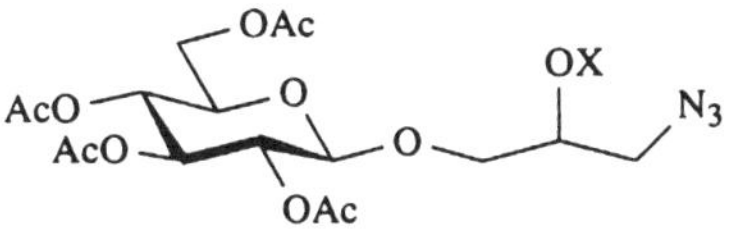

33 X = H
34 X = Ms

5 Epoxyalkyl C-Glycosides

For an enzyme inhibitor to have any use in the real world, it must be able to survive attack and degradation by "wild" enzymes in an in vivo situation. Epoxyalkyl glycosides are really not constituted for this purpose; apart from having a reactive epoxide moiety responsible for the mode of action of the inhibitor, the various glycosidic linkages within the inhibitor are too prone to hydrolysis. Therefore, over the years, chemists have replaced the labile glycosidic linkage with the carbon equivalent, namely a C-glycoside (a tetrahydropyran) [26].

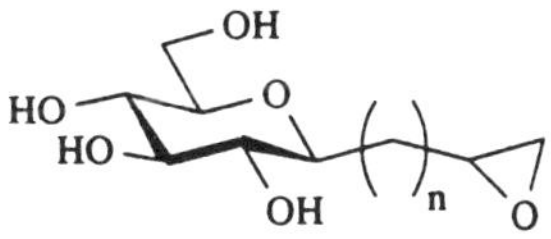

37 n = 1 - 3

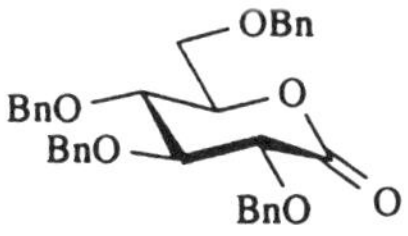

38

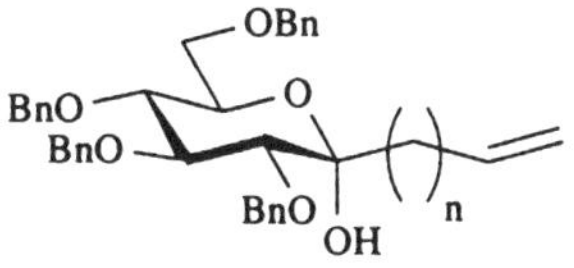

39 n = 1 - 3

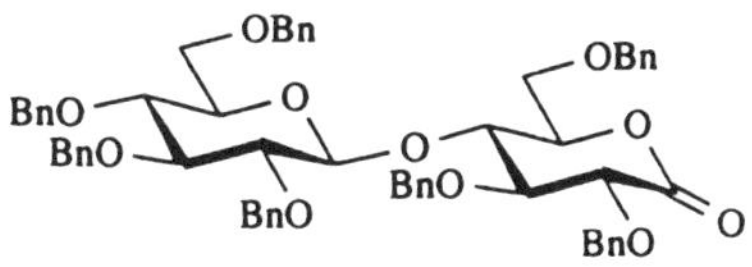

40

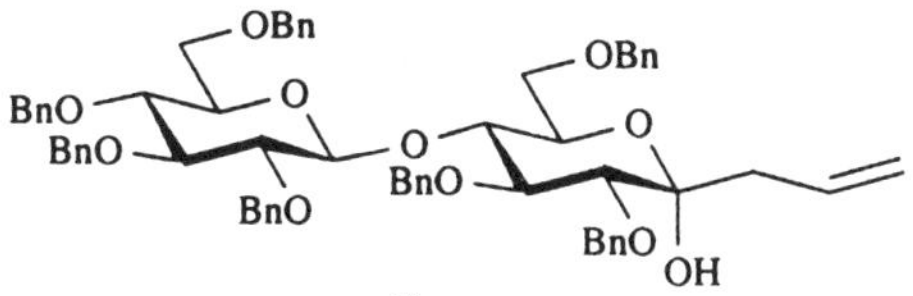

41

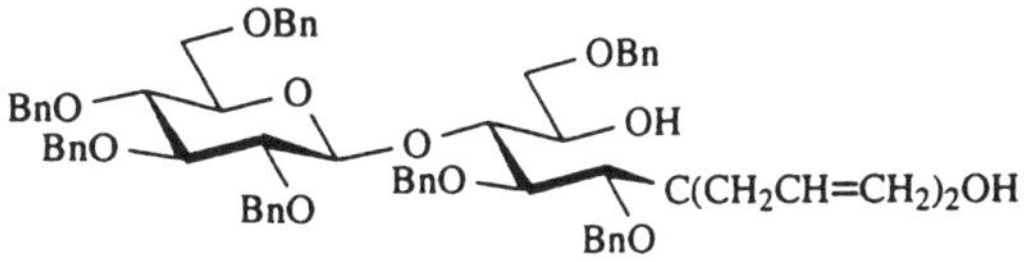

42

Some years ago we followed this lead by preparing the epoxyalkyl β-D-C-glucosides **37** [27], essentially following a literature precedent [28] whereby an unsaturated Grignard reagent was added to the lactone **38** to give the lactol **39**. Reduction, protecting group manipulation and epoxidation gave the desired epoxide **37** as a mixture of diastereoisomers. Similar success was not enjoyed in additions to the very unreactive cellobionolactone **40**; the allylmagnesium halides gave some of the desired lactol **41** but amounts of the "diallyl adduct" **42** were also formed. The butenyl- and pentenyl-magnesium bromide additions fared even worse [27].

We then considered an alternative approach to the epoxyalkyl C-glycosides **37**. The pioneering work of Danishefsky has made available the 1,2-anhydro-D-glucose derivative **43** [29]. Treatment of **43** with allylmagnesium bromide and zinc chloride in tetrahydrofuran gave the alcohol **44**, easily convertible to the known β-D-C-glucoside **45** [27]. The reaction was also catalyzed by ytterbium (III) triflate [30] and by lithium tetrachlorocuprate [31] and could be separately performed with lithium diallylcyanocuprate [32, 33]. However, no conditions have yet been found to add butenyl-and pentenyl-reagents to the epoxide **43**. Worse still, the 1,2-anhydrocellobiose **46** has been prepared (in the conventional manner from the glycal and dimethyldioxirane [29]) and is essentially unreactive even to the allylmagnesium bromide/catalyst system [33].

One final, but less direct, ploy is that we have prepared the nitromethyl β-D-C-glucoside **47** and hope to be able to alkylate this molecule on carbon [33, 34].

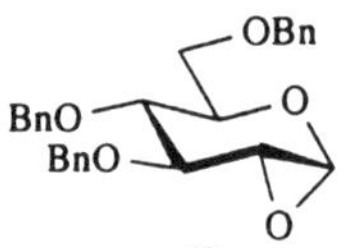

43

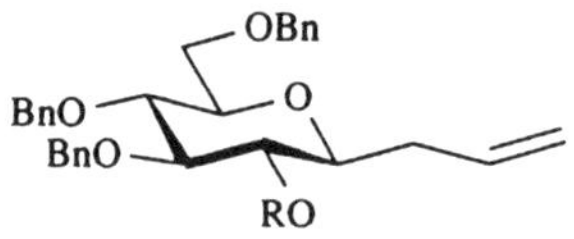

44 R = H
45 R = Bn

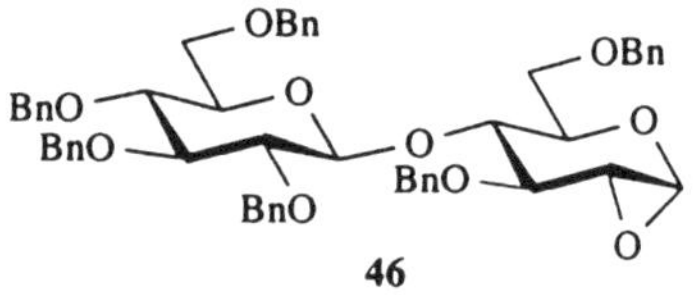

46

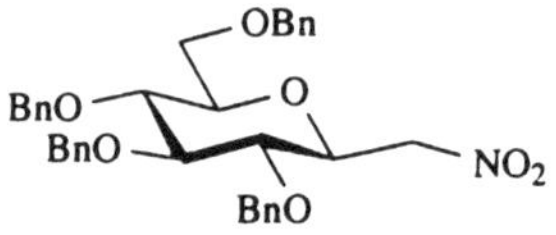

47

6
Tetrahydro-1,2-oxazines

Aza sugars have been recognized as potent competitive inhibitors of various enzymes [35, 36]. For example, nojirimycin **48** and deoxynojirimycin **49** have been found to inhibit the action of various glycosidases. Of some topical interest, the *N*-butyl derivative **50** of deoxynojirimycin has been found to show significant activity against HIV [37]. More recently, the "displaced" aza sugar isofagomine **51** has also been shown to be a powerful inhibitor of sweet almond β-glucosidase [38]. It struck us that a hybrid structure of a normal pyranose and an aza sugar, namely the tetrahydro-1,2-oxazine **52**, might prove to have interesting biological activity as an enzyme inhibitor. So began a saga!

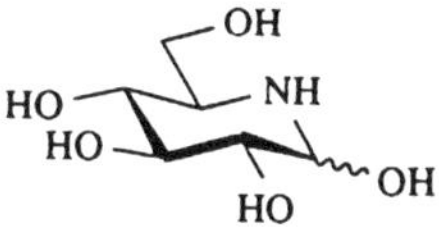

48 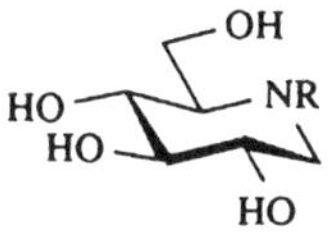 **49** R = H **50** R = Bu **51** 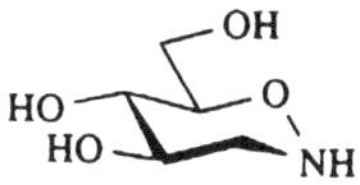 **52**

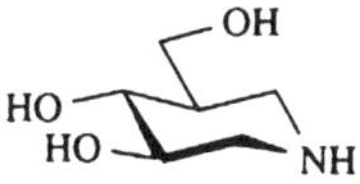

It was our intention to synthesize **52** by the reductive "amination" of the free sugar **53**; in actual fact, any such successful cyclization of **53** would yield the enantiomer of **52**. In the explorative stages it was cheaper to work with **53**, the precursor of which is D-xylose.

Benzyl β-D-xylopyranoside was converted into the alcohol **54** (a somewhat capricious isopropylidenation) [39] and a Mitsunobu inversion with *N*-hydroxyphthalimide, followed by protecting group removal, gave the hydroxylamine **55**. Transfer-hydrogenation (ammonium formate and palladium-on-charcoal in refluxing methanol) [40] then gave, on a small scale and in almost a quantitative yield, the enantiomer of the desired tetrahydro-1,2-oxazine **52**. We have never been able to repeat this result since! Figure 4 shows the NMR spectra acquired at the time [41].

Over the past years we have, off and on, repeated the transfer-hydrogenation of **55** with varying reagent, catalyst, solvent, temperature and pH. Classical hydrogenation under a variety of conditions has also been tried, but all to no avail; at different times, **55** has been converted into benzyl α-L-arabinopyranoside (N–O cleavage of the hydroxylamine), L-arabinose (N–O and glycoside cleavage) and perhaps even the amine **56** (N–O cleavage of the enantiomer of **52**!).

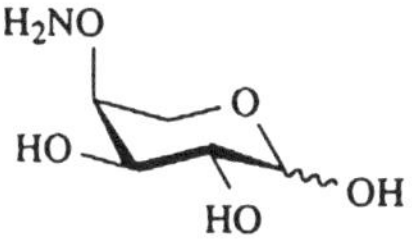

53 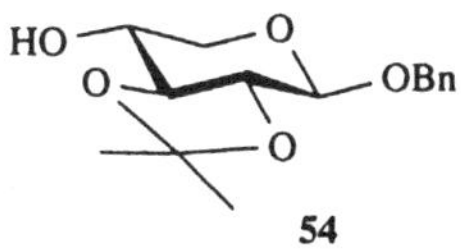 **54** 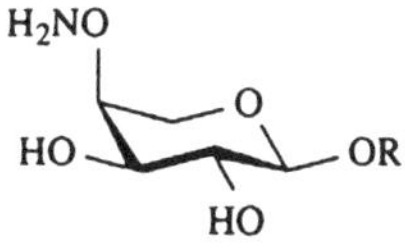 **55** R = Bn **57** R = $CH_2CH{=}CH_2$ **58** R = $CH_2CH_2SiMe_3$

We attempted to by-pass the above reductive amination by preparing alternative glycosides, namely the allyl- and 2-trimethylsilylethyl-β-D-xylopyranosides, **57** and **58**, respectively. However, to date, we have not been able to convert either **57** or **58** into the enantiomer of **52** or its expected precursor **59** [41].

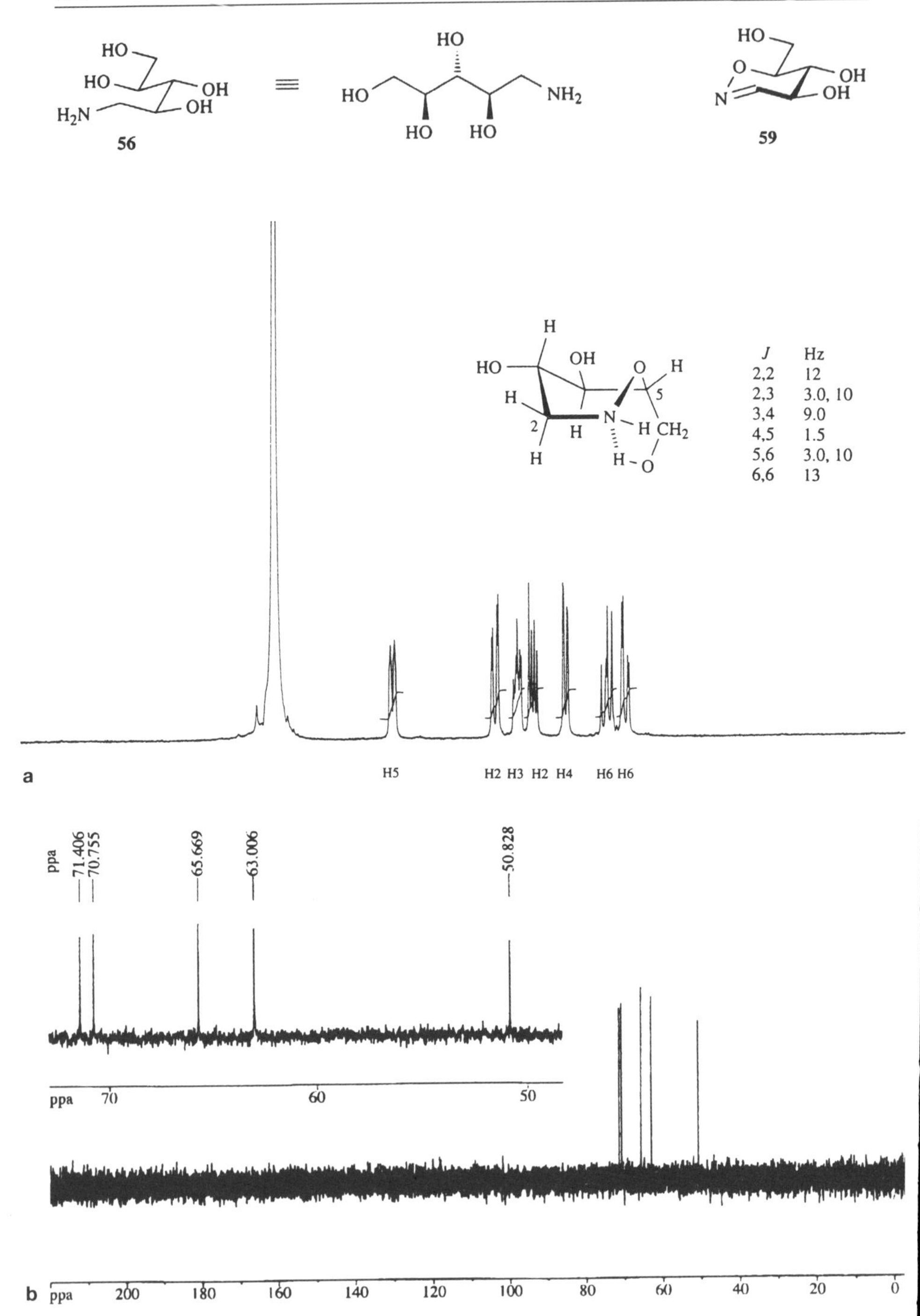

Fig. 4. **a** The ^{1}H NMR (500 MHz, D_2O) spectrum of the enantiomer of **52**. **b** The ^{13}C NMR (125 MHz, D_2O) spectrum of the enantiomer of **52**

7
Epoxyalkyl Glycosides of *N*-Acetyl-D-glucosamine

Chitin is an abundant biopolymer, especially in Perth where the thriving Western rock lobster industry produces tonnes of waste, a good proportion of which is exoskeleton and rich in chitin. It has been our aim for some years to put this waste chitin to some good use and so, initially, we have set out to prepare epoxyalkyl glycosides based on *N*-acetyl-D-glucosamine and its oligomers, for example **60** and **61**. It is hoped that molecules such as **60** and **61** will prove to be efficient inhibitors of chitinases, ubiquitous enzymes involved in many biological processes [42].

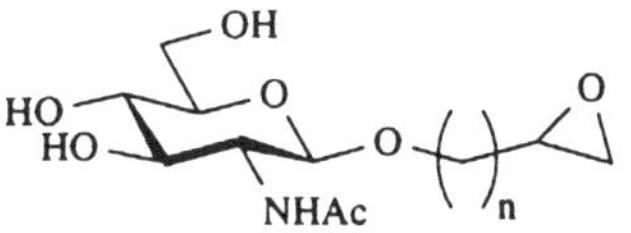

60 n = 1 - 4

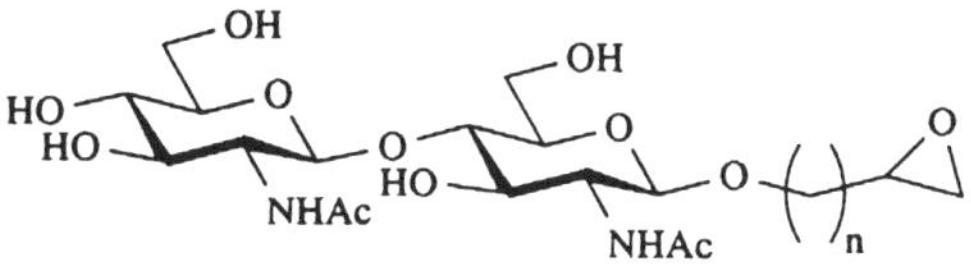

61

Our initial aim was to prepare the epoxyalkyl glycosides **60** as mixtures of diastereoisomers. The literature abounds with methods for such syntheses, ranging from Lewis-acid mediated condensations of the acetate **62** [43, 44] to acid-catalyzed additions to the oxazoline **63** [45, 46]. All of these transformations seemed rather cumbersome and long-winded to us, especially those involving the separate preparation of the oxazoline!

We have found that the secret to a successful synthesis of the alkenyl glycosides **64** lies in obtaining the pure β-acetate **65** [47]. Condensation of the acetate **65** with the appropriate alkenol, in the presence of a small amount (5–20 mol% depending on the scale) of trimethylsilyl triflate, gave the alkenyl glycosides **64** in excellent yield. Oxidation of the alkene **64** with dimethyldioxirane, followed by deacetylation, then gave the putative enzyme inhibitors **60**.

It was of obvious interest to prepare the inhibitors **60** as their pure diastereoisomers, **66** and **67**. Following on from our successful treatment of alkenyl β-D-glucosides under Sharpless asymmetric dihydroxylation conditions [21], we treated the alkenes **64** with the "α-AD"- and "β-AD"-mixes – the results are summarized in Table 2. In no case did we ever obtain a satisfactory diastereoisomeric excess of the diol **68** over the diol **69**, or vice versa. A similar lack of stereoselectivity was also obtained with the triol **70** and the amine **71** [48].

We finally reverted to the "tried and tested" method for the synthesis of **66** and **67** [19]. Although the acetate **65** could not be condensed with optically-pure

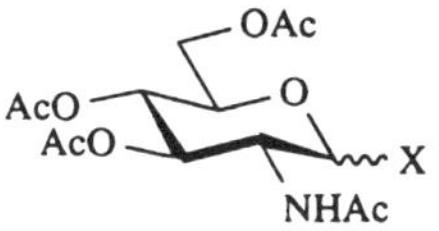

62 X = OAc
65 X = β-OAc

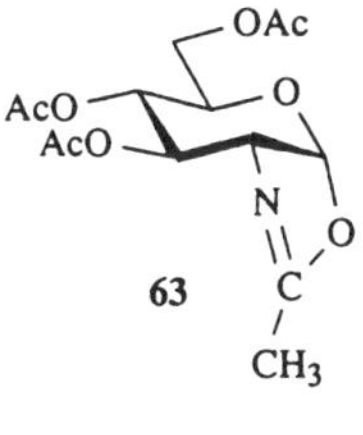

63

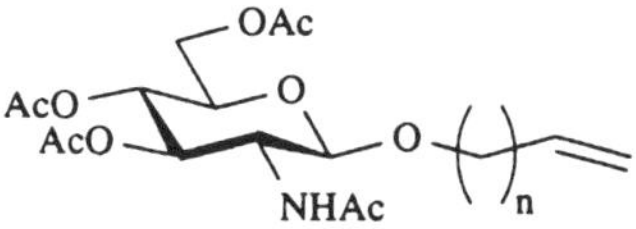

64 n = 1 - 4

Table 2. Diols **68** and **69** obtained by treatment of alkenes **64**

		Ligand	**68/69**[b]
64	n = 1	none	1:2.3
		Q[a]	1:1.6
		α[a]	1:1.7
		β[a]	1:2.3
	n = 2	Q	1:1
		α	1:1
		β	1.7:1
	n = 3	Q	1:1
		α	2.6:1
		β	1:3.2
	n = 4	Q	1:1
		α	1:3.5
		β	4:1

[a] (±)-3-Quinuclidinol, "α-AD mix", "β-AD mix"
[b] These ratios are only comparable for each individual compound and may be reversed between compounds

2,3-*O*-isopropylideneglycerol in the presence of a Lewis acid (pre-racemization of the glycerol was always observed) [49], condensation of the alcohol **72** [49] with **65** in the presence of the usual trimethylsilyl triflate gave the glycoside **73**. This glycoside could then be manipulated to form the epoxides **66** (n = 1) and **67**

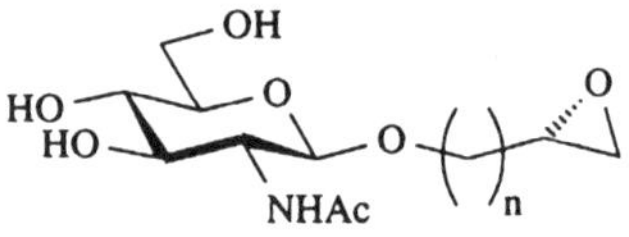

66 n = 1 - 3

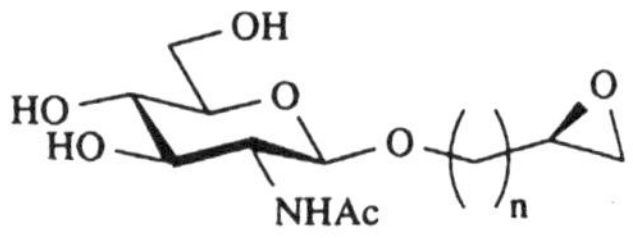

67 n = 1 - 3

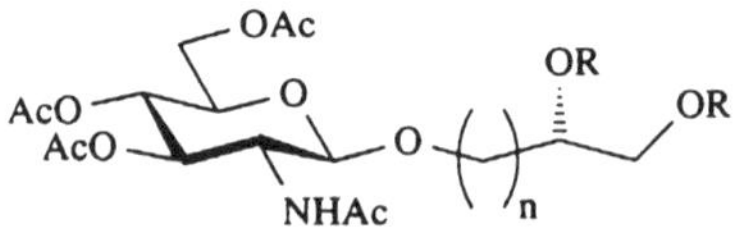

68 R = H, n = 1 - 4
73 R = Bn, n = 1

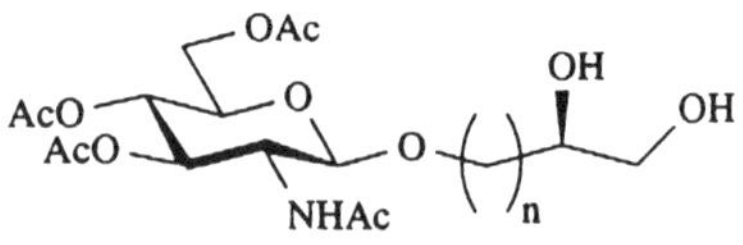

69 n = 1 - 4

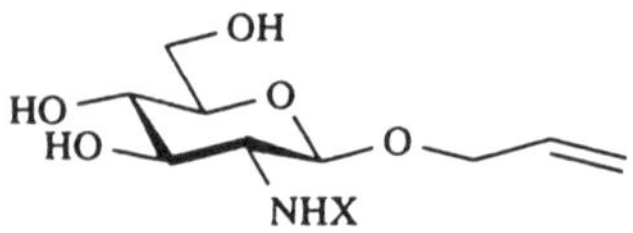

70 X = $COCH_3$
71 X = H

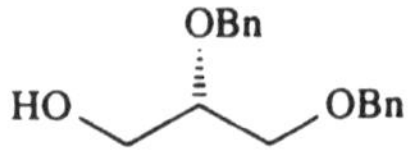

72

(n = 1). Similar methodology was used for the synthesis of **66** (n = 2,3) and **67** (n = 2,3) [48].

8
β-Acarbose: a Termite's Worst Nightmare?

At this stage of our work we had reached something of an impasse; what to do next! One of us [50], on noting the success of acarbose **74** (a carba-tetrasaccharide isolated from culture filtrates of actinomycetes in the *Actinoplanaceae* family [51, 52]) as an inhibitor of *α*-D-glucosidases, including intestinal sucrase and *α*-amylases [52], decided that "*β*-acarbose" **75** had the potential to be an inhibitor of *β*-D-glucosidases and *β*-glycan hydrolases, including cellulases. We therefore set about a synthesis of **75**.

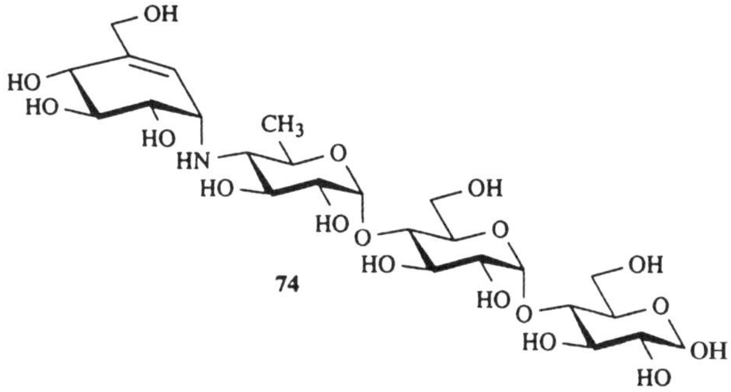

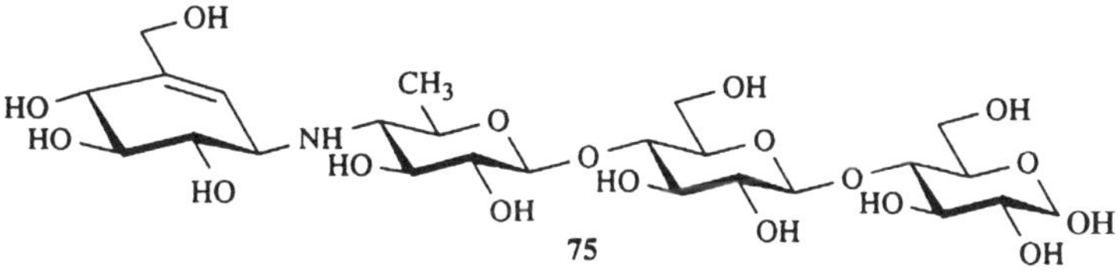

The main synthetic challenge in a synthesis of *β*-acarbose is the attachment of the polyhydroxylated cyclohexene unit, by a nitrogen atom, to the 6″-deoxy trisaccharide. If one accepts that molecular ammonia is not likely to be the ideal reagent to link the cyclohexene and trisaccharide units together, then one also realizes that only two approaches are left; either a trisaccharide amine may be attached to a cyclohexene unit or, conversely, a cyclohexenyl amine may be joined to a trisaccharide.

We did not explore the first of these two approaches too deeply. The synthesis of the epoxide **76**, an ideal partner for the alkylation of any trisaccharide amine, was daunting and seemingly difficult but there was available in the literature an excellent route to the enone **77** from tetra-*O*-benzyl-D-gluconolactone **38** [53]. However, earlier work by Kuzuhara suggested that any reductive amination of the enone **77** would probably proceed in low yield and certainly give a mixture of diastereoisomeric amines [54]. We did prepare the amine **78**, via the azide **79**, but the amine **78** would not condense with the enone **77** to give the required imine. We also prepared the phosphinimine **80** from the azide **79** but this showed no reactivity towards cyclohexanone, let alone the more stable enone **77** [55].

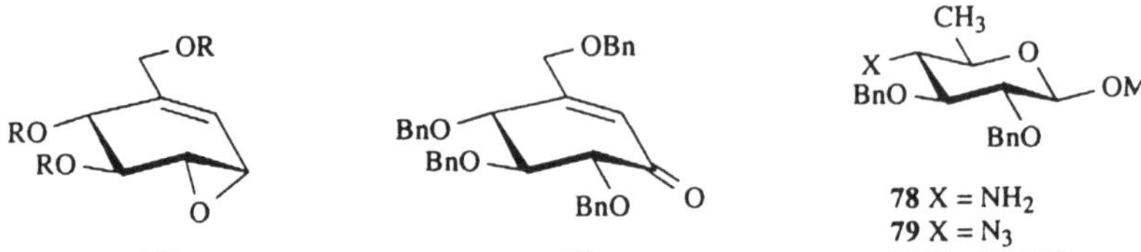

OR RO RO O **76**

OBn BnO BnO BnO O **77**

CH_3 X BnO O OMe BnO

78 X = NH_2
79 X = N_3
80 X = $N{=}PPh_3$

8.1 Synthesis

We now turned our attention to the second approach, the alkylation of a cyclohexenyl amine by some trisaccharide derivative. The logical choice of amine was the 1-epivalienamine derivative **81**, a molecule that was readily available through various literature procedures. We chose to utilize the excellent synthesis described by Panza et al. to produce multi-gram amounts of the amine **81** [56].

In our early work, we attempted the alkylation of the amine **81** with various carbohydrate triflates, namely **82**–**84** [57]. Somewhat disappointingly, the main products isolated were the alkenes, e.g. **85** and **86**. Only a small amount of the desired *N*-linked carbasugar, e.g. **87** was ever obtained. At one stage we attempted an alkylation of the amine **81** with the triflate **82** and with the triflate **84** in the presence of potassium carbonate; to our surprise, *carbamat*es were formed, **88** and **89**! These two carbamates have subsequently been deprotected to give the methyl *β*-D-glucoside **90** and the free sugar **91**, interesting candidates for enzyme inhibition studies [57].

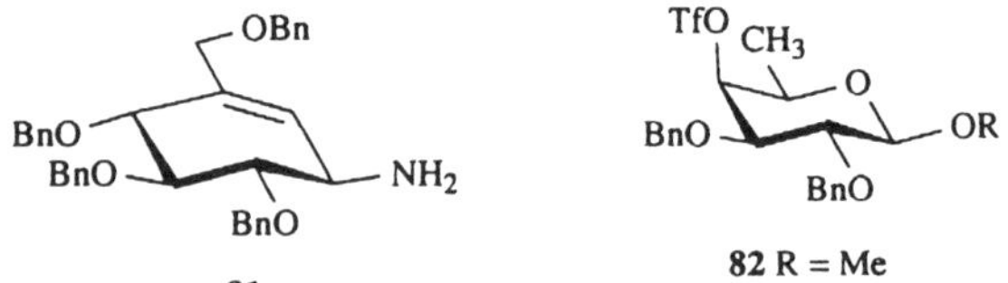

OBn BnO BnO NH_2 BnO **81**

TfO CH_3 O BnO OR BnO

82 R = Me
83 R = Bn

At a much later stage, we prepared the benzoylated triflate **92** and treated it with the amine **81**; the alkylated product **93** was obtained in a satisfactory yield, together with just one product of elimination, the alkene **94**. Deprotection of **93** gave the polyol **95**, a diastereoisomer of methyl acarviosin **96** [55, 58]. Although the change of protecting group at O-3 from benzyl **82** to benzoyl **92** had produced the desired effect, the resulting increase in yield of **93** and the sole formation of the alkene **94** were not entirely expected. In addition, it was forseen that any synthesis of the benzoate **97** (necessary for a similar approach to *β*-acarbose) would be long and tedious.

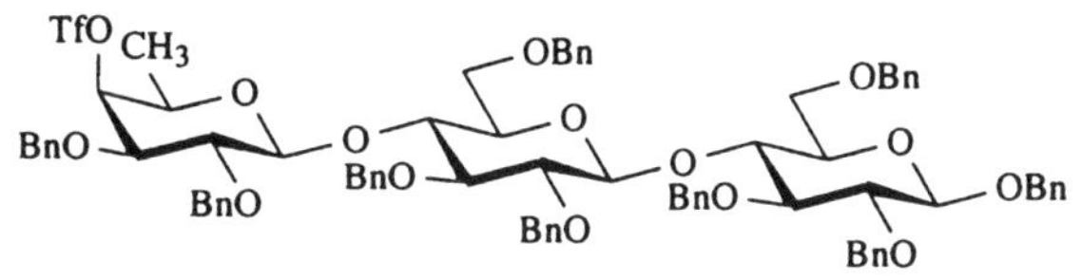

84

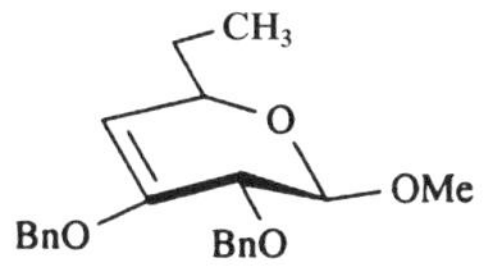

85

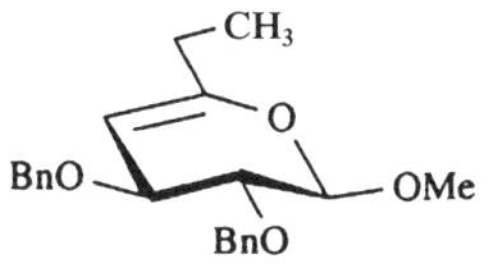

86

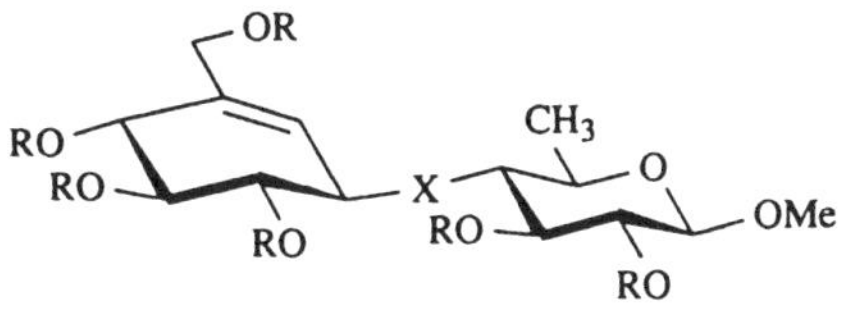

87 R = Bn, X = NH
88 R = Bn, X = NHCOO
90 R = H, X = NHCOO

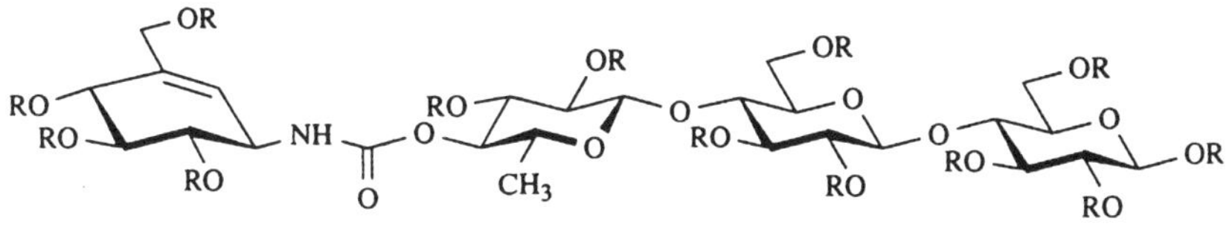

89 R = Bn
91 R = H

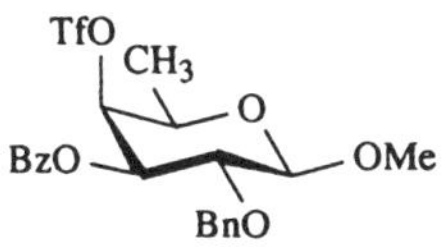

92

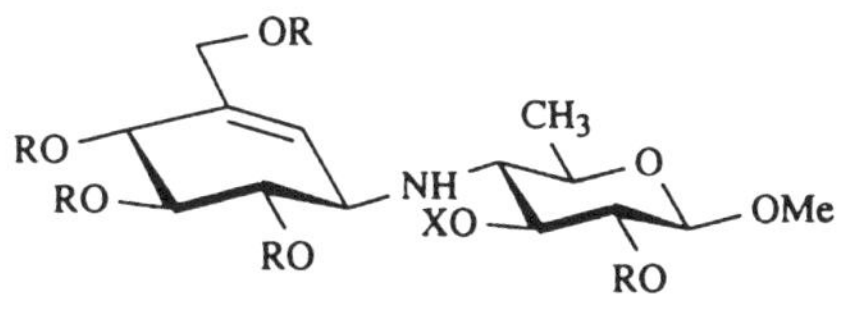

93 R = Bn, X = Bz
95 R = X = H

We next sought a different method to connect the amine **81** to some trisaccharide derivative and noted the success enjoyed by Shibata and Ogawa in their synthesis of acarbose [59] – treatment of the valienamine derivative **98** with an epoxide **99** gave some selectivity for the desired product **100**, but a related treatment of the β-D-*galacto* epoxide **101** with sodium azide was hopelessly in favour of opening at C-3 to give the undesired **102** [60]. Thus, although epoxides were obviously good alkylating agents for the valienamines, another strategy had to be devised for success with β-glycosides.

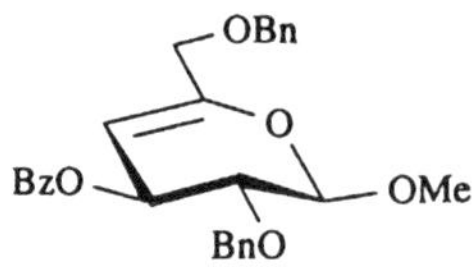

94

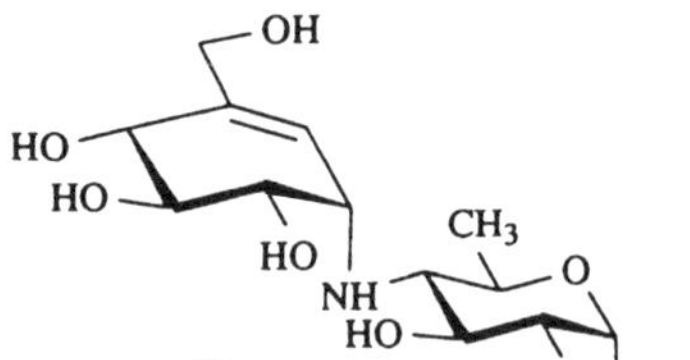

96

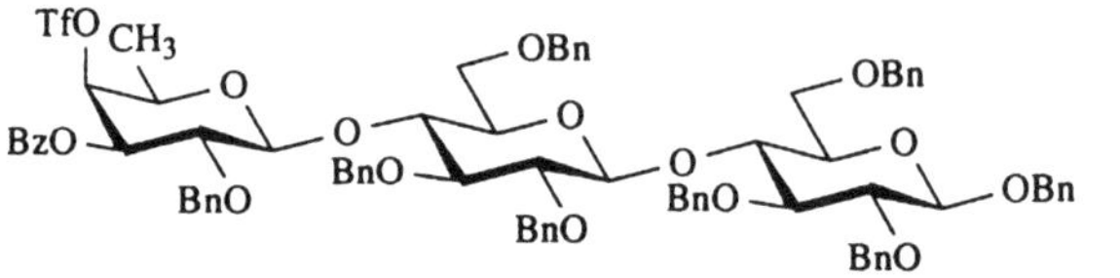

97

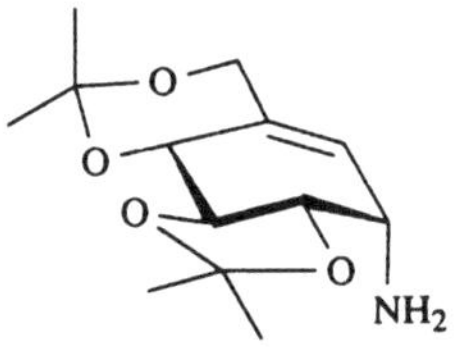

98

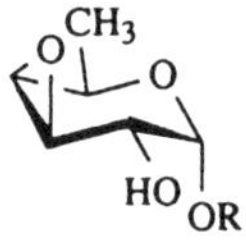

99

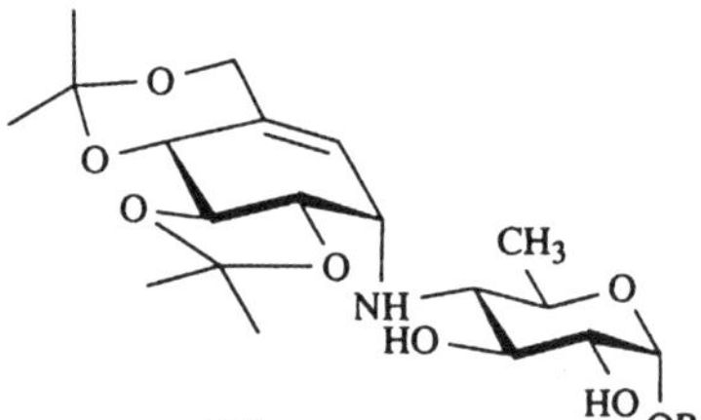

100

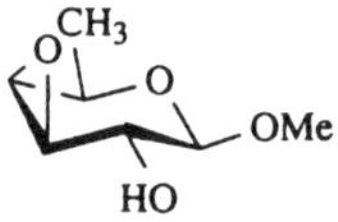

101

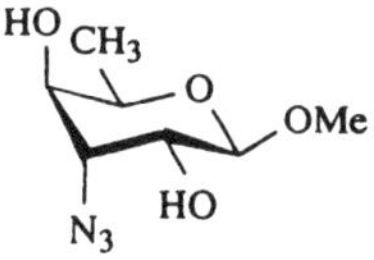

102

It occurred to us that the well known 1,6-anhydro sugars could provide a solution to our problem. Treatment of the epoxide **103** [61] with the amine **81** gave the amino alcohol **104** in good yield. Although a selective acetylation of **104** was not possible, the use of ethyl chloroformate or carbonyldiimidazole resulted in the formation of the cyclic carbamate **105**. Interestingly, the amino alcohol **104** and methanesulfonyl chloride gave the aziridine **106**. We have deprotected **106** to form the polyhydroxylated aziridine **107**, again an interesting candidate for enzyme inhibition studies [57].

With the nitrogen atom well protected as the cyclic carbamate **105**, all attempts to open the 1,6-anhydro ring under acidic conditions failed owing to the lability of the two allylic benzyl ethers present in the molecule. However, it was possible to debenzylate **105** with lithium in ammonia to form the new cyclic carbamate **108** and acetylation then gave **109**.

Treatment of **109** with triethylsilyl triflate and acetic anhydride gave the acetate **110** [62], convertible into the bromide **111** and thence the thioglycoside **112**. Of these three glycosyl donors, only the thioglycoside **112** was useful

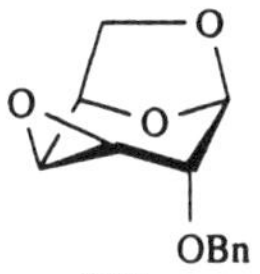

103

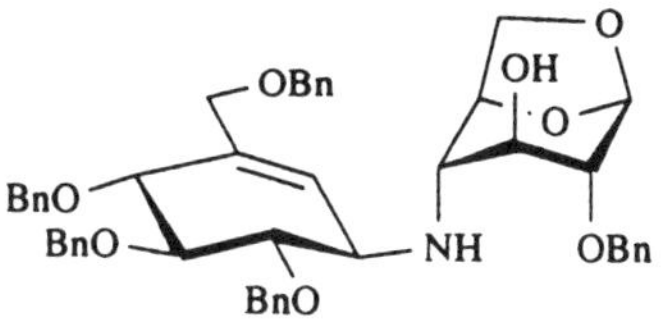

104

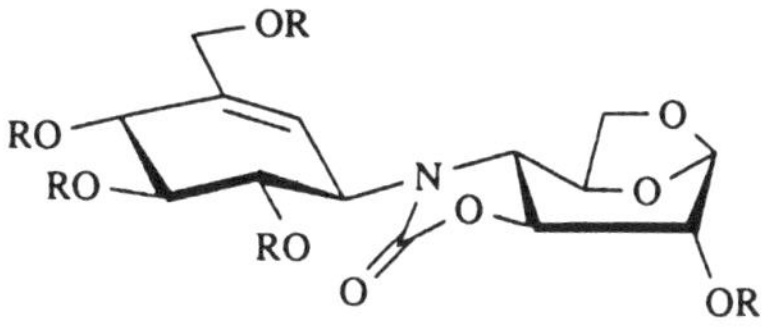

105 R = Bn
108 R = H
109 R = Ac

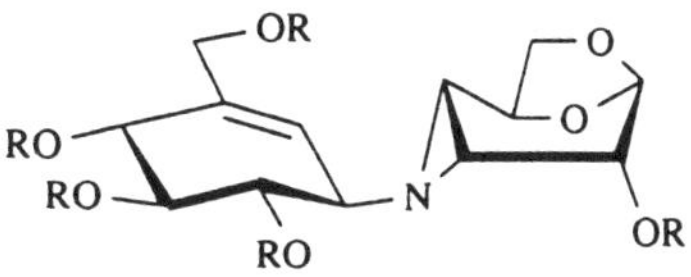

106 R = Bn
107 R = H

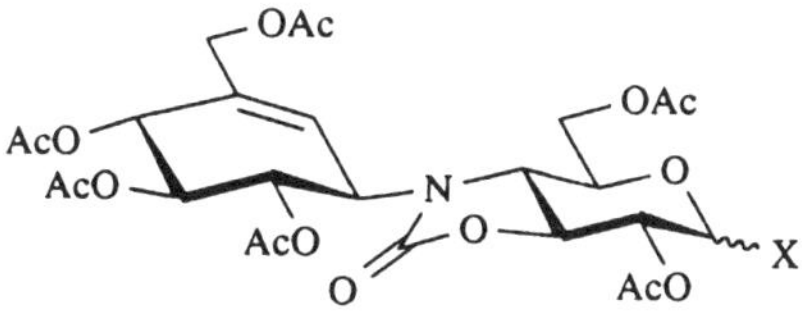

110 X = OAc
111 X = α-Br
112 X = β-SPh

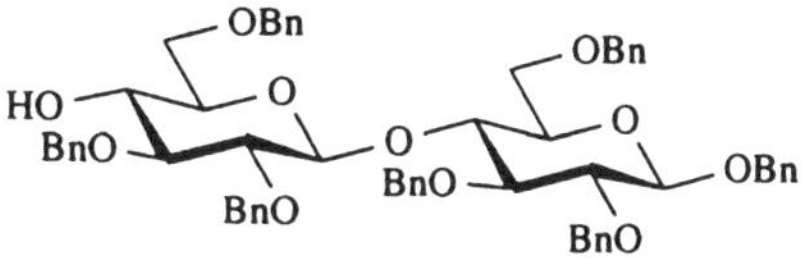

113

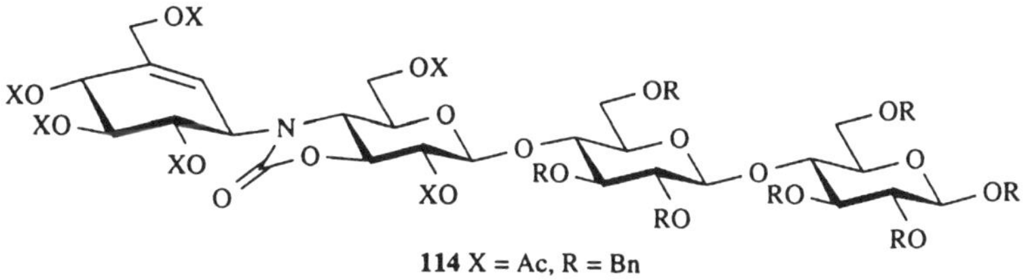

114 X = Ac, R = Bn

for converting the disaccharide alcohol **113** into the tetrasaccharide carbamate **114** [57].

We spent some time finding the right conditions for the successful removal of the protecting groups from **114**. In the end, base treatment of **114** gave the polyol **115** and subsequent reduction with lithium in ammonia gave the desired free sugar **116**. Although we have not actually prepared β-acarbose itself, **116** is the "all β" diastereoisomer of adiposin-2 **117**, itself a natural product possessing significant enzyme inhibitory activity [59, 63].

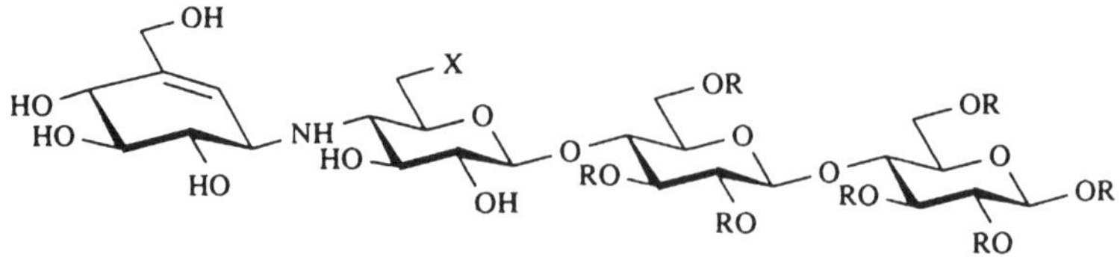

115 R = Bn, X = OH
116 R = H, X = OH
118 R = Bn, X = H

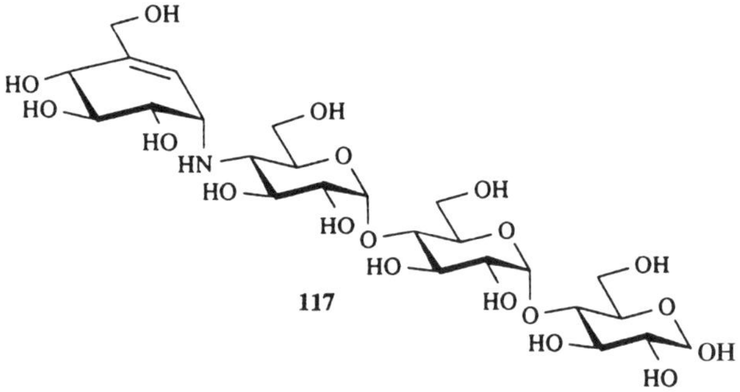

117

8.2
1,6-Epithio Sugars

It would, in principle, be possible to convert the polyol **115** into the 6″-deoxy sugar **118**, a direct precursor of β-acarbose. However, we decided that it would be better to reassess the whole synthesis and reasoned that a 1,6-epithio sugar such as **119** would be an ideal pivot for a synthesis of β-acarbose; the amine **81** could easily be introduced at C-4, the sulfur at C-1 allows for activation to produce a glycosyl donor ready for attachment to the disaccharide alcohol **113** and, finally, desulfurization at C-6 leads to the necessary methyl group.

We therefore needed a ready synthesis of the 1,6-epithio epoxide **119**. Normally, for such syntheses, the sulfur atom is introduced at an early stage [64]. However, it appeared to us that a suitable doubly-activated sugar such as **120** should be capable of direct conversion to the 1,6-epithio sugar **121** by treatment with some type of sulfur reagent. Indeed, the D-glucosyl bromide **122** was easily prepared and, upon treatment with either hydrogen sulfide and triethylamine in DMF or benzyltriethylammonium tetrathiomolybdate in chloroform, gave the 1,6-epithio sugar **121**. This success set the stage for a synthesis of the dimesylate **123** which was efficiently converted through to the sulfide **124** and thence the desired epoxide **125** [65].

Although we have not yet attempted the alkylation of the amine **81** with the epoxide **125**, preliminary experiments with cyclohexylamine and with sodium azide have allowed the synthesis of the model compounds **126** and **127** [65]. Additionally, we have converted the sulfide **121** into the sulfoxide **128** and the sulfone **129** (by oxidation with dimethyldioxirane) and have

119 X = Ac
125 X = H

120
122 X = Br, Y = OTs

121 X = Ac
124 X = Ms

123

126 X = $C_6H_{11}NH$
127 X = N_3

independently prepared the 1,6-episeleno sugar **130** by treatment of the bromo tosylate **122** with sodium hydrogen selenide [65, 66]. We are presently investigating methods for the synthesis of the telluride **131** and the disulfide **132**.

The bicyclic sugars **121** and **128–130** should all be capable of some sort of chemical activation to produce glycosyl donors. To date, we have investigated just the related sulfide **133**. Treatment of **133** with a range of alcohols, **134–138**, in the presence of *N*-iodosuccinimide and triflic acid, has given rise to intermediate disulfides, e.g. **139** in good yield and capable of reduction (Ra/Ni) to the 6,6′-dideoxy *β*-disaccharide **140** [66]. This is truly an excellent result in terms of a direct synthesis of *β*-acarbose.

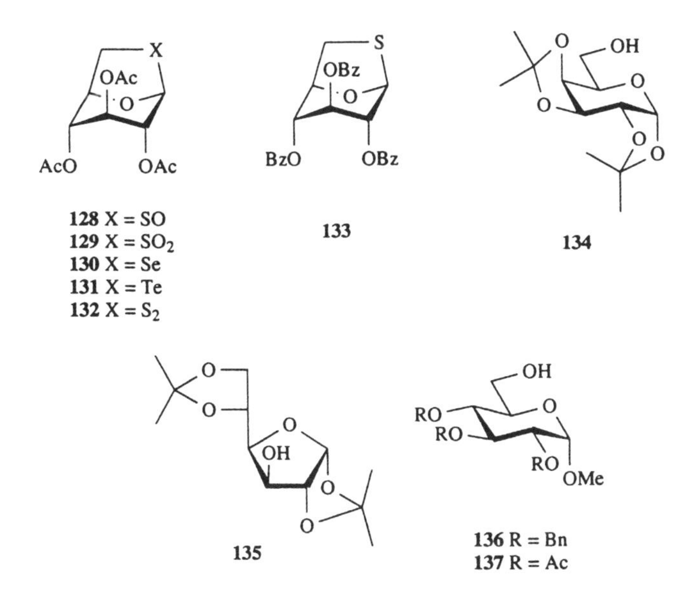

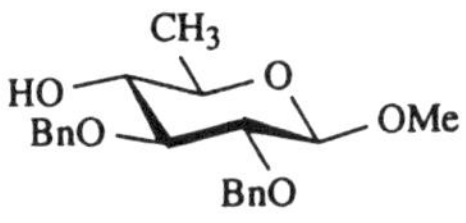

138

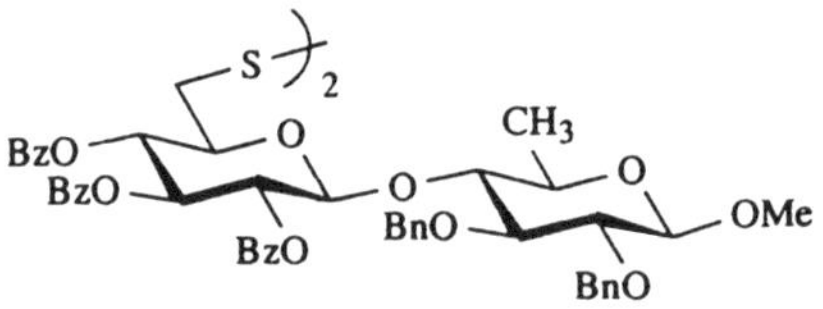

139

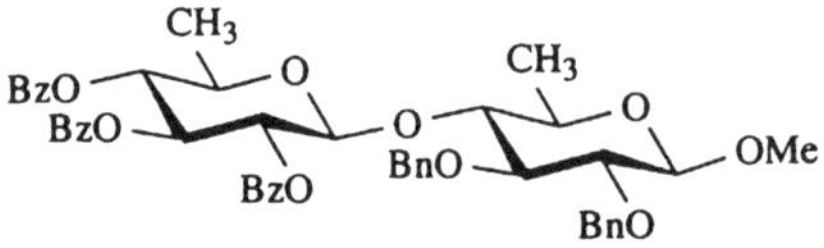

140

9
Enzyme Inhibition Studies

9.1
Epoxyalkyl Glycosides

The last aspect of this review of our work is to comment on the potential use of the various synthetic molecules in the inhibition of some plant and bacterial β-glucan endohydrolases. In 1989, we showed that the epoxylalkyl β-glycosides prepared early in our work were effective inhibitors of various β-glucan endohydrolases; the epoxylbutyl β-cellobioside **7** (n = 2) was a particularly good inhibitor of a *Bacillus subtilis* endohydrolase [18]. A little later, in 1991, we were able to use the optically-pure epoxides **7a** and **7b** to inhibit the action of two β-glucan endohydrolases isolated from *B. subtilis* and barley. The two enzymes, with identical substrate specificities but unrelated primary structures, were differentially inhibited by the two pure diastereoisomers, thus suggesting that the enzymes may employ sterically different mechanisms to achieve glycoside bond hydrolysis in their common substrate [67].

Another useful aspect of the epoxyalkyl β-glycosides was the development of an analytical technique (HPLC) for the identification of acidic, catalytic amino acids at the active site of a β-glucan endohydrolase. Proteolytic digestion of the complex formed from the β-glucan endohydrolase from *Bacillus amyloliquefaciens* and the epoxybutyl β-cellobioside **7** (n = 2) gave a labelled peptide which, using a novel HPLC analysis, identified Glu^{105} as a catalytic amino acid [68]. This novel HPLC technique has rendered the use of radiolabelled inhibitors essentially obsolete.

9.2
Epoxyalkyl C-Glycosides

The use of the epoxyalkyl β-D-C-glucosides **37** (n = 1–3) in further defining the active sites of β-glucan hydrolases has been investigated by Høj and coworkers

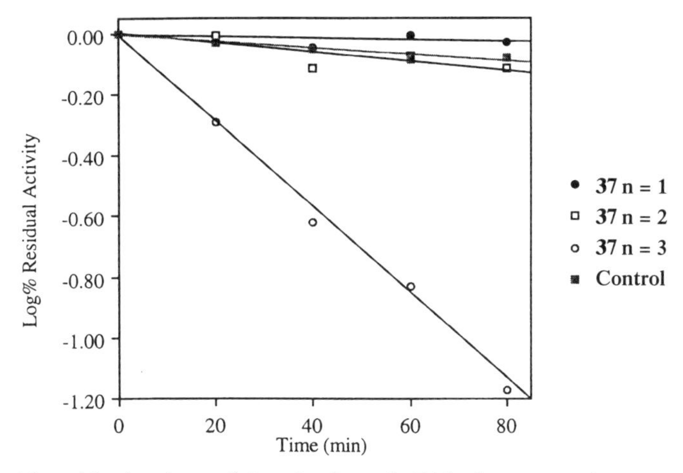

Fig. 5. Semi-logarithmic plots of inactivation of *Trichoderma reesei* QM9414 cellobiohydrolase I by the epoxides **37** (n = 1–3)

[69]. Figure 5 shows the results of inhibition experiments involving the D-glucosides **37** (n = 1–3) and a cellobiohydrolase from *Trichoderma reesei*. The best inhibition was achieved with the epoxypentyl β-D-C-glucoside, a result which corresponds nicely to the *O*-glycoside series where the epoxybutyl β-D-glucoside **6** (n = 2) was the most effective inhibitor.

9.3
Crystallographic Investigations

Varghese and coworkers have identified the active site nucleophiles of two crystalline β-D-glucan endohydrolases from germinating barley [70]. In more recent work, Høj and coworkers were able to introduce 2,3-epoxypropyl-β-laminaribioside **8** (n = 1) into the active site of the crystalline 1,3-β-D-glucanase from barley [71]. A subsequent X-ray crystallographic investigation gave invaluable information about the active site of this enzyme and the point of attachment between enzyme and inhibitor was confirmed.

10
Epilogue

The first ten years of this work has produced a plethora of novel, interesting and biologically active compounds. The focus has been on gaining an understanding of the mechanism of action of various β-glucan endohydrolases. We hope, in the next decade, to be able to transfer this experience and knowledge to other important enzymes, such as glycosyl transferases.

Acknowledgements. I would like to thank the host of Honours and Ph. D. students who performed most of the experiments described in this review. In particular, I would like to thank Dr Wayne Best and Dr Matthew Tilbrook (who, incidentally, drew all of the chemical structures contained in the review) for their invaluable collaboration; they are, indeed, a pair of excellent organic chemists.

I would also like to thank Professor Norm Lahey, to whom this review is dedicated, for dragging me out of the depths of the Government Chemical Laboratories (Queensland) to start a Ph. D. in 1967 and so lead to a life in academia.

The Australian Research Council has supported this work from its inception in 1984.

References

1. Hughes GK, Lahey FN, Price JR, Webb LJ (1948) Nature (London) 162:223
2. Macdonald PL, Robertson AV (1966) Aust J Chem 19:275
3. Colegate SM, Dorling PR, Huxtable CR (1979) Aust J Chem 32:2257
4. Humphries MJ, Matsumoto K, White SL, Molyneux RJ, Olden K (1988) Cancer Res 48:1410
5. Dorling PR, Huxtable CR, Colegate SM (1980) Biochem J 191:649
6. Hohenschutz LD, Bell EA, Jewess PJ, Leworthy DP, Pryce RJ, Arnold E, Clardy J (1981) Phytochemistry 20:811
7. Saul R, Chambers JP, Molyneux RJ, Elbein AD (1983) Arch Biochem Biophys 221:593
8. Furneaux RH, personal communication
9. Cannon JR, Joshi KR, McDonald IA, Retallack RW, Sierakowski AF, Wong LCH (1975) Tetrahedron Lett 2795
10. Decosterd LA, Parsons IC, Gustafson KR, Cardellina II JH, McMahon JB, Cragg GM, Murata Y, Pannell LK, Steiner JR, Clardy J, Boyd MR (1993) J Am Chem Soc 115:6673
11. von Itzstein M, Wu W-Y, Kok GB, Pegg MS, Dyason JC, Jin B, Phan TV, Smythe ML, White HF, Oliver SW, Colman PM, Varghese JN, Ryan DM, Woods JM, Bethell RC, Hotham VJ, Cameron JM, Penn CR (1993) Nature (London) 363:418
12. Thomas EW, McKelvy JF, Sharon N (1969) Nature (London) 222:485
13. Legler G, Bause E (1973) Carbohydr Res 28:45
14. Sinnott ML (1990) Chem Rev 90:1171
15. Legler G (1990) Adv Carbohydr Chem Biochem 48:319
16. Rodriguez EB, Stick RV (1990) Aust J Chem 43:665
17. Adam W, Bialas J, Hadjiarapoglou L (1991) Chem Ber 124:2377
18. Høj PB, Rodriguez EB, Stick RV, Stone BA (1989) J Biol Chem 264:4939
19. Rodriguez EB, Scally GD, Stick RV (1990) Aust J Chem 43:1391
20. Kolb HC, VanNieuwenhze MS, Sharpless KB (1994) Chem Rev 94:2483
21. Stick RV, Tilbrook DMG, Winslade ML, unpublished results
22. Gao Y, Hanson RM, Klunder JM, Ko SY, Masamune H, Sharpless KB (1987) J Am Chem Soc 109:5765
23. Best WM, Stick RV, Tilbrook DMG (1997) Aust J Chem, in press
24. Hudlický M (1988) Org React (NY) 35:513
25. Ermert P, Vasella A (1993) Helv Chim Acta 76:2687
26. Shulman ML, Shiyan SD, Khorlin AYa (1974) Carbohydr Res 33:229
27. Best WM, Ferro V, Harle J, Stick RV, Tilbrook DMG, unpublished results
28. Lewis MD, Cha JK, Kishi Y (1982) J Am Chem Soc 104:4976
29. Halcomb RL, Danishefsky SJ (1989) J Am Chem Soc 111:6661
30. Chini M, Crotti P, Favero L, Macchia F, Pineschi M (1994) Tetrahedron Lett 35:433
31. Tamura M, Kochi J (1971) Synthesis 303
32. Lipshutz BH, Crow R, Dimock SH, Ellsworth EL, Smith RAJ, Behling JR (1990) J Am Chem Soc 112:4063
33. Harle J, Stick RV, Tilbrook DMG, unpublished results
34. Martin OR, Lai W (1993) J Org Chem 58:176
35. Hughes AB, Rudge AJ (1994) Nat Prod Rep 11:135
36. Karpas A, Fleet GWJ, Dwek RA, Petursson S, Namgoong SK, Ramsden NG, Jacob GS, Rademacher TW (1988) Proc Natl Acad Sci USA 85:9229
37. Jacobs GS (1995) Current Opinion Structural Biology 5:605

38. Jespersen TM, Dong W, Sierks MR, Skrydstrup T, Lundt I, Bols M (1994) Angew Chem Int Ed Engl 33:1778
39. Takeo K, Murata Y, Kitamura S (1992) Carbohydr Res 224:311
40. Bieg T, Szeja W (1985) Synthesis 76
41. Best WM, Stick RV, Tilbrook DMG, unpublished results
42. Shahabuddin M, Kaslow DC (1993) Parasitology Today 9:252
43. Peter MG, Boldt P-C, Petersen S (1992) Liebigs Ann Chem 1275
44. Ogawa T, Beppu K, Nakabayashi S (1981) Carbohydr Res 93:C6
45. Matsuoka K, Nishimura S-I, Lee YC (1995) Bull Chem Soc Jpn 68:1715
46. Nishimura S-I, Matsuoka K, Kurita K (1990) Macromolecules 23:4182
47. Westphal O, Holzmann H (1942) Ber Deutsch Chem Ges 75:1274
48. Fairweather JK, Stick RV, Tilbrook DMG, unpublished results
49. McAdam DP, Perera AMA, Stick RV (1987) Aust J Chem 40:1901
50. Stone BA, private thought
51. Schmidt DD, Frommer W, Junge B, Müller L, Wingender W, Truscheit E, Schäfer D (1977) Naturwissenschaften 64:535
52. Truscheit E, Frommer W, Junge B, Müller L, Schmidt DD, Wingender W (1981) Angew Chem Int Ed Engl 20:744
53. Fukase H, Horii S (1992) J Org Chem 57:3651
54. Hayashida M, Sakairi N, Kuzuhara H (1989) Carbohydr Res 194:233
55. McAuliffe JC, Stick RV, Tilbrook DMG, Watts AG unpublished results
56. Nicotra F, Panza L, Ronchetti F, Russo G (1989) Gazz Chim Ital 119:577
57. McAuliffe JC, Stick RV, unpublished results
58. Shibata Y, Ogawa S, Suami T (1990) Carbohydr Res 200:486
59. Shibata Y, Ogawa S (1989) Carbohydr Res 189:309
60. Čapek K, Staněk Jr J, Čapková J, Jarý J (1975) Collect Czech Chem Commun 40:3886
61. Paulsen H, Röben W (1985) Liebigs Ann Chem 974
62. Zottola M, Venkateswara Rao B, Fraser-Reid B (1991) J Chem Soc Chem Commun 969
63. McAuliffe JC, Stick RV, Stone BA (1996) Tetrahedron Lett 37:2479
64. Lowary TL, Bundle DR (1994) Tetrahedron: Asymm 5:2397
65. Driguez H, McAuliffe JC, Stick RV, Tilbrook DMG, Williams SJ (1996) Aust J Chem 49:343
66. Stick RV, Tilbrook DMG, Williams SJ, unpublished results
67. Høj PB, Rodriguez EB, Iser JR, Stick RV, Stone BA (1991) J Biol Chem 266:11628
68. Høj PB, Condron R, Traeger JC, McAuliffe JC, Stone BA (1992) J Biol Chem 267:25059
69. Claeyssens M, Ferro V, Høj PB, Stick RV, Stone BA, unpublished results
70. Varghese JN, Garrett TPJ, Colman PM, Chen L, Høj PB, Fincher GB (1994) Proc Natl Acad Sci USA 91:2785
71. Chen L, Garrett TPJ, Fincher GB, Høj PB (1995) J Biol Chem 270:8093

Heparinoid Mimetics

Hans Peter Wessel

Pharma Division, Preclinical Research, F. Hoffmann-La Roche Ltd, CH-4070 Basel, Switzerland

Heparinoid polysaccharides such as heparan sulfate and heparin are able to interact with numerous proteins and influence vital biological processes. Heparinoid mimetics were prepared to reduce the structural complexity of heparinoids and to obtain selectivities. This article summarizes the development of heparinoid mimetics of different classes including representative syntheses and biological activities. Largely simplified compounds with regard to structure and synthetic access are described which maintain or exceed the activity of heparinoid polysaccharides. One of the recipes to increase binding or modify pharmacokinetic parameters was the introduction of hydrophobic groups.

Table of Contents

Topics in Current Chemistry, Vol. 187

List of Abbreviations and Symbols

Ac	=	acetyl
aFGF	=	acidic fibroblast growth factor
AIDS	=	acquired immune deficiency syndrome
APTT	=	activated partial thromboplastin time
AT III	=	antithrombin III
β-CD	=	β-cyclodextrin
bFGF	=	basic fibroblast growth factor
Bn	=	benzyl
CMV	=	cytomegalovirus
CR*S*-heparin	=	carboxyl-reduced sulfated heparin
DMF	=	*N*,*N*-dimethylformamide
DMSO	=	dimethylsulfoxide
DS	=	degree of sulfation
EGF	=	epidermal growth factor
Et	=	ethyl
FGF	=	fibroblast growth factor
HBGF	=	heparin-binding growth factor
HC II	=	heparin cofactor II
HIV	=	human immunodeficiency virus
HSV	=	herpes simplex virus
Lev	=	levulinoyl
Me	=	methyl
PDGF	=	platelet derived growth factor
PF4	=	platelet factor 4
Ph	=	phenyl
r_i	=	relative inhibitory activity compared to heparin in an SMC cell proliferation assay
RT	=	room temperature
SMC	=	smooth muscle cell
TMU	=	tetramethylurea
tol	=	toluene

1 Introduction

Heparan sulfate and heparin are sulfated polysaccharides of the glycosaminoglycan family. Both are biosynthetically related being generated from a common polysaccharide with β-D-glucuronic acid-α-(1$\rightarrow$4)-*N*-acetyl-D-glucosamine repeat sequence on a proteoglycan chain [1]. The enzymatic maturation process involves de-*N*-acetylation and *N*-sulfation; the following transformations, epimerization at C-5 of the uronic acid and *O*-sulfation at various specific positions, depend on *N*-sulfation. Heparin is processed to a higher degree and carries more L-iduronic acid residues, more *N*-sulfate and *O*-sulfate groups,

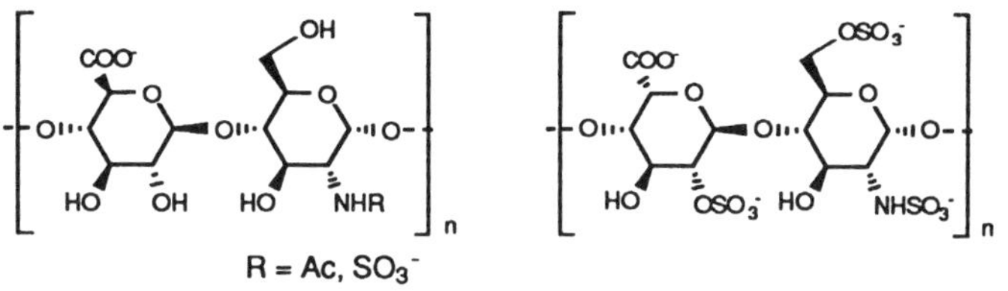

$R = Ac, SO_3^-$

1 2

Structures 1. Characteristic structural motifs in heparan sulfate and heparin

and less *N*-acetyl groups than heparan sulfate. Although characteristic repeating units (Structures 1) can be attributed to heparan sulfate (**1**) and heparin (**2**), these polysaccharides are characterized by great structural heterogeneity stemming from different substitution patterns, and 19 different naturally occurring uronic acid – glucosamine disaccharides have been identified so far. In addition, the preparations show dispersion of molecular weight after cleavage from the proteoglycan core. Emphasizing their common features, heparin and heparan sulfate are referred to as "*heparinoids*" [2].

Starting from bioactive proteins, peptides with similar activity or binding properties could frequently be identified; the development of peptide mimetics from those peptides is a classical exercise in medicinal chemistry. The analogous approach for carbohydrates, the development of carbohydrate mimetics from bioactive polysaccharides, is in its infancy.

In the case of heparinoids, the development of mimetics is desirable to reduce the enormous complexity of the polysaccharide mixtures to arrive at compounds of lower molecular weight that can be synthesized in a reasonable number of steps. As delineated in more detail in the next section, heparinoids are associated with a multitude of biological properties, and this raises the question of selectivity. Mimetics can be investigated in this respect and may be valuable tools to study heparinoid – protein interactions and their implications. As well established for peptide mimetics, heparinoid mimetics may also be tailored to modify pharmacokinetic parameters. Whereas heparinoid – protein interactions are mainly based on ionic forces, hydrophobic groups in heparinoid mimetics may lead to non-charged interactions and enhance the binding. In addition, since heparinoids are isolated from natural sources, e.g. commercial heparin presently mainly from porcine intestine, preparations may vary with respect to their heterogeneity and thus their composition; contamination with viruses or prions may also become an issue of purity. A heparinoid mimetic would ideally maintain only one pharmacological activity, be structurally less complex than heparin and be easily prepared.

Low molecular weight heparins, heparin fractions, or other sulfated polysaccharides still have the same or a similar complexity as heparin and are, therefore, not regarded as heparinoid mimetics; only compounds with one defined carbohydrate backbone serving as a template for sulfates will be discussed in this context.

2
Biological Properties of Heparinoids

As polyanionic saccharides, heparinoids interact with a high number of proteins, the so-called heparin binding proteins. Heparinoid – protein interactions have usually been described for in vitro systems, and their physiological relevance is not always clear. On the other hand, heparinoids can influence biological systems, but in most cases the molecular basis is not fully understood. For this reason, interactions with proteins and biological properties of heparinoids will be discussed separately in the following. As heparinoid – protein interactions have been reviewed [3] the focus will be in rather short form on those areas in which heparinoid mimetics have been investigated.

2.1
Heparinoid Interactions with Proteins

Interactions of heparinoids with the most diverse proteins such as enzymes and enzyme inhibitors, cytokines, and adhesion molecules have been described. To date, many more than a hundred heparin binding proteins are known. A number of heparin binding proteins are members of the serpin family of serine protease inhibitors. The best described example is antithrombin [4]. Antithrombin III (AT III) is able to inhibit various serine proteases involved in the blood coagulation process by formation of stable, equimolar complexes. Binding of heparin to AT III accelerates the kinetics of this complex formation by several orders of magnitude. This has been the basis for the successful clinical use of heparin as an anticoagulant for nearly sixty years.

Degradation of heparin followed by affinity chromatography on immobilized AT III led to the identification of a pentasaccharide unit with high antithrombin affinity. This pentasaccharide **3** is characterized by a unique highly sulfated central glucosamine unit with a 3-*O*-sulfate group (Structures 2).

This pentasaccharide sequence induces a conformational change in AT III which probably causes the complex to be more accessible to the active site of the proteases. The most relevant protease affected by the pentasaccharide **3** is factor Xa, but factor XIIa and plasma kallikrein activities can also be potentiated. Sequence **3** occurs in heparin as well as in various heparan sulfate proteoglycans of different origin including the vascular endothelium.

Other members of the coagulation cascade such as factor IIa (also referred to as thrombin), factor IXa, and factor XIa require a longer heparin unit of 13 or

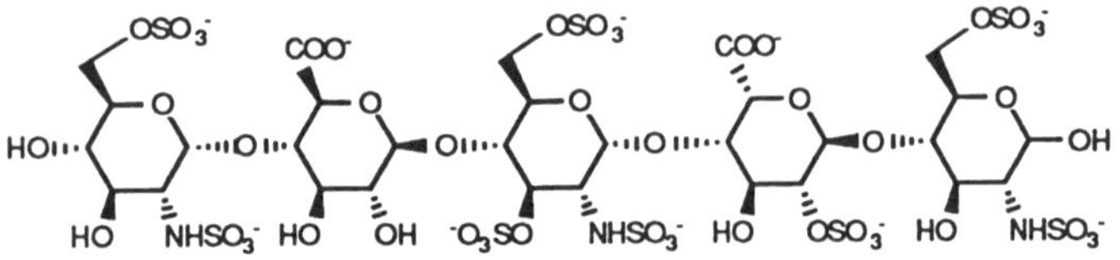

3

Structures 2. Structure of the heparin pentasaccharide with high antithrombin affinity

more saccharide units in addition to the pentasaccharide to accelerate the antithrombin – protein reaction. A template model was postulated in which a heparinoid binds to antithrombin plus the second protein in a ternary complex. To understand the conformational change of AT III on a molecular level, the protein – pentasaccharide binding site was investigated by various methods [4] including spectroscopic [5] as well as molecular modelling [6] studies; whereas first modelling experiments were based on α1-antitrypsin, crystal structures of cleaved AT III [7] and of active dimeric antithrombin [8] have since been reported. In addition to the AT III conformational change, the conformational flexibility of the L-iduronic acid unit of heparin, also present in the heparin pentasaccharide, has been postulated to play an important role in the protein interaction [9].

Another serine protease inhibitor of the α1-antitrypsin family (serpin) is heparin cofactor II (HC II), which also forms a 1:1 complex with thrombin, but does not react with factor Xa [4, 10]. The rate of inhibition of thrombin is not only increased by heparinoids but also by the related glycosaminoglycan dermatan sulfate. The identification of an inhibitor variant and site-directed mutagenesis studies on HC II cDNA led to the understanding that the binding sites for heparin and dermatan sulfate may be overlapping but not identical. Further proteinase inhibitors interacting with heparinoids are tissue factor pathway inhibitor and protease nexin-1.

Relevant heparin-binding enzymes not involved in the coagulation cascade are, for example, elastase, cathepsin G, superoxide dismutase, lipoprotein lipase and other lipases. The plasma clearing properties of heparin are associated with its binding to lipoprotein lipase and hepatic lipase when the enzymes are released from the surface of endothelial cells [11] and have been studied in view of a potential impact on the regulation of atherosclerosis.

Members of the fibroblast growth factor (FGF) family [12] bind heparin with high affinity and were also termed "heparin-binding growth factors" (HBGF). The most thoroughly characterized members are acidic FGF (aFGF, FGF-1) and basic FGF (bFGF, FGF-2); they are mitogenic for a wide variety of cells and support chemotactic migration and induction of proteases that could facilitate tissue remodelling. Heparinoids can stabilize the growth factors and prevent inactivation by proteolysis. According to binding studies, the binding site of heparinoids for bFGF and aFGF was estimated to be composed of four to eight monosaccharide units rich in 2-*O*-sulfated iduronic acid [13]. The involvement in heparinoid binding of the bFGF basic amino acids identified by various methods has been confirmed very recently by crystal structures of bFGF complexes with heparin tetra- and hexasaccharides, the first X-ray structures of heparinoid fragments [14]. It is interesting to note, however, that in a study of the bFGF/heparin interaction the amount of non-electrostatic interactions, i.e. hydrogen bonding and van der Waals packing, has been estimated at 70% [15]. The heparinoid sequence required to elicit a biological response (signal transduction, mitogenicity) is longer than the binding sequence mentioned above, only dodecasaccharides being equivalent to heparin. It was hypothesized that the additional saccharides are needed to bind the receptor or another bFGF molecule to facilitate receptor dimerization [16].

Other cytokines binding to heparinoids include platelet factor 4 (PF4) and interleukin-8 [17].

2.2 Selected Biological Properties of Heparinoids

Heparin and heparan sulfate inhibit the migration and proliferation of smooth muscle cells (SMC) in vitro and intimal thickening in vivo [18]. The high antiproliferative activity of heparan sulfates [19] particularly suggested that endogenous heparan sulfates function as specific regulators of SMC growth, and other cell types were only minimally inhibited. The antiproliferative effect of heparinoids was not dependent on the AT III-mediated anticoagulant activity [20]. Size fractionation of heparin demonstrated that a dodecasaccharide was required to obtain the full heparin effect, and oversulfation of the fractions slightly increased the activity [21]. The molecular mechanism for the heparinoid antiproliferative activity is not known yet; it seems that the inhibitory action is not exerted by binding of serum mitogens such as platelet derived growth factor (PDGF) or epidermal growth factor (EGF). While it has been shown that heparin is internalized by SMC [22] it is not clear whether heparinoids act intracellularly or through a specific site on the cell surface [23]. Excessive SMC growth plays an important role in diseases such as arteriosclerosis [24]. Clinical trials with heparins against vascular restenosis, a renarrowing of the arterial lumen which occurs with high incidence after angioplasty, an often used procedure to restore blood flow in stenotic coronary arteries, have not shown convincing efficacy, however, only low dosages were possible due to imminent bleeding complications.

Most mammalian cells carry heparan sulfates as plasma membrane substituents. Cell surface heparan sulfates have been shown to mediate the initial binding of enveloped viruses such as herpes simplex virus (HSV), cytomegalovirus (CMV), and pseudorabies virus to the cell [25]. Equally, it was shown that heparan sulfate proteoglycan mediates human immunodeficiency virus (HIV) infection of T-cell lines [26]. Thus, heparan sulfate and heparin, but also other polysulfated saccharides such as dextran sulfate, show anti-viral activity [27, 28]. Heparin fragments were significantly less active than heparin, and fragments below a mean molecular weight of 7 kD lost activity by more than a factor of ten [27]. HIV replication is believed to be inhibited by the inhibition of virus adsorption to the cell surface heparan sulfates, the mechanism being binding of sulfated polysaccharides to the virus to inhibit the binding of the virions to $CD4^+$ cells and subsequent giant cell (syncytium) formation. Studies with specific antibodies confirmed that polysulfated molecules interact with the positively charged amino acids concentrated in the V3 loop of gp 120 of HIV-1. Sulfated polysaccharides are also inhibitory to HIV-1 reverse transcriptase, but only at concentrations that are approximately 100-fold higher than those required to inhibit virus – cell binding [29]. Disappointingly, neither of the polysaccharides dextran sulfate nor the pentosan sulfate were clinically efficacious after oral or parenteral dosage although high plasma levels were achieved by continuous intravenous infusion [30].

Considering the many interactions of heparinoids with the various proteins and the multitude of resulting biological activities, including effects on tumour cell metastasis, angiogenesis, asthma, and immune cell migration in inflammation, it is clear that heparinoids and especially heparinoid mimetics may be useful not only in anticoagulation but also in a series of other indications [31].

3 Derivatives of Heparin Saccharides

Analogues of the heparin pentasaccharide **3** have been prepared by various authors [32] with the main contributions from the Sanofi and Organon groups; their achievements have been competently reviewed recently [33]. This work had the main goal of developing selective inhibitors of factor Xa as antithrombotics. Important for the development of synthetically simplified derivatives were the findings that an α-methyl glycoside had the same biological activity as

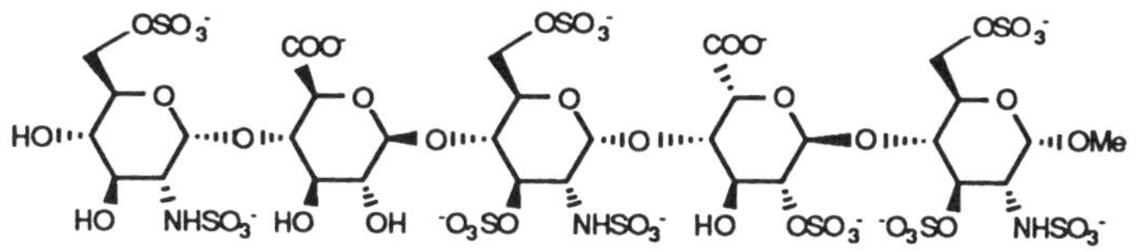

4

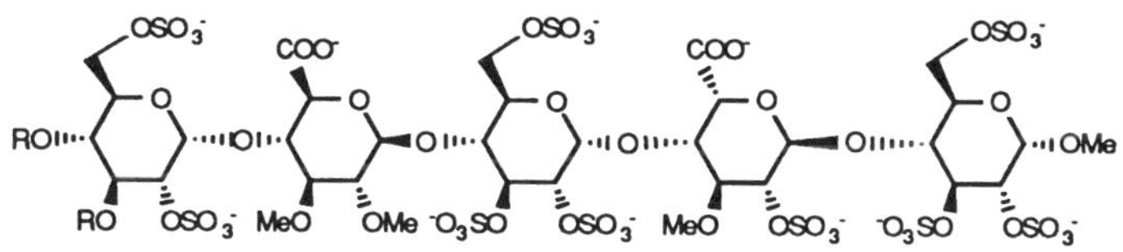

5 R = CH_3
6 R = $(CH_2)_3CH_3$

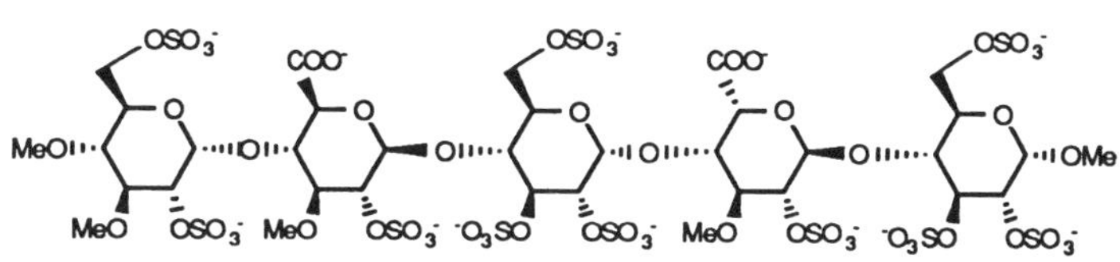

7

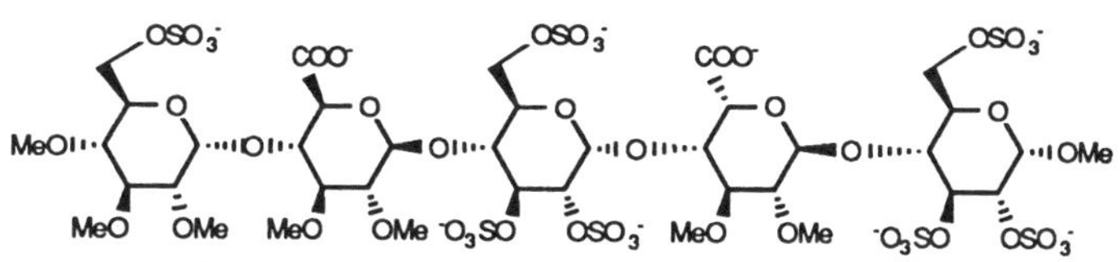

8

Structures 3. Derivatives of heparin pentasaccharides

Table 1. Activities of heparin pentasaccharide **3** and derivatives

	3	4	5	6	7	8
anti-Xa activitiy [U mg^{-1}]	700	1250	1323	1529	1159	1611
K_D [nM]		22	16	2	139	13

the parent pentasaccharide **3** with free anomeric center, and that replacement of *N*-sulfates by *O*-sulfates did not decrease the activity.

A series of modifications was based on the pentasaccharide **4** (Structures 3), which carries an additional *O*-sulfate group at the reducing terminus and was more active than the reference compound **3**. The methylated analogue **5** was synthesized and proved to have a similar activity. Replacement of the terminal methyl groups by n-butyl ethers (**6**) increased the anticoagulant activity, whereas longer alkyl chains led to a reduction in activity. Moreover, the hydrophobic groups increased the affinity towards AT III (Table 1); an increase in affinity (a lower K_D) was, however, not necessarily linked to higher anti-Xa activity: high affinity heparin saccharide derivatives with only low anti-Xa activity are known [33].

The very long syntheses (the first synthesis of pentasaccharide **3** required 75 steps) could be facilitated with the introduction of two disaccharide units that differ only in the configuration of the uronic acid as in **7** and **8** (Structures 3). The inversion of an L-iduronic acid disaccharide into a D-glucuronic acid analogue saved the independent construction of another disaccharide and thus an appreciable number of synthetic steps (Scheme 1) [34]. It is interesting to note that the substitution of the 2-*O*-sulfate in the L-iduronic acid moiety by an *O*-methyl group again increased the anti-Xa activity (Table 1), indicating that hydrophobic interactions contribute to AT III binding and compensate for the loss of an electrostatic sulfate interaction. The high number of hydrophobic groups led to an increase in elimination half-life in rat of up to more than 10 h in the case of pentasaccharide **8** [34].

For the inhibition of AT III-mediated thrombin activity, derivatives of **8** were prepared containing a flexible spacer of around 50 atoms plus a sulfated maltooligosaccharide or another AT III binding pentasaccharide [35].

While all these heparin saccharide derivatives are structurally close to the natural active pentasaccharide, a disadvantage is the complex synthesis comprising more than 30 steps even for the simplified analogues (Scheme 1) which renders a full development of these compounds less likely.

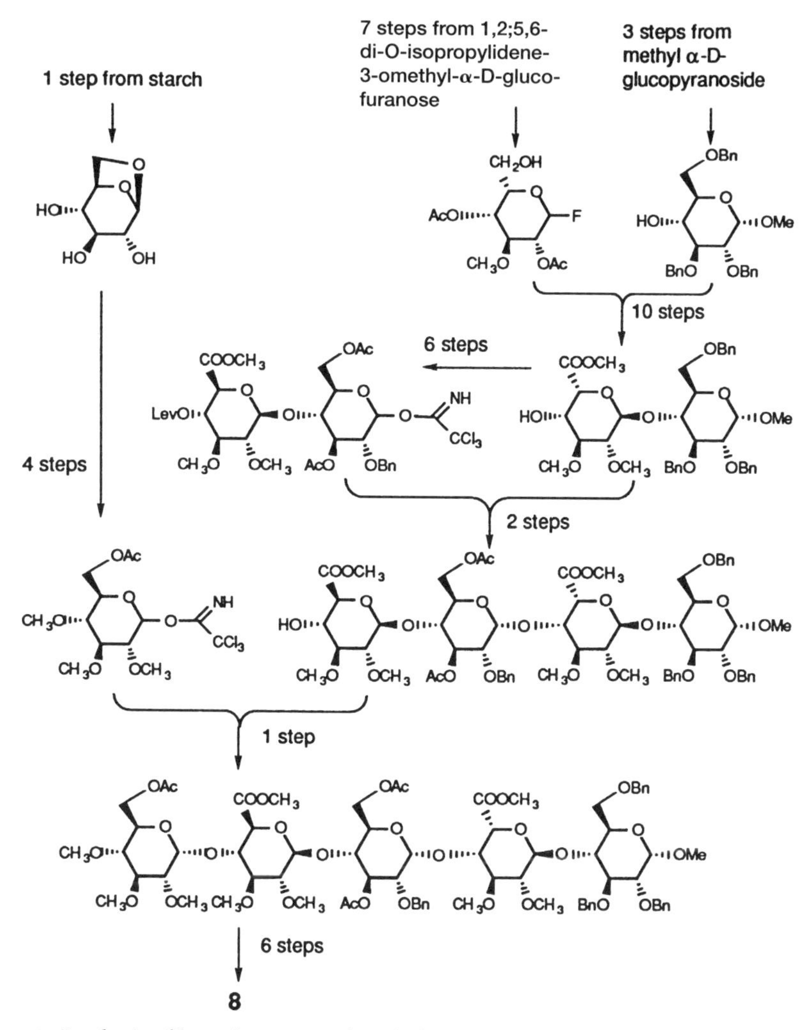

Scheme 1. Synthesis of heparin pentasaccharide derivative **8**

4 Sulfated Oligosaccharides

4.1 Sulfated Linear Oligosaccharides

The work on the first sulfated oligosaccharides was based on investigations of the polymeric heparinoids: derivatives of heparin had been prepared to obtain information on the structural requirements of heparin for SMC antiproliferative activity. A clue for further studies was brought about by carboxyl-reduced sulfated heparin (CRS-heparin, **9**) in which all carboxyl groups were reduced and the resulting primary hydroxyl groups were selectively sulfated (Structures 4). CRS-Heparin showed high antiproliferative activity but basically no AT III-mediated anticoagulant effect [36].

Consequently, sulfated non-uronic, mainly (1→4)-linked oligosaccharides were prepared. Trestatin A sulfate (**10**, Structures 4) emerged as highly active compound devoid of AT III-mediated anticoagulant activity [36]. Trestatin A, a pseudo-nonasaccharide obtained from strains of *Streptomyces dimorphogenes*, is a potent α-amylase inhibitor. As the molecule assumes an amylose-like helical tertiary structure it was obvious to compare the activity of sulfated malto-oligosaccharides. Very surprisingly, the sulfated amylose substructures had only low antiproliferative activity. Molecular modelling studies then indicated that the trehalose moiety of Trestatin A bends out from the helical conformation, so that Trestatin substructures containing the trehalose end were synthesized. The all-α-linked oligosaccharides were available in block syntheses from a trehalose glycosyl acceptor and a suitable maltooligoosyl donor [37], the critical α-D-glu-

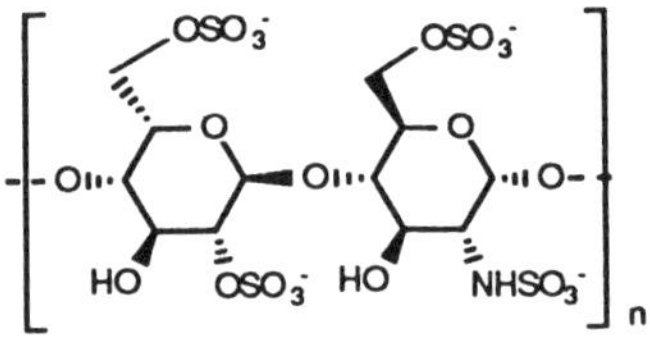

9

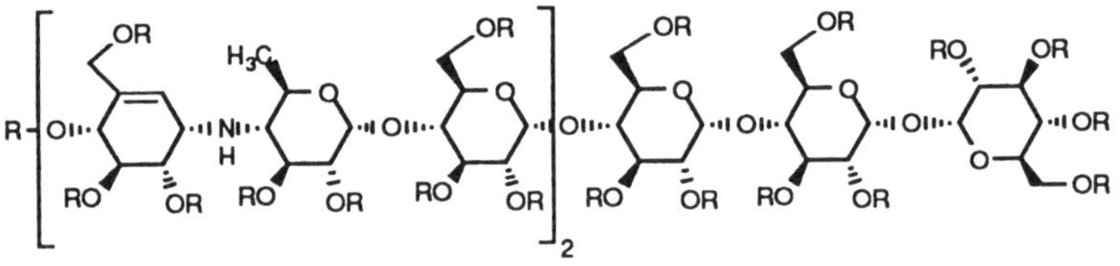

10 DS ≈ 2.4

11 n = 1 DS ≈ 3.0

12 n = 2 DS ≈ 2.9

13 n = 3 DS ≈ 2.4

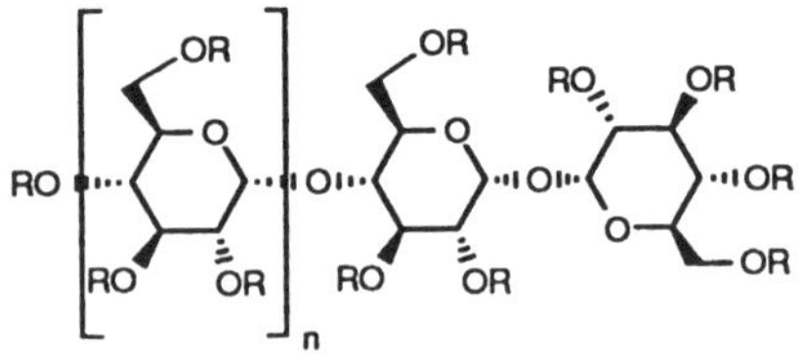

14 n = 0 DS ≈ 3.0

15 n = 1 DS ≈ 2.8

16 n = 2 DS ≈ 2.6

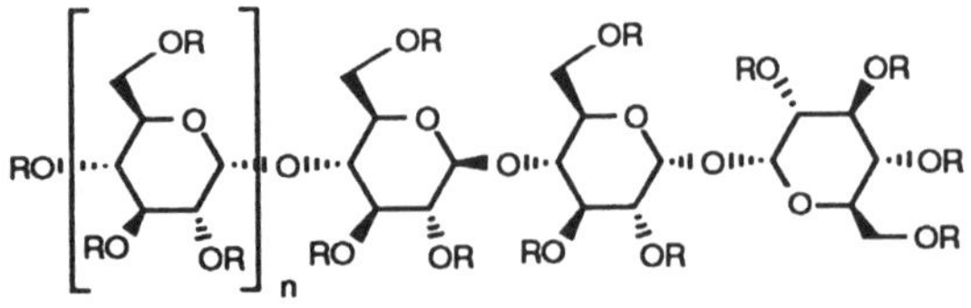

Structures 4. Structures of CRS-heparin, Trestatin A sulfate, substructures thereof, and β-linked analogues. DS denotes the degree of sulfation, defined as average number of sulfate groups per monosaccharide unit

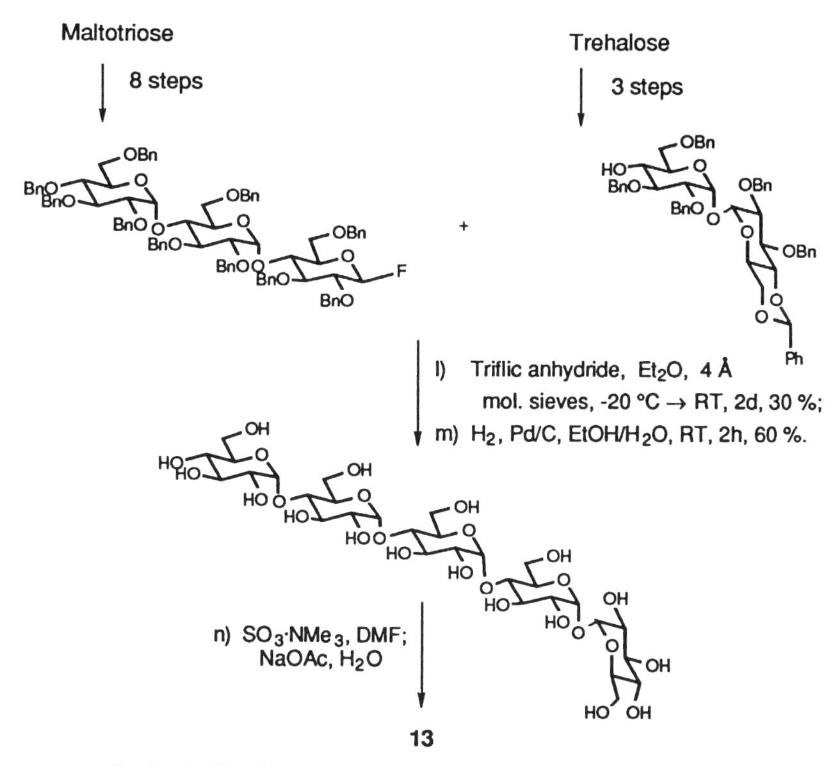

Scheme 2. Synthesis of sulfated Trestatin A pentasaccharide substructure **13**

cosyl linkage being achieved with high stereospecificity using triflic anhydride as a new promoter (Scheme 2) [38]. While the tri- and tetrasaccharide substructures **11** and **12** exhibited no appreciable activity, the highly sulfated maltotriosyl trehalose pentasaccharide **13** (Structures 4), a pentasaccharide substructure of Trestatin with an additional terminal hydroxyl group, had an antiproliferative activity comparable to heparin [39].

In an investigation of analogues of these substructures also the corresponding β-D-linked oligosaccharides were prepared; in this series, the sulfated β-D-(1→4)-linked maltosyl trehalose trisaccharide **14** had already appreciable SMC antiproliferative activity, and the activity of tetrasaccharide **15** was comparable to heparin; the activity of the β-D-(1→4)-linked pentasaccharide **16** was only slightly increased over the tetrasaccharide activity [39]. Tetrasaccharide **15** is the smallest sulfated carbohydrate with high antiproliferative activity known so far. The β-D-linkage of *gluco*-configurated building blocks is achieved easier than an α-D-linkage as in **13** since the stereodirecting influence of neighbouring groups can be exploited in the glycoside synthesis; thus the synthesis of **15** was short (9 steps) and proceeded with high yields (Scheme 3) [40].

With regard to the distinctly different activities of the α- and β-D-linked maltosyl trehalose tetrasaccharides, it is interesting to note that highly sulfated α-D-linked glucosides were conformationally stable whereas the β-D-linked glu-

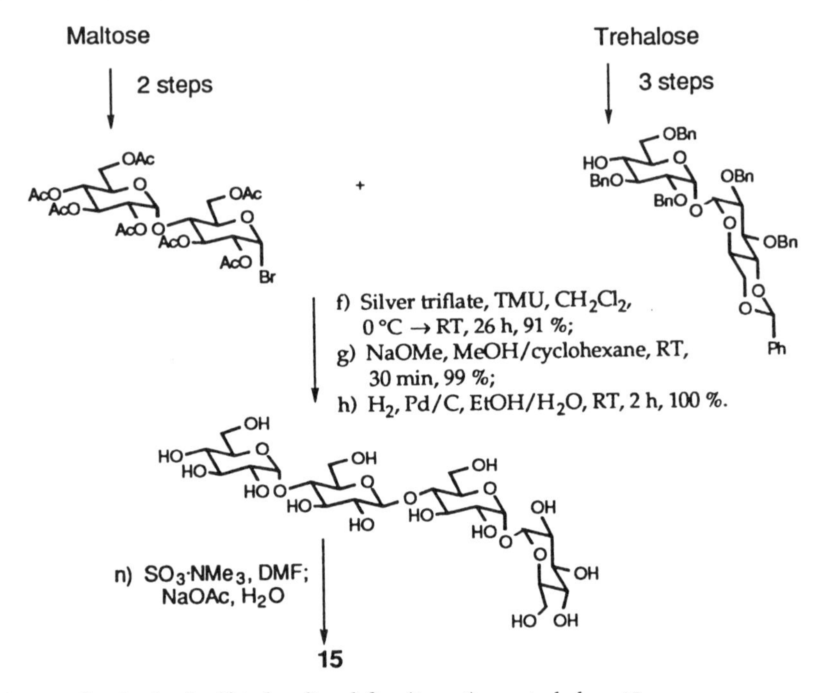

Scheme 3. Synthesis of sulfated maltosyl β-D-(1 $\rightarrow$ 4)-α,α-trehalose 15

cosides occurred in conformational equilibrium, which was, for example, shown for D-glucuronic acids and methyl D-glucopyranosides as model compounds [41]. This is a parallel to the conformational equilibrium of L-iduronic acid in heparin which was hypothesized to play a role in biological activity (see above). On the other hand, the different activities of the α- and β-D-linked maltosyl trehalose tetrasaccharides hinted at a specific interaction of the sulfated carbohydrate with other biomolecules; the requirement for the localization of sulfates at specific positions was supported in an investigation on tetrasaccharide analogues of **15** [42], in which the maltosyl moiety was replaced by different equatorially linked disaccharides. Whereas only the conformationally similar isomaltose analogue had a comparable activity, other modifications, such as the β-maltose replacement by β-cellobiose, led to a decrease in activity of up to 70% [43].

The sulfated oligosaccharides discussed above were highly, but not completely, sulfated so that mixtures of compounds with a slightly different sulfation pattern resulted. An attempt to understand which sulfation site was important for the biological activity of tetrasaccharide **15** was made by the synthesis of mono- and oligodeoxygenated analogues of **15**. These tetrasaccharides were prepared in a block synthesis approach comparable to the synthesis of **15**, and the deoxy functions were essentially introduced on the di- or tetrasaccharide level. A representative synthetic example is depicted in Scheme 4 [44].

Indeed, some single deoxygenations – and thus the absence of a sulfate group at this very position – led to a decrease in antiproliferative activity of more than

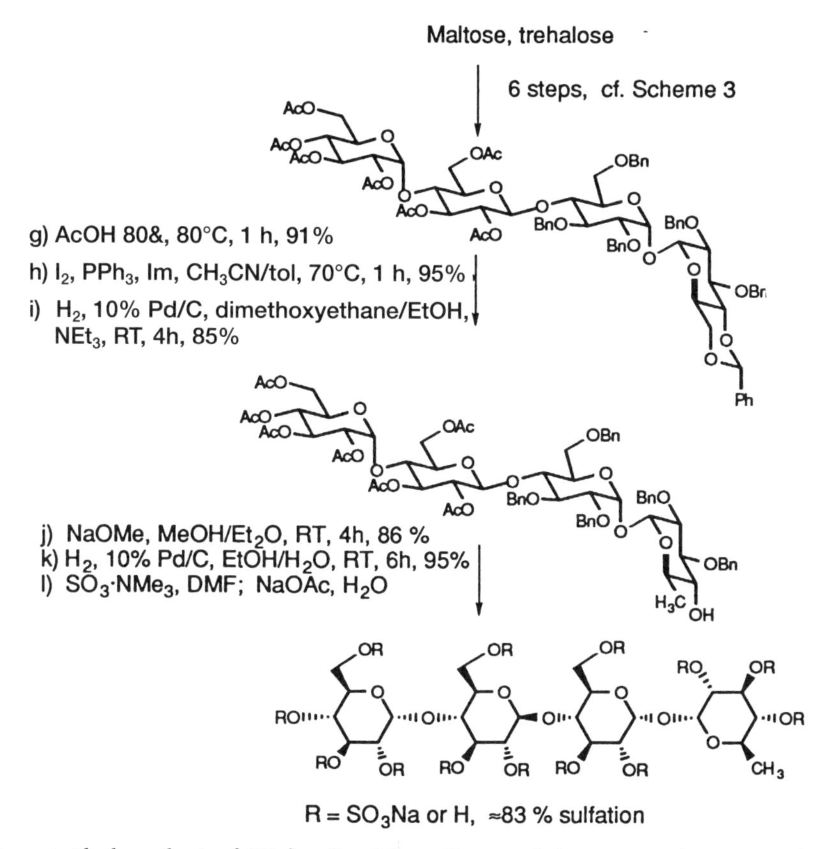

Scheme 4. Block synthesis of 6H-β-maltosyl-(1 $\rightarrow$ 4)-α,α-trehalose, a monodeoxygenated analogue of **15**

30% (Scheme 5) [45]. The sulfates contributing the most to the antiproliferative activity seem to be located at the both ends of the tetrasaccharide molecule.

Apart from its potent antiproliferative activity, tetrasaccharide **15** was effective in blocking human complement in vitro and inhibited the release of heparan sulfate from cardiac microvascular endothelial cells. To overcome hyperacute rejection, the tetrasaccharide has been investigated in a guinea pig to rat cardiac xenotransplantation model and significantly prolonged the survival of heart recipients when compared to control and heparin treated groups [46].

While the size requirement for SMC antiproliferative activity had been investigated for heparin (cf. Sect. 2.2), a similar study on fractions of highly active heparan sulfate has not been carried out. After the discovery of active sulfated tetrasaccharides it seemed possible that relatively small heparan sulfate substructures with the antiproliferative activity of heparin could be identified. Model heparan sulfate oligosaccharides were devised in which the *N*-sulfate groups were replaced by *O*-sulfates and the glucuronic acids were replaced by glucoses, formally carboxyl-reduced and sulfated glucuronic acids – in parallel to CRS-heparin (see above). With this approach the heparan sulfate backbone

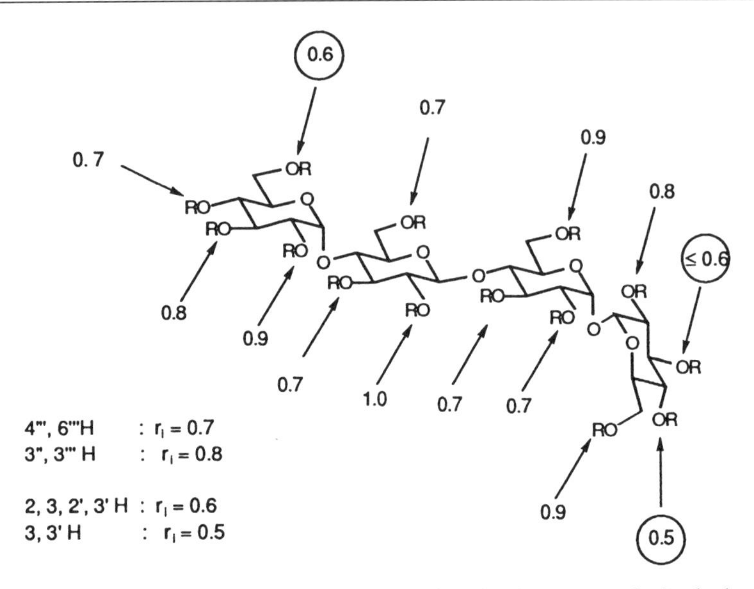

Scheme 5. Summary of antiproliferative activities of sulfated mono- and oligohydroxylated β-maltosyl-(1 $\rightarrow$ 4)-trehalose tetrasaccharides. Arrows indicate positions of monodeoxygenation, activities are given as r_i; values, the relative inhibition compared to heparin. Sites on which monodeoxygenation led to more than 30% loss of activity are circled

was reduced to an α,β-(1$\rightarrow$4)-glucan, i.e. alternating α-D- and β-D-(1$\rightarrow$4)-linked glucoses. Sulfated mono- to hexasaccharide substructures were synthesized as methyl glycosides; two series of compounds were investigated, namely methyl α-D-glucosides such as **17** and **18**, and the frame-shifted methyl β-D-glucosides represented by **19**–**21** (Structures 5).

Given those simplifications the oligosaccharide substructures were readily available in block syntheses using glucosyl or maltosyl building blocks as exemplified for the synthesis of pentasaccharide **20** (Scheme 6); the analogous [2+2+2] block synthesis furnished the hexasaccharide substructure **21** [47]. Pentasaccharide **20** and hexasaccharide **21** reached the antiproliferative activity of heparin (Table 2) [48]. The data suggested that three β-D-linked glucosyl units are required to elicit the full heparin activity. These heparan sulfate model saccharides were thus more active than oversulfated heparin fragments of equivalent size.

A series of alkyl malto- and laminaro α-oligosaccharides has been investigated for anti-HIV activity. An active representative (anti HIV EC_{50} = 0.2 mg ml^{-1}) is

Table 2. Relative antiproliferative activities of heparin and heparan sulfate model oligosaccharides

Compd.	2	17	18	19	20	21
r_i	1.0	0.7	0.7	0.8	1.0	1.0

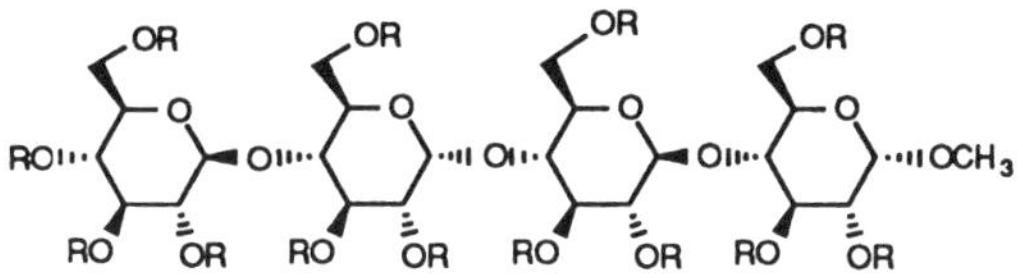

17 R = SO_3Na or H, DS ≈ 2.8

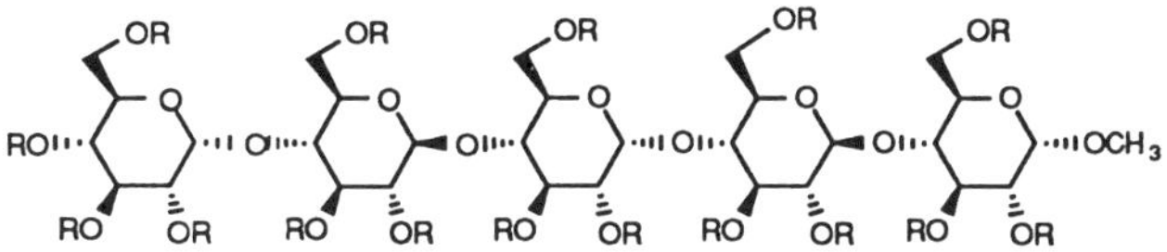

18 R = SO_3Na or H, DS ≈ 2.8

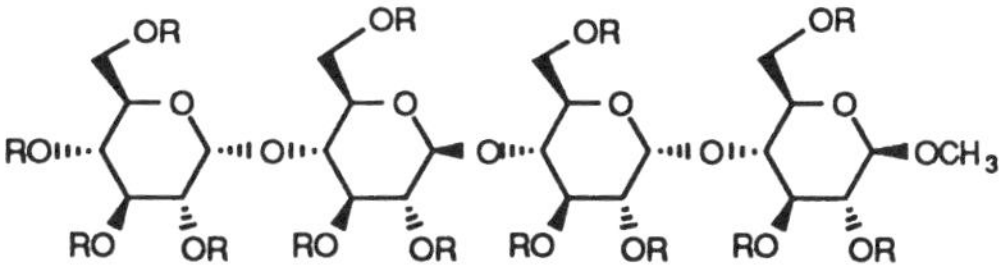

19 R = SO_3Na or H, DS ≈ 2.7

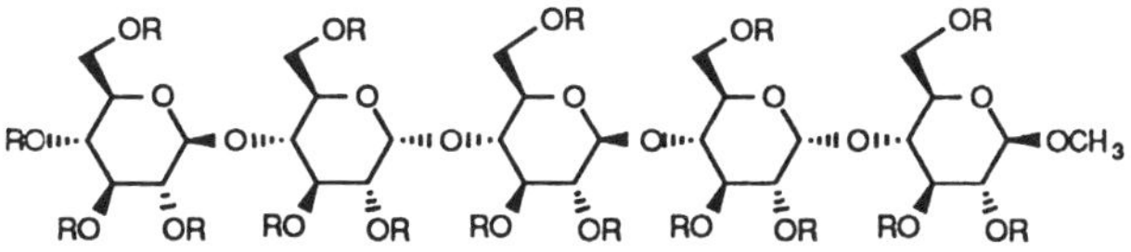

20 R = SO_3Na or H, DS ≈ 2.7

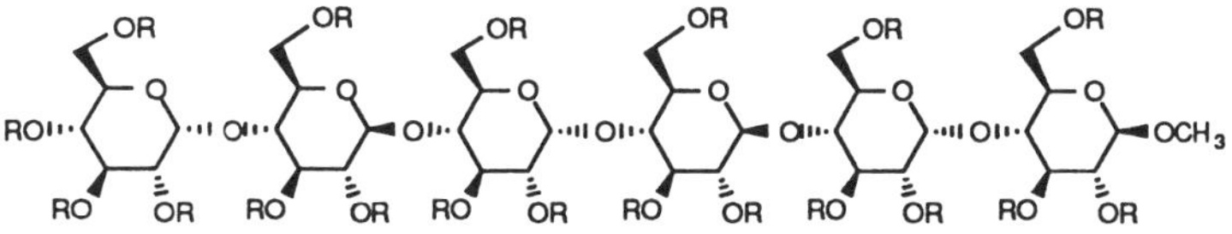

21 R = SO_3Na or H, DS ≈ 2.7

Structures 5. Structures of heparan sulfate model oligosaccharides. DS=degree of sulfation, cf. Structures 4

the sulfated dodecyl laminaropentaoside **22** with concomitantly low anticoagulant activity. These compounds probably have surface-active properties [49].

Sucrose octasulfate (**23**, Structures 6) bound to FGF and was able to induce, like heparin, a conformational change in the peptide, but showed very low mitogenic activity for an F32 cell line. The basic aluminium salt of sucrose octasulfate, which is used as a drug (Sucralfate, Carafate) for the treatment of duodenal ulcers was claimed to exert part of its activity through the stabilization of FGF

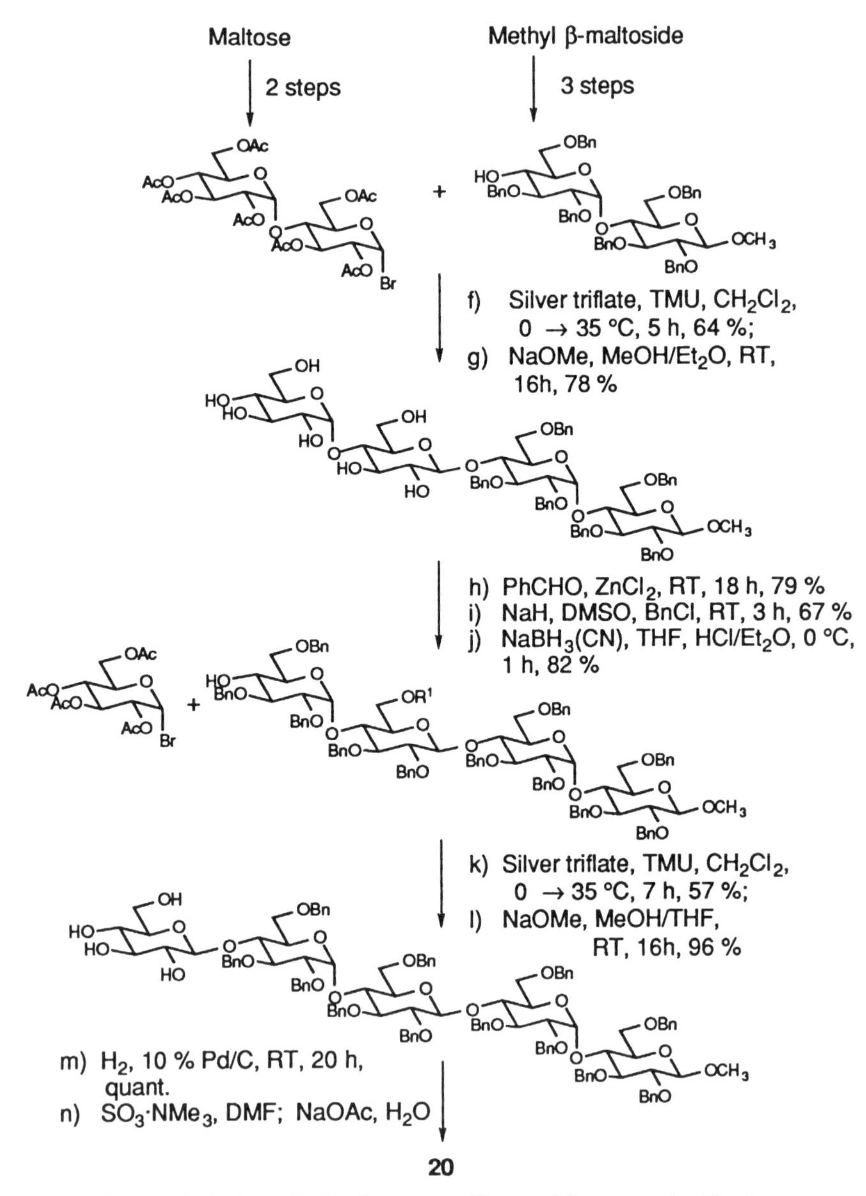

Scheme 6. [2+2+1] Block synthesis of heparan sulfate model pentasaccharide **20**

in the stomach [50, 51]. An X-ray of a sucrose octasulfate – aFGF complex was published with eight different aFGF molecules in the asymmetric unit cell in which the sulfate groups and sugar rings of sucrose octasulfate adopted different conformations [52].

In keeping with a high amount of non-ionic contact in the bFGF/heparin interaction is the recent and interesting finding that non-sulfated heparan-derived

22

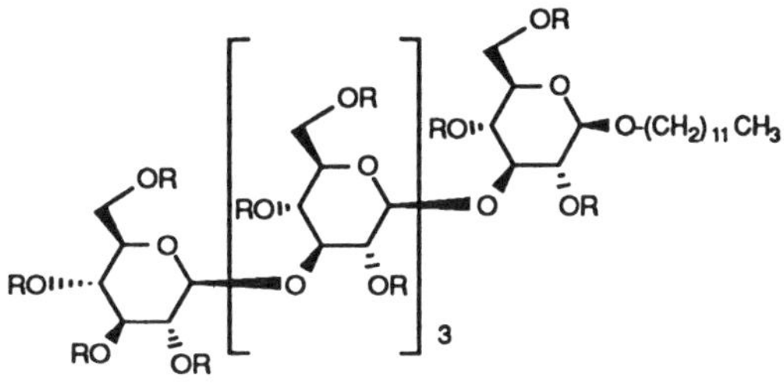

R = SO_3Na or H, DS ≈ 3.0

23

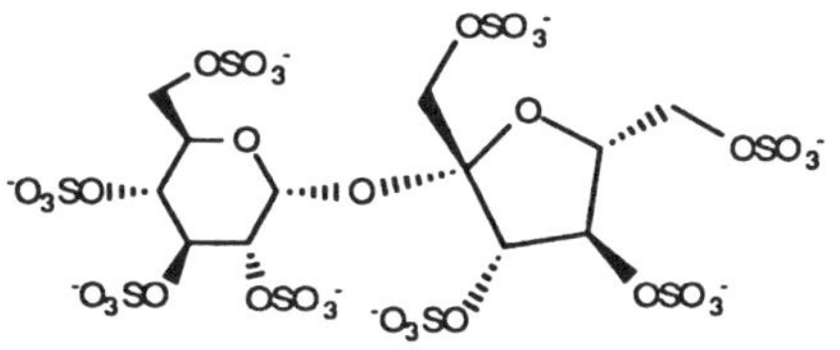

Structures 6. Structure of a laminaro oligosaccharide and sucrose octasulfate

di- and trisaccharides containing one or two uronic acids also elicited heparinoid activity binding to FGF and activated the FGF signalling pathway [51].

4.2
Sulfated Cyclic Oligosaccharides

Cyclodextrins are cyclic maltooligosaccharides available by enzymatic syntheses from starch. Due to their easy access, sulfated cyclodextrins had been investigated back in the early 1960s for their plasma clearing activity [53]. More recently, the activities of β-cyclodextrin (cycloheptamaltaose, β-CD) tetradecasulfate (**24**) were evaluated in more detail. It was prepared by sulfation of β-CD which led to the sulfation of the primary and one of the secondary hydroxyl groups per pyranose ring so that a mixture of differently sulfated compounds resulted (Structures 7).

β-CD Tetradecasulfate inhibited the migration of smooth muscle cells at high concentrations, and the effect on the proliferation of human and rabbit SMC was comparable to that of heparin [54]. The same preparation was also active in an angioplasty model in rats upon continuous infusion, although the heparin effect is not reached even at very high doses (50 mg kg^{-1} d^{-1}) [55]. Sulfated β-CD exhibited practically no anti-Xa activity but some APTT (activated partial thromboplastin time) clotting activity, although reduced vs heparin. Very remarkably, sulfated β-CD was also active in the rabbit model after oral dosage [56]. "Polymeric" sulfated β-CD was shown to have an affinity for basic FGF [57].

Furthermore, sulfated β-CD mimicked heparin in the augmentation of anti-angiogenic activity of angiostatic steroids, but used alone promoted blood vessel cell growth (angiogenesis) [58].

Whereas sulfated β-CD and γ-CD exerted considerable activity against the human immunodeficiency viruses HIV-1 and HIV-2 [59], the modification with hydrophobic groups could increase the in vitro activity determined by the inhi-

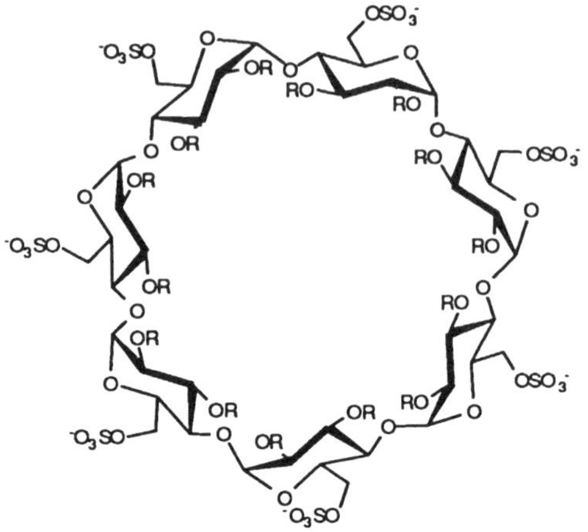

24 R = SO_3^- or H, 50 % sulfation on secondary hydroxyls

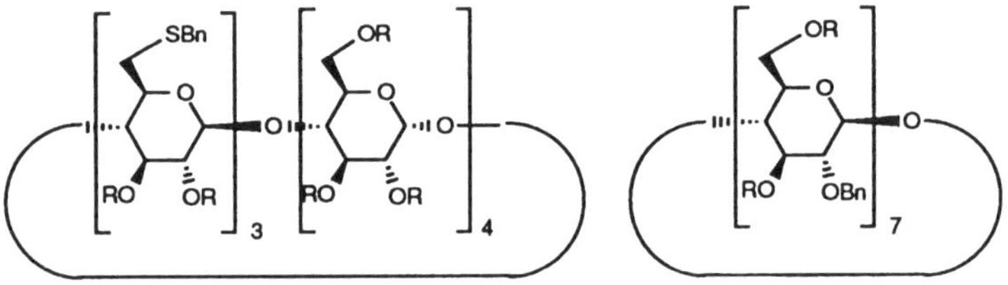

25 R = SO_3K or H (number of SO_3K = 16) mCDS11

26 R = SO_3Na mCDS71

Structures 7. Structures of β-CD tetradecasulfate and of modified cyclodextrins

bition of the cytopathic effect in MT-4 cells. Useful substitutions were the introduction of three *S*-benzyl groups in the primary positions of β-CD tetradecasulfate to give **25** (mCDS11, Structures 7) [60] or, particularly, *O*-benzylation of all 2-positions to result in **26** (mCDS71, Structures 7) [61]. This compound was prepared from β-CD by silyl protection of the primary hydroxyl groups and regioselective benzylation of the secondary hydroxyl groups in position 2 followed by desilylation and sulfation. As a persulfation was possible, **25** turned out to be a chemically defined compound. Advantageously, the substance was orally available and led to appreciable plasma levels in the rat [61].

27

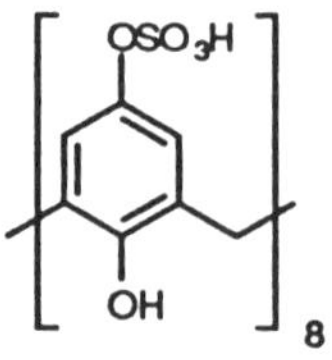

Structure 8. Structure of calixarene 27

The sulfonated calix[8]arene **27** (Structures 8), which can be viewed as a simplified γ-CD with the carbohydrate backbone replaced by an aryl moiety, mimicked heparin in the stimulation of heparan sulfate synthesis from cultured endothelial cells [62].

5 Sulfated Spaced Sugars

5.1 Sulfated Spaced Oligosaccharides

Based on the finding that sulfates in the center part of the tetrasaccharide **15** may not be of utmost importance for SMC antiproliferative activity, sulfated bis-glycosides of aromatic bis-hydroxyl compounds have been prepared; they were called sulfated 'spaced oligosaccharides' as formally a part of an oligosaccharide was replaced by a spacer group. The advantage of these structures was that the non-charged spacers avoided the charge interactions which in the oligosaccharides did not allow full sulfation; as expected from sulfation studies on disaccharides the spaced oligosaccharides could be persulfated, if the saccharide moieties were not larger than disaccharides, and thus yielded chemically defined compounds. High antiproliferative activities could be achieved in this series of compounds, e.g. compound **28** (Structures 9) was as active as heparin, and **29** was even 20% more active than heparin, but no AT III – mediated anticoagulant effects were observed [63]. These high activities pointed at a binding contribution of the hydrophobic spacer.

28

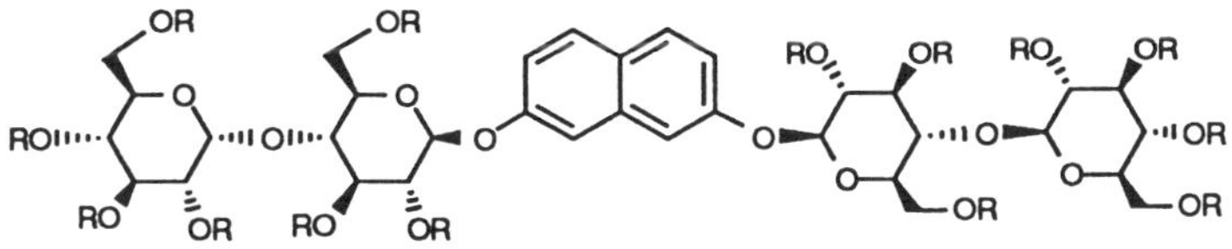

29

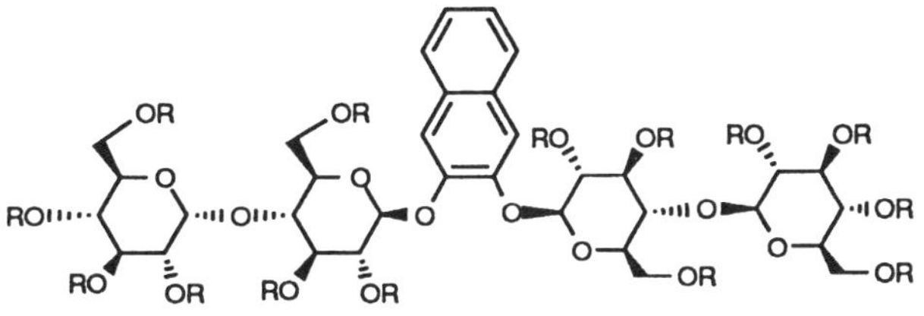

R = SO_3Na

Structures 9. Structures of sulfated spaced oligosaccharides

5.2
Sulfated Spaced Open Chain Sugars

Open chain C-6 sugars in extended conformation have approximately the same spatial extension as a disaccharide such as maltose. Sulfated spaced open chain sugars were therefore investigated in analogy to the sulfated spaced oligosaccharides. Glycamines, and particularly commercially available glucamine (1-amino-1-deoxy-D-glucitol), could be readily linked to various aromatic spacers via linker groups such as amines, sulfonamides, ureas, and amides as exemplified by compounds **30**–**33** (Structures 10). Other linker groups included ethers, thioureas, and inverted amides. All compounds were completely sulfated and thus chemically well defined. The synthetic access to this type of compound was straightforward and extremely short; in the simplest case unprotected glucamine could be linked to a bis-funtionalized spacer and be sulfated, and in some cases the intermediate protection of hydroxyl groups was advisable to facilitate purification.

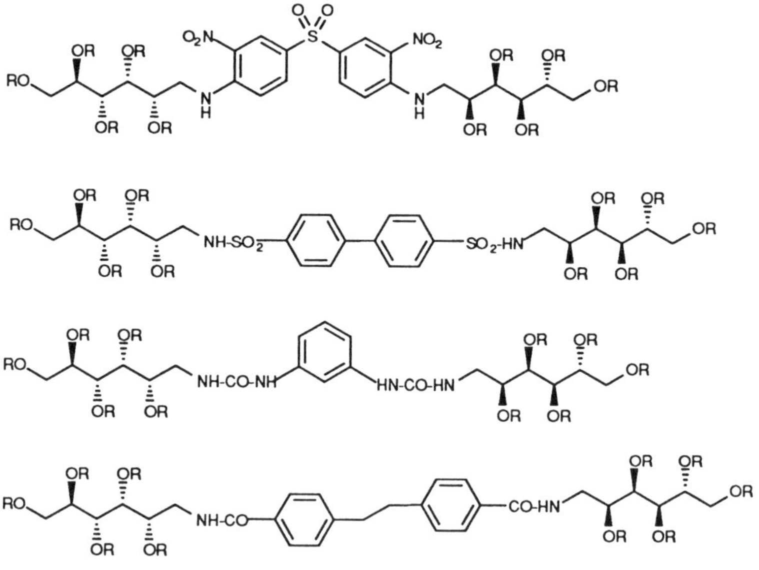

R = SO_3Na

Structures 10. Structures of sulfated spaced glucamines with aromatic spacers

The investigation of the antiproliferative activity on rat SMC culminated in the identification of highly active compounds with 50–60% more activity than heparin (Table 3) [63].

In an APTT clotting assay, which reflects not only the AT III – mediated events but also HC II and other factors of the coagulation cascade besides thrombin and factor Xa, sulfated spaced open chain sugars reached selectivities against

Table 3. Activities of sulfated spaced open chain sugars

Compound	**2**	**30**	**31**	**32**	**33**
Relative antiproliferative activity r_i	1.0	1.0	1.0	1.5	1.6
Anticoagulant activity					
anti-IIa/anti-Xa IC_{50} [µg ml^{-1}]	2.2/2.7	>1000/ >1000	>1000/ >1000	>1000/ >1000	>1000/ >1000

Table 4. APTT selectivities of different sulfated sugar derivatives. IC_{50} values are compound concentrations leading to a clotting time of twice the control

Compound	Compound class	r_i	APTT IC_{50} [µg ml^{-1}]
2	Polysaccharide	1.0	1.3
9	Polysacch. derivative	1.2	11
10	Nonasaccharide	1.2	7
15	Tetrasaccharide	0.9	12
28	Spaced saccharide	1.0	77
33	Spaced open chain sugar	1.5	>1000

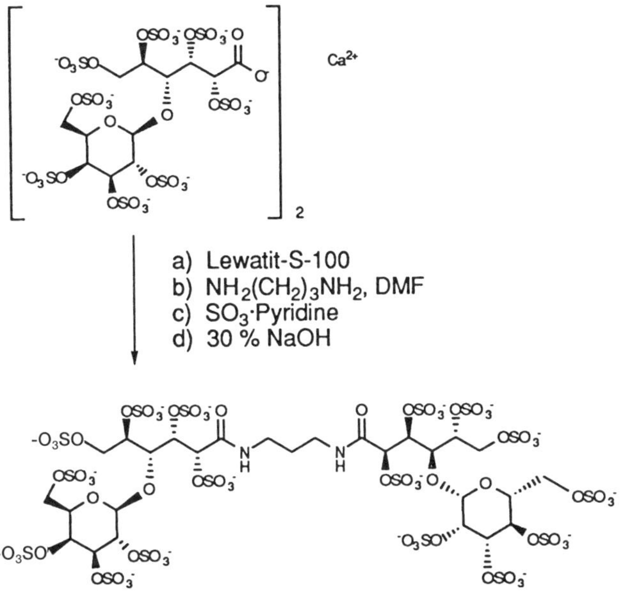

Ca^{2+}

a) Lewatit-S-100
b) $NH_2(CH_2)_3NH_2$, DMF
c) SO_3·Pyridine
d) 30 % NaOH

34

Scheme 7. Synthesis and structure of Aprosulate

heparin of more than three orders of magnitude depending on the choice of spacer (Table 4).

From a series of sulfated bis-aldonic acid amides with different alkyl spacer length, compound **34** was chosen for further evaluation as an antithrombotic. This compound was synthetically available in four steps from lactobionic acid (Scheme 7). Compound **34** had relatively high APTT values (42 U/mg) and antithrombotic activity, both of which decreased gradually when the number of methylene groups in the spacer was increased. It was thought to act via HC II and multiple sites in the intrinsic pathway of the coagulation cascade [64].

Under the name of Aprosulate compound **34** underwent phase I clinical trials. Whereas the drug was well tolerated and efficacious, it was associated with increased transaminase levels [65]. The development of Aprosulate sodium was suspended in March 1995.

6 Non-Carbohydrate Compounds

For obvious reasons of structural analogy to heparinoids the focus of this review is on sulfated carbohydrate derivatives. While it is not in all cases clear that these compounds really mimic the physiological activity of heparinoids, it is even less so for non-carbohydrate sulfates or sulfonates. Examples of the latter class include suramin and the simple 1,3-propanediol disulfate. Suramin is a sulfonated bis-naphthalene derivative used as a drug to treat African trypanosomiasis and onchocerciasis (a filarial infection); it was also tested in a number of other indications including adrenocortical carcinomas and AIDS. A wider use is, however, restricted by various toxic effects [66]. 1,3-Propanediol disulfate reduced inflammation-associated amyloid progression in vivo after oral administration which may be relevant to the treatment of Alzheimer's disease [67].

7 Concluding Remarks

This compilation of heparinoid mimetics that have evolved over the last years demonstrates that there is a tendency to largely simplified compounds with regard to structure and synthetic access; a minimum requirement for further pharmaceutical development seems to be chemical homogeneity. One of the recipes to increase binding and modify pharmacokinetic parameters was the introduction of hydrophobic groups.

In some of the carbohydrate mimetics discussed above the glycosidic linkage was replaced by other linkers such as, for example, an amide group (cf. Sect. 5.2). In an interesting approach towards new mimetics, carbohydrate amino acids have been employed as building blocks which can be easily coupled using peptide technology, including solid phase synthesis. These mimetics provide templates for the positioning of sulfates and maintain parts of the carbohydrate epitope [68].

With regard to drug development, the question of selectivity will remain a critical issue and the main challenge to avoid undesired side effects. However, the high selectivity accomplished already, for example, with the sulfated spaced open chain sugar **33** against the numerous heparinoid-binding proteins involved in the coagulation process, may signal that achievement of selectivity will not be an insurmountable task. The final answers will be given in the complex in vivo situation.

References

1. (a) Lindahl U, Kjellen L (1987) Biosynthesis of heparin and heparan sulfate. In: Wight TN, Mechem RP (eds) Biology of proteoglycans. Academic Press, New York, p 59; (b) Lindahl U (1989) Biosynthesis of heparin and related polysaccharides. In: Lane DA, Lindahl U (eds) Heparin – chemical and biological properties, clinical applications. Edward Arnold, London, p 159; (c) Yanagishita M, Hascall VC (1992) J Biol Chem 267: 9451
2. Conrad HE (1993) Pure Appl Chem 65:787
3. (a) Jackson RL, Busch SJ, Cardin AD (1991) Physiological Reviews 71: 481; (b) Kjellen L, Lindahl U (1991) Annu Rev Biochem 60:443; (c) Zhou F, Höök T, Thompson JA, Höök M (1992) Heparin protein interactions. In: Lane DA et al (eds) Heparin and related polysaccharides. Plenum Press, New York, p 141
4. Bourin M-C, Lindahl U (1993) Biochem J 289:313
5. (a) Lellouch AC, Lansbury PT (1992) Biochemistry 31:2279; (b) Horne A, Gettins P (1992) Biochemistry 31:2286
6. (a) Grootenhuis PDJ, van Boeckel CAA (1991) J Am Chem Soc 113:2743; (b) van Boeckel CAA, Grootenhuis PDJ, Meuleman D, Westerduin P (1995) Pure Appl Chem 67:1663
7. Mourey L, Samama J-P, Delarue M, Petitou M, Choay J, Moras D (1993) J Mol Biol 232:223
8. Carrell RW, Stein PE, Fermi G, Wardell MR (1994) Structure 2:257
9. (a) Casu B, Choay J, Ferro DR, Gatti G, Jacquinet J-C, Petitou M, Pravasoli A, Ragazzi M, Sinaÿ P, Torri G (1986) Nature (London) 322:215; (b) Casu B, Petitou M, Pravasoli A, Sinaÿ P (1988) Trends Biochem Sci 13:221
10. Tollefsen DM (1995) Thromb Haemost 74:1209
11. Olivecrona T, Bengtsson-Olivecrona G (1989) Heparin and lipases. In: Lane DA, Lindahl U (eds) Heparin – chemical and biological properties, clinical applications. Edward Arnold, London, p 335
12. (a) Burgess WH, Maciag T (1989) Annu Rev Biochem 58:575; (b) Bobik A, Campbell JH (1993) Pharmacol Rev 45:1; (c) Middaugh CR, Volkin DB, Thomas KA (1994) Curr Opin Invest Drugs 2:991
13. (a) Turnbull JE, Fernig DG, Ke Y, Wilkinson MC, Gallagher JT (1992) J Biol Chem 267: 10337; (b) Ishai-Michaeli R, Svahn CM, Weber M, Chajek-Shaul T, Korner G, Ekre H-P, Vlodavsky I (1992) Biochemistry 31:2080; (c) Habuchi H, Suzuki S, Saito T, Tamura T, Harada T, Yoshida K, Kimata K (1992) Biochem J 285:805; (d) Mach H, Volkin DB, Burke CJ, Middaugh CR, Linhardt RJ, Fromm JR, Loganathan D, Mattsson L (1993) Biochemistry 32:5480; (e) Tyrrell DJ, Ishihara M, Rao N, Horne A, Kiefer MC, Stauber GB, Lam LH, Stack RJ (1993) J Biol Chem 268: 4684; (f) Maccarana M, Casu B, Lindahl U (1993) J Biol Chem 268:23898
14. Faham S, Hileman RE, Fromm JR, Linhardt RJ, Rees DC (1996) Science 271:1116
15. Thompsen LD, Pantaliano MW, Springer BA (1994) Biochemistry 33:3831
16. (a) Ornitz DM, Yayon A, Flanagan JG, Svahn CM, Levi E, Leder P (1992) Mol Cell Biol 12: 240; (b) Guimond S, Maccarana M, Olwin BB, Lindahl U, Rapraeger AC (1993) J Biol Chem 268:23906; (c) Mason IJ (1994) Cell 78:547
17. Miller MD, Krangel MS (1992) Crit Rev Immunol 12:17
18. (a) Clowes AW, Karnovsky MJ (1977) Nature 265:625; (b) Hoover RL, Rosenberg R, Haering W, Karnovsky MJ (1980) Circ Res 47:578; (c) Castellot JJ, Addonizio ML, Rosenberg R, Karnovsky MJ (1981) J Cell Biol 90:372; (d) Majack RA, Clowes AW (1984) J Cell Physiol 118:253

19. (a) Fritze LMS, Reilly CF, Rosenberg RD (1985) J Cell Biol 100:1041; (b) Benitz WE, Kelley RT, Anderson CM, Lorant DE, Bernfield M (1990) Am J Respir Cell Mol Biol 2:13
20. Guyton JR, Rosenberg RD, Clowes AW, Karnovsky MJ (1980) Circ Res 46:625
21. Castellot JJ, Beeler DL, Rosenberg RD, Karnovsky MJ (1984) J Cell Physiol 120:315
22. Castellot JJ, Wong K, Herman B, Hoover RL, Albertini DF, Wright TC, Caleb BL, Karnovsky MJ (1985) J Cell Physiol 124:13
23. Au JPT, Kenagy RD, Clowes MM, Clowes AW (1993) Haemostasis 23 (suppl 1):177
24. Ross R (1993) Nature 362:801
25. (a) WuDunn D, Spear PG (1989) J Virol 63:52; (b) Sawitzky D, Hampl H, Habermehl K-O (1990) J Gen Virol 71:1221; (c) Lycke E, Johansson M, Svennerholm B, Lindahl U (1991) J Gen Virol 72:1131
26. Patel M, Yanagishita M, Roderiquez G, Bou-Habib DC, Oravecz T, Hascall VC, Norcross MA (1993) AIDS Res Hum Retrovir 9:167
27. Baba M, Pauwels R, Balzarini J, Arnout J, Desmyter J, De Clercq E (1988) Proc Natl Acad Sci USA 85:6132
28. (a) Bagasra O, Lischner HW (1988) J Infect Dis 158: 1084; (b) Neyts J, Snoeck R, Schols D, Balzarini J, Esko JD, van Schepdael A, De Clercq E (1992) Virology 189:48
29. (a) De Clercq E (1988) Chimioterapia 7: 357; (b) De Clercq E (1989) Antiviral Res 12:1
30. Flexner C, Barditch-Crovo PA, Kornhauser DM, Farzadegan H, Nerhood LJ, Chaisson RE, Bell KM, Lorentsen KJ, Hendrix CW, Petty BG, Lietman PS (1991) Antimicrob Agents Chemother 35:2544
31. (a) Lindahl U, Lidholz K, Spillmann D, Kjellén L (1994) Thromb Res 75: 1; (b) Tyrrell DJ, Kilfeather S, Page CP (1995) Trends Pharmacol Sci 16:198
32. (a) Ichikawa Y, Monden R, Kuzuhara H (1988) Carbohydr Res 172:37; (b) Wessel HP, Labler L, Tschopp TB (1989) Helv Chim Acta 72:1268
33. van Boeckel CAA, Petitou M (1993) Angew Chem Int Ed Engl 32:1671
34. Westerduin P, van Boeckel CAA, Basten JEM, Broekhoven MA, Lucas H, Rood A, van der Heijden H, van Amsterdem RGM, van Dinther TG, Meuleman DG, Visser A, Vogel GMT, Damm JBL, Overklift GT (1994) Bioorg Med Chem 2:1267
35. Westerduin P, Basten JEM, Broekhoven MA, de Kimpe V, Kuijpers WHA, van Boeckel CAA (1996) Angew Chem Int Ed Engl 35:331
36. Wessel HP, Hosang M, Tschopp TB, Weimann B-J (1990) Carbohydr Res 204:131
37. (a) Wessel HP, Englert G, Stangier P (1991) Helv Chim Acta 74:682; (b) Wessel HP, Mayer B, Englert G (1993) Carbohydr Res 242:141
38. (a) Wessel HP (1990) Tetrahedron Lett 31:6863; (b) Wessel HP, Ruiz N (1991) J Carbohydr Chem 10:901; (c) Dobarro-Rodriguez A, Trumtel M, Wessel HP (1992) J Carbohydr Chem 11:255
39. Wessel HP, Tschopp TB, Hosang M, Iberg N (1994) Bioorg Med Chem Lett 4:1419
40. Wessel HP, Niggemann J (1995) J Carbohydr Chem 14:1089
41. (a) Wessel HP (1992) J Carbohydr Chem 11: 1039; (b) Wessel HP, Bartsch S (1995) Carbohydr Res 274:1
42. (a) Wessel HP, Englert G (1994) J Carbohydr Chem 13: 1145; (b) Wessel HP, Englert G (1995) J Carbohydr Chem 14:179
43. Wessel HP, Vieira E, Trumtel M, Tschopp TB, Iberg N (1995) Bioorg Med Chem Lett 5:437
44. Wessel HP, Trumtel M, Minder R (1996) J Carbohydr Chem 15:523
45. Wessel HP, Iberg N, Trumtel M, Viaud M-C (1996) Bioorg Med Chem Lett 6:27
46. Deng S, Pascual M, Lou J, Bühler L, Wessel HP, Grau G, Schifferli J, Morel P (1996) Transplantation 61:1300
47. (a) Wessel HP, Minder R, Englert G (1995) J Carbohydr Chem 14:1101; (b) Wessel HP, Minder R, Englert G (1996) J Carbohydr Chem 15:201
48. Wessel HP, Iberg N (1996) Bioorg Med Chem Lett 6:427
49. Katsuraya K, Ikushima N, Takahashi N, Shoji T, Nakashima H, Yamamoto N, Yoshida T, Uryu T (1994) Carbohydr Res 269:51
50. (a) Szabo S, Vattay P, Scarbrough E, Folkman J (1991) Am J Med 91 (Suppl. 2A):158S; (b) Prestrelski SJ, Fox GM, Arakawa T (1992) Arch Biochem Biophys 293:314

51. Ornitz DM, Herr AB, Nilsson M, Westman J, Svahn C-M, Waksman G (1995) Science 268: 432
52. Zhu X, Hsu BT, Rees DC (1993) Structure 1:27
53. Berger L, Lee J (1960) Cycloamylose sulfates and derivatives thereof. USA: Hoffmann - La Roche
54. Okada SS, Kuo A, Muttreja MR, Hozakowska E, Weisz PB, Barnathan ES (1995) J Pharmacol Exp Ther 273:948
55. Reilly CF, Fujita T, McFall RC, Stabilito II, Eng W-S, Johnson RG (1993) Drug Dev Res 29:137
56. Herrmann HC, Okada SS, Hozakowska E, LeVeen RF, Golden MA, Tomaszewski JE, Weisz PB, Barnathan ES (1993) Arterioscl Thromb 13:924
57. Shing Y, Folkman J, Weisz PB, Joullié MM, Ewing WR (1990) Anal Biochem 185:108
58. Folkman J, Weisz PB, Joullié MM, Li MW, Ewing WR (1989) Science 243:1490
59. Schols D, De Clerc E, Witvrouw M, Nakashima H, Snoeck R, Pauwels R, van Schepdael A, Claes P (1991) Antiviral Chem Chemother 2:45
60. Moriya T, Kurita H, Matsumoto K, Otake T, Mori H, Morimoto M, Ueba N, Kunita N (1991) J Med Chem 34:2301
61. Moriya T, Saito K, Kurita H, Matsumoto K, Otake T, Mori H, Morimoto M, Ueba N, Kunita N (1993) J Med Chem 36:1674
62. Pinhal MAS, Dietrich CP, Nader HB, Jeske W, Walenga JM, Hoppensteadt D (1993) Thromb Haemostasis 69:1247
63. Wessel HP, Chucholowski A, Fingerle J, Iberg N, Märki HP, Müller R, Pech M, Pfister-Downar M, Rouge M, Schmid G, Tschopp T (1996) From glycosaminoglycans to heparinoid mimetics with antiproliferative activity. In: Chapleur Y (ed) Carbohydrate mimics: concepts and methods. Verlag Chemie, Weinheim, in press
64. (a) Klauser RJ, Meinetsberger E, Raake W (1991) Semin Thromb Hemost 17 (Suppl): 118; (b) Raake W, Klauser RJ, Meinetsberger E, Zeller P, Elling H (1991) Semin Thromb Hemost 17 (Suppl.): 129; (c) Ofosu FA, Fareed J, Smith LM, Anvari N, Hoppensteadt D, Blajchman MA (1992) Eur J Biochem 203:121
65. Papoulias UE, Wyld PJ, Haas S, Stemberger A, Jeske W, Hoppensteadt D, Kämmereit A (1993) Thromb Res 72:99
66. Horne MK, Stein CA, LaRocca RV, Myers CE (1988) Blood 71:273
67. Kisilevsky R, Lemieux LJ, Fraser PE, Kong X, Hultin PG, Szarek WA (1995) Nature Med 1:143
68. (a) Wessel HP, Mitchell CM, Lobato CM, Schmid G (1995) Angew Chem Int Ed Engl 34: 2712; (b) Müller C, Kitas E, Wessel HP (1995) J Chem Soc Chem Commun 2425; (c) Suhara Y, Hildreth JEK, Ichikawa Y (1996) Tetrahedron Lett 37:1575

Recent Developments in the Rational Design of Multivalent Glycoconjugates

René Roy

Department of Chemistry, University of Ottawa, Ottawa, Ontario,
Canada K1N 6N5. e-mail RRoy@oreo.uottawa.ca

This chapter describes the syntheses and recent applications of some novel neoglycoconjugates with an emphasis on a rather narrow number of key carbohydrate structures of special interest to modern glycobiology. The description is limited to those carbohydrates used as ligands by an increasing number of endogenous plant and animal lectins involved in recognition processes. A brief introduction to carbohydrate-protein interactions will be given to stress the need for multivalent neoglycoconjugates. Since the members of the neoglycoconjugate family are steadily increasing as a result of the demand for higher affinity and better specificity, an overview of new glycoforms covering the last ten years is presented with emphasis on work originating from our laboratory. Thus, glycopolymers, telomers, glycodendrimers glycopeptoids, and more recent calix[4]arenes will be discussed. Particular attention is given to sialic acid and sialyloligosaccharides, together with mannosides, and galactosides as these carbohydrate residues represent immunodominant structural elements.

Table of Contents

Topics in Current Chemistry, Vol. 187

1 Introduction

The chemistry and applications of neoglycoconjugates have been flourishing since the initial use of bacterial capsular polysaccharide-protein conjugates as vaccines some 60 years ago [1]. Even monosaccharide-protein conjugates are now known to be immunogenic. These findings have good as well as adverse implications. The positive aspect is that simple oligosaccharide neoglycoproteins such as those identified as cancer markers (e.g. T, T_N, and sialyl T_N) have reached clinical phase evaluation as potential human vaccines [2]. Moreover, it is also foreseeable to envisage entirely synthetic vaccines made from carbohydrate haptens conjugated to immunomodulatory peptidolipids [3] or to T-cell peptide (< 15 mer) epitopes recognized by MHC class II molecules [4, 5]. The negative aspect is that the same neoglycoproteins immunogenicity makes them therapeutically inappropriate as antiadhesive agents for the clearance of bacterial, viral, and parasitic infections. Therefore, another repertoire of multivalent glycan carriers are being investigated, among which, water-soluble glycopolymers are emerging.

The chemistry of neoglycoproteins has been well reviewed in the seminal paper by Stowell and Lee [6] and newer neoglycoconjugate syntheses are reported in books edited by Lee and Lee [7, 8]. A general review documenting the chemistry of neoglycoconjugates [9] and more specialized reference articles on glycopolymers [10–15], vaccines [16, 17], neoglycolipids [18, 19], affinity supports [20, 21], glycodendrimers [22], and drug carriers [23–25] have also been recently prepared. The need for original neoglycoconjugates results from old as well as relatively unexplored applications spanning from better defined vaccines, more specific affinity supports, immunodiagnostics and immunomodulators, cell specific drug delivery vectors, antiinflammatory and antiadhesin agents, cell culture media, and glycoprobes for radioimaging and immunostaining. The relevance of neoglycoconjugates for histology and pathology has also been reviewed [26]. This chapter will cover some of our work published in this field during the last few years and will thus be an update of a previous article [27].

This review will focus on novel multivalent neoglycoconjugates that have been synthesized in response to fundamental questions raised by glycobiologists. For obvious reasons, carbohydrate-protein interactions most widely studied currently are those involving macrophages and asialoglycoprotein Gal/GalNAc receptors (hepatic) together with their homologous galectins.

Serum-type mannose binding proteins and mannoside receptors of some fimbriated bacteria are appealing for those studying immune defense mechanisms. Finally, the increasing family of sialoside receptors has attracted considerable research activity since its members are implicated in immune cell interactions (sialoadhesins) and in viral (influenza, sendai, and rota-viruses), bacterial (*Helicobacter pilori*, *Bordetella pertussis*, *E. coli*, *Vibrio cholerae*, *Pseudomonas aeruginosa*), and parasitic infections (*Plasmodium falciparum*) [reviewed in 28]. Moreover, some of the cascade events at the origin of inflammatory diseases have implicated a series of three sialyloligosaccharide (sialyl Lewis X) binding proteins collectively known as selectins [29]. A review describing the syntheses and biological activity of multivalent sialosides as well as sialyl Lewis X mimetics has appeared recently [30].

Although it has been generally admitted that carbohydrate-protein interactions are of low affinity, the full consequences of these findings are only beginning to be well appreciated. Compensating binding strategies have been set by nature to override these weak interactions. In many biologically relevant recognition events, both carbohydrate ligands and carbohydrate recognition domains (CRDs) of lectins are organized as multivalent clusters. Consequently, cooperative associative forces help to link them together [31–32]. A large number of synthetic and semi-synthetic carbohydrate clusters have been prepared to address the above issues properly. This review highlights some of the synthetic endeavors toward the design of potent multivalent carbohydrate ligands. Again, glycopolymers and glycodendrimers represent striking examples.

2 Neoglycoconjugates

The chemistry of neoglycoconjugates encompasses a wide range of synthetic carbohydrate derivatives having their own particular characteristics and advantages. Most widely used neoglycoconjugates still belong to the family of neoglycoproteins and neoglycolipids since they constitute tools with which glycobiologists are already familiar with. However, glycopolymers and sophisticated clusters are emerging as novel classes of neoglycoconjugates of defined architectures and carbohydrate valencies and some of them have been prepared with added probes and effector molecules. In many cases, it has been strategically appealing to synthesize the necessary glycan precursors in forms suitable for the preparation of both neoglycoproteins and glycopolymers [33].

The latter approach has been followed in our laboratory. For instance, in studies related to the synthesis of potent multivalent sialoside inhibitors of influenza virus hemagglutinin [34], it became of interest to evaluate the immunogenicity of neoglycoproteins containing sialosides as sole immunodominant hapten. To this end, acetochloroneuraminic acid (**1**) was transformed into allyl glycoside **2** (Scheme 1) which upon reductive ozonolysis afforded aldehyde **3** in excellent overall yields [35]. Reductive amination of **3** to bovine serum albumin (BSA) and tetanus toxoid provided immunogenic vaccines from which specific rabbit IgG anti-sialic acid antibodies were obtained [36]. In order

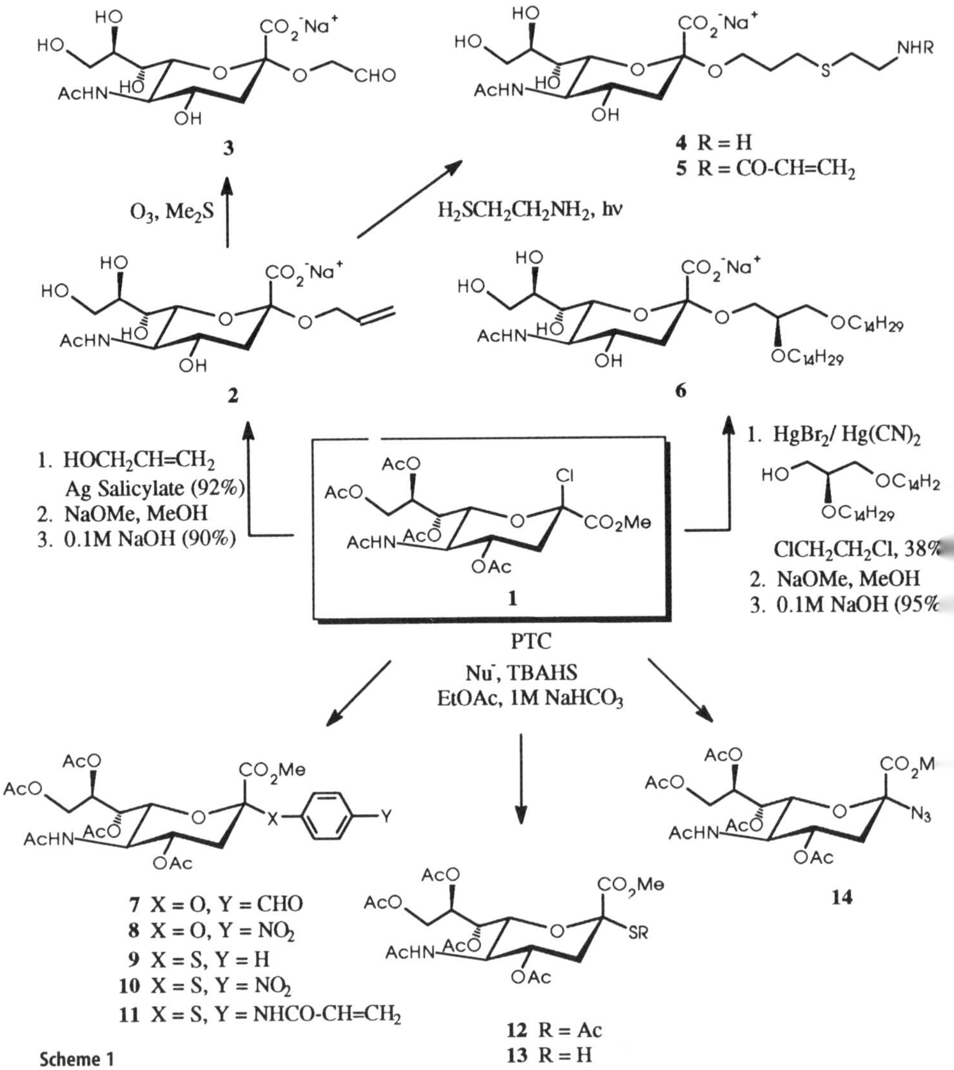

3

O_3, Me_2S

$H_2SCH_2CH_2NH_2$, hν

4 R = H
5 R = CO-CH=CH_2

2

6

1. $HOCH_2CH=CH_2$
Ag Salicylate (92%)
2. NaOMe, MeOH
3. 0.1M NaOH (90%)

1

1. $HgBr_2$/ $Hg(CN)_2$
$ClCH_2CH_2Cl$, 38%
2. NaOMe, MeOH
3. 0.1M NaOH (95%

PTC
Nu^-, TBAHS
EtOAc, 1M $NaHCO_3$

7 X = O, Y = CHO
8 X = O, Y = NO_2
9 X = S, Y = H
10 X = S, Y = NO_2
11 X = S, Y = NHCO-CH=CH_2

12 R = Ac
13 R = H

14

Scheme 1

to facilitate antibody combining site mapping and to avoid cumbersome cross-reactive antibodies from protein carriers, polymeric sialosides were simultaneously prepared and used as screening antigens. Thus, allyl sialoside **2** was directly transformed into water-soluble copolyacrylamide (acrylamide, H_2O, ammonium persulfate) [35]. Although the resulting glycopolymer had strong binding properties to a plant lectin such as wheat germ agglutinin, it failed to react to the rabbit antibodies, presumably because of the short spacer used. Consequently, treatment of **2** with cysteamine hydrochloride under photolysis conditions afforded the anti-Markovnikov adduct **4** in 60% yield which was readily transformed into the extended acrylamide monomer **5** (CH_2=CHCOCl,

MeOH, Et_3N, 88%) [37]. The copolyacrylamide polymer derived from **5** and acrylamide was found useful as coating antigen in solid phase immunoassays (ELISA) [36] and happened to be a very potent antiadhesin inhibitor against influenza virus hemagglutinin [34]. It is also noteworthy to mentioned that neoglycolipid **6** was equally effective as a coating antigen on microtiter plates [38], thus suggesting that complete ganglioside syntheses are not essential for screening assays.

It was later realized that acrylamido glycosides such as **5** could serve as precursors for both neoglycoprotein and polymer syntheses. Indeed, treating **5** directly with BSA under basic conditions (carbonate buffer, pH 10–10.5, 37 °C, 3 days) (conjugate addition) afforded neoglycoproteins having various glycan contents [33, 39, 40]. This strategy is now routinely used in our laboratory to generate both immunogenic carbohydrate derivatives and the required screening glycopolymers for monoclonal antibody production. The procedure has been instrumental in the preparation of human monoclonal antibodies against a breast cancer marker composed of T-antigen (Galβ-(1,3)-GalNAcα-) [41]. Allyl α-glycoside of the T-antigen disaccharide **15**, prepared under conventional glycosylation chemistry, was initially copolymerized with acrylamide to give copolymer **16** (Scheme 2). However, because of its short spacer arm, copolymer **16** failed to bind to human anti-T antibodies. To provide better accessibility to the carbohydrate moiety, allyl glycoside **15** was further extended to acrylamide

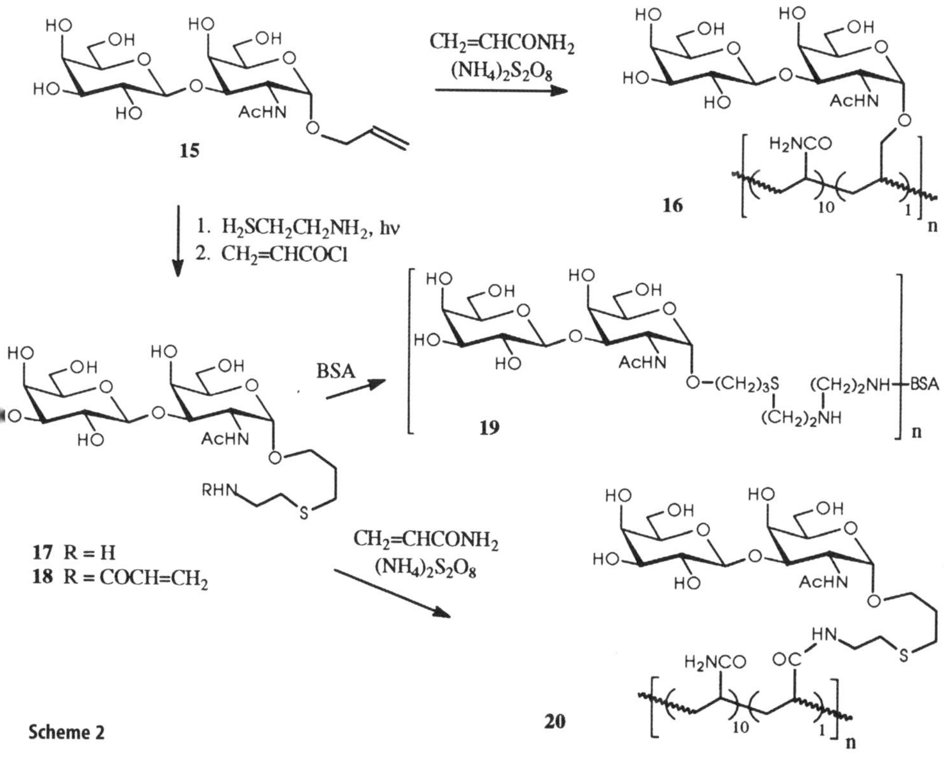

HO OH O HO O HO AcHN O
15
CH_2=CHCONH$_2$
$(NH_4)_2S_2O_8$
16 H_2NCO 10 1 n
1. $H_2SCH_2CH_2NH_2$, hv
2. CH_2=CHCOCl
BSA
19 O–$(CH_2)_3S$ $(CH_2)_2NH$–BSA $(CH_2)_2NH$ n
RHN S
17 R = H
18 R = COCH=CH$_2$
CH_2=CHCONH$_2$
$(NH_4)_2S_2O_8$
20 HN OC H_2NCO 10 1 n

Scheme 2

derivative **18** through *N*-acryloylation of intermediate amine **17**. Conjugation of **18** to BSA (**19**) or polymerization into **20** was then accomplished as for the sialic acid conjugates. Both conjugates were useful in immunological assays [41].

To facilitate accesses to suitably functionalized sialic acid derivatives and complex sialyloligosaccharides for other useful neoglycoconjugates, phase transfer catalysis (PTC) has been exploited extensively [for reviews see 42]. This process provided a wide range of carbohydrate derivatives under essentially clean S_N2 transformations. In the case of β-acetochloroneuraminic acid **1**, the PTC reactions always provided inverted α-sialic acid derivatives [43]. *para*-Formylphenyl sialoside **7** [44], together with many other sialoside derivatives such as **8–10** [43], including thioacetate **12** [45] and azide **14** [46], were thus obtained (Scheme 1). Aldehyde **7** and similar glycosides are of particular interest since they could be directly conjugated to protein by reductive amination after suitable deprotection [44].

Another noticeable strategy for the synthesis of neoglycoproteins has been recently described by Kamath et al. [47]. The procedure relies on a bifunctional coupling reagent, ethyl squarate (**23**) (Scheme 3), and allows successive transformation of aminated glycans such as **22** into monoethyl squarate intermediates. Derivatives **22** are readily obtained from 8-methoxycarbonyloctyl glycosides (**21**) by aminolysis with ethylenediamine (70 °C, neat, 2 d) which upon treatment with protein (lysyl ε-amino groups) provided neoglycoproteins (**24**). The same strategy has been used to prepare other useful heterobifunctional conjugates [48].

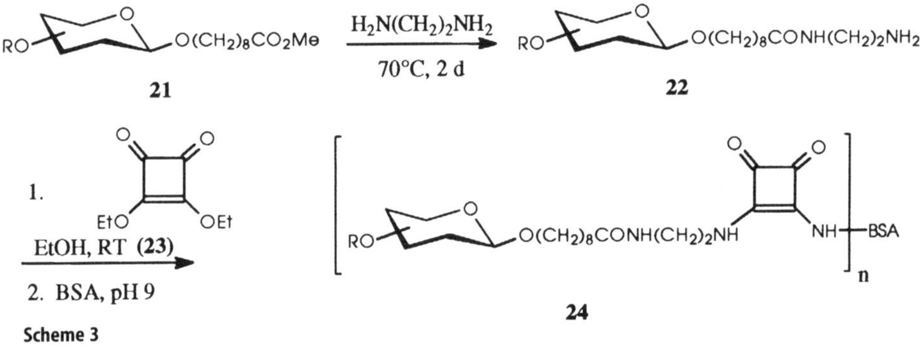

RO–[sugar]–$O(CH_2)_8CO_2Me$ (**21**) $\xrightarrow[70°C,\ 2\ d]{H_2N(CH_2)_2NH_2}$ RO–[sugar]–$O(CH_2)_8CONH(CH_2)_2NH_2$ (**22**)

1. EtO, OEt squarate, EtOH, RT (**23**)
2. BSA, pH 9

[RO–[sugar]–$O(CH_2)_8CONH(CH_2)_2NH$–squarate–NH]$_n$–BSA (**24**)

Scheme 3

3 Glycopolymers

Numerous polymerization methods exist for the synthesis of water-soluble as well as insoluble glycopolymers [reviewed in 9]. This section will focus on soluble polymers showing strong protein binding interactions. Most extensive applications of this family of polymers have centered around immunodiagnostic reagents and those having antiadhesin properties. As mentioned above, polysialosides have been extensively studied as influenza virus hemagglutinin inhibitors [for the most recent paper see 49]. Polymers with *O*-[35, 37, 40, 50],

S- [51], and C- [49, 52] *α*-sialosides have been prepared using copolymerization or "grafting" methods. More complex sialyloligosaccharide-containing polymers have also been prepared, including GM_3 [53], sialyl Lewis^x [54], sialyl Lewis^a [55], and a related 3′-sulfo-Lewis^X analog [56, 57], including polymerized liposomes [58].

3.1 Copolymerization

Following the first syntheses of acrylamide copolymers of allyl glycosides by Horejši et al. [59] and later extended to pentenyl glycosides by Matsuoka and Nishimura [for the latest ref. see 60], researchers have concentrated their activity on *N*-acryloylated carbohydrate monomers because of their better reactivity. In principle, any glycan derivatives ending with amine groups or their functional analogs (azide, nitro, etc.) can be used as polymer precursors. A landmark preparation has been the transformation of readily available reducing sugars such as GM_3 saccharide **25** into glycosylamine **26** using ammonium bicarbonate (Scheme 4) [40, 61]. Quenching unstable amine **26** with acryloyl chloride afforded monomer **27** [40]. As stated above, acrylamide **27** has been transformed into both neoglycoproteins and glycopolymers [39, 40]. Interestingly, libraries of glycosylamines such as **26** have been efficiently transformed into active monosuccinimidyl suberate **28** using excess disuccinimidyl suberate and hydroxybenzotriazole (HOBt) in DMSO [62].

Analogous GM_3 copolymers have also been synthesized using a new "active and latent" glycosylation strategy developed in our group [43, 53, 63]. Thus, minimally protected *p*-nitrophenylthio lactoside **29** ("latent" glycosyl acceptor) was stereo-and regio-chemically glycosylated with "active" phenylthio sialosyl donor **9** using *N*-iodosuccinimide (NIS) and triflic acid (TfOH) promotors at low temperature (47% yield) [53]. After conventional deprotection, the resulting trisaccharide **30** could be transformed into an active glycosyl donor by transforming its electron withdrawing nitro group into an electron donating *N*-acetamido group (1: Zn-HOAc reduction, 2: Ac_2O), thus reactivating sulfur's nucleophilicity toward electrophilic promotors. Alternatively, the *p*-nitrophenylthio precursor **30** was transformed into amine **31** (Zn, HOAc, MeOH, 90%) which was copolymerized into **33** (50–70%) after *N*-acryloylation to **32** (CH_2=CHCOCl, Et_3N, MeOH, 85%).

The next example represents a useful entry into this class of polymers since it is based on other widely available oligosaccharide bearing 8-methoxycarbonyloctyl glycosides (Lemieux's spacer). The synthesis of a very potent polymeric E- and L-selectin antagonist is illustrated in Scheme 5 [56]. 8-Methoxycarbonyloctyl glycoside of 3'-sulfo-Lewis X analog **34** was first hydrazinolyzed into hydrazide **35** (5 eq. H_2NNH_2, EtOH, Δ, quant) which after *N*-acryloylation to **36** (CH_2=CHCOCl, 90%) and copolymerization gave **38** (CH_2=$CHCONH_2$, $(NH4)_2S_2O_8$, heat, 15 min, 56%) (Scheme 5). Noteworthy was the transformation of acrylamide **36** into nucleophilic thiol **37** (1: AcSH, Et_3N, DMF, 81%; 2: NaOMe, MeOH, 0 °C, 5 min, quant). Compound **37** represents a useful precursor for the synthesis of multiantennary glycodendrimers (see Sect. 4 below).

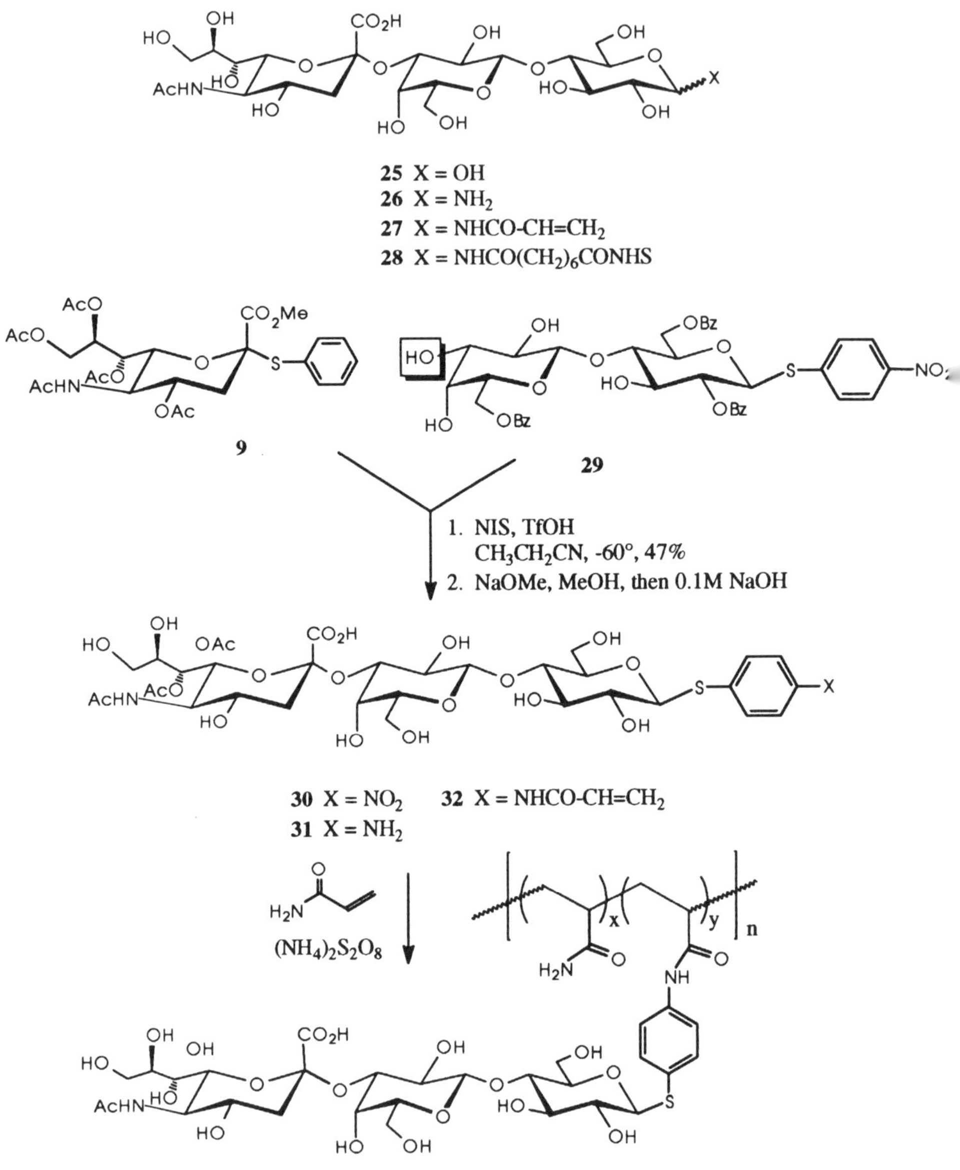

25 X = OH
26 X = NH_2
27 X = NHCO-CH=CH_2
28 X = NHCO$(CH_2)_6$CONHS

9

29

1. NIS, TfOH
   CH_3CH_2CN, -60°, 47%
2. NaOMe, MeOH, then 0.1M NaOH

30 X = NO_2 **32** X = NHCO-CH=CH_2
31 X = NH_2

$(NH_4)_2S_2O_8$

33

Scheme 4

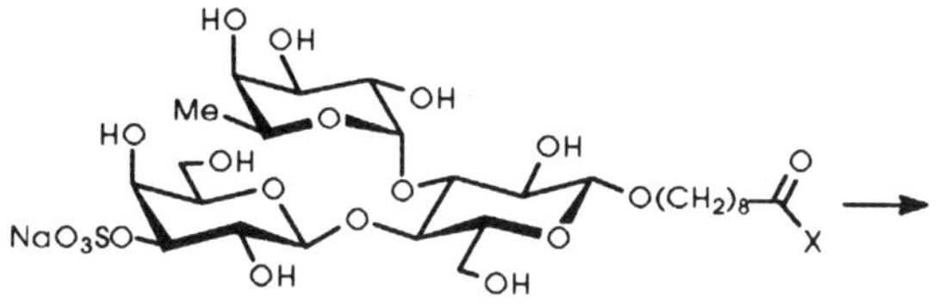

34 X = OMe
35 X = $NHNH_2$
36 X = $NHNH\text{-}CO\text{-}CH{=}CH_2$
37 X = $NHNH\text{-}CO\text{-}CH_2CH_2SH$

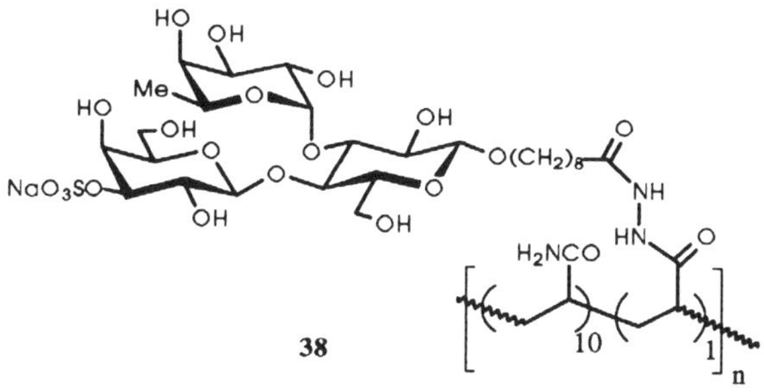

38

Scheme 5

3.2 Graft Conjugation

A useful alternative procedure exists to complement the copolymerization reactions discussed above. The method depends on preformed polymers having reactive functionality. Examples include poly-L-glutamic acid [64] and poly-L-lysine [24, 65] onto which oligosaccharides terminated with amine or acid groups can be "grafted". In the case of poly-L-lysine, coupling strategies have also included direct conjugation to reducing sugars and traditional reactions with *p*-isothiocyanatophenyl glycosides. We have similarly explored this approach using 1,4-conjugate addition of poly-L-lysine onto available *N*-acryloylated monomers. In the first case, *p*-acrylamidophenylthio sialoside **39** was treated with poly-L-lysine in carbonate buffer (pH 10–10.5, 72 h, 37 °C, 80%) to provide grafted conjugate **40** having desired molar ratios of sialoside incorporated (Scheme 6) [65]. Additionally, poly-α-(2,8) sialic acid (colominic acid/Group B *Neisseria meningitidis* capsular polysaccharide) was transformed into amine **41** by reductive amination of its reducing end group (NH_4HCO_3, $NaBH_3CN$, 72 h, 37 °C, 95%). Further derivatization of **41** with acryloyl chloride furnished Michael acceptor **42** in 74% yield. Finally, direct conjugate addition of poly-L-lysine, BSA or tetanus toxoid onto acrylamido polysaccharide **42** gave the corresponding neoglycoproteins. Moreover, treating poly-L-lysine with **42** and subsequently with an acrylamido biotin derivative afforded polymeric probe **43** in 70% yield.

Instead of preparing carbohydrate monomers, activated polymers of known molecular weight can also be synthesized [11]. The strategy offers the distinct

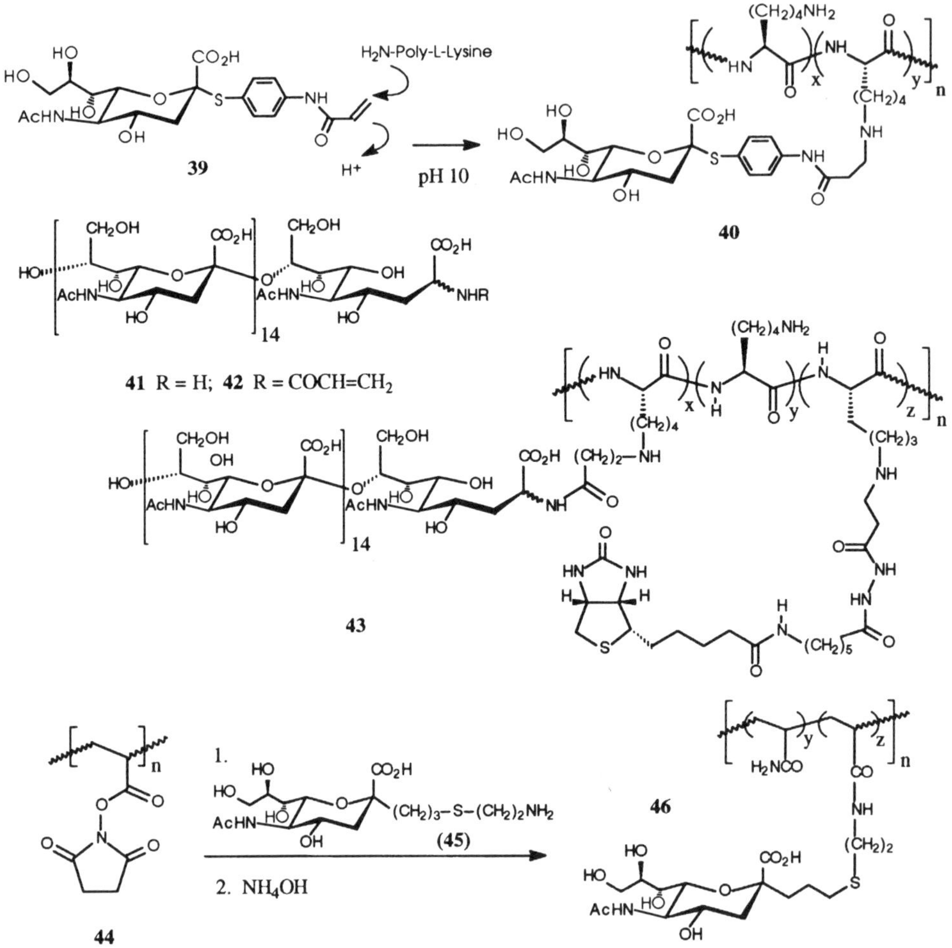

Scheme 6

advantage of providing constant molecular weight for a given family of glycopolymers. Initially developed by Russian chemists [11] with poly(*p*-nitrophenylacrylate), the chemistry has been recently extended to poly[*N*-(acryloyloxy)succinimide] **44** [41, 49]. The procedure has been used for the synthesis of copolyacrylamides containing a C-glycoside of sialic acid [49]. Amine derivative **45** was added to prepolymer **44** (Et_3N, DMF) to provide copolymer **46** after quenching excess active ester with aqueous ammonia. The sialoside content in the polymer can be adjusted with the initial concentration of amine **45** used. The strategy is appealing since it offers the possibility of adding an aminated probe to produce terpolymers. We have also used both activated polyesters to construct T-antigen polymers derived from amine **17** (Scheme 2) [41] and GM_3 saccharide similar to **26** (Scheme 4)having spacer arms of different size and hydrophobicity [66].

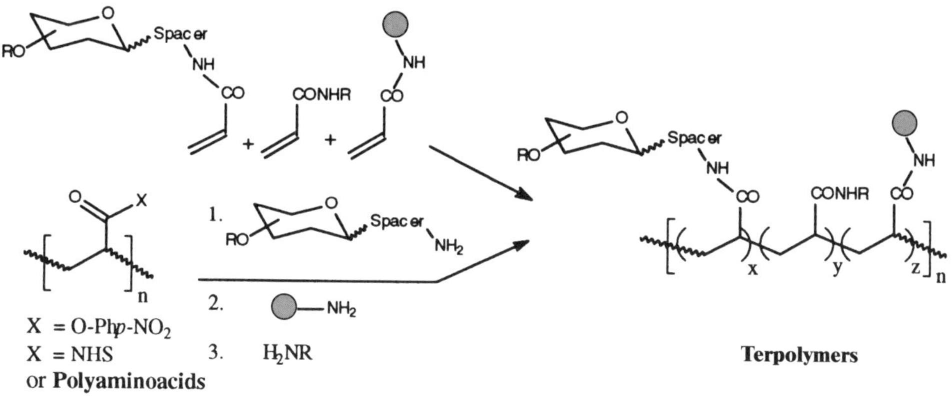

1. RO–sugar–Spacer–NH_2
2. ●–NH_2
3. H_2NR

$X = O\text{-}Php\text{-}NO_2$
$X = NHS$
or **Polyaminoacids**

Terpolymers

Scheme 7

3.3 Terpolymers

Terpolymers are copolymers made of three different components. Here again, they can be synthesized by copolymerization of three monomers or simply, by adding the carbohydrate, probe, drug or effector molecules to preactivated polymer backbones (Scheme 7). These neoglycoconjugates are valuable tools for targeting specific cells, tissues or organs. The glycan moieties serve as targeting devices, the polymer backbones are adjusted to confer desired physical or biophysical properties, while the effectors can be used to stain or label the cells or to deliver a given drug on site. Duncan and Kopecek [23] have described numerous applications of terpolymers derived from polyhydroxyethyl methacrylamide, adryamycin, and D-galactosamine or other aminosugars.

As shown by structure **43** above (Scheme 6) [65] rather complex neoglycoconjugates can be built by grafting amine derivatives onto poly-L-lysine and other polymers [11] or using three different *N*-acrylamide derivatives [49, 67–68]. Some of the examples discussed below were selected to provide an overall appreciation of the variety of procedures involved. Thus, terpolymer such as the T-antigen/biotin copolyacrylamide **47** (Scheme 8) was synthesized using acrylamide, T-antigen derivative **18** (Scheme 2), and an *N*-acryloylated biotin hydrazide residue [41]. Tichá and Kocourek [69] have similarly made an active copolymer by reacting allyl α-D-galactopyranoside, acrylamide, and allylamine. The resulting aminated polymer was further treated with fluorescein isothiocyanate to provide **48.** In a related approach, Monsigny et al. [24] have described an appealing poly-L-lysine conjugate bearing mannopyranoside residues, gluconamide solubilizing moieties, and the AZT-drug (**49**). The copolymer was used to target HIV-infected macrophages through the intermediary of mannoside receptors. As mentioned above [64] poly-L-glutamic acid as also been used for terpolymer syntheses. 8-Aminooctyl mannopyranoside and other monosaccharide analogs were coupled to the acid functionality of the polymer using carbodiimide chemistry. 2-(4-Hydroxyphenyl)ethylamine, as an iodinatable

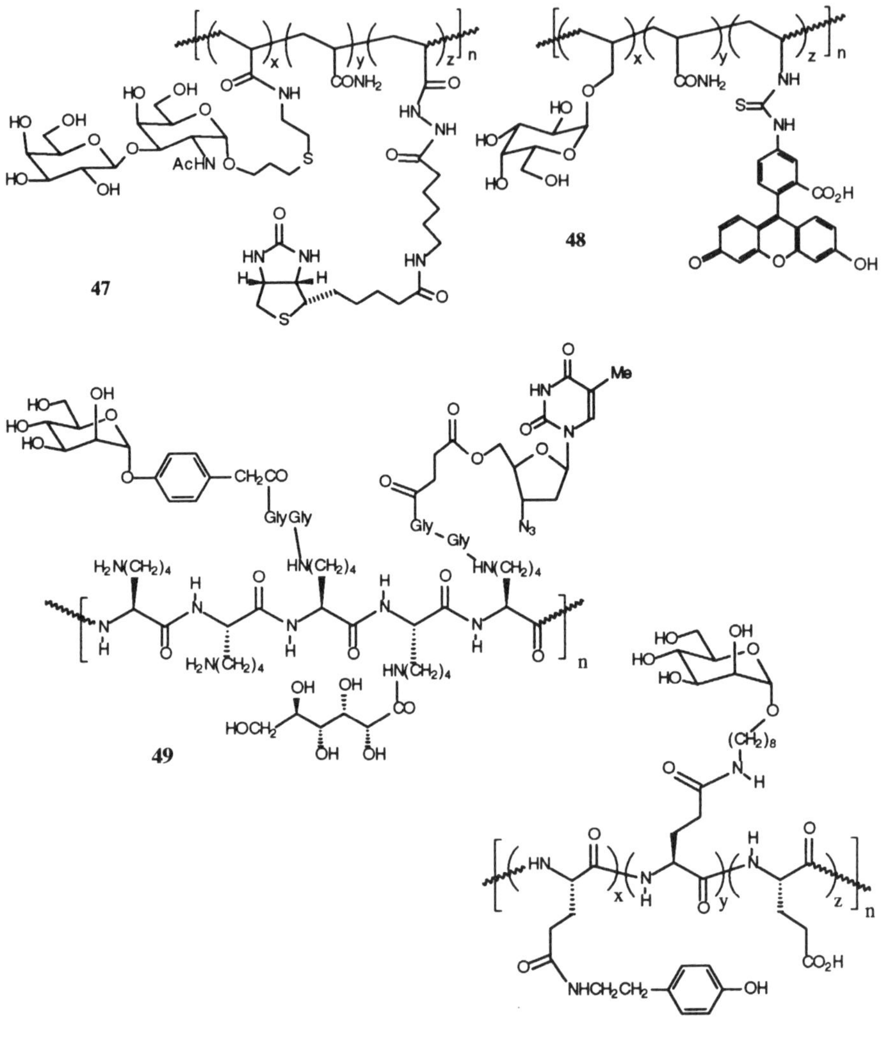

47

48

49

Scheme 8

50

unit, was subsequently added to provide copolymer **50** which was used to study the plasma elimination rates of ^{125}I-labeled poly-L-glutamic acid neoglycoconjugates.

3.4 Telomers

Telomers (or oligomers) are short homopolymers of less than approximately 10 ~ 12 residues long. They are to polymers what oligosaccharides are to polysaccharides. Only limited examples of glycotelomers have been described so far and this is a somewhat surprising finding given the fact that they are families of "mini clusters" readily available in single step reactions. Indeed, they can be prepared by quenching polymerization reactions with radical scavengers called telogens and thiols are representative examples.

Having readily available strategies for the syntheses of carbohydrate monomers, we first used lactose as a model. It was chosen because mammalian hepatic asialoglycoprotein receptors (ASGP-R) are well-studied endogenous lectins having high affinities toward non-reducing Gal/GalNAc glycoconjugates [70, 71]. Moreover, the routing of Gal/GalNAc-bearing clusters for the purpose of liver-specific drug delivery has been already demonstrated as viable anticancer and antiviral therapies [72]. Furthermore, lactosylated clusters are known to inhibit lung colonization by metastatic cancer cells [73].

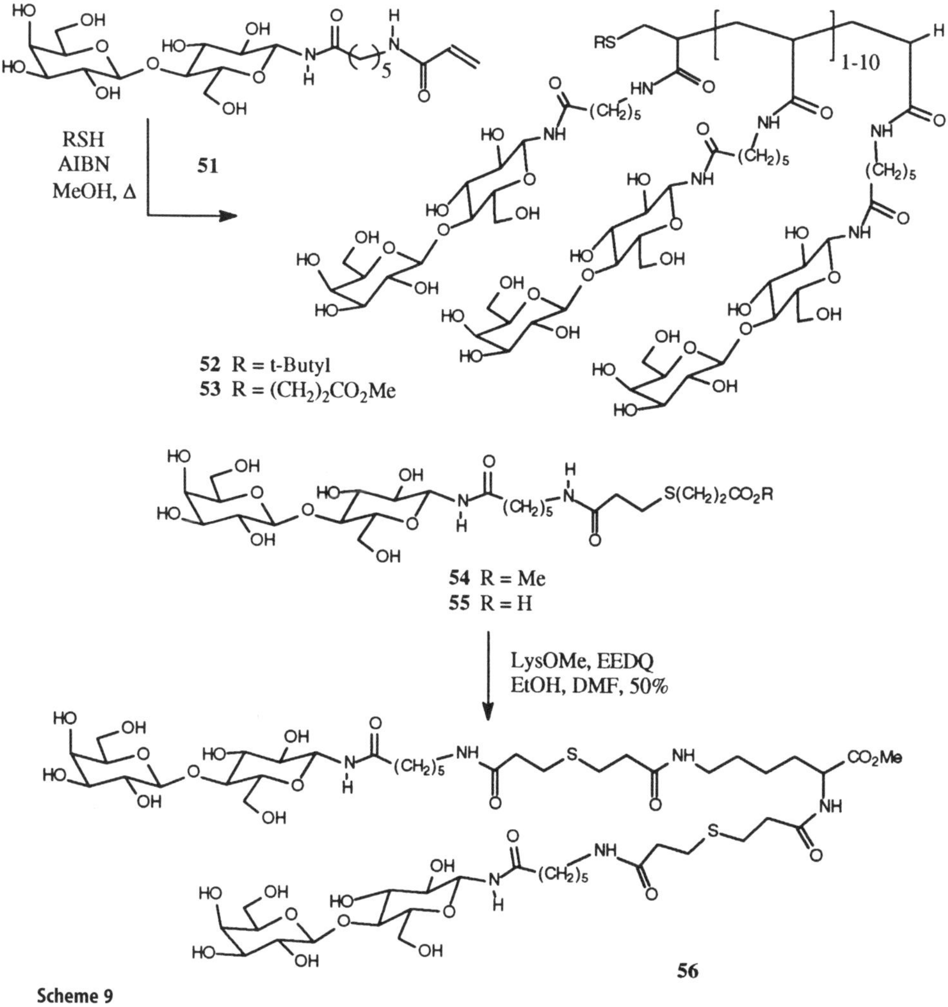

51

RSH
AIBN
MeOH, Δ

52 R = t-Butyl
53 R = $(CH_2)_2CO_2Me$

54 R = Me
55 R = H

LysOMe, EEDQ
EtOH, DMF, 50%

56

Scheme 9

6-*N*-Acrylamidohexanoyl *β*-lactosylamine **51** was telomerized (AIBN, MeOH) in the presence of either *tert*-butylmercaptan or methyl mercaptopropionate to provide families of discrete telomers **52** and **53** separable by size-exclusion chromatography (Scheme 9) [74, 75]. Using 5.5 equivalents of *t*-BuSH per monomer gave monoadduct (27%), dimer (14%), trimer (1%), and higher telomers **52** (45%) in 87% yield. When methyl mercaptopropionate was used (0.43 eq), monoadduct **54** (26%), dimer (22%), trimer (12%), tetramer (7%), and higher telomers **53** ($n \geq 5$, 31%) were obtained.

To further exploit the potential usefulness of this new family of clusters, monoadduct **54** was saponified into **55** (0.05 M NaOH, quant) and condensed to L-lysine methyl ester using 2-ethoxy-1-ethoxycarbonyl-1,2-dihydroquinoline (EEDQ) to give extended dimer **56** in 50% yield together with monoadduct in 15% yield [75]. Additionally, *tert*-butyl thioethers **52** could be transformed into thiols by a two step process involving 2-nitrobenzenesulfenyl chloride (2-NO_2-PhSCl, HOAc, r.t., 3 h, 84%) followed by disulfide reduction with 2-mercaptoethanol (60%). Curiously, attempts to directly obtain these thiolated telomers by reaction with thioacetic acid failed. These telomers were slightly better ligands then lactose in inhibition of binding of peanut lectin to a polymeric lactoside [76].

4 Glycodendrimers

Between high molecular weight neoglycoconjugates and small clusters, there has been a gap of structural moieties of intermediate size and valency that has only been partly filled with natural multiantennary oligosaccharides. As these are cumbersome to characterize and isolate in large quantities, it was felt that there was a need to create novel glycoconjugates with comparable properties. In 1993, the first synthesis of this novel class of neoglycoconjugates, coined "glycodendrimers", appeared [77]. The term dendrimer or starburst dendrimer has its origin in the fact that the molecules are multibranched and have tree-like structures (Latin: arbor-ol; Greek: dendron) [78], exactly like the tip of glycoproteins.

A review has been published about the synthesis and lectin binding properties of glycodendrimers produced by our group [22]. This section describes newer developments published since then. Conceptually it was deemed preferable to anchor desired carbohydrate units during the last stage of the syntheses since it provides opportunities to access dendrimer cores with different sugars incorporated. Commercially available polyamidoamine (PAMAM) and other dendrimers derived from poly-L-lysine, gallic acid, phosphorus, and 3,3′-iminobispropylamine have now been made accessible. As most glycodendrimer syntheses involve multiple attachment of carbohydrate derivatives on preformed structures, the conjugation chemistry ought to be efficient and high yielding. A description of these glycodendrimers is discussed below.

4.1 PAMAM Based Dendrimers

Spherical starburst dendrimers made of polyamidoamine (PAMAM, **57**) have already attracted a number research groups. These readily available amine-terminated molecules have been used to attach a number of carbohydrate derivatives using disaccharide lactones of lactose (*O*-β-D-galactopyranosyl-(1,4)-D-glucono-1,5-lactone) (**58**) and maltose (*O*-α-D-glucopyranosyl-(1,4)-D-glucono-(1,5)-lactone) via amide bond formation (Scheme 10) [79]. Dendrimers with 12, 24, and 48 glycan residues have been conjugated and shown to bind strongly and reversibly to Concanavalin A (Glc) and peanut (Gal) lectins. PAMAM dendrimers bearing *p*-thiophenyl urea of α-D-mannopyranoside (**59**) have been successfully prepared [80]. Using peracetylated glycosyl isothiocyanates of β-D-glucose, α-D-mannose (**60**), β-D-galactose, β-cellobiose, and β-lactose, Lindhorst and Kieburg [81] synthesized tetra-, hexa-, and octa-valent PAMAM dendrimers. The isothiocyanate coupling reaction was shown to be very efficient in both cases. Alternatively, in hopes to obtain a cancer vaccine

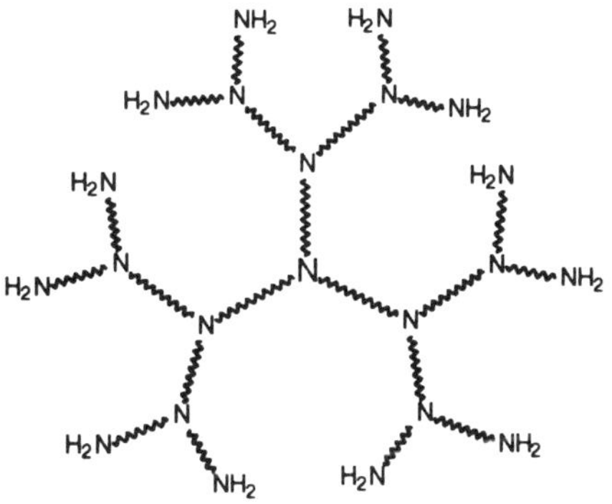

57 (PAMAM)

N~~N = $N(CH_2)_2NHCO(CH_2)_2N$

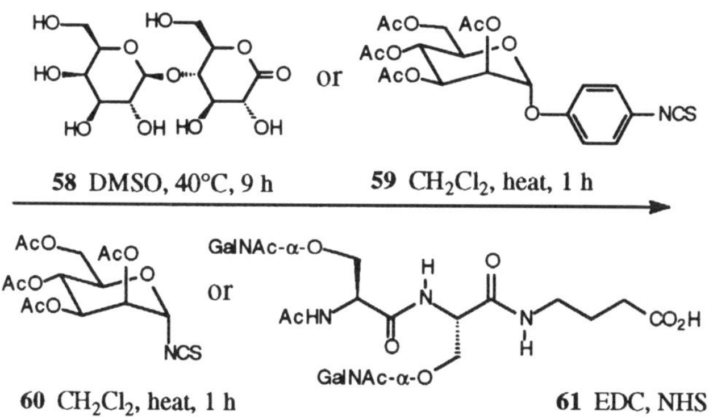

58 DMSO, 40°C, 9 h or **59** CH_2Cl_2, heat, 1 h

60 CH_2Cl_2, heat, 1 h or **61** EDC, NHS

Scheme 10

of the T_N-antigen, Toyokuni and Singhal [82] conjugated dimeric T_N-antigen (α-D-GalNAc-Ser) peptide (**61**) to 5th generation PAMAM dendrimers (48 amines) using *N*-hydroxysuccinimide ester. The material was however not immunogenic.

4.2
Dendrimers Based on Poly-L-Lysine

In order to mimic multiantennary oligosaccharide topology with greater variety of glycan structures and geometry, Roy et al. [77] used solid phase peptide syn-

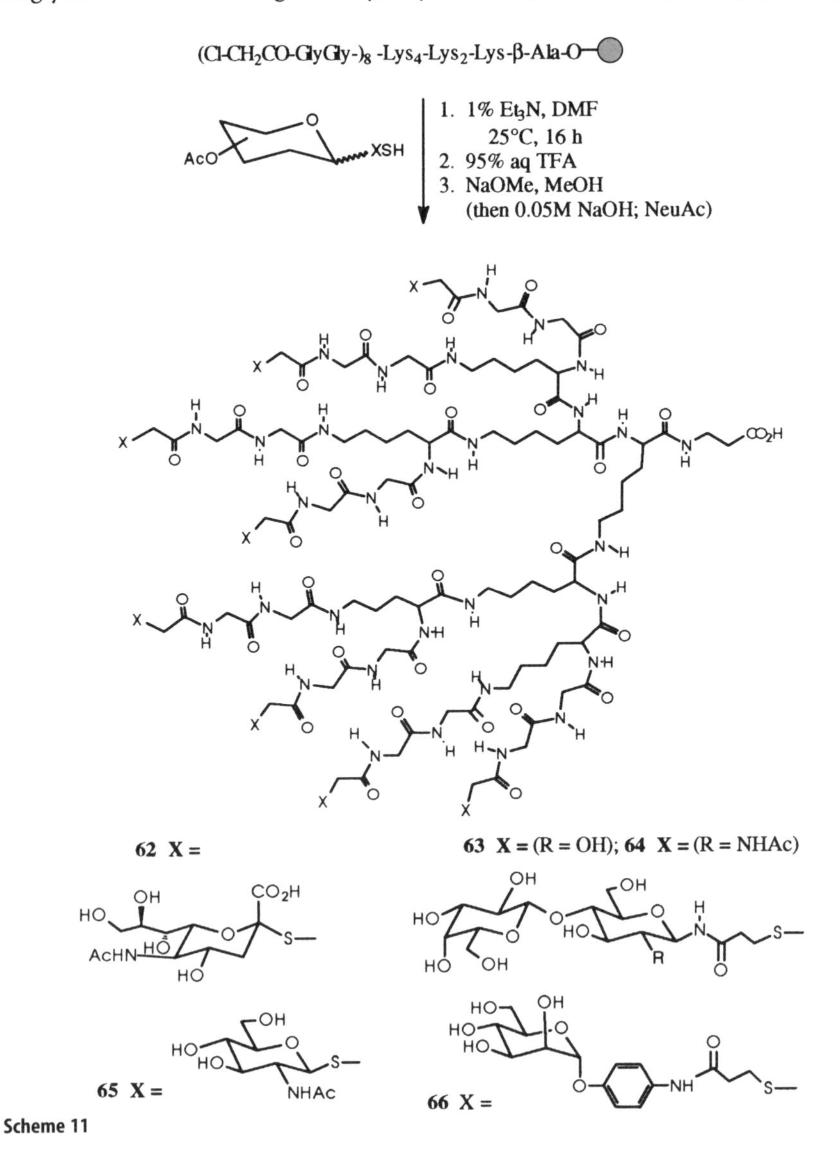

Scheme 11

thesis (Wang resin) to generate a multibranched L-lysine core of up to 16 available amine residues using 9-fluorenylmethoxycarbonyl (Fmoc) and hydroxybenzotriazole (HOBt) chemistry. All the terminal amino groups were then transformed into electrophilic *N*-chloroacetyl groups by treatment with *N*-chloroacetylglycylglycine hydroxybenzotriazolyl ester. Peracetylated glycosyl derivatives bearing thiol groups were added to the solid phase for nucleophilic substitution of the *N*-chloroacetyl groups. After sequential hydrolysis from the solid support by aqueous trifluoroacetic acid (95% TFA, 1.5 h) and saponification of the carbohydrate esters, glycodendrimers were obtained bearing α-thiosialosides **62** [77], β-lactosides **63** and *N*-acetyllactosaminides **64** [83], 1-thio β-D-*N*-acetylglucosamine **65** [84], α-D-mannosides **66** [85], T-antigen [86] and a 3′-sulfo-Lewis^X-Glc analog (not shown) [87]. In the later situation, unprotected and thiolated 3′-sulfo-Lewis^X-Glc derivative **37** (Scheme 5) was used and therefore, the conjugation was effected in homogenous solution (Et_3N, DMSO, r.t., 16 h) to provide di-, tetra-, and octavalent dendrimers in 41–80% yield. Representative octameric glycodendrimers are illustrated in Scheme 11.

A chemoenzymatic strategy was also applied for an alternative synthesis of *N*-acetyllactosamine-containing dendrimers. Thus, peracetylated 1-*S*-acetyl-1-thio β-D-*N*-acetylglucosamine was initially anchored to the *N*-chloroacetylated core dendrimers described above. Trifluoroacetolysis and ester hydrolysis provided GlcNAc-based dendrimers such as **65** which upon treatment with UDP-glucose, GlcNAc β–1,4-galactosyltransferase, and UDP-glucose 4′-epimerase provided dendritic *N*-acetyllactosaminide dendrimers with 2, 4, and 8 residues in good yields [84]. All of the above glycodendrimers exhibited fair to excellent enhanced binding affinity toward their respective binding proteins (L and E selectins, wheat germ agglutinin, Concanavalin A, peanut, and pea lectins, influenza virus hemagglutinin).

4.3 Dendrimers Based on Gallic Acid

With glycodendrimers of 2^n valencies in hands (2, 4, 8, 16), we became interested to get access to dendrimers with valencies of 3^n (3, 9, 27). To this end, gallic acid was chosen as core structure. Thus methyl gallate (**67**) was etherified with hydrophilic tetraethyleneglycol azido tosylate spacer **68** (K_2CO_3, DMF, 68%) (Scheme 12). The resulting azido ester was then reduced (H_2, 10% Pd-C, 62%) and treated with chloroacetic anhydride ($(ClCH_2CO)_2O$, Et_3N, EtOH, 72%) to generate trivalent *N*-chloroacetylated dendrimer **70** as its methyl ester [88]. Nucleophilic substitution with thiolated lactosylamide derivative **69** afforded dendrimer **71** after complete deprotection (Et_3N, CH_3CN, then 1 M NaOH in EtOH, 98%). As for L-lysine-based 3′-sulfo-Lewis^X-Glc dendrimer discussed above, reacting *N*-chloroacetylated methyl ester **70** with thiol derivative **37** (Et_3N, DMSO, r.t., 16 h) afforded trivalent cluster **72** in 80% yield [87]. To further demonstrate that higher generation dendrimers could be generated through this approach, azido acid **73** was coupled to amino ester **74** (EDC, HOBt, DIPEA, 25 °C, 3 h) to provide nonavalent azido ester **75** in 83% yield (Scheme 13) [88]. Using the same sequence of reactions described above for the first generation

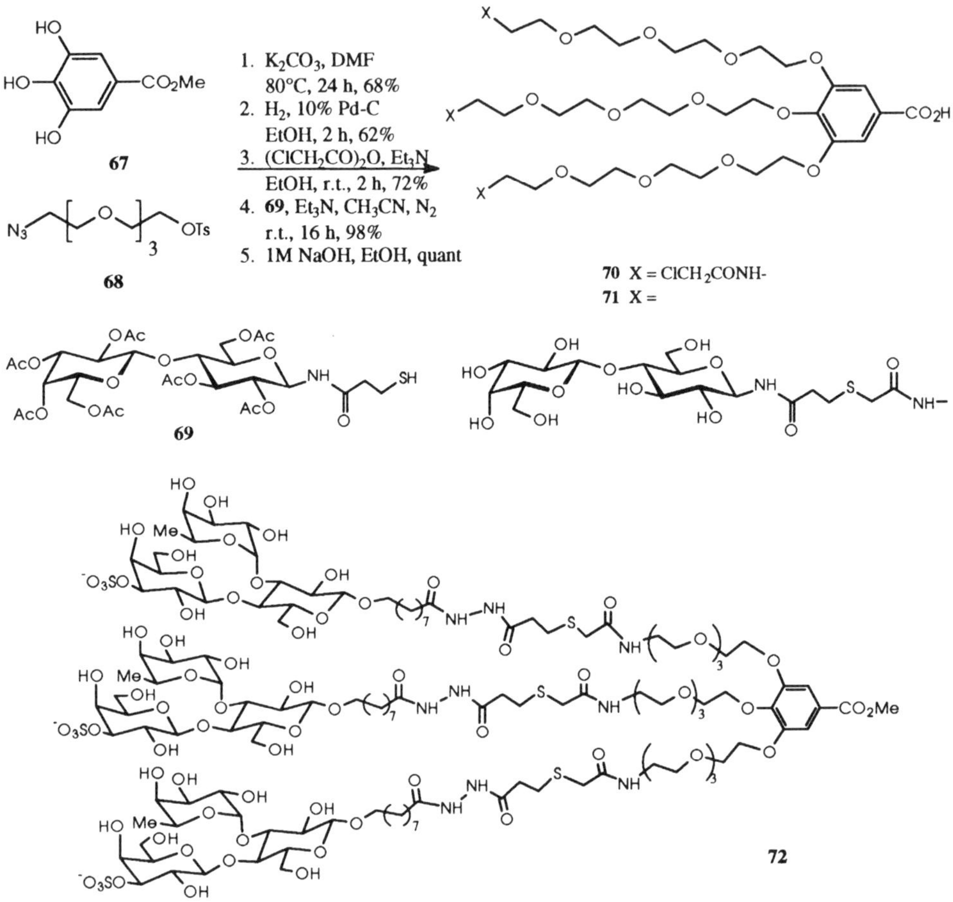

1. K_2CO_3, DMF
80°C, 24 h, 68%
2. H_2, 10% Pd-C
EtOH, 2 h, 62%
3. $(ClCH_2CO)_2O$, Et_3N
EtOH, r.t., 2 h, 72%
4. **69**, Et_3N, CH_3CN, N_2
r.t., 16 h, 98%
5. 1M NaOH, EtOH, quant

67

68

70 X = $ClCH_2CONH$-
71 X =

69

72

Scheme 12

(1: H_2, 10% Pd-C; 2: $(ClCH_2CO)_2O$, EtOH, Et_3N, 25 °C, 3 h, 70%; 3: **69** (11 eq), Et_3N, CH_3CN, DMSO, 25 °C, 16 h, 88%; 4: 1 M NaOH, EtOH, quant) nonavalent lactosylated dendrimer **75** was obtained.

4.4 Dendrimers Based on Phosphotriester Backbones

The chemistry and geometry of phosphodiesters and triesters were considered appealing factors for their use as new glycodendrimer scaffolds. The strategy discussed below was based on the synthesis of key phosphoramidite building blocks bearing either *N*-chloroacetyl end groups as above or alcohol functionality for further branching [89]. Diethylene glycol spacers were initially transformed into the mono *tert*-butyldiphenylsilyl derivative **77** which upon treatment with *N*, *N*-diisopropylphosphoramidous dichloride ($Cl_2PN(iPr)_2$, DIPEA,

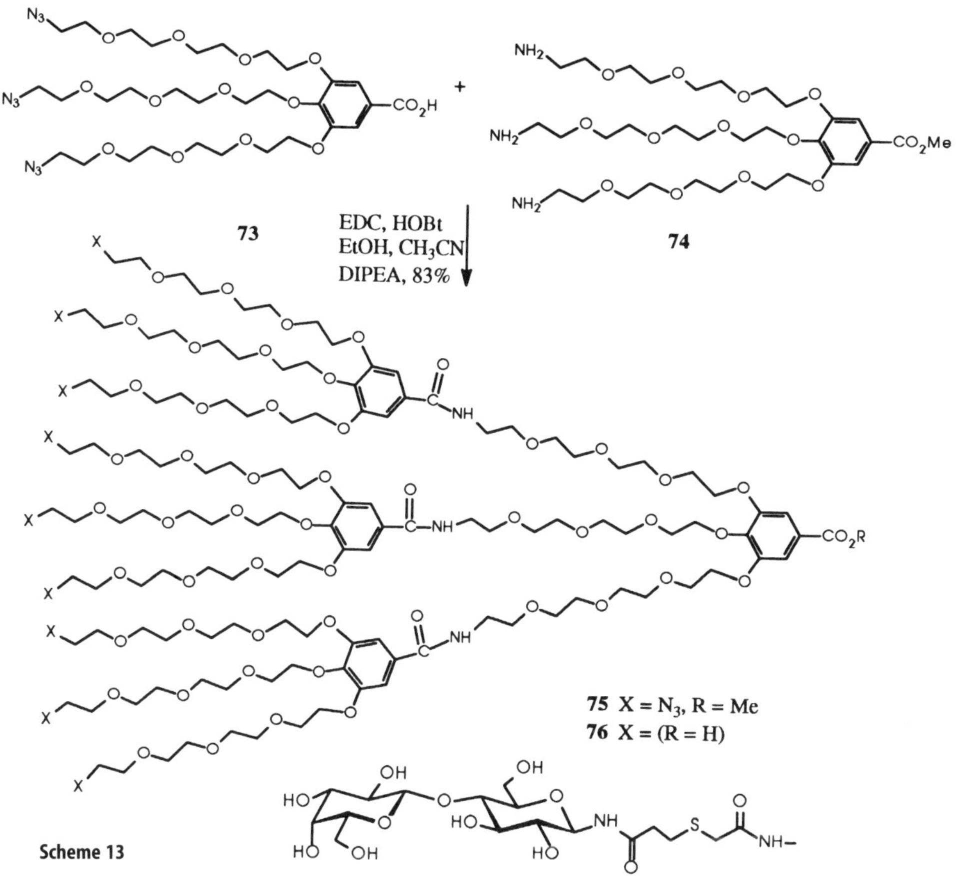

73 + 74

EDC, HOBt
EtOH, CH_3CN
DIPEA, 83%

75 $X = N_3$, R = Me
76 X = (R = H)

Scheme 13

DCM) provided the essential building block *N*, *N*-diisopropyldiphosphoramidite **79** in 84% yield (Scheme 14). Alternatively, mono *N*-chloroacetylated diethylene glycol derivative **78**, prepared from commercially available 2-(2-aminoethoxy)ethanol ($ClCH_2CO_2H$, EEDQ, 45 °C, 4 h, 82%) was similarly treated to provide phosphoramidite **80** (68%). Further couplings of **79** or **80** with alcohols **77** or **78** in the presence of 1 *H*-tetrazol gave phosphotriesters **81** or **83** after oxidation with *tert*-butylperoxide (70–71%). Coupling *N*-chloroacetyl derivative **83** (MeOH, Et_3N) with thiolated *N*-acetylgalactosaminide **85** afforded the first generation phospho-glycodendrimer **84** (63%) having three α-D-GalNAc moieties. Thiol **85** was prepared from allyl α-D-*N*-acetylgalactosaminide by reaction with thioacetic acid (HSAc, MeOH, AIBN, reflux, 10 h, 77%) followed by Zemplén de-*S*-acetylation.

tert-Butyldiphenylsilyl protected phosphotriester **81** was deprotected with fluoride anions (TBAF, THF, 78%) to give triol **82** which was further branched with building block **80** as above to provide a 2nd generation phosphotriester **88** (57%) having six *N*-chloroacetyl residues after phosphite oxidation. Further coupling of **82** with phosphoramidite building block **79** and oxidation gave **86**

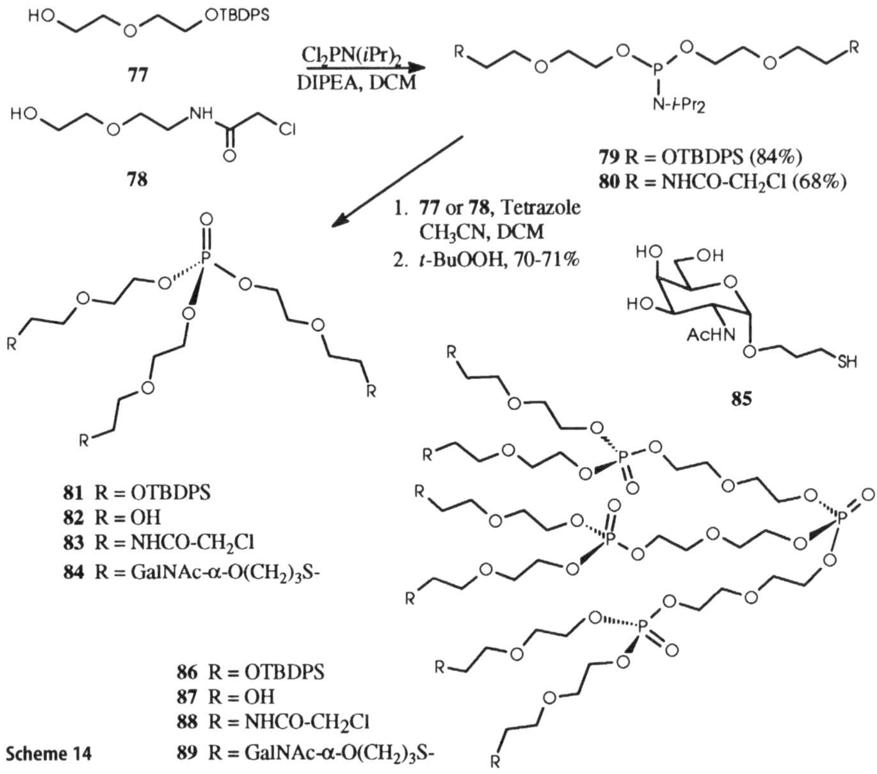

77 **78**

$Cl_2PN(iPr)_2$
DIPEA, DCM

79 R = OTBDPS (84%)
80 R = NHCO-CH_2Cl (68%)

1. **77** or **78**, Tetrazole
CH_3CN, DCM
2. *t*-BuOOH, 70-71%

85

81 R = OTBDPS
82 R = OH
83 R = NHCO-CH_2Cl
84 R = GalNAc-α-$O(CH_2)_3S$-

86 R = OTBDPS
87 R = OH
88 R = NHCO-CH_2Cl
89 R = GalNAc-α-$O(CH_2)_3S$-

Scheme 14

(83%) which after silyl ether deprotection as above gave hexaol **87** (85%) useful as precursor for the next higher generations using simple reiterative coupling strategies (up to 12 GalNAc residues were incorporated, not shown). Nucleophilic displacement of hexavalent *N*-chloroacetylated precursor **88** by thiolated *N*-acetylgalactosaminide **85**, as above, provided the hexavalent dendrimer **89** (52%). Further processing of the synthetic sequence, coupled with tethering strategies with various spacers, furnished access to a family of dendrimers having many different valencies. All the GalNAc phospho-dendrimers thus produced were tested as inhibitor of *Vicia villosa* binding to asialoglycophorin. The results indicated 3–10-fold (hexamer) enhanced affinity, supporting once again the cluster effect.

5 Clusters

Once multivalent carbohydrate-protein interactions are firmly established with the assistance of neoglycoconjugates such as those described above, further focus toward fine-tuned geometry and valency requirements becomes necessary for a thorough understanding of the binding interactions involved. Until now, these investigations have been more or less dependent on trial and error which required substantial time investment from the synthetic chemist. A large number of clusters of different size and shapes are needed to define the topography of any given receptors. From a crude estimation, it is considered that unraveling rat hepatocyte Gal/GalNAc receptor's optimum trivalent ligand has necessitated ten years of research [70]. It is likely that dendritic glycoconjugates will serve as tools for rapidly assessing multivalent interactions. The family of carbohydrate ligands described below will serve to illustrate the usefulness of glycoclusters in various biological investigations.

A number of carbohydrate recognition domains (CRDs) of animal lectins are known to exist as clusters. ASGP-Rs exist as bundle of hexamers, mannose 6-phosphate receptor and galectin-3 are known as dimers, while mannose binding proteins and receptors form multiple CRDs [31].

5.1 Galactosides

As stated above, asialoglycoprotein receptors, macrophages, and galectins are well studied Gal/GalNAc binding receptors. Classical investigations by Lee and Lee [31, 70] on rat and rabbit hepatocytes have culminated in the design of a very potent trivalent GalNAc ligand (**92**, YEE(GalNAcAH)$_3$) [90]. It has been demonstrated that **92** has a K_d as low as 0.2 nM as determined by the inhibition of binding of asialoorosomucoid to isolated as well as intact hepatocyte receptors. *N*-Acetylgalactosaminide **92** was readily prepared by coupling 6-aminohexyl 2-acetamido-2-deoxy-β-D-galactopyranoside (**90**) to benzyloxycarbonyl-protected L-tyrosyl-L-glutamyl-L-glutamic acid (**91**) according to Scheme 15.

Similar investigations by Kichler and Schuber [71] resulted in the syntheses of a number of β-D-galactopyranoside clusters built on L-lysyl-L-lysine. Trimer **95** was shown to be more potent than mono- or di-valent clusters. Moreover, structures analogous to **95** having shorter or longer spacers resulted in less efficient inhibitors. Ligand **95** was synthesized from pseudothiourea **93** which after treatment with 2-(2-(2-iodoethoxy)ethoxy)ethanoic acid under basic conditions afforded thiogalactoside **94** in 65% yield. Transformation of acid **94** to an active ester and further coupling to L-lysine dimer provided trivalent ligand **95** in 25% overall yield (Scheme 15). The above studies confirmed that tri-antennary oligosaccharides having Gal/GalNAc moieties situated at the apexes of a triangle whose sides are 1.5, 2.2, and 2.5 nm constituted optimum geometry [91]. Toyokuni and Hakomori [73] have used L-lysine dimer as a scaffold for the syntheses of related trivalent ligands used as inhibitors of lung colonization by metastatic cells.

Natural and semi-synthetic multiantennary oligosaccharides derived from glycoproteins are obvious candidates for studies related to cluster effects. The various branches of multiantennary oligosaccharides can be used to scan clustered receptors and many investigations are based on this premise. Recent studies with di- and tetra-antennary oligosaccharides bearing terminal α-D-(1,3)-galactopyranoside residues were shown to be effective inhibitors of sperm adhesions to mouse gametes [92]. The artificial oligosaccharides were prepared by trimming existing structures and reconstituting them by enzyma-

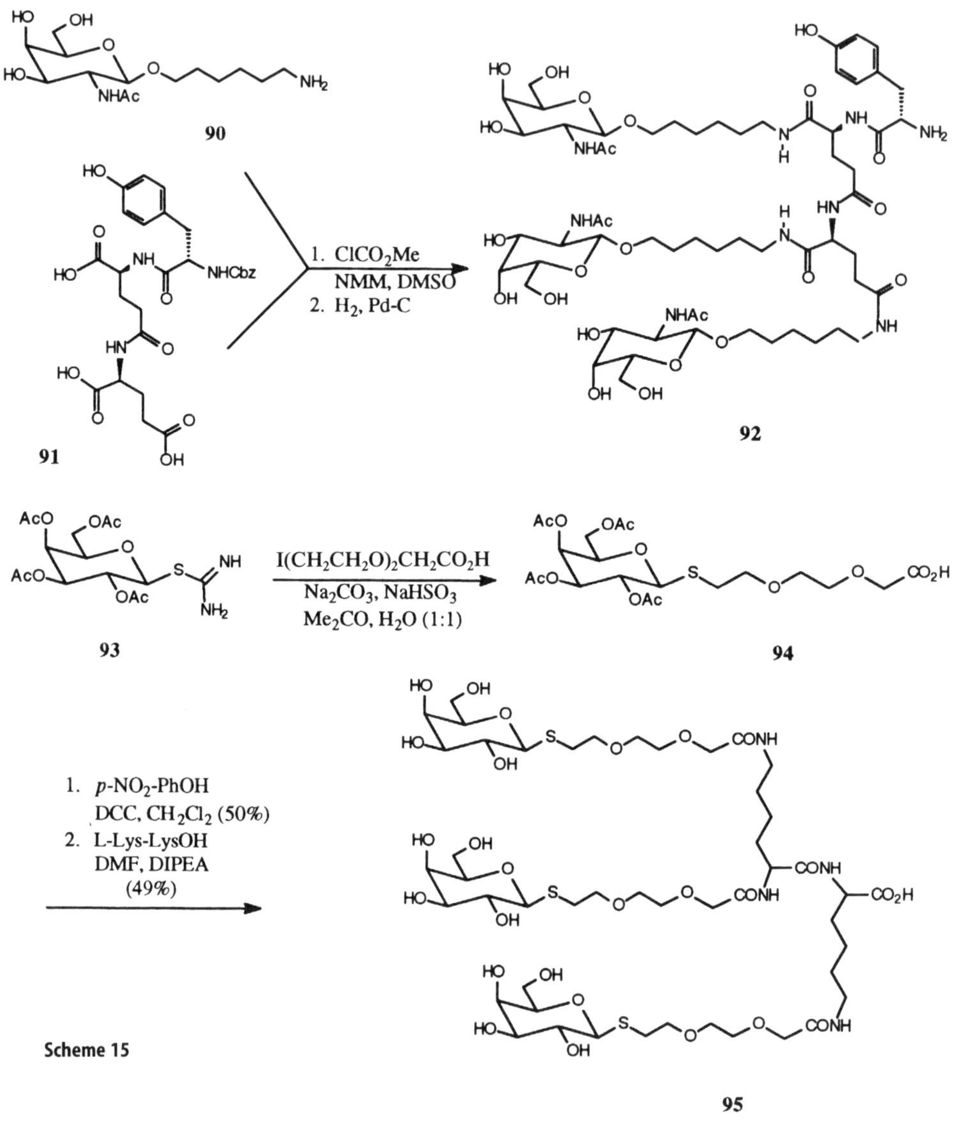

1. $ClCO_2Me$
NMM, DMSO
2. H_2, Pd-C

90

91

92

$I(CH_2CH_2O)_2CH_2CO_2H$
Na_2CO_3, $NaHSO_3$
Me_2CO, H_2O (1:1)

93

94

1. *p*-NO_2-PhOH
DCC, CH_2Cl_2 (50%)
2. L-Lys-LysOH
DMF, DIPEA
(49%)

95

Scheme 15

tic glycosylation. Again, a cluster effect has been demonstrated. The strategy has been also extended to study mannopyranoside and sialoside receptors.

5.2 Mannosides

Systematic investigations of mannose-binding proteins with clustered mannosides other than natural multiantennary oligosaccharides have been rather scarce until recently. However, using 6′-*O*-phosphorylated mannose disaccharide scaffolded on peptide templates using combinatorial glycopeptide libraries,

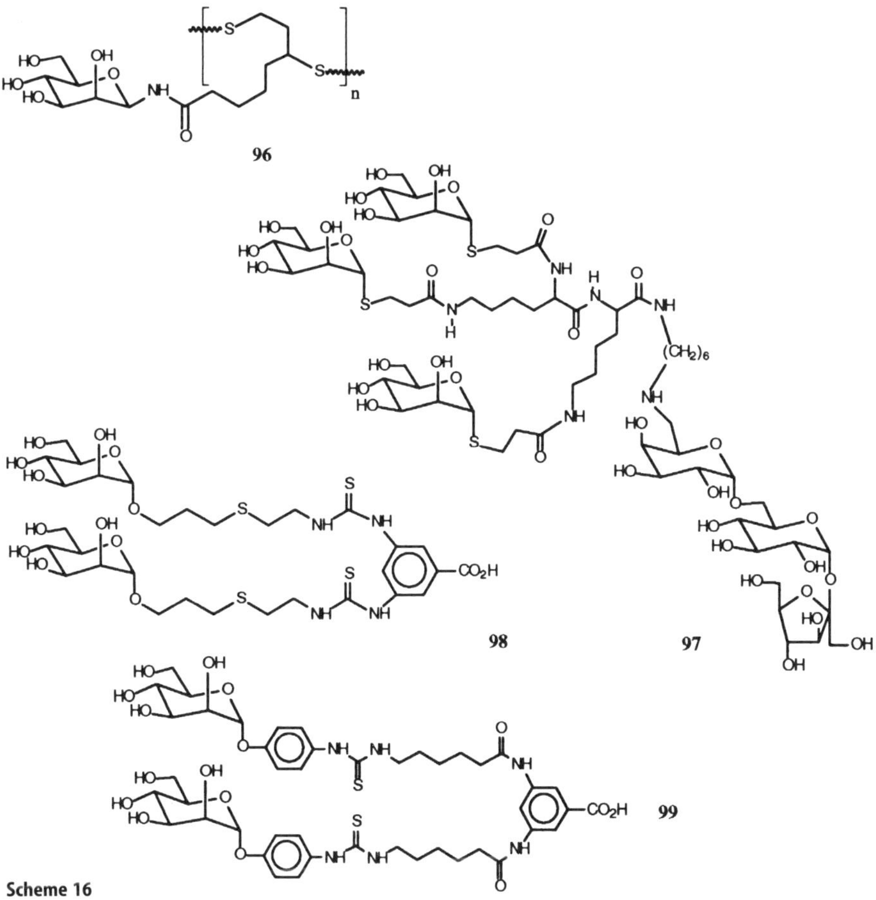

Scheme 16

Christensen et al. [93] successfully obtained a divalent cluster having ~600–1500 enhanced binding properties. We have also demonstrated that dendritic mannosides such as **66** (Scheme 11) can provide up to 100-fold higher affinity (on a per-mannoside basis) when used to inhibit the binding of plant lectins (Concanavalin A and pea lectins) to yeast mannan [85]. As these interactions are at the origin of host infections by fimbriated bacteria, mannoside dendrimers can form the basis of novel antiadhesin molecules.

Previous investigations on mannose macrophage receptors by Ponpipom et al. [94,95] have demonstrated the strong binding properties of multivalent mannose-containing polymers (**96**), liposomes, and trivalent mannoside built from raffinose dilysyl conjugates **97** (Scheme 16). The construction of heterobifunctional trimannoside **97** was based on the regioselective C-6 Gal oxidation of raffinose using galactose oxidase. The trimannoside L-lysyl-lysine moieties was then introduced by reductive amination ($NaBH_3CN$) of its 6-aminohexanoyl derivative.

To further investigate multivalent mannose-binding interactions, we have recently begun the systemic syntheses of mannoside clusters of various shapes, size, and valencies. Dimannoside clusters such as **98** and **99** [96], and analogous hexamer [97] were synthesized. They had binding properties almost as high as dendritic mannosides. Using turbidimetric experiments, they were shown to efficiently and reversibly cross-link di-and tetra-valent lectins.

5.3 Sialosides

Sialic acid and sialyloligosaccharide clusters, like their other neoglycoconjugate counterparts (proteins, polymers, liposomes), have so far been mainly used as inhibitors of influenza virus hemagglutination and for selectins-dependent erythrocytes adhesion to inflamed tissues. Analogously to the above two cases, natural and semi-synthetic multiantennary oligosaccharides have constituted topological models for the scaffolding of sialoside clusters. In recent examples, enzymatically prepared di- [98] and tetra-valent [99] sialyl Lewis^X oligosaccharides (**100**) were shown to be potent inhibitors (IC_{50} 0.15 μM and < 50 nM, respectively) of L-selectin-mediated lymphocyte-endothelium interactions (Scheme 17). The above *tetramer* was approximately 60-fold better than the corresponding monomer, suggesting again multivalent interactions. In fact, recent studies with electron microscopy have unequivocally demonstrated the clustering of L-selectins on the tips of leukocyte microvilli [100]. Other sialyl Lewis^X clusters and their structural mimetics (3′-sulfo-Lewis^X-Glc) have been synthesized, either enzymatically or chemically, using various tethering units. Thus, 1,4-butanediol and 1,5-pentanediol (**101**) [101], gallic acid (**72**) [30, 102], multibranched L-lysine dendrimers [30, 102], galactoside (**102**) [101], nitromethane-trispropionic acid (**103**) [103], and cyclic peptide [104] have been successfully used as scaffolding elements to further demonstrate multivalent interactions in selectin binding processes (reviewed in [30]).

By carefully adjusting the distances between two sialoside residues in a number of divalent clusters, Glick and Knowles [105] have obtained dimer **104** having the two sialic acid 5.7 nm apart. Compound **104** was 100-fold more potent than methyl *α*-sialoside (Neu5Ac*α*2Me) in influenza virus inhibitions and 500-fold more potent in the case of polyomia virus. Alternatively, sialyl-*α*-(2,6)-*β*-LacNAc dimers (**105**) branched at different positions of synthetic peptides, including compact glycine-rich and helical proline-rich peptides, afforded clusters which were only 8- and 4-fold more potent, respectively, than the corresponding monovalent trisaccharide [106].

All of the above evidence suggests that cooperative binding interactions depend on both the overall number of carbohydrate ligands and their relative positioning from one each other. Residues too close or too far apart in any given clusters may contribute negatively to the binding interactions. Moreover, each clustered protein receptor will impose its own geometrical constraints. From a purely fundamental point of view, isolated single receptors would give false perspective of cluster effects. For instance, in ongoing activities in our laboratory, it has become apparent that soluble di- and tetra-valent lectins would require only optimum tetra- and di-valent clusters respectively for maximum efficacy in forming stable cross-linked lattices.

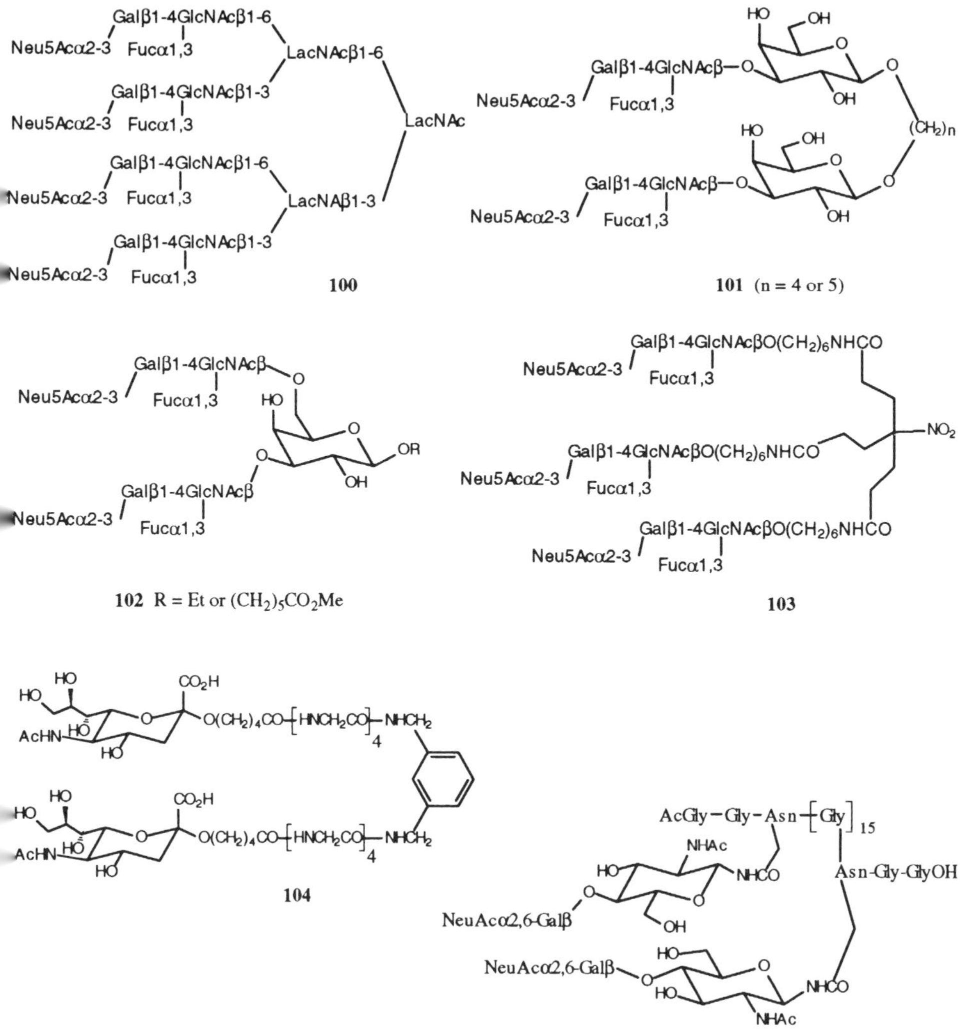

100

101 (n = 4 or 5)

102 R = Et or $(CH_2)_5CO_2Me$

103

104

105

Scheme 17

6
Novel Multivalent Neoglycoconjugates

As better understanding of multivalent carbohydrate-protein interactions is gained, it is becoming evident that clusters of different shapes and size would furnish potent ligands from which critical information about receptors topography will be obtained. It is similarly obvious that the geometry, conformation, and valency of the necessary clusters will need to be known. In order to address these issues, we initiated systematic developments of novel multivalent neoglycoconjugates. The following section will described recent developments toward this goal.

6.1 Calix[4]arenes

In many ways, calix[n]arenes are structurally related to cyclodextrins. As the syntheses of cyclodextrins fully substituted with carbohydrate residues have been described [107], this topic will not be covered here. Calix[n]arenes offer a certain number of non-negligible advantages over their cyclodextrin counterparts. They can be readily prepared with various ring sizes encompassing four, six or eight phenoxy groups at the upper rim. Moreover, the *para*-phenyl positions can be chemically modified to incorporate a wide range of functionality thus allowing further connections to other molecules and probes. The chemistry by which the phenoxy group can be modified is more straightforward than that of cyclodextrins. They can also form inclusion complexes with aromatic compounds, thus offering the extra benefit of drug targeting. They can also be trapped in different conformations, thus further allowing desired orientations of the carbohydrate ligands.

Marra et al. [108] have been the first to synthesize carbohydrate-linked calix[4]arenes. However, the resulting glycoconjugates were not water-soluble and were deprived of any spacer arms necessary for efficient interactions with protein receptors. In a recent paper, we described the synthesis of *p-tert*-butylcalix[4]arene fixed in its cone conformation and substituted with four α-thiosialoside residues at the tip of *N*-chloroacetylated 1,4-butanediamine spacer [109]. Scheme 18 describes the synthesis of the first water-soluble and biologically active "glyco-calix[4]arene" (**111**). Thus, *p-tert*-butylcalix[4]arene **106** was initially transformed into the known tetraacid chloride **107** which was treated with a slight excess of mono-Boc protected 1,4-butanediamine to afford derivative **108** in 62% yield. The Boc-protecting groups were removed (20% TFA in CH_2Cl_2, quant.) and the resulting tetraamine **109** was transformed into key tetra-*N*-chloroacetylated derivative **110** ($(ClCH_2CO)_2O$, Et_3N, CH_2Cl_2, 63%). Using the coupling strategy already developed for glycodendrimer syntheses, electrophile **110** was treated with nucleophilic 2-thio-α-sialic acid derivative **13**, freshly prepared from thioacetate **12** by chemoselective de-*S*-acetylation, to give fully protected calix[4]arene intermediate. Protecting group hydrolysis provided **111**. Sialylated derivative **111** was shown to bind strongly and reversibly to wheat germ agglutinin (WGA) and to form insoluble cross-linked lattices as demonstrated by turbidimetric experiments.

6.2 Glycopeptoids

As illustrated above, peptides have been used as tethering elements for the suitable positioning of carbohydrate moieties. Since the chemistry of glycopeptides is well under control, the approach may appear valuable, at least a priori. This is likely to be true for research investigations. However, as neoglycopeptides should eventually developed into useful therapeutics, the synthesis of novel and metabolically stable neoglycopeptidomitics was undertaken [110–113]. The approach chosen was commensurate to that used for the synthesis of peptoids.

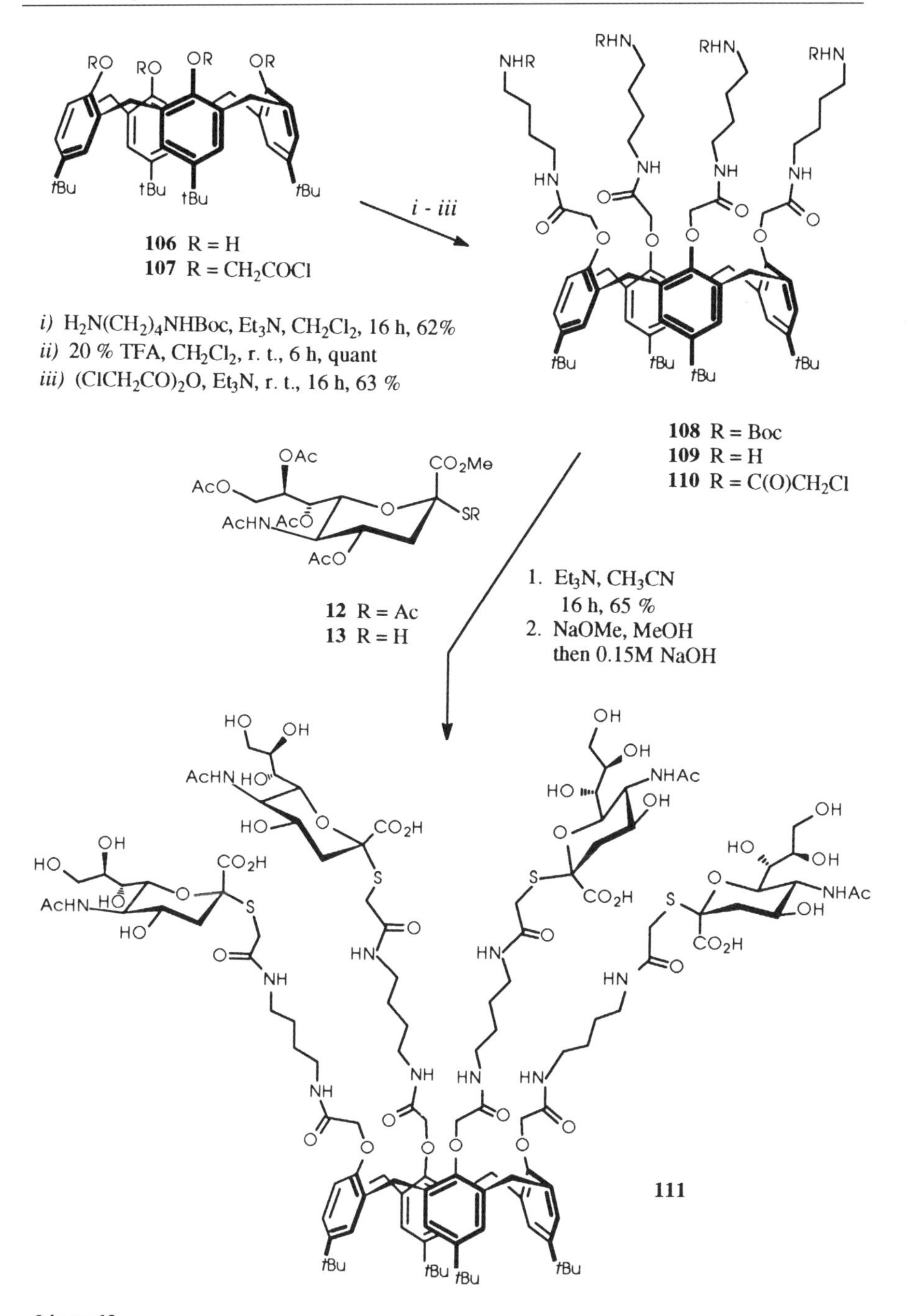
RO RO OR OR
tBu tBu tBu tBu
106 R = H
107 R = CH2COCl
i - iii
i) H2N(CH2)4NHBoc, Et3N, CH2Cl2, 16 h, 62%
ii) 20 % TFA, CH2Cl2, r. t., 6 h, quant
iii) (ClCH2CO)2O, Et3N, r. t., 16 h, 63 %
NHR RHN RHN RHN
HN NH NH NH
108 R = Boc
109 R = H
110 R = C(O)CH2Cl
OAc
CO2Me
AcO
AcHN AcO
SR
AcO
12 R = Ac
13 R = H
1. Et3N, CH3CN
16 h, 65 %
2. NaOMe, MeOH
then 0.15M NaOH
HO OH
AcHN HO
HO
CO2H
S
NHAc
CO2H
111

Scheme 18

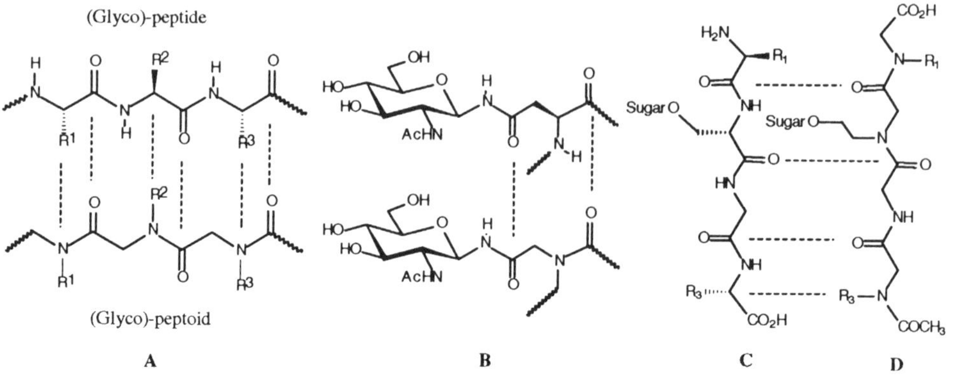

Scheme 19

Peptoids are *N*-substituted oligoglycines whose carbonyl and side-chain residues are superimposable to those of natural peptides. Scheme 19 illustrates the structural similarities between both *N*- and *O*-linked glycopeptoids in relation to their analogous glycopeptides.

After the syntheses of model glycopeptoids was undertaken, it was anticipated that oligomeric units, having pre-established spacers between the carbohydrate moieties and the peptoid backbone and inbetween the branching residues, could be readily prepared by a convergent blockwise approach using orthogonally protected derivatives [112]. By virtue of the secondary amide linkages at every branching point, it has been possible to generate libraries of conformational rotamers (*E/Z*) which can be used to screen a wide area of multivalent receptor *loci*. The example below illustrate the synthesis of sialic acid oligomers spaced from the backbone with 6-aminocaproic acid. All the *N*-linked sialosides were themselves interspaced by one glycine residue [114].

α-Sialosyl azide **14** [115] was derivatized into the protected amine derivative **112** which upon hydrogenolysis gave **113** (Scheme 20). Amine **113** was transformed into *N*-substituted key building block **114** by treatment with *tert*-butyl bromoacetate (75%) followed by *N*-bromoacetylation using bromoacetic anhydride (92%). The orthogonally protected dimer **115** was then prepared by alkylating **114** with **113** (DIPEA, CH_3CN, r.t., 3 h, 74%). The secondary amine of **115** was then *N*-acetylated (AcCl, DIPEA, CH_2Cl_2, 91%) to provide end-group dimer **116** or alternatively transformed into *N*-benzyloxycarbonyl-protected derivative **117** (CbzCl, DIPEA, CH_2Cl_2, 89%). Sequential deprotection of *tert*-butyl ester and Cbz-group followed by amide coupling (DCC) of the resulting amino acids afforded trimer, tetramer, hexamer and octamer such as **118** after protecting group removal under standard basic conditions. The resulting "sialopeptoids" were suitable for biological testing.

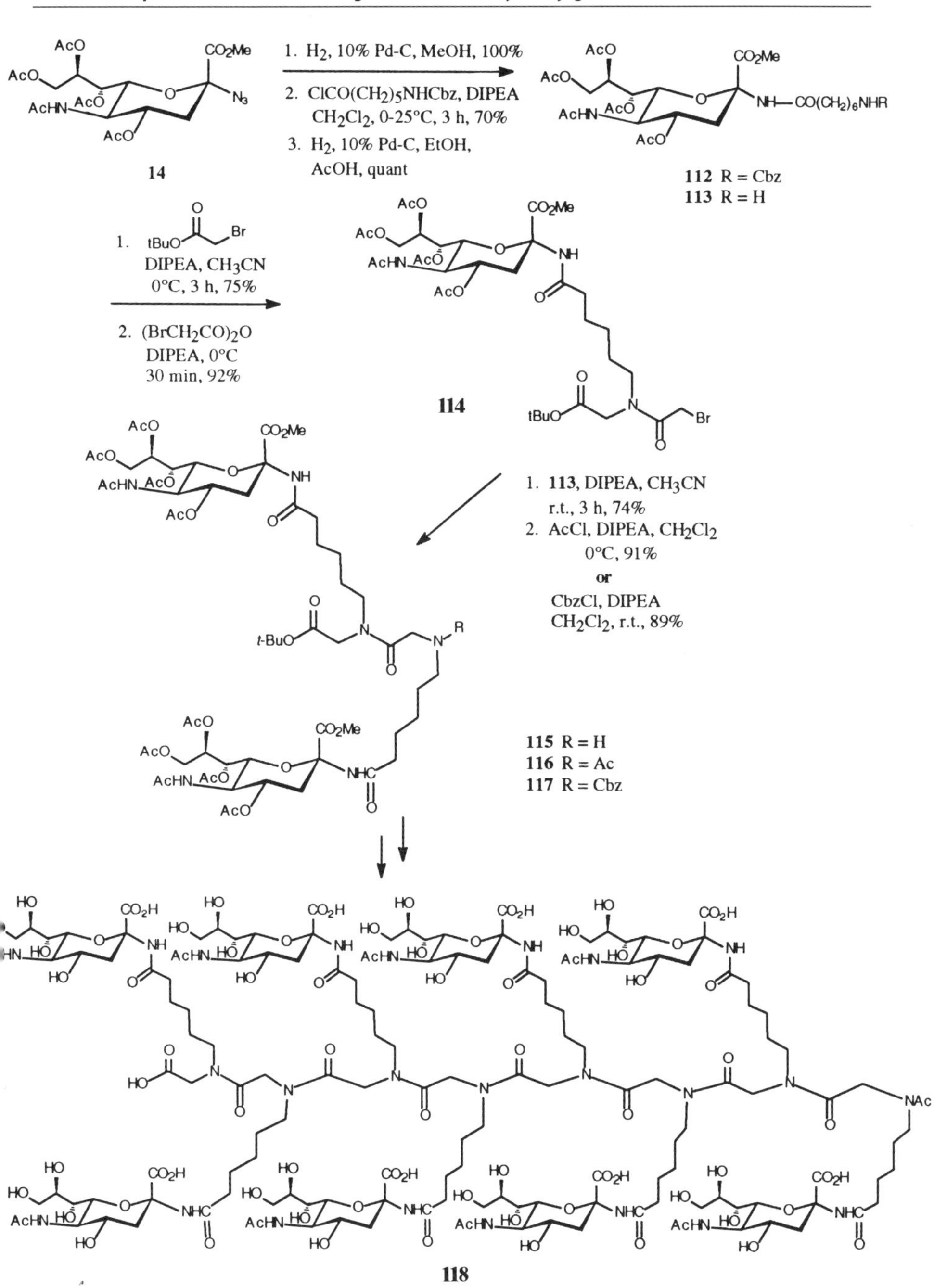

1. H2, 10% Pd-C, MeOH, 100%
2. ClCO(CH2)5NHCbz, DIPEA CH2Cl2, 0-25°C, 3 h, 70%
3. H2, 10% Pd-C, EtOH, AcOH, quant
14
112 R = Cbz
113 R = H
1. tBuO Br DIPEA, CH3CN 0°C, 3 h, 75%
2. (BrCH2CO)2O DIPEA, 0°C 30 min, 92%
114
1. 113, DIPEA, CH3CN r.t., 3 h, 74%
2. AcCl, DIPEA, CH2Cl2 0°C, 91%
or
CbzCl, DIPEA CH2Cl2, r.t., 89%
115 R = H
116 R = Ac
117 R = Cbz
118

Scheme 20

7
Heterobifunctional Neoglycoconjugates

Carbohydrate chemists have been creative in their design of novel glycotools to address a wide range of biological investigations. A large number of carbohydrate derivatives have been synthesized with added effector molecules such as drugs and probes. Particular attention has been devoted to biotinylated and fluoresceinylated heterobifunctional neoglycoconjugates. The examples are numerous and their individual description would require a chapter on its own. The unique example described below has taken into consideration both multivalent interactions and insertion of a fluorescent probe [116]. *N*-Acetylglucosamine (GlcNAc) was also chosen because the resulting compound may be further glycosylated using known enzymatic transformations. The target compound incorporates a novel glycosidic *O-N*-linkage such a that found in antitumor antibiotics calicheamicin γ_1 and esperamycin A_1. It has been stereoselectively introduced using phase transfer catalysis (PTC) [42] by treating 2-acetamido-3,4,6-tri-*O*-acetyl-2-deoxy-α-D-glucopyranosyl chloride **119** with

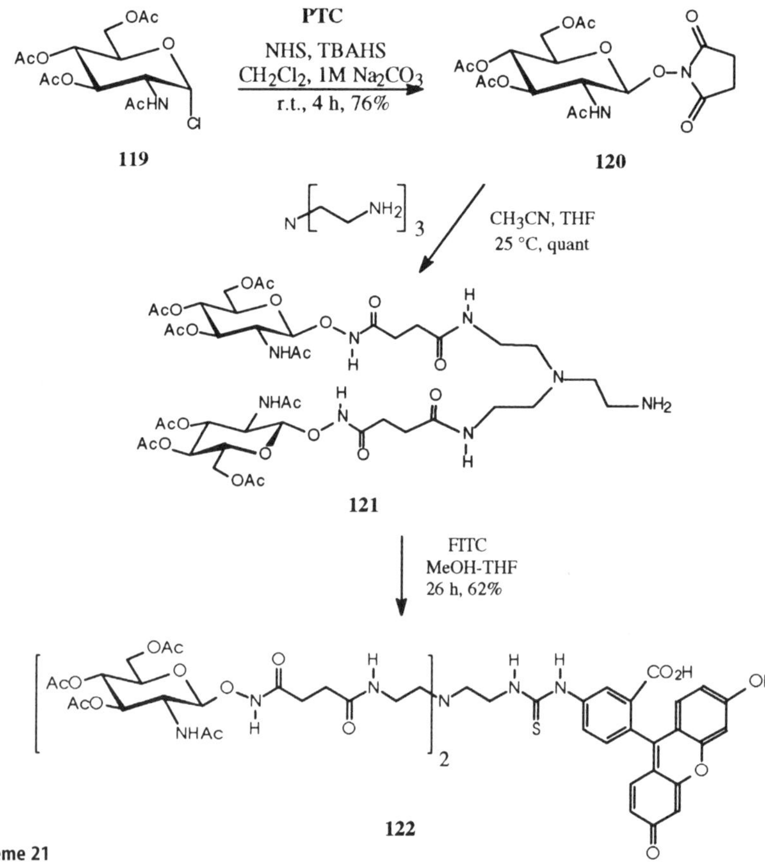

Scheme 21

N-hydroxysuccinimide (NHS) (TBAHS, 1 M Na_2CO_3, CH_2Cl_2, r.t., 4 h) which afforded glycosyloxysuccinimide derivative **120** in 76% yield (Scheme 21). Interestingly, reacting excess of **120** with tris(2-aminoethyl)amine gave exclusively and in quantitative yield divalent intermediate **121**. Further treatment of amine **121** with fluorescein isothiocyanate (FITC) provided heterobifunctional neoglycoconjugate **122** in 62% yield.

References

1. Goebel WF, Avery OT (1929) J Exp Med 50:521
2. Apostopoulos V, McKenzie IFC (1994) Crit Rev Immunol 14:293
3. Toyokuni T, Dean B, Cai S, Boivin D, Hakomori S-I, Singhal AK (1994) J Am Chem Soc 116:395
4. Meldal M, Mouritsen S, Bock K (1993) ACS Symposium Ser 519:19
5. Harding CV, Kihlberg J, Elofsson M, Magnusson G, Unanue ER (1993) J Immunol 151:2419; Ishioka GY, Lamont AG, Thomson D, Bulbow N, Gaeta FCA, Sette A, Grey HM (1992) J Immunol 148:2446
6. Stowell CP, Lee YC (1980) Adv Carbohydr Chem Biochem 37:225
7. Lee YC, Lee RT (1994) Neoglycoconjugates: Preparation and applications. Academic Press, San Diego
8. Lee YC, Lee RT (1994) Methods Enzymol, vol 242 and 246. Academic Press, San Diego
9. Roy R (1996) The chemistry of neoglycoconjugates. In: Boons GJ (ed) Carbohydrate chemistry. Chapman and Hall, Glasgow, (in press)
10. Roy R (1996) Trends Glycosci Glycotechnol 8:79
11. Bovin NV, Korchagina EYu, Zemlyanukhina TV, Byramova NE, Ivanov AE, Zubov VP, Mochalova LV (1993) Glycoconjugate J 10:142
12. Bovin NV, Gabius H-J (1995) Chem Soc Rev 24:413
13. Magnusson G, Chernyak AYa, Kihlberg J, Kononov LO (1994) Synthesis of neoglycoconjugates. In: Lee YC, Lee RT (eds) Neoglycoconjugates: preparation and applications. Academic Press, San Diego, p 53
14. Chernyak AYa (1994) ACS Symposium Series 560:133
15. Kochetkov NK (1984) Pure Appl Chem 56:923
16. Dick Jr WE, Beurret M (1989) Glycoconjugates of bacterial carbohydrate antigens. In: Cruse JM, Lewis Jr RE (eds) Contrib Microbiol Immunol. Karger, Basel, vol 10, p 48
17. Jennings HJ, Sood RK (1994) Synthetic glycoconjugates as human vaccines. In: Lee YC, Lee RT (eds) Neoglycoconjugates: preparation and applications. Academic Press, San Diego, p 325
18. Feizi T, Childs RA (1994) Methods Enzymol 242:205
19. Magnusson G (1986) Synthetic neo-glycoconjugaters. In: Lark D (ed) Protein-carbohydrate interactions in biological systems. Academic Press, London, p 215
20. Schnaar RL (1984) Anal Biochem 143:1
21. Pazur J (1981) Adv Carbohydr Chem Biochem 39:405
22. Roy R (1996) Polymers News 21:226
23. Duncan R, Kopecek J (1984) Adv Polymer Sci 57:51
24. Monsigny M, Roche, A-C, Midoux P, Mayer R (1994) Adv Drug Deliv Rev 14:1
25. Seymour LW (1994) Adv Drug Deliv Rev 14:89; Molema G, Meijer DKF (1994) Adv Drug Deliv Rev 14:25
26. Danguy A, Kayser K, Bovin NV, Gabius H-J (1995) Trends Glycosci Glycotechnol 7:261
27. Roy R (1996) Design and syntheses of glycoconjugates. In: Khan SH, O'Neil R (eds) Modern methods in carbohydrate synthesis. Harwood Academic, Switzerland, p 378
28. Karlsson K-A (1995) Curr Opin Struct Biol 5:622
29. Lasky LA (1995) Ann Rev Biochem 64:113; Rosen SD, Bertozzi CR (1994) Curr Opin Cell Biol 6:663

30. Roy R (1996) Sialoside mimetics and conjugates as antiinflammatory agents and inhibitors of flu virus infections.In: Witczak ZJ (ed) Carbohydrates: targets for drug design. Marcel Dekker, New York, p 84
31. Lee RT, Lee YC (1994) Enhanced biochemical affinities of multivalent neoglycoconjugates. In: Lee YC, Lee RT (eds) Neoglycoconjugates: preparation and applications. Academic Press, San Diego, p 23
32. Matrosovich MN (1989) FEBS Letters 252:1; Spillmann D (1994) Glycoconjugate J 11:169
33. Roy R, Tropper FD, Morrison T, Boratynski J (1991) J Chem Soc Chem Commun 536
34. Gamian A, Chomik, Laferrière CA, Roy R (1991) Can J Microbiol 37:233
35. Roy R, Laferrière CA (1990) Can J Chem 68: 2045; Roy R, Laferrière CA, Gamian A, Jennings HJ (1987) J Carbohydr Chem 6:161
36. Roy R, Laferrière CA, Pon RA, Gamian A (1994) Methods Enzymol 247:351
37. Roy R, Laferrière CA (1988) Carbohydr Res 177:C1
38. Roy R, Romanowska A, Andersson FO (1994) Methods Enzymol 242:198
39. Romanowska A, Meunier SJ, Tropper FD, Laferrière CA, Roy R (1994) Methods Enzymol 242:90
40. Roy R, Laferrière CA (1990) J Chem Soc Chem Commun 1709
41. Roy R, Baek MG, Filion L, Ogunnaike S, (unpublished data)
42. Roy R (1997) Phase transfer catalysis in carbohydrate chemistry. In: Sasson Y, Neumann R (eds) Handbook of phase transfer catalysis, Chapman and Hall, Glasgow, (in press); Roy R, Tropper FD, Cao S, Kim JM (1997) ACS Symposium Series, (in press)
43. Cao, S, Meunier SJ, Andersson FO, Letellier M, Roy R (1994) Tetrahedron: Asymm 5:2303
44. Roy R, Tropper FD, Romanowska A, Letellier M, Cousineau L, Meunier SJ, Boratynski J (1991) Glycoconjugate J 8:75
45. Park WKC, Meunier SJ, Zanini D, Roy R (1995) Carbohydr Lett 1:179
46. Tropper FD, Andersson FO, Braun S, Roy R (1992) Synthesis 618
47. Kamath VP, Diedrich P, Hindsgaul O (1996) Glycoconjugate J 13:315
48. Hällgren C, Hindsgaul O (1995) J Carbohydr Chem 14:453
49. Sigal GB, Mammen M, Dahmann G, Whitesides GM (1996) J Am Chem Soc 118:3789
50. Byramova NE, Mochalova LV, Belyanchikov IM, Matrosovich MN, Bovin NV (1991) J Carbohydr Chem 10:691; Lees WJ, Spaltenstein A, Kingery-Wood JE, Whitesides GM (1994) J Med Chem 37:3419
51. Roy R, Andersson FO, Harms G, Kelm S, Schauer R (1992) Angew Chem Int Ed Engl 31:1478
52. Charych DH, Nagy JO, Spevak W, Bednarski MD (1993) Science 261:585; Spevak W, Nagy JO, Charych DH, Schaefer ME, Gilbert JH, Bednarski MD (1993) J Am Chem Soc 115:1146; Nagy JO, Wang P, Gilbert JH, Schaefer ME, Hill TG, Callstrom MR, Bednarski MD (1992) J Med Chem 35:4501; Sparks MA, Williams KW, Whitesides GM (1993) J Med Chem 36:778; Mammen M, Dahmann G, Whitesides GM (1995) J Med Chem 38:4179
53. Cao S, Roy R (1996) Tetrahedron Lett 37:3421
54. Nifant'ev NE, Tsvetkov YuE, Shashkov AS, Tuzikov AB, Maslennikov IV, Popova IS, Bovin NV (1994) Russ J Bioorg Chem 20:311
55. Nifant'ev NE, Shashkov AS, Tsvetkov YuE, Tuzikov AB, Abramenko IV, Gluzman DF, Bovin NV (1994) ACS Symposium Ser 560:267
56. Roy R, Park WKC, Srivastava OP, Foxall C (1996) Bioorg Med Chem Lett 6:1399
57. Zemlyanukhina TV, Nifant'ev NE, Shashkov AS, Tsvetkov YE, Bovin NV (1995) Carbohydr Lett 1:277
58. Spevak W, Foxall C, Charych DH, Dasgupta F, Nagy JO (1996) J Med Chem 39:1018
59. Horejší V, Smolek P, Kocourek J (1978) Biochim Biophys Acta 538:293
60. Matsuoka K, Nishimura S-I (1995) Macromolecules 28:2961
61. Kallin E, Lönn H, Norberg T, Elofsson M (1989) J Carbohydr Chem 8:597
62. Vetter D, Tate EM, Gallop MA (1995) Bioconjugate Chem 6:319
63. Roy R, Andersson FO, Letellier M (1992) Tetrahedron Lett 33:6053
64. Sugawara T, Susaki H, Nogusa H, Gonsho A, Iwasawa H, Irie K, Ito Y, Shibukawa M (1993) Carbohydr Res 238:163

65. Roy R, Pon RA, Tropper FD, Andersson FO (1993) J Chem Soc Chem Commun 264
66. Park WKC (1995) PhD thesis, University of Ottawa, Ottawa, Canada; Roy R, Gan Z, (unpublished data)
67. Roy R, Tropper FD, Romanowska A (1992) J Chem Soc Chem Commun 1611
68. Tropper FD, Romanowska A, Roy R (1994) Methods Enzymol 242:257
69. Tichá M, Kocourek J (1991) Carbohydr Res 213:339
70. Lee YC (1989) Binding modes of mammalian hepatic Gal/GalNAc receptors. In: Bock G, Harnette S. (eds) Carbohydrate recognition in cellular function. Ciba Found Symp. 145. Wiley, Chichester, p 80
71. Kichler A, Schuber F (1995) Glycoconjugate J 12:275
72. Hangeland JJ, Levis JT, Lee YC, Ts'o PO (1995) Bioconjugate Chem 6:695
73. Toyokuni T, Hakomori S-I (1994) Methods Enzymol 247:325
74. Park WKC, Aravind S, Romanowska A, Renaud J, Roy R (1994) Methods Enzymol 242:294
75. Aravind S, Park WKC, Brochu S, Roy R (1994) Tetrahedron Lett 35:7739
76. Roy R, Tropper FD, Romanowska A (1992) Bioconjugate Chem 3:256
77. Roy R, Zanini D, Meunier SJ, Romanowska A (1993) J Chem Soc Chem Commun 1869; Roy R, Zanini D, Meunier SJ, Romanowska A (1994) ACS Symposium Series 560:104
78. Tomalia DA, Durst HD (1994) Topics Curr Chem 165:193
79. Aoi K, Itoh K, Okada M (1995) Macromolecules 28:5391
80. Roy R, Pagé D, (unpublished data)
81. Lindhorst TK, Kieburg C (1996) Angew Chem Int Ed Engl 35:1953
82. Toyokuni T, Singhal AK (1995) Chem Soc Rev 231
83. Zanini D, Park WKC, Roy R (1995) Tetrahedron Lett 36:7383
84. Zanini D, Roy R (1996) Chemo-enzymatic synthesis of multivalent *N*-acetyllactosamine compounds as dendritic sialyl LewisX precursors. Proceedings of the XVIIIth International Carbohydrate Symposium, Milan, Italy, Juli 21–26, p 236
85. Pagé D, Zanini, D, Roy R (1996) Bioorg Med Chem 4:1949
86. Roy R, Baek MG, Zanini D, (unpublished data)
87. Roy R, Park WKC, Zanini D, Foxall C, Srivastava OP (unpublished data)
88. Roy R, Park WKC, WU Q, Wang S-N (1995) Tetrahedron Lett 36:4377
89. Zanini D, Park WKC, Meunier SJ, Wu Q, Aravind S, Kratzer B, Roy (1995) PMSE 73:82; Park WKC, Kratzer B, Zanini D, Wu Q, Meunier SJ, Roy R (1995) Glycoconjugate J 12:456
90. Lee RT, Lee YC (1987) Glycoconjugate J 4:317
91. Lee RT, Lin P, Lee YC (1984) Biochemistry 23:4255
92. Seppo A, Penttilä L, Niemellä R, Maaheimo H, Renkonen O, Keana A (1995) Biochemistry 34:4655
93. Christensen MK, Meldal M, Bock K, Cordes H, Mouritsen S, Elsner H (1994) J Chem Soc Perkin Trans 1 1299
94. Ponpipom MM, Bugianesi RL, Robbins JC (1982) Carbohydr Res 107:142
95. Ponpipom MM, Bugianesi RL, Robbins JC, Doebber TW, Shen TY (1981) 24:1388
96. Pagé D, Roy R (1996) Bioorg Med Chem Lett 6:1765
97. Pagé D, Aravind S, Roy R (1996) J Chem Soc Chem Commun 1913
98. Maaheimo H, Renkonen R, Turunen JP, Penttilä L, Renkonen O (1995) Eur J Biochem 234:616
99. Seppo A, Turunen JP, Penttilä L, Keane A, Renkonen O, Renkonen R (1996) Glycobiology 6:65
100. Von Andrian UH, Hasslen SR, Nelson RD, Erlandsen SL, Butcher EC (1995) Cell 82: 989
101. DeFrees SA, Kosch W, Way W, Paulson JC, Sabesan S, Halcomb RL, Huang D-H, Ichikawa Y, Wong C-H (1995) J Am Chem Soc 117:66
102. Zanini D, Roy R, Park WKC, Foxall C, Srivastava OP (1996) 3′-Sulfo-LewisX-Glc dendrimers as potent L- and E-selectin inhibitors. Proceedings of the XVIIIth International Carbohydrate Symposium, Milan, Italy, July 21–26, p 543
103. Kretzschmar G, Sprengard U, Kunz H, Bartnik E, Schmidt W, Toepfer A, Hörsch B, Krause M, Seiffge D (1995) Tetrahedron 51:13015

104. Sprengard U, Schudok M, Schmidt W, Kretzschmar G, Kunz H (1996) Angew Chem Int Ed Engl 35:321
105. Glick GD, Knowles JR (1991) J Am Chem Soc 113:4701
106. Unverzagt C, Kelm S, Paulson JC (1994) Carbohydr Res 251:285
107. De Robertis L, Lancelon-Pin C, Driguez H, Attioui F, Bonaly R, Marsura A (1994) Bioorg Med Chem Lett 4:1127
108. Marra A, Schermann M-C, Dondoni A, Casnati A, Minari P, Ungaro R (1994) Angew Chem Int Ed Engl 33:2479
109. Meunier SJ, Roy R (1996) Tetrahedron Lett 37:5469
110. Saha UK, Roy R (1995) Tetrahedron Lett 36:3635
111. Saha UK, Roy R (1995) J Chem Soc Chem Commun 2571
112. Roy R, Saha UK (1996) J Chem Soc Chem Commun 210
113. Kim JM, Roy R (1996) Carbohydr Lett 1:465
114. Saha UK, Kim JM, Roy R (1995) Syntheses of glycoforms of biological interests. Proceedings of the 8th European carbohydrate symposium, Seville, Spain, July 2–7, C IL-5; Roy R (1996) Novel classes of clusters for studies related to carbohydrate recognition processes. 2nd Euroconference on carbohydrate mimics, Lago di Garda, Italy, July 16–19, IL-7
115. Tropper FD, Andersson FO, Braun S, Roy R (1992) Synthesis 618
116. Cao S, Tropper FD, Roy R (1995) Tetrahedron 51:667

Amphiphilic Carbohydrates as a Tool for Molecular Recognition in Organized Systems

Paul Boullanger

Université Claude-Bernard-Lyon 1, C.P.E. Lyon, Laboratoire Chimie Organique 2, (U.M.R. C.N.R.S. 5622), 43 Bd 11 Novembre 1918, 69622 Villeurbanne cedex, France

Organized assemblies of carbohydrates are spontaneously formed when glycolipids or synthetically hydrophobized carbohydrates are dispersed in water (possibly in the presence of other lipids). Carbohydrate recognition of such organized systems (mono or bilayers) is somewhat different from recognition in isotropic media. The main differences arise from the kinetics and thermodynamics points of view (hindered approach and entropy changes at the surface, respectively). Furthermore, the conformation and the motion of the carbohydrate embedded at interfaces are strongly affected by the natures of its lipid anchor and that of the surrounding lipid components. The self-organization of amphiphilic carbohydrates can be rationalized by considering the geometry of the molecule; depending on the surface of the polar head, length and volume of the apolar tail, micelles, liquid crystals, monolayers, or vesicles can be formed. The recognition, by specific receptors, of carbohydrates assembled as monolayers or Langmuir-Blodgett films, can display quantitative information such as the thermodynamic parameters of binding or cluster effects, whereas the recognition at the surface of vesicles mainly affords a qualitative knowledge of the binding.

This paper is not a review covering the entire field of carbohydrate-recognition in any organized system. Many excellent papers have already been devoted to supramolecular systems such as cyclodextrins, podands, coronands or cryptants able to entrap carbohydrate molecules [1]. This article only deals with the molecular recognition of mono and oligosaccharides in organized self-assemblies of amphiphilic carbohydrates (possibly blended with other lipids) in aqueous medium; i.e. in assemblies mimicking the cell membrane.

Table of Contents

Topics in Current Chemistry, Vol. 187

List of Abbreviations

Cer	Ceramide
CHAPSO	3-[(CHolAmidopropyl)dimethylammonio]-2-hydroxyl-1-PropaneSulfOnate
cmc	Critical Micellar Concentration
Con A	Concavalin A
DMPC	L-α-DiMyristoylPhosphatidylCholine
EFO	Evanescent Fiber Optic
FITC	Fluorescein IsoThioCyanate
GbO_3	Galα(1–4)Galα(1–4)Glcβ-Cer
GbO_4	GalNAcβ(1–3)Galα(1–4)Galα(1–4)Glcβ-Cer
GbO_5	GalNAcα(1–3)GalNAcβ(1–3)Galα(1–4)Galα(1–4)Glcβ-Cer
GPI	GlycosylPhosphatidyl Inositol
H-	Hydrophobized
HLB	Hydrophilic Lipophilic Balance
IgG	Immunoglobulin G
IgM	Immunoglobulin M
Lac	Lactose
LB	Langmuir-Blodgett
LUV	Large Unilamellar Vesicle
Mal	Maltose
MLV	MultiLamellar Vesicle
PC	Phosphatidyl Choline
PE	Phosphatidyl Ethanolamine
QCM	Quartz Crystal Microbalance
RET	Resonance Energy Transfer
SEM	Scanning Electron Microscopy
SFA	Surface Force Apparatus

Sug-A	Aldonic acid (any carbohydrate in oxidized form)
Sug-ol	Sugar alditol (any carbohydrate in reduced form)
SUV	Small Unilamellar Vesicle
TBDMS	*tert*-ButylDiMethylSilyl
TMSOTf	TriMethylSilyl TriFluoromethanesulfonate
WGA	Wheat Germ Agglutinin
2D, 3D	Two-Dimensional, Three-Dimensional

1
Introduction

Important biological phenomena depend on carbohydrate/protein interactions. For example, blood group substances are written in terms of carbohydrates and blood transfusion is governed by the immune recognition of oligosaccharides as self or non-self. Bacterial antigens of Gram(+) and Gram(-) bacteria are oligosaccharides whose recognition by lymphocytes and immunoglobulins constitute the first activity of the invaded organism's defense-system. Selectins – proteins residing on the cell membranes – mediate the initial recognition of injured sites by binding specific carbohydrates. Many other examples could be mentioned since the recognition of carbohydrates by proteins is involved in most of the major biological events (inflammatory response, blood coagulation, cell differentiation and maturation, malignant transformations, information transfer between cells, cellular adhesion, etc...).

The "recognition" is the binding of a carbohydrate by a specific receptor, which is usually part of a protein (enzyme, immunoglobulin, lectin). The process is due to the sum of weak interactions (van der Waals-London, hydrogen bonds) that may results in high association constants.

The receptor-substrate binding takes place in water, which complicates the study of the thermodynamics of the reaction. The entropy change does not favor the binding, which can be explained by the loss of degree of freedom of the protein [2], the loss of conformational entropy of the ligand [3] and the structural changes in water [4]. Since the enthalpy change is in favor of the binding, the neat ΔG of the reaction is most often low. This enthalpy-entropy compensation has recently been analyzed by JP Carver [5].

Moreover, the carbohydrate portion effectively recognized is rather small (one to six monosaccharide units). Nevertheless the carbohydrate-ligand is most often "presented" to the receptor-protein as a high molecular weight conjugate (glycolipid or glycoprotein that can itself be embedded at the surface of a cell). Therefore, the binding takes place at interfaces where the degrees of freedom of both the receptor and the ligand are reduced. The carbohydrate and the protein cannot move freely in a three-dimensional homogenous and isotropic medium but they have restricted motions in macromolecular or supramolecular lattices (Fig. 1).

At the surface of cells, carbohydrates are encountered as organized systems such as glycoproteins (transmembrane protein **A** or surface proteins **C**) or glycolipids **B** anchored into the bilayer via their lipid moiety. Surface glycoproteins (**C**), are most often themselves anchored to the membrane, by a glycosylphosphatidylinositol (GPI) glycolipid.

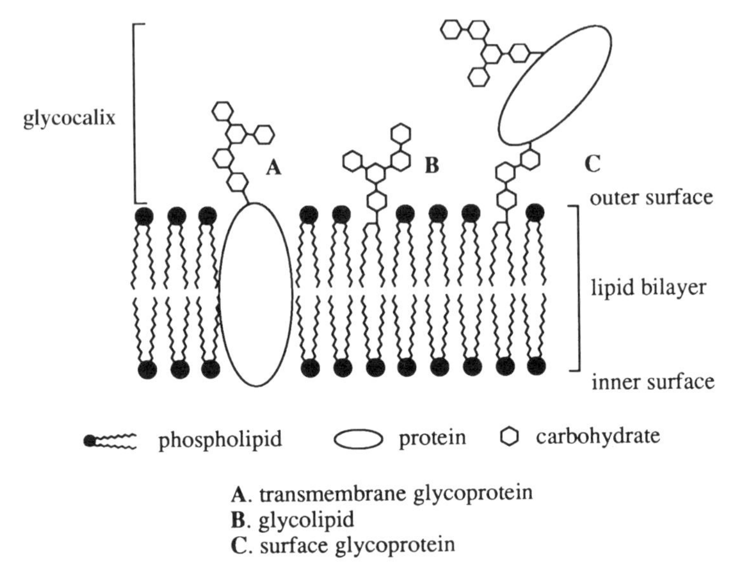

Fig. 1. Schematic representation of a cell surface

The thermodynamic parameters of binding are therefore strongly influenced by interfacial phenomena. Conformational constraints could be favorable to the binding in such organized systems. Indeed, an unfavored conformation of the carbohydrate at the surface of a carrier, or the structure of water around the system carrier-ligand-protein, which is less favorable than around the system ligand-protein, could afford a lower entropy variation at constant enthalpy (and therefore a higher ΔG of binding). The name "site-directed" presentation has been suggested for the enhancement of the free energy of binding in organized systems [5].

Moreover, the supramolecular presentation of the ligand could affect the kinetics of the reaction thus allowing, for example, selectins to immobilize leucocytes flowing in the blood-stream [6].

The high density of ligands in supramolecular or polymeric assemblies could also be favorable to the binding. This phenomenon was well examplified for immunoglobulin-antigen recognition [7] and carbohydrate-antigenicity enhancement [8]. More recently, cooperative effects were also suspected in the recognition of sialyl-Lewis^x antigen by E-selectin [9].

The main differences between recognition in isotropic media and organized systems can therefore be attributed to kinetic and thermodynamic parameters arising from: (1) entropy-effects due to changes in the structure of water and conformations of both the ligand and the protein in the course of binding, (2) cooperative effects from which a high ΔG of binding could result from the sum of low individual ΔGs.

The name "glycolipid" usually refers to the natural compounds, whereas more generally, the terms "amphiphilic carbohydrate", "neoglycolipid" or "hydrophobized carbohydrate" (*H*-carbohydrate) are used for synthetic unnatural structures. Glycolipids and neoglycolipids are able to self-organize in water, their supramolecular assemblies will be discussed in Sect. 2. Sections 3 and 4 will deal with their syntheses and their recognition by carbohydrate-receptors, respectively.

2 Supramolecular Assemblies of Amphiphilic Carbohydrates

Natural or synthetic glycolipids can be used as models that can form supramolecular assemblies ressembling the biological organizations. Depending on their structure, they are able to self-assemble in water as: micelles, vesicles or liposomes (3D systems). They can also organize as monolayers at air/water interfaces (2D systems). Mixtures of glycolipids and phospholipids (the natural constituents of the cell membranes) organize most often in the form of liposomes. The construction of well defined supramolecular models requires the knowledge of the main types of self-organization of glycolipids, which will be discussed below.

2.1 Supramolecular Assemblies, General Statements

The mesogenic structures of glycolipids are due to the occurrence, on the same molecule, of a hydrophilic and a hydrophobic moiety often referred to as "head" and "tail" respectively. As a result, glycolipids are able to self-organize into a large variety of mesophases also called liquid crystals (Fig. 2) [10]. Supramolecular assemblies of mesogenic compounds can be caused by a rise in temperature (thermotropic liquid crystals) or by the addition of water (lyotropic liquid crystals); they result from different responses of the carbohydrate and the alkyl chain to temperature or solvent (water), respectively.

When they are heated, mesogenic compounds do not melt directly from the highly ordered crystalline state to an isotropic liquid. They form instead, intermediate phases in which the molecules are orientated in a parallel direction and referred to as smectic (centers of the molecules organized in layers) or nematic (centers of the molecules distributed at random). Smectic and nematic mesophases are in turn divided into a variety of subgroups of thermotropic liquid crystals which will not be dealt with in detail in the present article.

When dispersed in water, amphiphilic molecules self-organize in assemblies, the structures of which depend on the nature and length of the lipid moiety and also on the concentrations. At low concentrations, the molecules spread at the air-water interface with the hydrophilic moiety located in the aqueous phase and the lipid chains in the air. This organization ensures the best thermodynamic stability. When the surface is saturated, the molecules form a monolayer. At higher concentration, aggregates are formed in water; their natures strongly depend on geometric factors. Hydrophobic chains assemble together under the effect of forces of attraction (van der Waals-London) whereas hydrophilic carbohydrate moieties are submitted to forces of repulsion (electrostatic and steric interactions). The competition between forces of attraction and repulsion results in an organization in which the hydrophobic parts are kept away from the

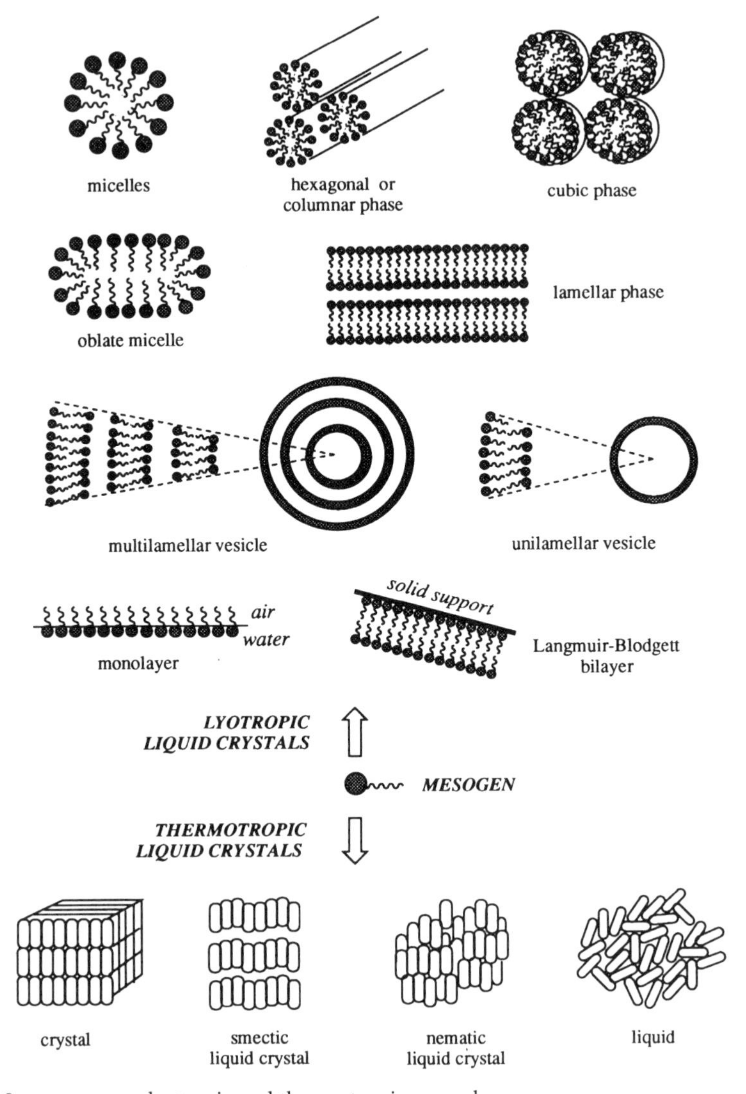

Fig. 2. Some common lyotropic and thermotropic mesophases

bulk of water whereas the hydrophilic parts are arranged at a more or less curved interface.

With short chain derivatives, the forces of repulsion are higher than the ones of attraction; the curvature is high and spherical micelles are formed at a concentration called the critical micellar concentration (cmc). This concentration can be detected by a change in the physico-chemical properties of the solution (e.g. surface tension, Fig. 3a). Above a characteristic temperature (referred as Krafft temperature), the tensio-active molecules are infinitely soluble in the form of micelles (Fig. 3b).

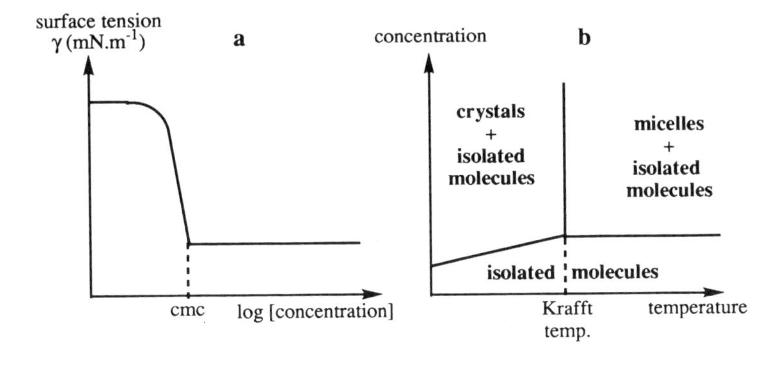

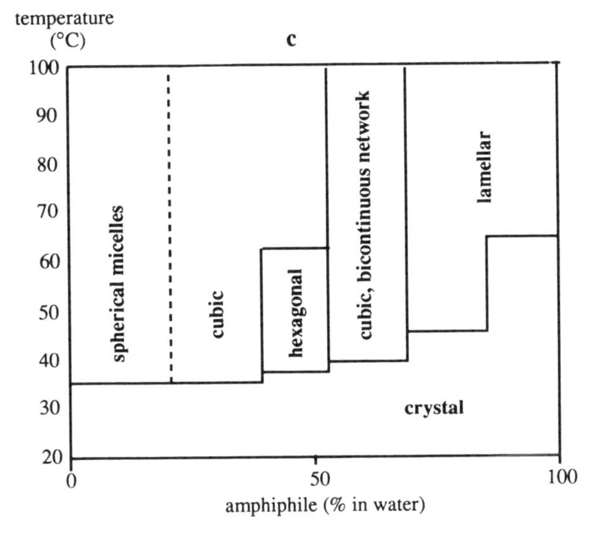

Fig. 3. Schematic diagrams: **a** surface tension versus concentration of a surfactant, **b** phase diagram of a surfactant near the Krafft temperature, **c** phase diagram of 3-*O*-dodecyl-D-glucitol [11]

With longer chain derivatives, the forces of attraction increase, the curvature decreases and micelles become oblate or form hexagonal (or columnar) phases. When the zero-curvature is reached, the flat oblate micelles can fold and close spontaneously, thus entrapping a volume of water and form vesicles that may contain one or several bilayers of the amphiphile.

At higher concentrations, micelles assemble in turn, to form hexagonal or cubic phases while longer chains or multi-chain compounds afford lamellar phases in which the amphiphilic derivative is arranged in parallel bilayers, separated by water. The succession of mesophases depending on temperature and concentration of the amphiphile can be visualized in a phase diagram (Fig. 3c).

The various supramolecular arrangements are related to the geometry of the molecules through the "packing criterion" [12]:

$$a_0 l_c / v$$

in which

- a_0 is the interfacial area per molecule (depends on the size and hydration of the polar head)
- l_c is the maximum length of the hydrophobic chain in extended conformation
- v is the volume of the alkyl chain (depends on the fluidity of the lipid tail and therefore on the temperature).

Supramolecular assemblies can be predicted from the values of the "packing criterion":

- $a_0 l_c/v \geq 3$ spherical micelles
- $3 > a_0 l_c/v > 2$ cylindrical micelles
- $2 \geq a_0 l_c/v \geq 1$ lamellar phases or vesicles
- $a_0 l_c/v < 1$ inverse micelles (the hydrophilic head are assembled in the core of the micelle).

2.2 Self-Assemblies of Glycolipids

2.2.1 *Micelles*

Single short chain glycolipids usually form micelles since $a_0 l_c/v > 1$. Octyl *β*-D-glucopyranoside **1a** (sug=Glc*p*, R=C_8H_{17}), methyl 6-*O*-heptylcarbamoyl-*β*-D-glucopyranoside (HECAMEG) **2b** [13] and alkanoyl-*N*-methylglucamide (e.g. MEGA-N, **3c**, sug=Glc-ol) [14] are the most widely used carbohydrate surfactants of biological interest. Alkyl glycosides **1a** and alkyl thioglycosides **1b** containing C_8 to C_{12} *n*-alkyl chains, have long been studied as nonionic detergents. Continuous interest has been shown in their micellar properties [15], aggregate sizes [16], distributions [17] and structures [18], in the thermodynamics [19] and kinetics [20] of micelle formation and in molecular dynamics simulations in surfactant/water/oil systems [21]. The 2-amino-2-deoxy derivatives **4a** have also been compared with their 2-hydroxy counterparts [22–24]. The cmc of alkyl glycosides are depending on several parameters: (1) they roughly are 5- to 10-fold decreasing for an increase of two CH_2 groups in the alkyl chain, (2) for the same C_8-chain length, they are following the order *β*-D-Glc*p* > *α*-D-Man*p* > α-D-Glc*p* > *β*-D-Gal*p*, (3) they are only weakly affected by the addition of one or two sugar residues or by the ring size (pyranose or furanose) of the carbohydrate.

A large series of amphiphilic carbohydrates was shown to form micelles in aqueous solution; only the more common of them will be quoted below. Alkyl glycosylamines (**1c**, sug=Glc*p*, Gal*p*, Mal and **4b**, R=C_nH_{2n+1}) [25–27], alkylaminolactitols **3c**, (sug=Glc-ol, Mal-ol) [28], or aldonamides **5** (sug=Lac-A, R′=H, R=C_9H_{18}-CH=CH_2) [29] are compounds in which the carbohydrate and lipid moieties are linked at C-1; in sugar thio ethers **2a** [30] or 6-amino-6-deoxy derivatives **2c**, the hydrophilic-hydrophobic linkage is located at C-6 [31]. More-

sug–O … X-R (R= alkyl or acyl)

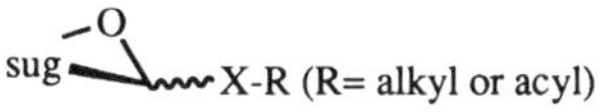

1a X= O
1b X= S
1c X= NR' (R'= H, alkyl, acyl)

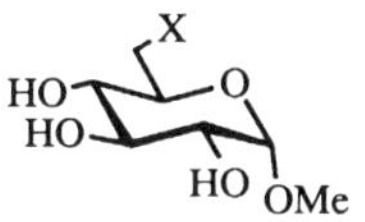

2a X= SC_nH_{2n+1}
2b X= $OCONHC_7H_{15}$ (HECAMEG)
2c X= $NHCOC_nH_{2n+1}$

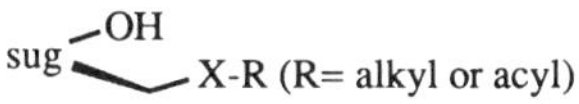

sug–OH … X-R (R= alkyl or acyl)

3a X= O
3b X= S
3c X= NR' (R'= H, alkyl, acyl)
e.g. MEGA-10: X= $N(CH_3)$, R= COC_9H_{19}
3d X= CH_2

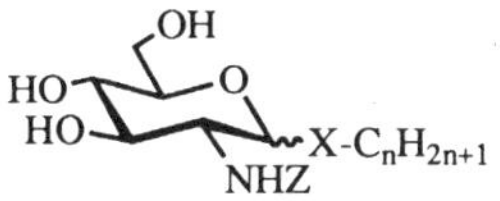

4a X= O; Z= H, Ac, or H-HCl
4b X= NH; Z= Ac
4c X= NHCO; Z= Ac
4d X= $O(CH_2CH_2O)_n$; Z= Ac

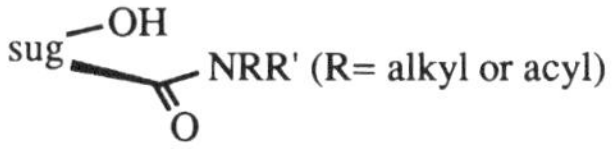

sug–OH … NRR' (R= alkyl or acyl)

5 R'= H, alkyl, acyl

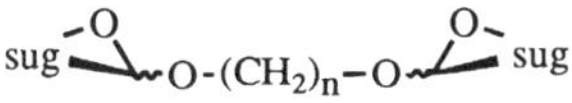

sug … $O-(CH_2)_n-O$ … sug

6a

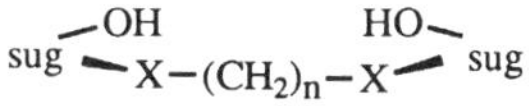

sug … $X-(CH_2)_n-X$ … sug

6b

C_mH_{2m+1}, C_mH_{2m+1} … $NHCO(CHOH)_nCH_2OH$, $NHCO(CHOH)_nCH_2OH$

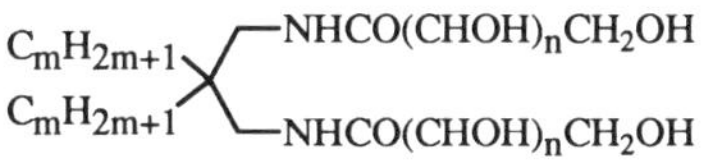

7

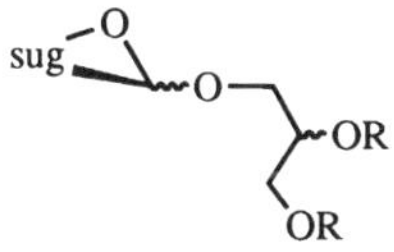

8a R=alkyl or acyl

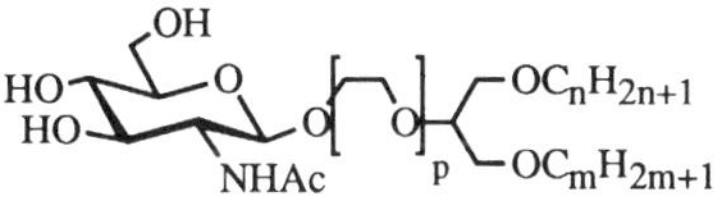

8b n=m=11; m=11, n=16; m=n=16; p=0-4

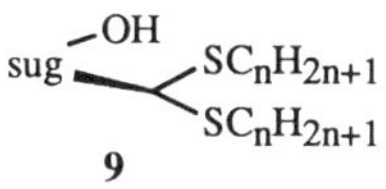

sug–OH … SC_nH_{2n+1}, SC_nH_{2n+1}

9

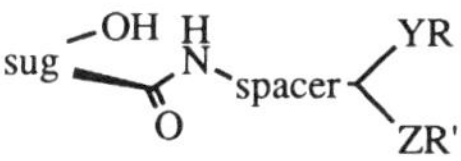

sug–OH … N-spacer … YR, ZR'

10 Y, Z= NH, NHCO, CH_2;
R,R'= alkyl, perfluoroalkyl
spacer= $(gly)_n$, $(CH_2)_n$

Compounds 1–10. 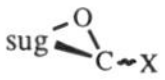 sug–O–C~X and 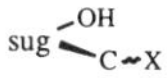 sug–OH–C~X represent O5 and C1 of cyclic and acyclic carbohydrate, respectively, in which the carbohydrate moiety may be any mono or oligosaccharide, α or β anomer (cyclic forms); C_nH_{2n+1} is a saturated aliphatic chain, while alkyl may be either a saturated or an unsaturated aliphatic chain

over, sugar derived perfluoroalkyl surfactants [32–34] were shown to be highly effective as emulsifiers of fluorocarbons able to carry oxygen in vivo [35, 36].

Bolaamphiphiles **6a** (sug=$GlcNH_2$, GlcNAc [37, 38], sug=Glc*f*, Gal*p*) or **6b** (sug=Glc-A, Lac-A, X=NH [39, 40]; DL-Xyl-ol [41]) or gemini surfactants **7** [42] are micelle-forming derivatives bearing two hydrophilic heads.

2.2.2
Lyotropic and Thermotropic Liquid Crystals

The structure and function of cell membranes have long been associated with lyotropic liquid crystalline phases. Since most of the glycolipids are amphitropic (both thermotropic and lyotropic) their was an increase of interest in the comparison of the structures of both types of mesophases formed by the same compound.

A paper by Jeffrey and Wingert [43] covers all the aspects of carbohydrate liquid crystals. This section will only give a brief overview of the topics, following the same classification as the aforementioned article. Since then, eighty different amphiphilic non-ionic carbohydrates, gleaned from the literature have been analyzed; excellent correlations were found between their lyotropic liquid crystal properties and their structures [44]. The roles of hydrogen bonding interactions in the formation of liquid crystals, including amphiphilic carbohydrates, has also been recently reported [45].

Cyclic carbohydrates with one alkyl chain generally form smectic S_A mesophases in which the thickness of the bilayer is 1.5 to 1.7 times the calculated length of the fully extended alkyl chain, which implies an overlapping of the latter. Despite the large number of papers devoted to alkyl glycoside mesophases, questions remain unanswered concerning thermotropic [46] and lyotropic [47] liquid crystals as well. For example, an unexpected formation of vesicular aggregates was observed in aqueous solutions of *n*-octyl 1-thio-α-D-talopyranoside (**1b**, sug=Tal, R=C_8H_{17}) [48]. Some very complete study on octyl α- and β-furanosides of D-glucose, D-galactose and D-mannose were recently published [49–51]. All of them exhibit thermotropic smectic S_A phases (except galactofuranosides) but only octyl β-D-glucofuranoside exhibits a full range of lyotropic phases. Alkyl and thioalkyl glucopyranosides **1a,b**, with C_{12} chains or longer, spontaneously display myelin figures (tubular structures formed at the interface with bulk water and consisting of 300–5000 bilayers constituting the lamellar phase) [52]. A complete set of ordered lyotropic phases of dodecyl β-maltoside (**1a**, sug=Mal, R=$C_{12}H_{25}$) and *N*-alkyllactosylamines (**1c**, sug=Lac, XR=NHC_nH_{2n+1}) has been determined by X-ray diffraction studies [53] and compared with open structures **3c**. Carbohydrate mesogens bearing the hydrophobic residue at positions other than the anomeric carbon have also been found to display lyotropic liquid crystals properties [51, 54–57].

Cyclic carbohydrates with two alkyl chains (e.g. 1,2-dialkyl (or 1,2-diacyl) glycerol **8a** (sug=Glc*p*, Gal*p*) present structural similarities with glycerophospholipids. They form complex mesophases such as bicontinuous cubic phases, inverted hexagonal phases or myelin figures [58–61]. Other dialkyl derivatives of carbohydrates bearing both alkyl chains on the same carbon [62] or at different positions have also been reported [62–64].

Acyclic carbohydrates with one alkyl chain are either derived from cyclitols or aldonic acids. Acyclic cyclitol derivatives **3** (X=O, S, NH, $N(CH_3)$, NHCO, $N(CH_3)CO$; R=C_6-C_{16}; sug=Glc-ol, Man-ol) form thermotropic liquid crystals [65]. 3-*O*-Alkyl derivatives of D-glucitol and D-mannitol also display a wide variety of mesophases (Fig. 3c, [11]). Predictive rules for the occurrence of various phases in relation with geometric parameters were established for **3c** [66]. Mesophases of related structures such as **3b** (sug=Gal-ol) [67] or **3d** (sug=L-Rib-ol, D-Lyx-ol) [68, 69] were also depicted recently.

Aldonamides **5** display smectic S_A thermotropic as well as lyotropic mesophases [70, 71]; they form helical tubes with diameters between 40 and 500 Å and lengths of several µm and have gelation properties [72–76].

Acyclic carbohydrates with two alkyl chains have been studied in detail in the case of dialkyldithioacetals **9**. It is one of the more complete series whose results have been cross-checked and compiled [43]. Their liquid crystalline properties were compared with those of dodecyl α- and β-D-glucofuranosides [50]; depending on their structures, they form either hexagonal disordered columnar phases or a new rectangular columnar phase [77].

Alkyl chain(s) with carbohydrates at both termini (bolaamphiphiles) have also been reported since they are potential building blocks for the construction of membrane mimetics with a single monolayer [78]. Bisgluconamide and lactobionamides **6b** (sug=Glc-A or Lac-A, X=NH) were studied for their crystalline properties and their arrangements in water [39, 40]. Alkyl-α,ω-dimannitol **6b** (sug=Man-ol, n=16–22) [66] or bolaamphiphiles with identical or different carbohydrates at both ends of the alkyl chain **6a** (sug=D-Glc*f*, D-Gal*p*, DL-Xyl-ol) were found to form micelles and lyotropic liquid crystals as well [41].

2.2.3 *Vesicles*

Many double-tailed amphiphilic carbohydrates tend to form vesicles when they are dispersed in an aqueous medium. This is due to a favorable geometry, also found in phospholipids, in which $a_0 l_c / v \approx 1$. Among structures which agree with this finding, we have prepared several 1,3-dialkylglycerol derivatives bearing *N*-acetyl-D-glucosamine at C-2 (**8b**). Their dispersion in water affords vesicles whose formation was confirmed by three different methods: 6-carboxyfluorescein encapsulation, gel filtration chromatography and quasi-elastic light scattering. The mean diameter of these assemblies is compatible with small unilamellar vesicles (SUV) [79]. Rhamnolipid A and B, both microbial and membrane constituents, bearing alkyl chains at C-2 and C-1 also form vesicles which can change their morphology to lamellae or micelles within a narrow pH range [80]. Lactobiouronamide and maltobiouronamide derivatives **10** (sug=Lac, Mal) linked to the double-tailed lipophilic moiety by a peptide-spacer form very stable vesicles when they are dispersed in an aqueous medium; their stability and morphology (unilamellar or multilamellar) depend on the nature and length of the spacer [81] as well as the possible occurrence of fluorine in the lipid moiety [81, 82]. The formation of vesicles with other amphiphilic carbohydrates is sometimes more difficult to understand with simple geometric considerations: e.g. dialkyl polyol derivatives **7** (C_{18}) [42], bolaamphiphiles **6b** (X=CONH) [39], *N*-octadecylchitosan [83], or *n*-octyl 1-thio-α-D-talopyranoside **1b** (sug=Tal, R=C_8H_{17}) [48].

3
Syntheses of Amphiphilic Carbohydrates

The syntheses of natural glycolipids and neoglycolipids have been the subject of intensive research over the last ten years. Most of the papers were stimulated by the biological and physico-chemical properties of amphiphilic carbohydrates.

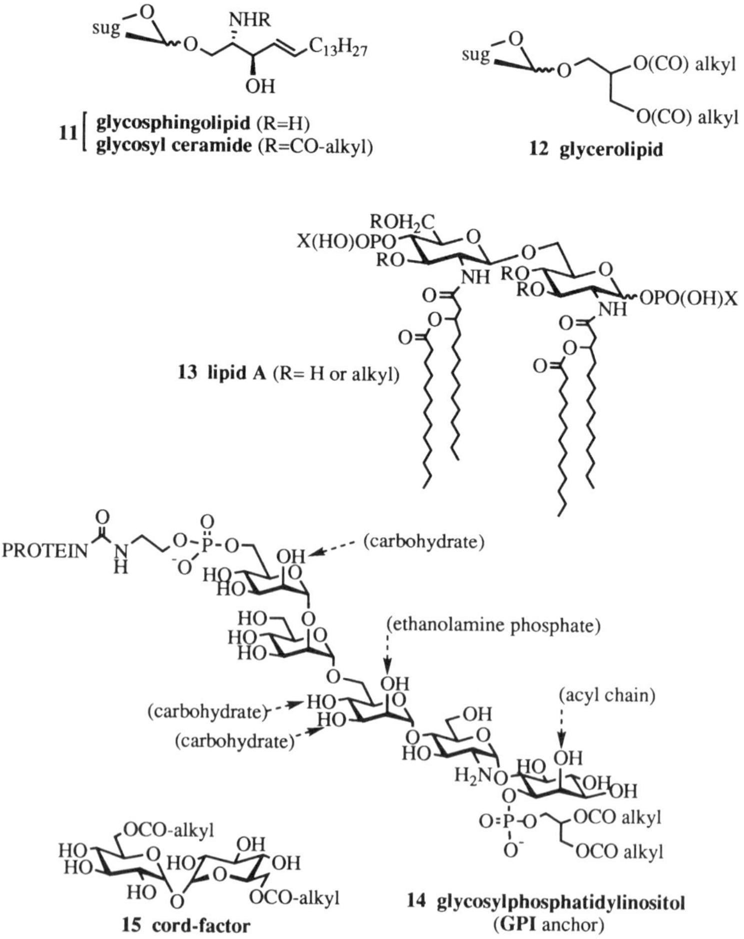

Fig. 4. Some classes of natural glycolipids (in parentheses are occasional substitutions)

Glycolipids are widely distributed in animals, plants, algae and bacteria where they occur as minor components of the lipid mixture. Their roles are not completely elucidated but it is well established that they are involved in the biosynthesis of glycoproteins and they also serve as biological markers and as ligands for toxins, lectins, antibodies, bacteria or viruses [84, 85]. A wide variety of structures are encountered in glycolipids, among them two important classes can be distinguished: (1) glycosphingolipids **11** (Fig. 4) are animal glycolipids, glycosylated at the primary OH of sphingosine. They are divided into subclasses such as glycosyl ceramides (or cerebrosides) acylated on the amino group by a fatty acid, or gangliosides containing one or more sialic acid residues in the oligosaccharide moiety. Glycosyl ceramides are mainly distributed in the central nervous system whereas gangliosides are found in many tissues. (2) Glycosyl glycerides **12** found in animals, plants, algae or bacteria are glycerol derivatives usually carrying two fatty acid esters (sometimes ethers [86]) and glycosylated at a primary position of the glycerol moiety.

Bacteria also contain a very rich variety of glycolipids with unusual structures. Lipid A **13** is the site of attachement of the O-specific chain of Gram (–) bacteria, which constitutes the antigenic lipopolysaccharide [87]. Other members of this family can be quoted, for example glycosyl glycerophospholipids in which the carbohydrate and glycerol moieties are linked by a phosphodiester bond (e.g. GPI anchor **14**) [88] or carbohydrate esters (e.g. cord-factor of mycobacteria **15**).

The key-step of the synthesis of glycolipids, and more generally of amphiphilic carbohydrates, is the covalent coupling of a hydrophilic carbohydrate with a lipophilic compound. A hydrophilic or hydrophobic spacer may be inserted between them in order to control the hydrophilic-lipophilic balance (HLB). This modulation allows to obtain variously organized systems with the same polar head and apolar tail.

3.1
Lipid Moiety

The lipids used in the syntheses of neoglycolipids are often commercially available; most of the long-chain carboxylic acids, *n*-alcanols, *n*-alcanethiols or *n*-alkylamines are readily available as well as α,ω-dicarboxylic acids, diols or diamines useful for the syntheses of bola-amphiphiles.

More complex structures, often related to natural products are prepared by organic synthesis. Among them can be mentioned (*R*)-3-hydroxytetradecanoic acid (the double-tail hydrophobic moiety of lipid A), sphingosine derivatives related to the ceramides or 1,2- and 1,3-dialkyl(acyl)glycerols related to glycoglycerolipids, glycerophospholipids, and GPI anchors of membrane proteins. The preparations of the above derivatives were reported several years ago but some improvements have been published more recently.

Acyl chains of lipid A are often branched double-chain derivatives. Thus 3-O-acyl-(*R*)-3-hydroxytetradecanoic acid methyl ester **16** was obtained in 85% e.e. by the enantioface-differentiating hydrogenation of methyl 3-oxotetradecanoate over (*R,R*)-tartaric acid-NaBr-modified nickel [89]. Pure (*R*)-3-hydroxytetradecanoic acid was obtained by recrystallization of the dicyclohexylammonium salt.

Sphingosine derivatives are intermediates in the syntheses of glycosyl ceramides. The occurrence of several reactive functions on the same lipid prescribes the use of protective groups before the coupling of the carbohydrate. Thus, azidosphingosine **17** was reported to be a valuable intermediate in the synthesis of sphingosine derivatives. The latter was synthesized in a few steps by a Wittig reaction between compound **18** (easily obtained from 4,6-*O*-benzylidene-D-galactose) and hexadecanylidene triphenylphosphorane. The condensation was followed by: (1) conversion of the alcohol to a triflate, (2) nucleophilic sustitution to the azido compound **19**, 3) removal of the acetal function, (4) protection of the secondary alcohol to afford **17**, suitable as acceptor in glycosylation reactions [90].

1,2-Dialkyl(acyl)glycerols **21** were prepared many years ago but their syntheses have seen recent improvements. Most of the methods reported to date use 1,2-*O*-isopropylidene-D-glycerol **22** (easily obtained from D-mannitol) as the starting material. After protection of OH-3, the acetal is removed to afford a 1,2-diol. Coupling of both alkyl chains by ester (or ether) bonds, followed by the deprotection of OH-3, concludes the synthesis of compound **21**. This pathway

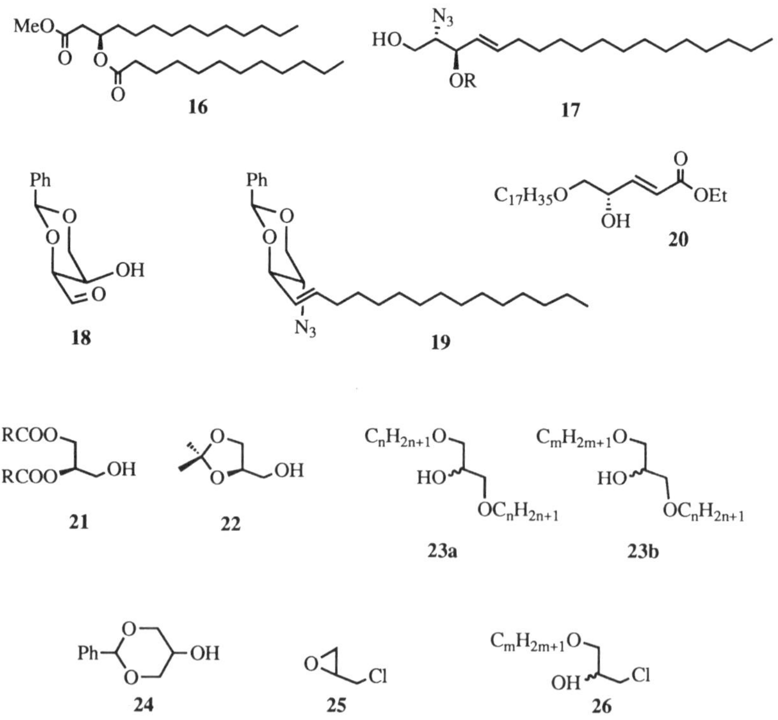

Compounds. 16–26

was recently used by Mioskowski et al. [91] and by Ogawa et al. [92] to prepare phospholipid and GPI anchor intermediates, respectively.

In 1,3-dialkyl(acyl)glycerols **23 a,b** the alkyl chains are separated from each other, which could enhance the formation of vesicles. Several methods are reported in the literature for preparing such derivatives: 1,3-*O*-benzylidene glycerol **24** or epichlorohydrin **25**, among others, can be used as starting materials. In the first instance, OH-2 has to be protected before the cleavage of the benzylidene acetal, followed by the coupling of the alkyl (or acyl) chains and deprotection of OH-2. Starting from epichlorohydrin **25** we have prepared both symetrical (**23 a**) and dissymetrical (**23 b**) derivatives. When **25** was treated with an excess of *n*-alkanol in a strongly alkaline medium, compound **23 a** was obtained in 70% yield. On the other hand, the opening of the epoxide ring of **25** by *n*-alkanols in a strongly acidic medium afforded the intermediate **26** which was able, in turn, to be reacted with another *n*-alkanol in alkaline medium, to afford compound **23 b** in good yield [93].

3.2 Covalent Coupling of the Lipid and Carbohydrate Moieties

If surfactant properties are searched for (gel forming derivatives, detergents, wetting agents, etc...), the coupling of low-cost carbohydrates and lipids (no protective group) has to be realized in the lowest possible number of steps and with a brief final purification (none if possible). The availability of large amounts of derivatives at low-cost is a prerequisite for competing with commercial surfactants. On the other hand, if biological applications or liquid crystal properties are searched for, the structure of the glycolipid may be submitted to more drastic requirements (e.g. regio and stereochemistry). Complex syntheses, involving protective groups and high-cost intermediates can then be considered. These two aspects will be covered in the following sections; they will be identified as short and multistep syntheses.

3.2.1 *Short Syntheses*

The *C-1 position* is the most usual site of anchorage of the lipid and carbohydrate moieties. Long chain alkyl glycosides **1a** cannot be prepared by a Fischer type reaction restricted to reactive, low molecular weight alcohols. Nevertheless, heptyl and octyl D-glucopyranosides **1 a** and 1-thio-D-glucopyranosides **1 b** were obtained by reaction of a free carbohydrate with the corresponding alcohol or thiol, using poly-(hydrogen fluoride)-pyridinium as both the solvent and reagent of the reaction [94]. Actually, short syntheses most often result in the preparation of open chain derivatives since the linkage at C-1, in cyclic compounds, requires the α/β stereocontrol if anomerically pure amphiphiles are desired. The only derivatives in which a cyclic carbohydrate is linked to the lipid moiety at the anomeric position with a reasonable β-stereoselectivity are glycosylamines **1 c**. The latter compounds (e.g. sug=Lac, XR=NHC_nH_{2n+1}) were prepared by condensation of *n*-alkylamines with lactose in a mixture 2-propanol/water [28]. Unfortunately these derivatives are unstable in water where they hydrolyze almost spontaneously. This difficulty can be overcome by reduction or acetylation. Reduction affords *N*-alkylaminoalditols **3c** (e.g. sug=Lac-ol, XR=NHC_nH_{2n+1}) [27,95] in which the terminal carbohydrate is in an open form. Acetylation of the amino group affords **1c** (e.g. sug=Lac, XR=$NAcC_nH_{2n+1}$) [26] in which the terminal carbohydrate is in a cyclic form. Structures **1c** (XR=$NHCOC_nH_{2n+1}$) and **4c** are less prone to hydrolysis. **1c** (sug=D-Glc*p*, D-Gal*p*, Lac, XR=$NHCOC_nH_{2n+1}$) and **4c** were prepared by reaction of a mixture of D-glycosylamine with *n*-alcanoic acids and a thiocarbamate used as the coupling agent [25]. The preparations of D-glyconamides **5** (NRR′=NHC_nH_{2n+1}) were also reported many years ago. An improvement was nevertheless published more recently in which D-glucono-γ-lactone was reacted with *n*-alkylamine in methanol to afford the expected open-chain derivatives in good yield [72].

Positions other than C1 can be achieved, taking advantage of classical carbohydrate intermediates that allows to discriminate one position among the others. Thus OH-6 of D-Gal can be discriminated via 1,2:3,4-di-*O*-isopropylidene-α-D-galactopyranose, OH-6 of most carbohydrates via their methyl glycopyranosides, OH-3 of D-Glc via 1,2:5,6-di-*O*-isopropylidene-α-D-glucofuranose, COOH of uronic acids or NH_2 of D-glycosamines via the differentiated chemical reactivities of these functions. On the other hand, acetals of long chain aldehydes can provide easy access to amphiphilic 4,6-acetals from unprotected carbohydrates.

Several surfactants were prepared in accordance with the aforementioned general principles. Only the recent examples will be discussed below. Hecameg **2b** was prepared in 90% yield by a single-step reaction between methyl α-D-glucopyranoside and *n*-heptyl isocyanate [13]. Methyl 6-*O*-*n*-alcanoyl-β-D-glucopyranosides were prepared by transesterification of methyl alcanoates at OH-6 of methyl D-glucopyranoside by means of lipases [96, 97]. D-Galacturonic acid esters, thioesters and amides were obtained by reaction of *n*-alkyl alcohols, thiols or amines respectively, on 1,2:3,4-di-*O*-isopropylidene-α-D-galactopyranuronosyl chloride [57]. 6-*S*-Alkyl derivatives of D-Gal*p*, 3-*S*-alkyl derivatives of D-Glc*f* **27** [30] as well as symetrical and dissymetrical bola-amphiphiles [41] were prepared from di-*O*-isopropylidene derivatives of α-D-Gal*p* and α-D-Glc*f*. New surfactants in which the carbohydrate and lipophilic moieties are linked at C-2 have also been reported by reaction of the activated acid function of the steroidal ursocholic acid on NH_2-free D-glucosamine [98].

4,6-Acetals of D-Glc*p*NAc **28a** (Z=NHAc) [99] and D-Glc*p* **28b** [55] were prepared by condensation of the appropriate aldehyde or aldehyde dimethyl acetal with the unprotected carbohydrate. An improvement of the classical methods of acetalation (microwave irradiation of montmorillonite) made it possible to synthesize various 5,6-*O*-*n*-alkylacetals of L-galactono-1,4-lactone [100].

Double-tail derivatives were sometimes prepared by short syntheses; thus, dithioacetals **9** were obtained by condensation of a free carbohydrate with the corresponding *n*-alkanethiol [50] whereas the 6,6-di-*O*-alkyl derivatives **29** were synthesized by the reaction of alkyl magnesium bromide with 1,2:3,4-di-*O*-isopropylidene-D-galacturonic acid [62].

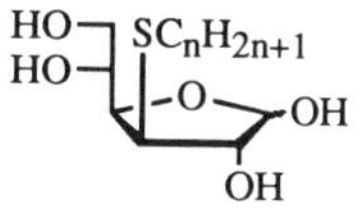

27

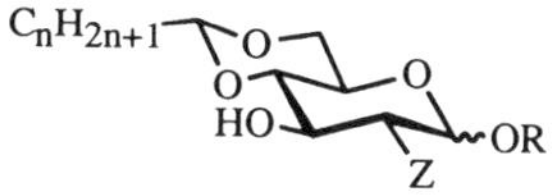

28a R= Me; Z= NHAc, $N(CH_3)_3^+I^-$, $NHCH_2COONa$

28b R= H; Z= OH

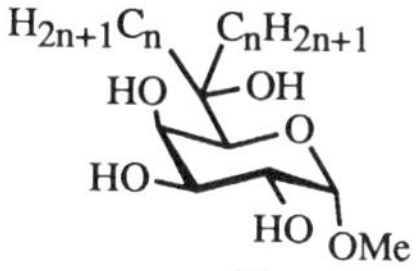

29

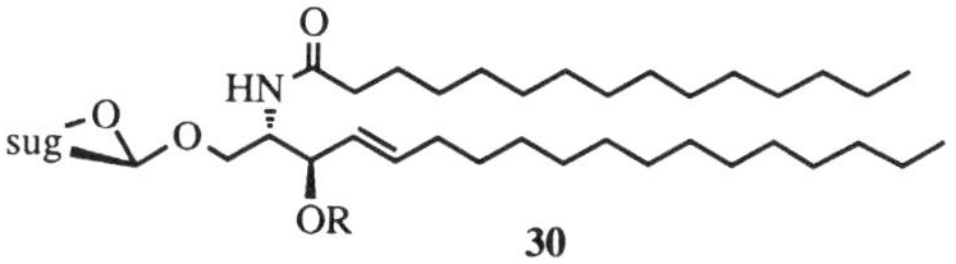

30

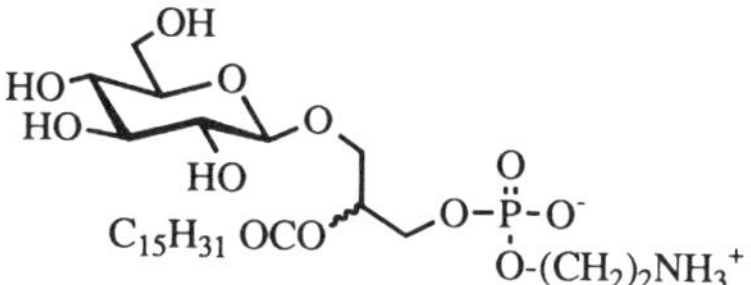

31

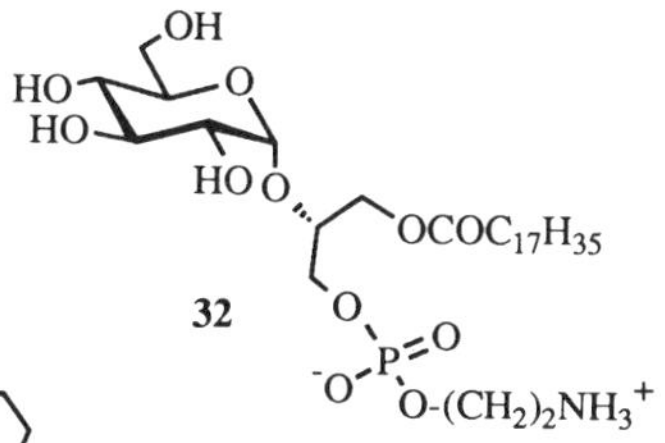

32

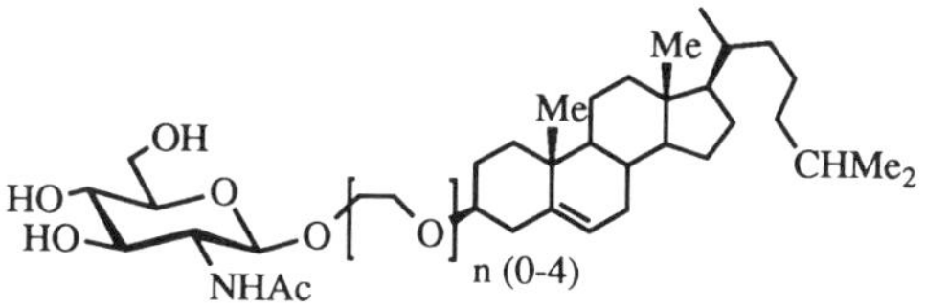

33 n (0-4)

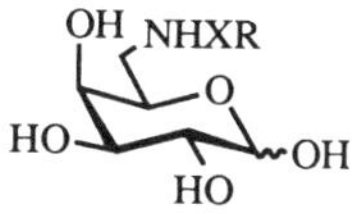

34 XR= COC_nH_{2n+1}, $SO_2C_nH_{2n+1}$

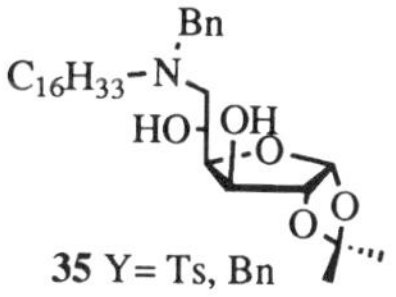

35 Y= Ts, Bn

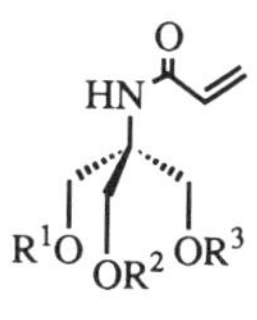

36 a R^1= R^2= R^3= β-D-Gal(OAc)$_4$

b R^1= R^2= β-D-Gal(OAc)$_4$, R^3= H

c R^1= β-D-Gal(OAc)$_4$, R^2= R^3= H

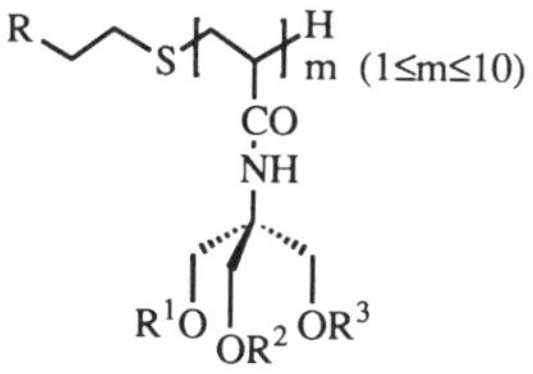

37 a-c R^1, R^2, R^3= β-D-Gal or H

R= C_nH_{2n+1} or C_nF_{2n+1}

Compounds 27–37. sug–O–C–X and sug–OH–C–X represent 05 and C1 of cyclic and acyclic carbohydrate, respectively, in which the carbohydrate moiety may be any mono or oligosaccharide, *α* or *β* anomer (cyclic forms); C_nH_{2n+1} is a saturated aliphatic chain, while alkyl may be either a saturated or an unsaturated aliphatic chain

3.2.2
Multistep Syntheses

Due to the biological roles of glycolipids, many papers have been devoted to their syntheses over the last ten years. The coupling of a fully protected carbohydrate donor to a lipid acceptor requires efficient and highly stereoselective glycosylation methods because lipid derivatives often have low reactivity. A few examples of glycosphingolipids syntheses will be discussed below as well as multistep preparations of other amphiphilic carbohydrates designed as biochemical mimetics, surfactants or liquid crystals.

Glycosphingolipids **11** and their synthetic analogs **30** require very sophisticated methods which are imposed by the structure of the oligosaccharide (which may contain more than ten sugar units) as well as by that of the lipid moiety (sphingosine). Two peracylated glycosyl fluoride donors were reported to react with the acceptor **17** (R=Bz) in poor glycosylation yields (20–40%) but the use of a β-fluorinated donor with silver trifluoromethane sulfonate as the promoter afforded better results (75% yield) [101]. Moreover, good glycosylation yields were reported if glycosyl fluorides were used as donors, compound **17** (R=TBDMS) as the acceptor and $AgClO_4$-$SnCl_2$ as the promoter [102].

The total synthesis of the *para*-Forsman glycolipid was achieved by the same reaction pathway, using either a methyl 1-thioglycoside or a D-galactose trichloroacetimidate as donor [103]. An improvement was suggested: the use of a 2-*O*-pivaloyl donors [104]. Nevertheless the use of trichloroacetimidate leaving groups with Lewis acid as promoters ($BF_3 \cdot Et_2O$ most often) generally constitutes the highest yielding method of widespread use for the synthesis of glycolipids [90, 105].

Glycerolipids **12** were synthesized many years ago but some improvements have been reported recently. For example, the preparations of two related glycoglycerophospholipids were described by independent groups. Compound **31** was synthesized as (2*R*) and (2*S*) epimers by reacting 2,3,4,6-tetra-*O*-benzyl-D-glucopyranose with the 3-*O*-triflate ester of 1,2-*O*-isopropylidene glycerol. The condensation was followed by: (1) removal of the acetal, (2) selective protection of the primary OH (TBDMS), (3) reaction of palmitic anhydride at the secondary OH, (4) deprotection of the primary OH, (5) reaction with 2-bromoethyl phosphorodichloridate quenched with benzyl alcohol, (6) reaction with Me_3N [106]. On the other hand, a C-2 analogue **32** was prepared by the reaction of 2,3,4,6-tetra-*O*-benzyl-α-D-glucopyranosyl fluoride with the synthetic intermediate **20** in the presence of $AgClO_4$-$SnCl_2$ at −15 °C (89% yield). After ozonolytic cleavage, the α,β-unsatürated ester afforded an aldehyde which was reduced to a primary alcohol; the phosphocholine group was introduced afterwards at C-1. After deprotections, compound **32** was obtained on the gram scale in eight reaction steps and an overall yield of 22% [107].

Glycosylphosphatidylinositols (GPI anchors **14**) are a class of naturally occuring glycophospholipids that do not only bind the *C*-termini of membrane protein but also mediate signal transduction. Several papers have been devoted to their syntheses. Most of them have a structure related to **14** (with various sidechains that are species-specific) but some yeast GPI have been reported to possess a sphingosine lipid moiety [108]. Their syntheses represent a challenge with regard to the complex structure of the carbohydrate moiety. The latter contains, most often, a pentasaccharide core [Manα(1–2)Man(α1–6)Maα(1–4) GlcNH$_2$(α1–6)*myo*-inositol] on which are grafted mono or oligosaccharide side-chains. Their syntheses are most often realized by separate preparations of the oligosaccharide and glycerophospholipid moieties which are linked together in the last steps of the reaction pathway. Thus, there have been recent reports of the preparations of the inner core of GPI [1L-6-*O*-(2-amino-2-deoxy-α-D-glucopyranosyl)-*myo*-inositol-*sn*-2,3-*di*-*O*-palmitoyl glycerol hydrogenphosphate] [109], parts of the GPI anchor *of Trypanosoma brucei* containing a tetragalactose side-chain [110, 111], and the preparation of a fully phosphorylated GPI of rat brain Thy-1, which represents the first example of a complete synthesis of such glycolipids [112].

Alkyl glycopyranosides **1a** are structurally simple derivatives; several of them have been synthesized recently in order to afford compounds with expected surfactant or liquid crystal properties. Since they were prepared in multistep reactions, involving protective groups, they were not reported in terms of short syntheses. These derivatives are difficult to obtain in an anomerically pure form. Their syntheses require classical glycosylation reactions in which the anomeric orientation is governed by the structures of donors, acceptors and promoters. For example, glycosides of 2-acetamido-2-deoxy-D-glucose **4a** (Z=H, Ac) were prepared by two independent groups. The first one reported on the Fischer type glycosylation of *N*-acetyl-D-glucosamine with *n*-alcanols (n=8–14) used as the solvent, and a cation exchange resin as the acid catalyst. Both anomers were formed in low yields (4–14% with a 2.5–3/1 α/β ratio) and were separated by ion exchange chromatography. We independently prepared the β-anomers by procedures related to the Koenigs-Knorr reaction. The glycosylation was realized either from a 2-acetamido or a 2-allyloxycarbonylamino peracetyl derivative of D-glucosamine, using iron (III) chloride (Kiso and Anderson procedure) or trimethylsilyl trifluoromethane sulfonate (TMSOTf) as promoters respectively [23]. The overall yields, after deprotection (60–70%) were much higher than those obtained by the single-step procedure discussed before. A series of alkyl β-D-mono- and oligosaccharides **1a** was also prepared by Koenigs-Knorr reaction using peracetylated glycosyl bromides as donors and silver salts as promoters [113, 114]. The corresponding α-D anomers were obtained by the Ferrier reaction (addition of the alcohol to a glycal under acidic catalysis) [113]. In the case of L-fucose, the glycosylation reaction with heavy alcohols was shown to be laborious. Thus 1,2,3,4-tetra-*O*-acetyl-L-fucopyranose reacts with *n*-octanol or *n*-decanol to afford a mixture of α and β-L-fucopyranoside and β-L-fucofuranoside in a 1:1:1 ratio [115].

Alkyl glycopyranosides with an oligoethylene glycol spacer between the carbohydrate and lipid moieties were synthezised because the intercalation of oligoethoxyethylene spacers between the hydrophilic and hydrophobic moieties allows HLB control. Furthermore, it favors the separation of the carbohydrate-head from the tail which can avoid non-specific interactions in molecular recognition [91, 116]. The glycosylation of alkyl oligoethylene glycols needs efficient methods because of their low reactivities. Thus, the preparations of 1,2-*trans*-

glycosides of the above alcohols were shown to require peracetylated donors and TMSOTf as the promoter in the D-*gluco*, D-*galacto*, D-*manno* and D-*xylo* series [117] as well as in the 2-acetamido-2-deoxy-D-*gluco* series [93]. 1,2-*cis*-Glycoside epimers were prepared with 2,3,4,6-tetra-*O*-benzyl donors and tetrabutyl ammonium bromide-$CoBr_2$ as the activators and promoters, but yields and stereoselectivities remain low [117]. We have also prepared a series of 2-acetamido-2-deoxy-β-D-glucopyranosides of cholesteryl oligoethylene glycols by the allyloxycarbonyl procedure. After deprotection and *N*-acetylation the expected β-glucosides **33** were obtained in high yields and with excellent stereoselectivities [118].

Alkyl (or acyl) derivatives of the 6-amino-6-deoxy carbohydrates are examples of derivatives in which the hydrophilic and hydrophobic moieties are linked at other positions than C-1. Thus 6-amino-6-deoxy-D-galactose derivatives **34** were prepared from 1,2:3,4-di-*O*-isopropylidene-6-*O*-tosyl-α-D-galactopyranose by the following reactions: (1) substitution of the leaving group at C-6 by a phthaloyl function, (2) hydrazinolysis to afford a 6-amino-6-deoxy intermediate, (3) reaction of acyl or sulfonyl chlorides at the amino function, (4) deprotection of the acetal rings to afford the expected glycolipid **34** [56].

Methyl 6-amino-6-deoxy-α-D-glucopyranoside derivatives **2c** were synthesized in our laboratory by a somewhat different procedure [31]. 6-*O*-Sulfonyl or 6-bromo-6-deoxy derivatives of methyl α-D-glucopyranoside were substituted at C-6 by sodium azide. The 6-azido-6-deoxy intermediate was then treated by acyl chlorides in the presence of triphenylphosphine (Staudinger reaction) to afford amido derivatives which were finally de-*O*-acetylated to give **2c**. The same reaction pathway allowed the preparation of 6-alkylamido-6-deoxy-D-glucopyranose derivatives, starting from D-glucose [31].

A third strategy was applied to the synthesis of a neighboring family of amphiphilic carbohydrates. The starting material was 5,6-anhydro-1,2-*O*-isopropylidene-α-D-glucofuranose, easily obtained in a few steps from D-glucose. The latter was reacted with *N*-benzyl-*n*-hexadecylamine to afford regioselectively the 6-alkylamino-6-deoxy compound **35** [119].

Telomeric glycolipids **37** were prepared in a somewhat different manner: the lipid chain was built up by polymerization, after the coupling of the carbohydrate moiety. Thus 2,3,4,6-tetra-*O*-acetyl-α-D-galactopyranosyl bromide was glycosylated with tris (hydroxymethyl)acrylamidomethane (or partly protected analogs) to afford fully acetylated mono, di or tri β-D-galactopyranosides **36a–c**. These derivatives were then polymerized in the presence of a telogen thiol ($C_{12}H_{25}SH$, $C_{16}H_{33}SH$, $C_6F_{13}C_2H_4SH$, $C_8F_{17}C_2H_4SH$) to afford, after deprotection, the telomeric amphiphiles **37 a–c** [120].

These few examples, which do not constitute an exhaustive coverage of the literature, were selected for their diversity and to illustrate the finding that, besides the amazing multisteps syntheses of natural glycolipids analogs, even the simplest structures could require complex preparations.

4
Molecular Recognition of Organized Systems by Protein and Cells

Apart from their being interesting as constituents of supramolecular assemblies, and because they are able to form such arrangements, glycolipids are also able to incorporate into phospholipid bilayers if their alkyl chain(s) is(are) properly selected. Mixed glycolipid/phospholipid supramolecular assemblies thus constitute mimetics of the cell membrane [121, 122] that can be used for the study of molecular recognition in organized systems. Arrangements such as monolayers, bilayers, mono or multilayer liposomes can be built up for that purpose. The vesicles also possess biological properties that can be exploited both in vitro and in vivo. For example, the coating of phospholipid bilayers with carbohydrates increases the stability of the colloidal system against low temperatures (cryoprotection) [123, 124] or in vitro agglutination [125–127]. This stabilization is also observed for egg phosphatidylcholine monolayers [128] or black lipid membranes [129] coated with *H*-polysaccharides. Actually, the proper choice of the carbohydrate and the hydrophobic anchor allows to influence most of the parameters governing the mechanical stability of a liposome (aggregation, membrane fluidity, permeability, etc...) [130] or the thermotropic behavior of mixed glycolipid/phospholipid vesicles [131]. Glycolipids protect liposomes from rapid in vivo uptake by the mononuclear phagocyte system [132]. It can be noted that the steric stabilization of vesicles both in vitro and in vivo is also achieved with other amphiphilic derivatives acting on colloidal stability. Thus alkyl polyethylene glycols anchored into the bilayer membranes can hide vesicles from the immune system and give rise to the so-called stealth liposomes [133, 134]. Amongst potential medical applications, specific drug targeting can be achieved by coating liposomes with *H*-carbohydrates [135]. In vivo distribution of such liposomes is mostly governed by the structures of surface-carbohydrates [136, 137]. Another medical application of such assemblies is found in immunology, thus glycolipids can be used as immunostimulating agents probably due to their propensity to form weakly immunogenic vesicular assemblies and able to act as adjuvants [138]. Furthermore, egg-PC liposomes coated with *H*-pullulan and containing anti-sialyl Le^x IgM fragments were shown to bind specifically with cells carrying the antigenic determinant [139].

4.1
Parameters That Govern Interfacial Molecular Recognition

The recognition of a ligand at the surface of a natural or artificial membrane by a specific receptor is governed by kinetic and thermodynamic parameters. In a non-isotropic system, perturbations occur to the classical kinetics and thermodynamics of the reaction:

$$\text{m.ligands} + \text{n.receptors} \rightleftarrows [\text{receptor}]_m - [\text{ligand}]_n$$

As was discussed in the introduction, kinetic parameters of the reaction are disturbed by the hindered approach of the partners whereas thermodynamic

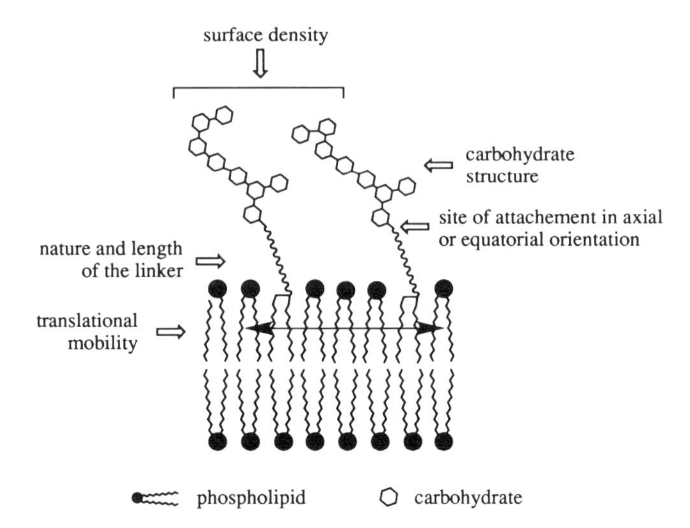

Fig. 5. Parameters that govern interfacial recognition

parameters are modified by entropy changes in the close vicinity of the surface. In the recognition of *H*-carbohydrates at the surface of a (phospho)lipid layer, an appropriate choice of the neoglycolipid constituents (carbohydrate and lipid moieties) may favor the recognition process. Several structural parameters affecting the recognition are schematically represented in Fig. 5.

The nature of the carbohydrate, the site of attachment with the lipid moiety (most often in the anomeric position) and the axial/equatorial orientation of the latter are of obvious major importance to the presentation of the carbohydrate to the receptor, and therefore to the specificity of the recognition. These structural parameters are quite usual in solution reactions and do not need further discussion. What will be discussed below are the specific interactions induced by molecular recognition at interfaces. The nature and length of the linker that joins the carbohydrate and the lipid anchor can interfere with the orientation and conformation of the ligand or with the steric hindrance of the latter with the surface. What is here called a "linker" can either be a spacer (e.g. polyoxyethylene or polypeptide) or it can be part of the hydrophilic or hydrophobic moieties, depending on the HLB of the glycolipid. The nature and length of the lipid anchor themselves are responsible for membrane fluidity and therefore translational mobility of the ligand at the surface of the layer. The structures of both the glycolipid and the phospholipid (in mixed glycolipid/phospholipid layers) are therefore responsible for the surface density of the carbohydrate, which must be taken in account in the binding.

4.1.1
Carbohydrate Density at the Interface

The amount of ligand adsorbed at the interface might control the binding with specific proteins or cells. Despite the large number of reports on the coating of surfaces with carbohydrates, very few of them deal with the measurement of the incorporated amounts. However, this measurement is of major importance when mixed glycolipid/phospholipid layers are studied. Thus for example, carboxymethyl chitin incorporated into PC bilayers was quantified in several steps: (1) extraction of the lipids with chloroform, (2) acid hydrolysis of the chitin, (3) colorimetric titration of the carbohydrate. The method is dependent on the extraction process, duration of hydrolysis and absorbances (420 nm) of mixtures of ferrocyanide/chitin, which are concentration-dependent [140]. This rather laborious and time-consuming method is nevertheless accurate and seems to be better than turbidimetry which was used for the same supramolecular system [141]. Both methods demonstrate that the amounts of carboxymethyl chitin adsorbed at the surface of liposomes increase linearly with increasing initial carboxymethyl chitin concentrations. Nevertheless the incorporated amounts, with regard to the initial carboxymethyl chitin concentrations were not reported.

Measurements of the quantities of glycolipids inserted into the membrane have also been reported by a technique based on the use of ^{14}C-labeled lipid anchors. In this method, the carbohydrate (α-D-Man) was covalently coupled to the anchor at the surface of a pre-formed vesicle. Indeed, the liposome structure was shown to remain intact in the treatment. Nevertheless, the measurement of the incorporated mannose was performed after separation of bound and unbound material by centrifugation. The yields of coupling were shown to increase with the increase of the initial mannose/^{14}C-anchor ratio, but non covalent insertions were displayed at high initial mannose concentrations. Therefore, the aforementioned method was not as accurate as could have been expected for the use of radioactive materials [142]. Radiolabeled phospholipids were also used for such determinations; thus the amounts of glycosphingolipids incorporated into liposomes were quantified by the use of ^{3}H-phospholipids whereas the amounts of glycolipids were determined by a sphingosine assay [143].

Another technique allowed to determine the composition of octyl glucoside/PC mixed micelles and vesicles: the resonance energy-transfer (RET) between two fluorescent lipid probes present in trace amounts in the membrane. The results were consistent with the assumption of ideal mixing of the two amphiphiles at high octyl glucoside concentrations (micellar region): the amount of phosphatidyl choline in the micelles appears to be a function of the total phosphatidyl choline concentration. At a low concentration of octyl glucoside (SUV region), the RET method was found to be invalid and did not allow such a determination [144]. A phase transition model (micelles <-> vesicles), developed for the same system, showed that the transition was driven by very large differences in spontaneous curvatures between lipid and surfactant [145].

Because the labeling of either the phospholipid or the glycolipid were not very efficient, we decided to measure the amounts of glycolipids and phospholipids into vesicles by labeling both of them. Thus, ^{3}H-glycerophospholipid and alkyl glycosides of *N*-acetyl-D-glucosamine ^{14}C-labeled on the acetamido group (**4a**, $R=C_nH_{2n+1}$, **8b, 33**) were evaporated with egg-PC to afford films that were suspended in water, and sonicated [93, 146]. Small unilamellar vesicles (SUV) thus prepared were separated from multilamellar vesicles (MLV) by centrifuga-

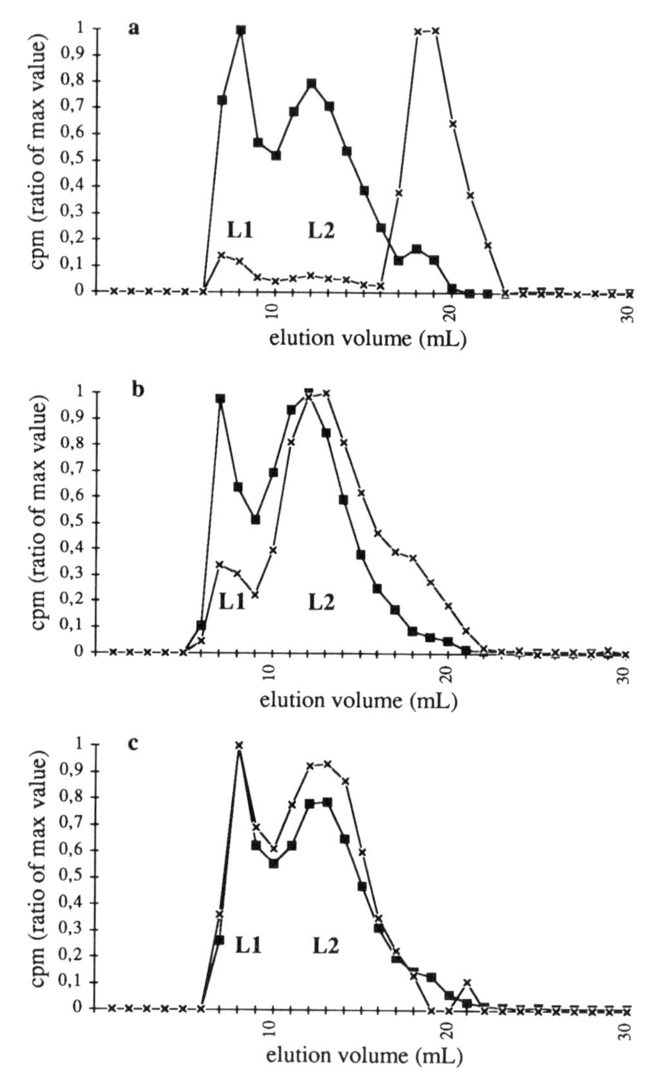

Fig. 6. Elution profiles of three alkyl 2-acetamido-2-deoxy-β-D-glucopyranosides **4a** (a. n=8; b. n=11; c. n=14) in gel filtration chromatography on sepharose 4B (exclusion 2.10^7 Da, exclusion volume 10 mL) monitored by ^3H and ^{14}C radioactivity countings (normed to 1.0 for reason of convenience). (■) ^3H countings of ^3H-labelled phosphatidyl choline; (x) ^{14}C countings of ^{14}C-labelled glycolipids **4a**

tion and then purified by gel filtration chromatography (Sepharose 4B). ^3H and ^{14}C were separately counted before and after centrifugation, and in each chromatographic fraction. The elution profiles were shown to depend on several parameters: (1) sonication duration, (2) nature of the glycolipid, (3) initial glycolipid/phospholipid ratio. Examples are shown in Fig. 6.

Three ^{14}C-labeled alkyl 2-acetamido-2-deoxy-β-D-glucopyranosides (**4a**, n=8, 11, 14) were incorporated into the bilayer of ^{3}H-labeled egg-PC. Two liposome populations were eluted successively: the first one (L1), excluded from the column could correspond with large unilamellar vesicles, LUV (70–80 nm as mean diameter determined by quasi-elastic light-scattering), the second one (L2) corresponds with SUV (30–40 nm as mean diameter). Both populations were eluted before low molecular weight aggregates, the natures of which were not fully determined. The relative abundances of L1 and L2 were found to be strongly dependent on the sonication conditions, longer periods can lead to the complete disappearance of L1. Figure 6a, b and c show the elution profiles of **4a**, (n=8, 11, 14 respectively). The shorter chain glucoside (C_8) is almost not incorporated and eluted after the liposomes, the C_{11}-glucoside is imperfectly recovered into the liposomes whereas the longest chain derivative (C_{14}) is totally incorporated into the liposome bilayers (^{3}H and ^{14}C curves are almost superimposable). The amounts of glycerophospholipids and glycolipids in the liposomes can be accurately quantified for different initial concentrations of both reactants. The results obtained with several glycolipids are recorded in Table 1.

Several conclusions can be drawn from Table 1. With single-chain derivatives **4a**, the maximum incorporation of glycolipids into the vesicles is close to 3–5%. The addition of higher initial amounts of glycolipids results in their elimination in the centrifugation pellets. With double-chain derivatives **8b** (p = 3), the incorporation is almost complete, whatever the initial glycolipid/phospholipid ratio. Furthermore, the initial glycolipid/PC ratio is recovered almost unchanged into the bilayers. The same observations can be made with the cholesteryl derivative **33** (n=0).

With a view to controlling the density of carbohydrates at the surface of vesicles, these results demonstrate the importance of measuring the amounts of both the glycolipid and the phospholipid in the vesicles. The initial glycolipid/phospholipid ratio, used for the preparation of liposomes, is not always recovered unchanged at the surface of the vesicles, as sometimes postulated in the literature. As shown in Table 1, the incorporation strongly depends on the nature of the lipid moiety; this assessement is in accordance with previously reported results, which showed that a portion of the glycolipid was lost during the formation of the vesicles [143]. Furthermore, we have shown that the cholesteryl glycoside **33** (n=0) incorporates into the phospholipid bilayer even at high concentrations, in contrast to previous reports [147]. Incorporations of cholesteryl derivatives bearing hydrophilic anchors **33** (n=1–4) have further confirmed this statement in our laboratory.

As discussed before, the surface density of carbohydrates is one of the parameters that might be controlled to prepare membrane mimetics. The driving force for binding of surface-carbohydrates by extracellular proteins is probably lying largely in cluster effects. As recently reported [148] several multivalent carbohydrate-assemblies are able to mimic biological recognition (e.g. several carbohydrates bound to the same linker or to a polymer backbone). For a multivalent ligand, the free energy of binding to a multivalent receptor is greater than the sum of the individual contributions; once the ligand has attached itself at one site, it is closer to other sites and will suffer a smaller entropy loss by bin-

Table 1. Incorporation of several glycolipids (*GS*) into egg phosphatidyl choline (*PC*). Results obtained with ^{3}H-labelled phospholipids and ^{14}C-labelled glycolipids

glycolipid	initial GL/PC ratio (mol %)	pellets: recovered GL (% of initial amount)	supernatants: recovered GL (% of initial amount)	GL/PC[a] in the liposomes (mol %)
4a (n=8, Z=Ac)	7.2	1.9	98.0	2.6 ± 0.1[b]
	27.1	4.0	93.6	0.4 ± 0.1[b]
4a (n=11, Z=Ac)	5.6	11.2	76.8	8.0 ± 2.0[b]
4a (n=14, Z=Ac)	3.2	0.4	98.2	4.5 ± 0.1
	16.5	81.4	16.5	3.1 ± 0.2
4a (n=18, Z=Ac)	3.8	0.3	91.8	3.7 ± 0.2
	31.9	76.6	12.9	2.3 ± 0.3
8b (m=n=11)	4.6	2.1	80.9	3.8 ± 0.2
(p=3)	8.6	1.3	94.9	7.3 ± 0.2
	52.0	1.0	86.2	43.0 ± 3.0
8b (m=11, n=16)	3.7	0.2	78.1	3.4 ±0.1
(p=3)	9.6	0.3	88.4	8.9 ± 0.2
	51.3	0.2	93.3	42.7± 0.1
33 (n=0)	6.9	0.3	98.6	6.4 ± 0.3
	32.7	0.1	97.1	28.0 ± 0.3

[a] mean ratios of incorporation in L1 and L2.
[b] results of low significance since the glycolipid is either not incorporated or forming other supramolecular assemblies than liposomes with PC.

ding to them. With a matrix bearing several carbohydrate groups, the binding affinity of a receptor to the ligands increases geometrically with a linear increase of the number of sugar residues [149].

This cooperative effect of several ligands at the surface of a liposome was recently demonstrated by the inhibition of agglutination of erythrocytes (natural vesicles covered with sialic acids) with the influenza virus by means of sialyl gangliosides. The lowest concentration of sialyl derivatives required for the inhibition was found to be 10 μM in solution whereas it was 20 nM at the surface of a vesicle. Actually, arrays of sialic acid at the surface of liposomes were found to be moderately more effective than sialic acid groups linked to a soluble polymer but as good or better than the best known natural inhibitors of hemagglutination (i.e. mucins and macroglobulins) [150].

4.1.2
Changes of Carbohydrate-Conformations at Interfaces

If supramolecular assemblies of amphiphilic carbohydrates can display cluster effects, the specific binding of the latter are also dependent on the presentation of the carbohydrate to the receptor, i.e. its orientation and conformation.

Carbohydrate-conformations at the surface of a membrane were studied from a static as well as from a dynamic point of view. Molecular mechanics

Table 2. Minimum energy conformations and conformer populations for a series of glucosphyngolipids and glucoglycerolipids. Values are obtained by MM3 minimizations at a value of dielectric constant ε=4 [151]

Glycolipid	φ (deg)	Ψ (deg)	θ_1 (deg)	β_1 (deg)	population (%)
β-D-Glc-O–CH$_2$–CH(NHCOC$_n$H$_{2n+1}$)–CH(OH)–CH=CH–C$_{13}$H$_{27}$	47	189	-64	141	46
	43	179	-179	145	28
	45	-90	167	96	19
β-D-Glc-O–CH$_2$–CH(NHCOCH(OH)C$_n$H$_{2n+1}$)–CH(OH)–CH=CH–C$_{13}$H$_{27}$	47	179	-63	125	48
	45	180	180	132	22
	44	-91	179	107	13
β-D-Glc-O–CH$_2$–CH(OCOC$_n$H$_{2n+1}$)–CH$_2$–OCOC$_n$H$_{2n+1}$	50	-180	179	132	41
	42	-89	168	106	21
	45	180	-56	132	19
β-D-Glc-O–CH$_2$–CH(OC$_n$H$_{2n+1}$)–CH$_2$–OC$_n$H$_{2n+1}$	49	179	177	163	40
	43	180	-57	165	35
	49	-178	70	162	12

(MM3) calculations of the preferred conformations of *β*-D-glucopyranosyl ceramides and *β*-D-glucopyranosyl diacyl and dialkyl glycerolipids were performed by imposing steric restrictions, accounting for the surrounding membrane-surface, and by using a low dielectric constant (ε=4) for the dielectric properties of a lipid bilayer. The nature of the lipid moiety was shown to strongly affect the preferred conformations around the linkage between the hydrophilic and hydrophobic parts of the molecule. Results reported in Table 2 display the rotation angles of the three major conformers if four different lipid anchors were considered [151].

For all the glycolipids, the surrounding membrane considerably reduces the range of possible conformations by comparison with a homogenous isotropic medium. Furthermore, sphingosine derivatives were shown to induce a more restricted rotation than glycerol derivatives. The differences of conformational preferences could also account for differences observed in biological activity [152] as well as phase behavior. The extended conformation of glyceroglycolipids could favor hexagonal phases whereas the larger packing cross section of glycosphingolipids could favor lamellar structures.

The orientation and dynamics of dodecyl *β*-D-glucopyranoside in phospholipid bilayers were studied by NMR in oriented samples [153, 154]. The fully ^{13}C-labeled glycolipid was embedded in a magnetically orientable membrane system (DMPC-CHAPSO) and examined by NMR. The analysis of the set of data allowed to determine the average orientation of D-glucose with respect to the bilayer; the ring is extended from the plane of the bilayer with vector C-2/C-6 approximately parallel to the surface, the middle-plane of the sugar ring is in the prolongation of the alkyl chain. Motions of the ring are restricted in an anisotropic manner to avoid placing OH-2 or OH-6 in the apolar region of the membrane. The same findings were roughly displayed with other alkyl glycosides [155]. The preferred conformations thus deduced, were compared with energy maps of the glycolipid torsion angles. The results illustrate the need to include a membrane interface in modeling glycolipid average orientation and range of motion.

The same procedure was applied to monogalactosyldiacylglycerol using both NMR in oriented membranes and molecular modeling (AMBER); a preferred set of low-energy conformations extends the galactose headgroup away from the membrane. Conformations in solution display a more extensive motional averaging than conformations of the glycolipid embedded in the lipid matrix [156]. It should be noted, nevertheless, that the inclusion of the membrane-interface in the energy modeling presents some limitations if oligoethoxyethylene groups are placed between the carbohydrate and lipid moieties [157].

^{2}H-NMR was also shown to give insights into the structure and dynamics of the membrane components. The use of ^{2}H-labeled lipid chains in quadrupolar echo NMR allowed to get access to the bilayer thickness of liposomes, oriented bilayers or even whole cells [158]. Relaxation times in deuterium NMR made it possible to determine the phase diagrams of lipids. From the angular dependence of the quadrupolar splitting in oriented bilayers one can determine the main rotation axis of any molecule inserted into the bilayer. Deuterium relaxation times give access to the dynamics of the membrane components. Thus, the orientation, conformation and motion of 3-*O*-(4-*O*-β-D-galactopyranosyl-β-D-glucopyranosyl)-1,2-di-*O*-tetradecyl-*sn*-glycerol (**8a**, sug=Lac, R=$C_{14}H_{29}$) relative to a phospholipid bilayer membrane was determined by ^{2}H-NMR [159]. The lactosyl headgroup is extended away from the surface and displays motion with an axial symmetry averaging in the bilayer normal. The conformations of the Gal-Glc linkage and rotamers about the C-5′/C-6′ bond of the glucose unit, calculated from ^{2}H-NMR, were shown to differ substantially from those determined by X-ray diffraction or by NMR in solution again suggesting interactions with the bilayer.

The conformations and motion properties of saccharides at the surface of model membranes might induce differences of recognition properties in biological media. The role of the matrix itself was recently displayed in molecular recognition. Thus three allelic variants of G-adhesins (lectins: $PapG_{J96}$, $PapG_{AD110}$ and $PrsG_{J96}$) were analyzed for their ability to agglutinate erythrocytes from various species and containing different glycosphingolipids on their surfaces [Galα(1–4)Galα(1–4)Glcβ-Cer: GbO_3; GalNAcβ(1–3)Galα(1–4)Galα(1–4)Glcβ-Cer: GbO_4; GalNAcα(1–3)GalNAcβ(1–3)Galα(1–4)Galα(1–4)Glcβ-Cer: GbO_5]. All three adhesin variants exhibit similar specificities for the globo-series when they are inserted onto artificial surfaces. However, important differences are observed when the ligands are fixed on erythrocytes [160]. This result suggests conformational differences induced by the nature of the matrix. The preferred conformations of the glycolipid at the surface of the membrane, modeled by MM3 calculations, display differences in accordance with the observed reactivities and cross-reactivities of the lectin variants (Fig. 7).

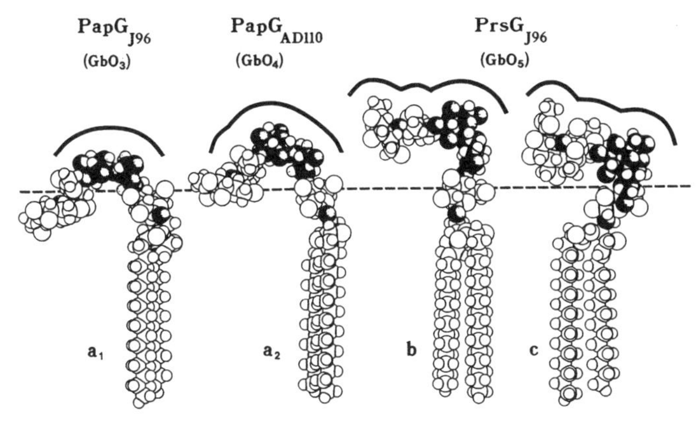

Fig. 7. Preferred conformations and epitope presentations of globoglycosphingolipids. Favored conformations are shown with the Galα(1–4)Gal unit and glycosidic oxygen shaded. The internal conformation of the saccharide chain is practically identical for all four conformers and is independent of the sugar residue distal to Galα(1–4)Gal unit. The four conformers differ in the conformation of the saccharide-ceramide linkage, resulting in different saccharide orientations with respect to the membrane surface (dashed line). Conformers a_1, b and c correspond to the three minimum energy conformations for the saccharide-ceramide linkage as calculated by MM3. The preferred isoreceptor for the various adhesins is indicated in parentheses. *PapG receptor-active conformers:* conformer a ($\varphi/\Psi/\theta$=45/–90/170) is a minimum energy conformer that most efficiently exposes the Galα(1–4)Gal epitope recognized by $PapG_{J96}$. In membranes, however, conformer a_1 can only be adopted by GbO_3, while it is sterically excluded in the case of GbO_4 and GbO_5. Conformer a_2 ($\varphi/\Psi/\theta$=45/–115/180) is a slightly strained modification of conformer a_1 with the GalNAcβ residue raised above the membrane surface. The existence of GbO_4 in this conformation may explain the recognition of GbO_4 in membranes of both PapG adhesins. *PrsG receptor-active conformers:* conformers b and c ($\varphi/\Psi/\theta$=45/180/180) in b and ($\varphi/\Psi/\theta$=45/180/–65) in c are minimum energy conformers that can be adopted by three globoglycosphingolipids. In these conformers the GalNAcα residue and its *N*-acetyl goup are efficiently exposed, while the Galα(1–4)Gal unit is not. These saccharide orientations favor recognition of GbO_5 by $PrsG_{J96}$ adhesin but result in a crypticity of GbO_5 for $PapG_{AD110}$ and particularly for $PapG_{J96}$ [160]

These latter results, together with those previously reported, clearly demonstrate differences between solution and surface conformations. In supramolecular assemblies, the conformation of a carbohydrate does not only results from the arrangement of atoms inside surrounding water or steric hindrance with the carrier (if any); the lipophilic moiety of amphiphilic carbohydrates and the lipidic environment also control the conformation of the hydrophilic moiety, and therefore the recognition.

4.2
Measurements of Interfacial Recognition

Several recent and often sophisticated methods allow to make measurements of interfacial recognition. These methods, performed either on Langmuir-Blodgett (LB) films or on vesicles allow the direct or indirect measurement of the binding (i.e. recognition). Several of them have been set up on systems in which the ligand is not necessarily a carbohydrate. Nevertheless, they will be briefly discussed since they can be extended (and have sometimes been extended) to carbohydrate-ligands.

Steroid hormones derivatized as glycerylphosphate dialkyl esters and spread at an air/water interface allowed the two-dimensional recrystallization of proteins bearing the specific receptors. The direct observation of 2D-proteins arrays by electron microscopy allowed to exhibit the parameters that govern the binding such as film fluidity or ligand accessibility, ordering and surface density [149].

Several methods have been used to investigate the high affinity binding of streptavidin with *H*-biotin both immobilized at the surface of lipid bilayers. Long range attractions were detected at distances up to 85 Å with a surface force apparatus (SFA) [161]. Surface pressure isotherms and fluorescence microscopy of several *H*-biotin spread at the air/water interface display two stages in the biological recognition: (1) specific binding at one site of streptavidin, (2) diffusion in the plane of the monolayer and binding with the second site of streptavidin [162]. The influence of *H*-biotin surface density and accessibility to avidin has also been reported at the surface of the tip of an evanescent fiber-optic sensor (EFO) able to conduct an excitation light and to collect the fluorescence emission of FITC-labeled avidin [163]. The molecular recognition by streptavidin, of *H*-biotin monolayers tightly bound to a gold electrode has also been observed by surface plasmon resonance spectroscopy (SPR) and microscopy. The results displayed a better binding when the electrode exhibited a low biotin packing density [164].

Several papers also deal with the antigen-antibody recognition in ordered systems. Fluorescence microscopy at air/water interfaces of dense monolayer packing allowed to conclude that non-specific adsorption and lateral aggregation leading to phase separated lipid areas had taken place[165]. The antigen recognition properties of antibody monolayers were also quantified by an enzyme-linked immunoassay [166] and EFO sensor, using IgG coupled with a fluorescent label [167].

A very useful method for the direct measurement of binding at interfaces is the quartz-crystal microbalance (QCM). LB films deposited on a QCM provides a very sensitive mass measuring device because the quartz resonance frequency decreases upon the increase of mass on the QCM electrode, at the nanogram level. Furthermore, QCM data can be taken in real time, thus making analysis of the kinetics of binding possible. Several ligand-receptors have been studied by this technique. For example, phospholipids deposited as LB films on the QCM electrode display in the presence of a phospholipase two successive phenomena. The first one is an increase in mass due to the binding of the enzyme and the second one a decrease corresponding to the hydrolysis of the phospholipid from which the complete clearing of the electrode results[168]. The kinetics of binding of antifluorescein antibodies with *H*-fluorescein/lipids deposited onto the gold electrode of a QCM have illustrated the roles of steric hindrance and that of a hydrophilic spacer in the binding [169].

Studies of recognition are often realized by covalently coupling LB films on gold surfaces by means of a SH function located either at the end of the alkyl chain or on the hydrophilic moiety. The latter occurrence confines water between the support and the lipid layer and allows their use in SPR, impedance measurements or cyclic voltametry [170].

Several of the aforementioned methods, and others, have been applied to the recognition of carbohydrates. The most widely used of them will be further discussed, distinguishing whether they were applied to monolayers (2D-arrays) or to vesicles (3D-systems).

4.2.1
Monolayers

Spreading a glycolipid at an air/water interface and reducing progressively the surface until obtaining an organized monolayer gives access to physical parameters such as the surface occupied by the polar heads at the interface or the inclination of the lipid moieties on the latter. When a protein, bearing specific receptors to the carbohydrate is injected into the subphase, the recognition at the interface is accompanied by a reorganization of the monolayer that can be quantified and related to the physico-chemical parameters of binding. Thus the recognition of lipid A (**13**) by specific antibodies (IgG) displays three successive steps: (1) diffusion of IgG at the interface, (2) specific binding, (3) aggregational and conformational rearrangements in the plane of the monolayer. The formation of the immune complex at the interface is a diffusion limited process controlled by the lipid A/IgG ratio. This ratio must be high enough to allow the binding, but not too high in order not to avoid the reorientation of the antibody molecules from the vertical to horizontal position [171].

A very recent device allowed the direct colorimetric visualisation of the binding of E. coli enterotoxin with gangliosides. The latter glycolipids were incorporated into a LB matrix of diacetylenic lipids. After polymerization of the monolayer, the membrane acquired a characteristic blue color. In the presence of the toxin, the membrane turned red. The specific response, that can be colorimetrically quantified, is due to a reorganization of the film, which induces changes in the effective conjugate length and therefore the chromatic transition. The ganglioside does not need to be covalently coupled to the membrane; a 5% ratio in the total amount of lipids afforded the best results with regard to membrane stability and detection sensitivity [172].

The recognition by Con A, of *H*-maltose and *H*-lactose at air/water interfaces was also monitored by QCM. The glycolipids were separately diluted in a PE matrix at concentrations ranging from 0% to 100%. No complexation with Con A was observed with lactose in contrast to maltose (which is in agreement with the specificity of Con A for D-glucose). The maximum binding amount of lectin was shown to be independent on the molar fraction of glycolipids in the matrix but the binding constant (k_1), the dissociation constant (k_{-1}) and the association constant ($K_a=k_1/k_{-1}$) were largely affected. Both k_1 and k_{-1} values decreased largely and then increased slightly with increasing concentrations of glycolipids. As a result, the association constant K_a displayed a maximum at 40% of glycolipid in the matrix. This phenomenon was explained by steric hindrance of the lectin at the surface of the monolayer and by the accessibility to one or two sites of the protein depending on the glycolipid concentration [173]. A related system (Con A/*H*-maltose or *H*-lactose) was also studied at the surface of a gold electrode to which the glycolipid has been coupled via a thiol group located at the hydrophobic end. The interaction of the lectin was measured by cyclic voltametry. At pH 8, the initial binding to maltose monolayers was about four times faster than to lactose monolayers, which was again in agreement with the binding specificity of Con A [174].

A carbohydrate biosensor was realized for the detection of uropathogenic bacteria, specific of Galα(1-4)Galβ. The *H*-digalactoside was coupled to a gold surface via a thiol function either located at the end of the lipid chain or belonging to a BSA molecule intercalated between the glycolipid and the gold surface. The binding of bacteria was observed by scanning electron microscopy (SEM). The data, thus collected, demonstrated that the direct covalent coupling (*H*-disaccharide/Au) affords the higher number of bound bacteria per surface unit but the lower selectivity by comparison with *H*-disaccharide/BSA/Au [175].

4.2.2
Vesicles

Several methods were also suggested for studying molecular recognition of glycolipids at the surface of liposomes or vesicles. 3D-systems are closer to the structures of cells and they constitute better mimetics of biological membranes than 2D-arrays. Nevertheless, the third dimension strongly complicates their study. The thermodynamic and kinetic parameters available with 2D-arrays are not all accessible with 3D-structures; mostly qualitative informations remain available. Thus, diacetylenic glycolipids, already mentioned for their use in monolayer systems [172], were incorporated as vesicle components (5–10%) in diacetylenic lipids. The vesicles, obtained by ultrasonic dispersion in water, were polymerized. The deep blue vesicles obtained after 5–10 min of UV treatment turned purple after a longer time (30 min). The colors, assumed to be due to different degrees of conjugation, changed from blue to pink and from purple to orange when the vesicle suspensions were treated with influenza virus bearing the specific receptors to the glycolipid (sialic acid) [176].

Molecular recognition of glycolipids at the surface of vesicles is highly specific and could be exploited in the field of biological sensors. Nevertheless caution should be exercised since non-specific interactions can produce similar effects. Thus, for example, the increase of fluorescence observed when liposomes internally loaded with 6-carboxy fluorescein were treated with Con A was observed, even in the absence of specific surface ligands. This means that the formation of protein-saccharide aggregates is not a prerequisite to the alteration of SUV membranes and that great care should be taken in interpreting the results [177].

The carbohydrate recognition at the surface of vesicles has also been monitored by agglutination measurements. Thus the binding of the carbohydrate moieties of two different glycolipids (galactosyl ceramide and cerebroside sulfate) was displayed which specifically interact with each other [178]. Both reagents were separately incorporated into SUV, composed of DMPC and cholesterol, varying their concentrations and the nature of their ceramide moieties. The densities of carbohydrates at the surface of the vesicles were estimated by the use of ^{3}H-phospholipids and by a fluorescence titration of sphingosine allowing the measurements of the lipid and carbohydrate contents separately. The surface density of carbohydrates in the vesicles was found to be lower than the initial ratio, which is in agreement with our proper results. The agglutination of the vesicles carrying either of the carbohydrates was quantified by the increase of optical density at 450 nm [143].

Model bilayers composed of dodecyl β-D-glucoside embedded in a mixture of phospholipid (DMPC) and detergent (CHAPSO) which orient in a magnetic field, were also used to study the recognition of *N*-acetyl-D-glucosamine by a specific lectin (WGA). The ^{13}C-labeled carbohydrate (GlcNAc) was hydrophobized with oligoethoxyalkyl aglycons (**4d**; n=0–4), which were then inserted into the artificial membrane. In ^{13}C-NMR, the glycolipids with the longer ethoxyethylene spacers (2–4) appear isotropic in the bilayers whereas those with shorter or no spacers were significantly ordered. When WGA was added to the samples, only the glycolipids with 3–4 ethoxyethylene spacers bind to the lectin. This result is in agreement with molecular modeling that displayed significant contacts between the lectin-carbohydrate complex and the membrane with glycolipids containing zero or one ethoxyethylene spacer [157]. This result was confirmed by agglutination experiments of DMPC-SUV in which 5% of *H*-GlcNAc were incorporated. In the presence of WGA, the turbidity of the solution was monitored by measurement of optical density at 450 nm. The longest ethoxyethylene (3–4) derivatives only were agglutinated at the aforementioned glycolipid/phospholipid ratio [157]. This result clearly demonstrates the role of the linker between the carbohydrate and the lipid moieties. Depending on the nature and size of this linker, the carbohydrate is more or less buried into the membrane and therefore more or less accessible to the binding.

5 Conclusion and Future Prospects

Several evidences, reported in the literature and briefly reviewed in the present article, demonstrate that the carbohydrate recognition at the surface of organized systems is somewhat different from that observed in isotropic media. These differences lie in: (1) the conformation of carbohydrate which is affected by hydrophobization and by the nature of the surrounding lipids, (2) cluster effects from which can result in high energies of binding and which are affected by the fluidity of the lipid system, (3) entropy changes at the surface of a supramolecular structure.

The molecular recognition of glycolipids at the surface of organized systems can be seen as a continuing process resulting from: (1) long range carbohydrate/receptor interactions, (2) monovalent binding, (3) lateral diffusion of the carbohydrate/receptor complex, (4) multivalent binding (for multivalent receptors). The four steps of the recognition process are strongly influenced by the supramolecular organization, which could explain most of the differences with isotropic recognition.

Some methods are now available to observe and to quantify the binding in such systems. 2D-Arrays (mainly LB films) can afford a lot of information concerning the thermodynamics and kinetics of the reaction, whereas 3D-supramolecular structures (liposomes and vesicles) most often do not allow such precise measurements. Nevertheless, many biological phenomena are governed by molecular recognition at the surface of cells. The knowledge of such intricate systems should proceed via the preliminary cognition of model organized assemblies.

New glycolipids have to be synthesized to get further insights into liquid crystal properties (mainly lyotropic liquid crystals), surfactant properties (useful in the extraction of membrane proteins), and factors that govern vesicle formation, stability and tightness. New techniques have to be perfected in order to allow to make precise measurements of thermodynamic and kinetic parameters of binding in 3D-systems and to refine those already avalaible with 2D-arrays. Furthermore, molecular mechanics calculations should also be improved to afford a better modeling of the conformations of carbohydrates at interfaces, in relation with physical measurements such as NMR.

These advancements will then be applied to biological systems both in vitro and in vivo. As already mentioned, several medical applications are in a close relationship to molecular recognition; the organized systems of glycolipids should find widespread use such as: drug delivery (glycolipids protect liposomes from the phagocyte system [132]), drug targeting (the carbohydrate moiety of glycolipids is able to target specific receptor cells [135–137]), immunology (immunostimulating agents [138], immunoliposomes [139], diagnostics), cancer therapy (changes at the surface of malignant cells are written in terms of carbohydrates), biosensors. The list is not and cannot be exhaustive, many applications may be anticipated, many others have yet to be discovered.

Acknowledgements. This work was made possible by the support of the Centre National de la Recherche Scientifique and the University of Lyon 1. The author warmly acknowledges his colleagues of the team "PASTAGA" who were involved in this work, namely Dr. D. Lafont, Dr. M.R. Sancho, S. Chierici, V. Maunier and P. Lemattre. Thanks are also addressed to colleagues from other laboratories for fruitful discussions, help and encouragement, namely Professor B. Roux, M. Gelhausen, Dr. A. Coleman, Dr. A. Baszkin and Dr. Y. Chevalier.

References

1. Vögtle F (1991) In: Supramolecular chemistry. Wiley, Chichester, UK
2. Spohr U, Paszkiewicz-Hnatiw E, Morishima N, Lemieux RU (1992) Can J Chem 70:254
3. Carver JP, Michnick SW, Imberty A, Cumming DA (1989) Carbohydrate Recognition in Cellular Function. In CIBA Fnd Symp 145 Wiley, Chichester UK, p 6
4. Lumry R, Rajender S (1970) Biopolymers 9:1125
5. Carver JP (1993) Pure Appl Chem 65:763
6. Alon R, Hammer JP, Springer TA (1995) Nature 374:539
7. Goldberg RJ (1952) J Amer Chem Soc 74:5715
8. Lemieux RU, Bundle DR, Baker DA (1975) J Amer Chem Soc 97:4076

9. Ichikawa Y, Halcomb RL, Wong CH (1994) Chem Brit 117
10. Ringsdorf H, Schlarb B, Venzmer J (1988) Angew Chem Intern Ed Engl 27:113
11. Raaijmakers HWC, Arnouts EG, Zwanenburg B, Chittenden GJF, van Doren HA (1995) Recl Trav Chim Pays-Bas 114:301
12. Israelachvili JN, Marcelja S, Horn RG (1980) Quat Rev Biophys 13:121
13. (a) Plusquellec D, Chevalier G, Talibart R, Wroblewski H (1989) Anal Biochem 179:145 (b) Plusquellec D, Wroblewski H (1993) US Patent 5,223,411
14. Hildreth JEK (1982) Biochem J 207:363
15. Kameyama K, Takagi T (1990) J Colloid Interface Sci 137:1
16. Lamesa C, Bonincontro A, Sesta B (1993) Colloid Polym Sci 271:1165
17. Roxby RW, Mills BP (1990) J Phys Chem 94:456
18. a) Focher B, Savelli G, Torri G, Vecchio G, McKenzie DC, Nicoli DF, Bunton CA (1989) Chem Phys Lett 158:491 (b) Focher B, Savelli G, Torri G (1990) Chem Phys Lipids 53:141
19. Antonelli ML, Bonicelli MG, Ceccaroni G, Lamesa C, Sesta B (1994) Colloid Polym Sci 272:704
20. Frindi M, Michels B, Zana R (1992) J Phys Chem 96:8137
21. van Buuren AR, Berendsen HJC (1994) Langmuir 10:1703
22. Matsumura S, Kawamura Y, Yoshikawa S, Kawada K, Uchibori T (1993) J Amer Oil Chem Soc 70:17
23. Boullanger P, Chevalier Y, Croizier MC, Lafont D, Sancho MR (1995) Carbohydr Res 278:91
24. Boullanger P, Chevalier Y (1996) Langmuir 12:1771
25. Plusquellec D, Brenner-Hénaff C, Léon-Ruaud P, Duquenoy S, Lefeuvre M, Wroblewski H (1994) J Carbohydr Chem 13:737
26. Costes F, Elghoul M, Bon M, Rico-Lattes I, Lattes A (1995) Langmuir 11:3644
27. Latgé P, Bon M, Rico I, Lattes A (1992) New J Chem 16:387
28. (a) Garelli-Calvet R, Latgé P, Rico I, Lattes A, Puget A (1992) Biochim Biophys Acta 1109:55 (b) van Doren HA, van der Geest R, de Ruijter CF, Kellogg RM, Wynberg H (1990) Liq Cryst 8:109
29. Boyer B, Lamaty G, Moussamou-Missima JM, Pavia AA, Pucci B, Roque JP (1991) Tetrahedron Lett 32:1191
30. (a) Postel DG, Ronco GL, Villa P (1989) FR 2,630,445 (b) Postel DG, Ronco GL, Villa P (1989) FR 2,630,446
31. Maunier V, Boullanger P, Lafont D, Chevalier Y. Carbohydr Res, in press
32. Elghoul M, Escoula B, Rico I, Lattes A (1992) J Fluorine Chem 59:107
33. Jouani MA, Szonyi F, Trabelsi H, Cambon A (1994) Bull Soc Chim Fr 131:173
34. Boyer B, Lamaty G, Moussamou-Missima JM, Pavia AA, Pucci B, Roque JP (1991) Eur Polym J 27: 359
35. Pavia AA, Pucci B, Riess JG, Zarif L (1992) Makromol Chem 193:2505
36. Myrtil E, Zarif L, Greiner J, Riess JG, Pucci B, Pavia AA (1995) J Fluorine Chem 71:101
37. Ats SC, Lehmann J, Petry S (1994) Carbohydr Res 252:325
38. Lafont D, Boullanger P, Chevalier Y (1995) J Carbohydr Chem 14:533
39. Garelli-Calvet R, Brisset F, Rico I, Lattes A (1993) Synthetic Commun 23:35
40. Müller-Fahrnow A, Saenger W, Fritsch D, Schnieder P, Fuhrhop JH (1993) Carbohydr Res 242:11
41. Gouéth P, Ramiz A, Ronco G, Mackenzie G, Villa P (1995) Carbohydr Res 266:171
42. Briggs CBA, Newington IM, Pitt AR (1995) J Chem Soc Chem Commun 379
43. Jeffrey GA, Wingert LM (1992) Liq Cryst 12:179
44. van Doren HA, Wingert LM (1994) Recl Trav Chim Pays-Bas 113:260
45. Paleos C, Tsiourvas D (1995) Angew Chem Intern Ed Engl 34:1696
46. Dorset DL (1990) Carbohydr Res 206:193
47. Sakya P, Seddon JM, Templer RH (1994) J Phys II 4:1311
48. van Doren HA, Galema SA, Engberts JBFN (1995) Langmuir 11:687
49. Goodby JW, Haley JA, Mackenzie G, Watson MJ, Plusquellec G, Ferrieres V (1995) J Mater Chem 5:2209

50. Tietze LF, Boge K, Vill V (1994) Chem Ber 127:1065
51. Raaijmakers HW, Zwanenburg B, Chittenden GJF (1994) Recl Trav Chim Pays-Bas 113:79
52. Sakurai I, Suzuki T, Sakurai S (1989) Biochim Biophys Acta 985:101
53. Auvray X, Petipas C, Anthore R, Rico-Lattes I, Lattes A (1995) Langmuir 11:433
54. Dahlhoff WV, Riehl K, Zugenmaier P (1993) Liebigs Ann Chem 1063
55. Thiem J, Vill V, Miethchen R, Dietmar P (1991) J Prakt Chem 333:173
56. Jeschke U, Vogel C, Vill V, Fischer H (1995) J Mater Chem 5:2073
57. Vogel C, Jeschke U, Vill V, Fischer H (1992) Liebigs Ann Chem 1171
58. Mannock DA, Lewis RNAH, McElhaney RN (1990) Biochemistry 29:7790
59. Sen A, Hui SW, Mannock DA, Lewis RNAH, McElhaney RN (1990) Biochemistry 29:7799
60. Hinz HJ, Kuttenreich H, Meyer R, Renner M, Fründ R, Koynova R, Boyanov AI, Tenchov BG (1991) Biochemistry 30:5125
61. Turner DC, Wang ZG, Gruner SM, Mannock DA, Mcelhaney RN (1992) J Phys II:2039
62. Miethchen R, Schwarze M, Holz J (1993) Liq Cryst 15:185
63. Miethchen R, Holz J, Prade H (1993) Colloid Polymer Sci 271:404
64. Langlois V, Williams JM (1994) J Chem Soc Perkin Trans 1:2103
65. Dalhoff WV (1990) Liebigs Ann Chem 811:1025
66. Vill V, Kelkenberg H, Thiem J (1992) Liq Cryst 11:459
67. Dahlhoff WV, Radkowski K, Riehl K, Zugenmaier P (1995) Z Naturf B 50:1079
68. Dahlhoff WV (1992) Liebigs Ann Chem 109
69. Praefcke K, Kohne B, Diele S, Pelzl G, Kjaer A (1992) Liq Cryst 11:1
70. Goodby JW, Marcus MA, Chin E, Finn PL, Pfannemüller B (1988) Liq Cryst 3:1569
71. van Doren HA, Wingert LM (1991) Mol Cryst Liq Cryst 198:381
72. (a) Fuhrhop JH, Schnieder P, Rosenberg J, Boekema E (1987) J Amer Chem Soc 109:3387 (b) Fuhrhop JH, Schnieder P, Boekema E Helfrich W (1988) J Amer Chem Soc 110:2862
73. Jeffrey GA, Maluszynska H (1990) Carbohydr Res 207:211
74. Taravel FR, Pfannemüller B (1990) Makromol Chem 191:3097
75. Frankel DA, O'Brien DF (1991) J Amer Chem Soc 113:7436
76. Svenson S, Koning J, Fuhrhop JH (1994) J Phys Chem 98:1022
77. Fischer H, Vill V, Vogel C, Jeschke U (1993) Liq Cryst 15:733
78. Fuhrhop JH, Abch R (1992) Adv Supramolecular Chem 2:25
79. Boullanger P, Sancho MR, Chevalier Y, unpublished results
80. Ishigami Y, Gama Y, Nagahora H, Yamaguchi M, Nakahara H, Kamata T (1987) Chem Lett 763
81. Guedj C, Pucci B, Zarif L, Coulomb C, Riess JG, Pavia AA (1994) Chem Phys Lipids 72:153
82. Zarif L, Gulikkrzywicki T, Riess JG, Pucci B, Guedj C, Pavia AA (1994) Colloids Surf. A 84:107
83. Wakita M, Hashimoto M (1995) Kobunshi Ronbunshu 52:589
84. Sharon N, Lis H (1993) Scientific Amer 74
85. Feizi T (1993) Curr Opin Struct Biol 3:701
86. Falk KE, Karlsson KA, Samuelsson BE (1980) Chem Phys Lipids 27:9
87. Lüderitz O, Tanamoto KI, Galanos C, Westphal O, Zähringer U, Rietschel ET, Kusumoto S, Shiba T (1983) Amer Chem Soc Symp Ser 231:3
88. Robinson PJ (1991) Immunology Today 12:35
89. Tai A, Nakahata M, Harada T, Izumi Y, Kusumoto S, Inage M, Shiba T (1980) Chem Lett 1125
90. (a) Zimmermann P, Bommer R, Bär T, Schmidt RR (1988) J Carbohydr Chem 7:435 (b) Schmidt RR, Zimmermann P (1986) Tetrahedron Lett 27:481
91. Lebeau L, Oudet P, Mioskowski C (1991) Helv Chim Acta 74:1697
92. Murakata C, Ogawa T (1992) Tetrahedron Lett 32:101
93. Sancho MR (1994) Thesis, Lyon-France
94. Defaye J, Gadelle A, Pedersen C (1989) Fr Pat Appl 8 910 301
95. Rico-Lattes I, Garrigues JC, Perez E, André-Barrès C, Madelaine-Dupuich C, Lattes A, Linas MD, Aubertin AM (1995) New J Chem 19:341

96. Abe Y, Harata K, Fujiwara M, Ohbu K (1995) Carbohydr Res 269:43
97. Björkling F, Godtfredsen SE, Kirk O (1989) J Chem Soc Chem Commun 934
98. Azéma J, Chebli C, Bon M, Rico-Lattes I, Lattes A (1995) J Carbohydr Chem 14:805
99. Kida T, Yurugi K, Masuyama A, Nakatsuji Y, Ono D, Takeda T (1995) J Amer Oil Chem Soc 72:773
100. Csiba M, Cleophax J, Loupy A, Malthete J, Gero SD (1993) Tetrahedron Lett 34:1787
101. Ito Y, Sato S, Mori M, Ogawa T (1988) J Carbohydr Chem 7:359
102. Nicolaou KC, Caulfield T, Kataoka H, Kumazawa T (1988) J Amer Chem Soc 110:7910
103. Nunomura S, Mori M, Iti Y, Ogawa T (1989) Tetrahedron Lett 30:5619
104. Sato S, Nunomura S, Nakano T, Ito Y, Ogawa T (1988) Tetrahedron Lett 29:4097
105. Toepfer A, Schmidt RR (1990) Carbohydr Res 202:193
106. Bierer DE, Dubenko LG, Litvak J, Gerber RE, Chu J, Thai DL, Tempesta MS, Truong TV (1995) J Org Chem 60:7646
107. Mickeleit M, Wieder T, Buchner K, Geilen C, Mulzer J, Reutter W (1995) Angew Chem Int Ed Engl 34:2667
108. Mayer TG, Kratzer B, Schmidt RR (1994) Angew Chem Intern Ed Engl 33:2177
109. Cottaz S, Brimacombe JS, Ferguson MAJ (1993) J Chem Soc Perkin Trans 1:2945
110. Murakata C, Ogawa T (1992) Carbohydr Res 235:95
111. Khiar N, Martin-Lomas M (1995) J Org Chem 60:7017
112. Campbell AS, Fraser-Reid B (1995) J Amer Chem Soc 117:10387
113. Böcker T, Thiem J (1989) Tenside Surf Det 26:318
114. Ma YD, Takada A, Sugiura M, Fukuda T, Miyamoto T, Watanabe J (1994) Bull Chem Soc Jpn 67:346
115. Vill V, Lindhorst TK, Thiem J (1991) J Carbohydr Chem 10:771
116. Lebeau L, Regnier E, Schultz P, Wang JC, Mioskowski C, Oudet P (1990) FEBS Lett 267:38
117. Wilhelm F, Chatterjee SK, Rattay B, Nuhn P, Benecke R, Ortwein J (1995) Liebigs Ann Chem1673
118. Lafont D, Boullanger P, Chierici S, Gelhausen M, Roux B (1996) New J Chem 20:1093
119. Sharma L, Singh S (1995) Carbohydr Res 270:43
120. Polidori A, Pucci B, Maurizis JC, Pavia AA (1994) New J Chem 18:839
121. Kunitake T (1992) Angew Chem Int Ed Engl 31:709
122. Fuhrhop JH, Mathieu J (1984) Angew Chem Int Ed Engl 23:100
123. Sakai H, Takisada M, Takeoka S, Tsuchida E (1993) Chem Lett 1891
124. Rodin VV, Izmailova VN (1994) Colloid J 56:745
125. Crowe JH, Crowe LM, Carpenter JF, Rudolph AS, Wistrom CA, Spargo BJ, Anchordoguy TJ (1988) Biochim Biophys Acta 947:367
126. Takeoka S, Sakai H, Takisada M, Tsuchida E (1992) Chem Lett 1877
127. Takeoka S, Sakai H, Ohno H, Yoshimura K, Tsuchida E (1992) J Colloid Interf Sci 152:351
128. Baszkin A, Rosilio V, Albrecht G, Sunamoto J (1991) J Colloid Interf Sci 145:502
129. Moellerfeld J, Prass W, Ringsdorf H, Hamazaki H, Sunamoto J (1986) Biochim Biophys Acta 857:265
130. Sunamoto J, Sato T, Taguchi T, Hamazaki H (1992) Macromol 25:5665
131. Bach D, Miller IR, Barenholz Y (1993) Biophys Chem 47:77
132. Allen TM (1994) Adv Drug Delivery Rev 13:285
133. Virden JW, Berg JC (1992) J Colloid Interf Sci 153:411
134. Lasic DD (1994) Angew Chem Int Ed Engl 33:1685
135. (a) Sunamoto J, Iwamoto K (1986) CRC Crit Rev Therapeutic Drug Carrier Systems 2:117 (b) Sunamoto J, Sato T (1989) In: Tsuruta T, Nakajima J (eds) Multiphase Biomedical Materials. VSP, p 167
136. Szoka Jr FC, Mayhew E (1983) Biochim Biophys Res Commun 110:140
137. Ghosh PC, Bachhawat BK (1992) J Liposome Res 2:369
138. Lockhoff O (1991) Angew Chem Int Ed Engl 30:1611
139. Sunamoto J, Sato T, Hirota M, Fukushima K, Hiratani K, Hara K (1987) Biochim Biophys Acta 898:323
140. Nishiya T, Robibo D (1992) Carbohydr Res 235:239

141. Dong C, Rogers JA (1991) J Microencapsulation 8:153
142. Weissig V, Lasch J, Gregoriadis G (1989) Biochim Biophys Acta 1003:54
143. Stewart RJ, Boggs JM (1993) Biochemistry 32:10666
144. Eidelman O, Blumenthal R, Walter A (1988) Biochemistry 27:2839
145. Andelman D, Kozlov MM, Helfrich W (1994) Europhysics Lett 25:231
146. Sancho MR, Boullanger P, Létoublon R (1993) Colloids Surfaces B: Biointerfaces 1:373
147. (a) Chabala JC, Shen TY (1978) Carbohydr Res 67: 55 (b) Ponpipom MM, Bugianesi RL, Shen TY (1980) Can J Chem 58:214
148. Kiessling LL, Pohl NL (1996) Chem and Biol 3:71
149. (a) Lee RT, Ichikawa Y, Fay M, Drickamer K, Shao MC, Lee YC (1991) J Biol Chem 266:4810 (b) Lee YC, Lee RT (1995) Acc Chem Res 28:321
150. Kingery-Wood JE, Williams KW, Sigal GB, Whitesides GM (1992) J Am Chem Soc 114:7303
151. Nyholm PG, Pascher I (1993) Biochemistry 32:1225
152. Nyholm PG, Pascher I (1993) Int J Biol Macromol 15:43
153. Sanders CR, Prestegard JH (1991) J Amer Chem Soc 113:1987
154. Hare BJ, Howard KP, Prestegard JH (1993) Biophys J 64:392
155. Sanders CR, Prestegard JH (1992) J Amer Chem Soc 114:7096
156. Howard KP, Prestegard JH (1995) J Amer Chem Soc 117:5031
157. Hare BJ, Rise F, Aubin Y, Prestegard JH (1994) Biochemistry 33:10137
158. Augé S, Marsan MP, Czaplicki J, Demange P, Muller I, Tropis M, Milon A (1995) J Phys Chem 92:1715
159. Renou JP, Giziewisz JB, Smith ICP, Jarell HC (1989) Biochemistry 28:1804
160. Strömberg N, Nyholm PG, Pascher I, Normark S (1991) Proc Natl Acad Sci USA 88:9340
161. Leckband DE, Israelachvili JN, Schmitt FJ, Knoll W (1992) Science 255:1419
162. Blankenburg R, Meller P, Ringsdorf H, Salesse C (1989) Biochemistry 28:8214
163. Zhao S, Reichert WM (1992) Langmuir 8:2785
164. Häussling L, Ringsdorf H, Schmitt FJ, Knoll W (1991) Langmuir 7:1837
165. Piepenstock M, Losche M (1992) Thin Solid Films 210:793
166. Ahluwalia A, Derossi D, Schirone A (1992) Thin Solid Films 210:726
167. Turko IV, Lepesheva GI, Chashchin VL (1992) Anal Chim Acta 265: 21
168. Okahata Y, Ebara Y (1992) J Chem Soc Chem Commun 116
169. Ebato H, Gentry CA, Herron JN, Müller W, Okahata Y, Ringsdorf H, Suci PA (1994) Anal Chem 66:1683
170. Lang H, Duschl C, Vogel H (1994) Langmuir 10:197
171. Ivanova M, Panaiotov I, Eshkenazy M, Tekelieva R, Ivanova R (1986) Colloids Surfaces 17:159
172. Charych D, Cheng Q, Reichert A, Kuziemko G, Stroh M, Nagy JO, Spevak W, Stevens RC (1996) Chem Biol 3:113
173. Ebara Y, Okahata Y (1994) J Am Chem Soc 116:11209
174. Niwa M, Mori T, Nishio E, Nishimura H, Higashi N (1992) J Chem Soc Chem Commun 547
175. Nilsson KGI, Mandenius CF (1994) Bio/Technology:1376
176. Reichert A, Nagy JO, Spevak W, Charych D (1995) J Am Chem Soc 117:829
177. Tauskela JS, Thompson M (1992) Anal Chim Acta 264:185
178. Hakomori S, Igarashi Y, Kojima N, Okoshi K, Handa B, Fenderson B (1991) Glycoconjugate J 8:178

Springer and the environment

At Springer we firmly believe that an international science publisher has a special obligation to the environment, and our corporate policies consistently reflect this conviction.

We also expect our business partners – paper mills, printers, packaging manufacturers, etc. – to commit themselves to using materials and production processes that do not harm the environment. The paper in this book is made from low- or no-chlorine pulp and is acid free, in conformance with international standards for paper permanency.

Printing: Saladruck, Berlin
Binding: Buchbinderei Lüderitz & Bauer, Berlin